The Michigan Historical Reprint Series

Reprints from the collection of the University of Michigan University Library

This volume is produced from digital images created through the University of Michigan University Library's preservation reformatting program. The Library seeks to preserve the intellectual content of items in a manner that facilitates and promotes a variety of uses. The digital reformatting process results in an electronic version of the text that can both be accessed online and used to create new print copies. This book and thousands of others can be found in the digital collections of the University of Michigan Library. The University Library also understands and values the utility of print, and makes reprints available through its Scholarly Publishing Office.

For access to the University of Michigan Library's digital collections, please see http://www.lib.umich.edu.

The Scholarly Publishing Office seeks to disseminate high-quality, cost-effective scholarly content through both print and electronic publishing. Information about the Scholarly Publishing Office can be found at http://spo.umdl.umich.edu.

The Scholarly Publishing Office
the University of Michigan
University Library

MAN AND NATURE;

OR,

PHYSICAL GEOGRAPHY

AS MODIFIED BY HUMAN ACTION.

BY

GEORGE P. MARSH.

"Not all the winds, and storms, and earthquakes, and seas, and seasons of the world, have done so much to revolutionize the earth as MAN, the power of an endless life, has done since the day he came forth upon it, and received dominion over it."—H. BUSHNELL, *Sermon on the Power of an Endless Life.*

NEW YORK:
CHARLES SCRIBNER & CO., No. 654 BROADWAY.
1869.

JOHN F. TROW & CO.
PRINTER, STEREOTYPER, AND ELECTROTYPER,
46, 48, & 50 Greene St., New York.

PREFACE.

THE object of the present volume is: to indicate the character and, approximately, the extent of the changes produced by human action in the physical conditions of the globe we inhabit; to point out the dangers of imprudence and the necessity of caution in all operations which, on a large scale, interfere with the spontaneous arrangements of the organic or the inorganic world; to suggest the possibility and the importance of the restoration of disturbed harmonies and the material improvement of waste and exhausted regions; and, incidentally, to illustrate the doctrine, that man is, in both kind and degree, a power of a higher order than any of the other forms of animated life, which, like him, are nourished at the table of bounteous nature.

In the rudest stages of life, man depends upon spontaneous animal and vegetable growth for food and clothing, and his consumption of such products consequently diminishes the numerical abundance of the species which serve his uses. At more advanced periods, he protects and propagates certain esculent vegetables and certain fowls and quadrupeds, and, at the same time, wars upon rival organisms which prey upon these objects of his care or obstruct the increase of their numbers. Hence the action of man upon the organic world tends to subvert the original balance of its species, and while it reduces the numbers of some of them, or even extirpates them altogether, it multiplies other forms of animal and vegetable life.

The extension of agricultural and pastoral industry involves an enlargement of the sphere of man's domain, by encroachment upon the forests which once covered the greater part of the earth's surface otherwise adapted to his occupation. The felling of the woods has been attended with momentous consequences to the drainage of the soil, to the external configuration of its surface, and probably, also, to local climate; and the importance of human life as a transforming power is, perhaps, more clearly demonstrable in the influence man has thus exerted upon superficial geography than in any other result of his material effort.

Lands won from the woods must be both drained and irrigated; river banks and maritime coasts must be secured by means of artificial bulwarks against inundation by inland and by ocean floods; and the needs of commerce require the improvement of natural, and the construction of artificial channels of navigation. Thus man is compelled to extend over the unstable waters the empire he had already founded upon the solid land.

The upheaval of the bed of seas and the movements of water and of wind expose vast deposits of sand, which occupy space required for the convenience of man, and often, by the drifting of their particles, overwhelm the fields of human industry with invasions as disastrous as the incursions of the ocean. On the other hand, on many coasts, sand hills both protect the shores from erosion by the waves and currents, and shelter valuable grounds from blasting sea winds. Man, therefore, must sometimes resist, sometimes promote, the formation and growth of dunes, and subject the barren and flying sands to the same obedience to his will to which he has reduced other forms of terrestrial surface.

Besides these old and comparatively familiar methods of material improvement, modern ambition aspires to yet grander achievements in the conquest of physical nature, and projects are meditated which quite eclipse the boldest enterprises hitherto undertaken for the modification of geographical surface.

The natural character of the various fields where human industry has effected revolutions so important, and where the multiplying population and the impoverished resources of the globe demand new triumphs of mind over matter, suggests a corresponding division of the general subject, and I have conformed the distribution of the several topics to the chronological succession in which man must be supposed to have extended his sway over the different provinces of his material kingdom. I have, then, in the Introductory chapter, stated, in a comprehensive way, the general effects and the prospective consequences of human action upon the earth's surface and the life which peoples it. This chapter is followed by four others in which I have traced the history of man's industry as exerted upon Animal and Vegetable Life, upon the Woods, upon the Waters, and upon the Sands; and to these I have added a concluding chapter upon Probable and Possible Geographical Revolutions yet to be effected by the art of man.

I have only to add what, indeed, sufficiently appears upon every page of the volume, that I address myself not to professed physicists, but to the general intelligence of educated, observing, and thinking men; and that my purpose is rather to make practical suggestions than to indulge in theoretical speculations properly suited to a different class from that to which those for whom I write belong.

GEORGE P. MARSH.

December 1, 1863.

BIBLIOGRAPHICAL LIST

OF WORKS CONSULTED IN THE PREPARATION OF THIS VOLUME.

Amersfoordt, J. P. Het Haarlemmermeer, Oorsprong, Geschiedenis, Droogmaking. Haarlem, 1857. 8vo.

Andresen, C. C. Om Klitformationen og Klittens Behandling og Bestyrelse. Kjöbenhavn, 1861. 8vo.

Annali di Agricoltura, Industria e Commercio. Pubblicati per cura del Ministero d'Agricoltura, Industria e Commercio. Fasc i–v. Torino, 1862–'3. 8vo.

Arago, F. Extracts from, in Becquerel, Des Climats.

Arriani, Opera. Lipsiæ, 1856. 2 vols. 12mo.

Asbjörnsen, P. Chr. Om Skovene og om et ordnet Skovbrug i Norge. Christiania, 1855. 12mo.

Aus der Natur. Die neuesten Entdeckungen auf dem Gebiete der Naturwissenschaften. Leipzig, various years. 20 vols. 8vo.

Avé-Lallemant, K. C. B. Die Benutzung der Palmen am Amazonenstrom in der Oekonomie der Indier. Hamburg, 1861. 18mo.

Babinet. Études et Lectures sur les Sciences d'Observation. Paris, 1855–1863. 7 vols. 18mo.

Baer, von. Kaspische Studien. St. Petersburg, 1855–1859. 8vo.

Barth, Heinrich. Wanderungen durch die Küstenländer des Mittelmeeres. V. i. Berlin, 1849. 8vo.

Barth, J. B. Om Skovene i deres Forhold til Nationalœconomien. Christiania, 1857. 8vo.

Baude, J. J. Les Côtes de la Manche, Revue des Deux Mondes, 15 Janvier, 1859.

Baumgarten. Notice sur les Rivières de la Lombardie; in Annales des Ponts et Chaussées, 1847, 1er sémestre, pp. 129–199.

Beckwith, Lieut. Report in Pacific Railroad Report, vol. ii.

Becquerel. Des Climats et de l'Influence qu'exercent les Sols boisés et non-boisés. Paris, 1853. 8vo.

——— Éléments de Physique Terrestre et de Météorologie. Paris, 1847. 8vo.

Belgrand. De l'Influence des Forêts sur l'écoulement des Eaux Pluviales; in Annales des Ponts et Chaussées, 1854, 1er sémestre, pp. 1, 27.

Berg, Edmund von. Das Verdrängen der Laubwälder im Nördlichen Deutschlande durch die Fichte und die Kiefer. Darmstadt, 1844. 8vo.

Bergsöe, A. F. Greve Ch. Ditlev Frederik Reventlovs Virksomhed som Kongens Embedsmand og Statens Borger. Kjöbenhavn, 1837. 2 vols. 8vo.

Berlepsch, H. Die Alpen in Natur-und Lebensbildern. Leipzig, 1862. 8vo.

Bianchi, Celestino. Compendio di Geografia Fisica Speciale d'Italia. Appendice alla traduzione Italiana della Geog.-Fisica di Maria Somerville. Firenze, 1861. (2d vol. of translation.)

Bigelow, John. Les États Unis d'Amérique en 1863. Paris, 1863. 8vo.

Blake, Wm. P. Reports in Pacific Railroad Report, vols. ii and v.

Blanqui. Mémoire sur les Populations des Hautes Alpes; in Mémoires de l'Académie des Sciences Morales et Politiques, 1843.

——— Voyage en Bulgarie. Paris, 1843. 12mo.

——— Précis Élémentaire d'Économie Politique, suivi du Résumé de l'Histoire du Commerce et de l'Industrie. Paris, 1857. 12mo.

Boitel, Amédée. Mise en valeur des Terres pauvres par le Pin Maritime. 2d edition. Paris, 1857. 8vo.

Bonnemère, Eugène. Histoire des Paysans depuis la fin du Moyen Age jusqu'à nos jours. Paris, 1856. 2 vols. 8vo.

Böttger, C. Das Mittelmeer. Leipzig, 1859.

Boussingault, J. B. Économie Rurale considerée dans ses Rapports avec la Chimie, la Physique, et la Météorologie. 2d edition. Paris, 1851. 2 vols. 8vo.

Brémontier, N. T. Mémoire sur les Dunes; in Annales des Ponts et Chaussées, 1833, 1er sémestre, pp. 145, 223.

Brincken, J. von den. Ansichten über die Bewaldung der Steppen des Europæischen Russland. Braunschweig, 1854. 4to.

Büttner, J. G. Zur Physikalischen Geographie; in Berghaus, Geographisches Jahrbuch, No. iv, 1852, pp. 9–19.

Caimi, Pietro. Cenni sulla Importanza e Coltura dei Boschi. Milano, 1857. 8vo.

Cantegril, and others. Extracts in Comptes Rendus à l'Académie des Sciences. Paris, 1861.

Castellani. Dell' immediata influenza delle Selve sul corso delle acque. Torino, 1818, 1819. 2 vols. 4to.

Census of the United States for 1860. Preliminary Report on, Washington, 1862. 8vo.

Cerini, Giuseppe. Dell' Impianto e Conservazione dei Boschi. Milano, 1844. 8vo.

Champion, Maurice. Les Inondations en France depuis le VIme Siècle jusqu'à nos jours. Paris, 1858, 1862. Vols. i–iv, 8vo.

Chateauvieux, F. Lullin de. Lettres sur l'Italie. Seconde edition, Genève, 1834. 8vo.

Chevandier. Extracts in Comptes Rendus à l'Académie des Sciences. Juillet-Decembre, 1844. Paris.

Clavé, Jules. Études sur l'Économie Forestière. Paris, 1862. 12mo.

——— La Forêt de Fontainebleau ; Revue des Deux Mondes, 1 Mai, 1863.

Cooper, J. G. The Forests and Trees of Northern America; in Report of the Commissioner of Patents for the year 1860, pp. 416-445.

Cotta, Bernhard. Deutschlands Boden. Leipzig, 1858. 2 vols. 8vo.

——— Vorwort zu Paramelle's Quellenkunde. See *Paramelle.*

——— Die Alpen. Leipzig, 1851. 8vo.

Coultas, Harland. What may be Learned from a Tree. New York, 1860. 8vo.

Courier, Paul-Louis. Œuvres Complètes. Bruxelles, 1833. 8vo.

Dana, James D. Manual of Geology. Philadelphia, 1863. 8vo.

Delamarre, L. G. Historique de la Création d'une Richesse Millionaire par la culture des Pins. Paris, 1827. 8vo.

D.Héricourt, A. F. Les Inondations et le livre de M. Vallès; Annales Forestières, December, 1857, pp. 310, 321. Paris.

Diggelen, B. P. G. van. Groote Werken in Nederland. Zwolle, 1855. 8vo.

Dumas, M. J. La Science des Fontaines. 2me edition, Paris, 1857. 8vo.

Dumont, Aristide. Des Travaux Publics dans leurs Rapports avec l'Agriculture. Paris, 1847. 8vo.

Dwight, Timothy. Travels in New England and New York. New Haven, 1821. 4 vols. 8vo.

Emerson, George B. A Report on the Trees and Shrubs growing naturally in Massachusetts. Boston, 1850. 8vo.

Emory, Wm. H., Col. Report of Commissioners of the United States and Mexican Boundary Survey, vol. i, 1857.

Escourrou-Miliago, A. L'Italie à propos de l'Exposition Universelle de Paris. Paris, 1856. 8vo.

Evelyn, John. Silva; or, a Discourse of Forest Trees. With Notes by A. Hunter. York, 1786. 2 vols. 4to.

——— Terra, a Philosophical Discourse of Earth. York, 1786. 4to. in vol. ii of Silva.

Féraud-Giraud, L. J. D. Police des Bois, Défrichements et Reboisements,

Commentaire pratique sur les lois promulguées en 1859 et 1860. Paris, 1861. 8vo.

Ferrara, Francesco. Descrizione dell' Etna. Palermo, 1818. 8vo.

Feuillide, C. de. L'Algérie Française. Paris, 1856. 8vo.

Figuier, Louis. L'Année Scientifique et Industrielle. Paris, 1862–'3. 12mo.

Finnboga Saga hins rama. Kaupmannahöfn, 1812. 4to.

Foissac, P. Meteorologie mit Rücksicht auf die Lehre vom Kosmos, Deutsch von A. H. Emsmann. Leipzig, 1859. 8vo.

Forchhammer, G. Geognostische Studien am Meeres-Ufer; in Leonhard und Bronn's Neues Jahrbuch für Mineralogie, Geognosie, Geologie, etc. Jahrgang, 1841, pp. 1–38.

Fossombroni, Vittorio. Memorie Idraulico-Storiche sopra la Val-di-Chiana. Montepulciano, 3za edizione, 1835. 8vo.

Fraas, C. Klima und Pflanzenwelt in der Zeit. Landshut, 1847. 8vo.

Frisi, Paolo. Del Modo di regolare i Fiumi e i Torrenti. Lucca, 1762. 4to.

Fuller, Thomas. The History of the Worthies of England. London, 1662. Folio.

Gilliss, J. M., Capt. United States Naval Astronomical Expedition to the Southern Hemisphere. Washington, 1855. 2 vols. 4to.

Giorgini. Paper by; in Salvagnoli-Marchetti, Rapporto sul Bonificamento delle Maremme, App. v.

Girard et Parent-Duchatelet. Rapport sur les Puits forés dits Artésiens; Annales des Ponts et Chaussées, 1833, 2me sémestre, 313–344.

Graham, J. D., Lieut.-Col. A Lunar Tidal Wave in the North American Lakes demonstrated. Cambridge, 1861. 8vo. *pamphlet.* Also in vol. xiv, Proc. Am. Ass. for Adv. of Science for 1860.

Hakluyt, Richard. The Principal Navigations, Voyages, &c., of the English Nation. London, 1598–'9. 3 vols. folio.

Harrison, W. An Historicall Description of the Iland of Britaine; in Holinshed's Chronicles. Reprint of 1807, vol. i.

Hartwig, G. Das Leben des Meeres. Frankfurt, 1857. 8vo.

Haxthausen, August von. Transkaukasia. Leipzig, 1856. 2 vols. 8vo.

Henry, Prof. Joseph. Paper on Meteorology in its connection with Agriculture; in United States Patent Office Report for 1857, pp. 419–550.

Herschel, Sir J. F. W. Physical Geography. Edinburgh, 1861. 12mo.

Heyer, Gustav. Das Verhalten der Waldbäume gegen Licht und Schatten. Erlangen, 1852. 8vo.

Hohenstein, Adolph. Der Wald sammt dessen wichtigem Einfluss auf das Klima, &c. Wien, 1860. 8vo.

Humboldt, Alexander von. Ansichten der Natur. Dritte Ausgabe, Stuttgart und Tübingen, 1849. 2 vols. 12mo.

Hummel, Karl. Physische Geographie. Graz, 1855. 8vo.

Hunter, A. Notes to Evelyn, Silva, and Terra. York, 1786. See *Evelyn.*

Jacini, Stefano. La Proprietà Fondiaria e le Popolazioni agricole in Lombardia. Milano e Verona, 1857. 8vo.

Joinville. Histoire de Saint-Louis. Nouvelle Collection des Mémoires pour servir à l'Histoire de France, par Michaud et Poujoulat. Tome i. Paris, 1836. 8vo.

Josselyn, John. New England Rarities. London, 1672. 12mo.

Knorr, E. A. Studien über die Buchen-Wirthschaft. Nordhausen, 1863. 8vo.

Kohl, J. G. Alpenreisen. Dresden und Leipzig, 1849. 3 vols. 8vo.

——— Die Marschen und Inseln der Herzogthümer Schleswig und Holstein. Dresden und Leipzig, 1846. 3 vols. 8vo.

Kramer, Gustav. Der Fuciner-See. Berlin, 1839. 4to.

Krause, G. C. A. Der Dünenbau auf den Ostsee-Küsten West-Preussens. 1850. 8vo.

Kremer, Alfred von. Ægypten, Forschungen über Land und Volk. Leipzig, 1863. 2 vols. 8vo.

Kriegk, G. L. Schriften zur allgemeinen Erdkunde. Leipzig, 1840. 8vo.

Ladoucette, J. C. F. Histoire, Topographie, Antiquités, Usages, Dialectes des Hautes Alpes. Seconde édition, 1834. 1 vol. 8vo. and Atlas.

Læstadius, Lars Levi. Om Möjligheten och Fördelen af allmänna Uppodlingar i Lappmarken. Stockholm, 1824. 12mo.

Læstadius, Petrus. Journal för första året af hans Tjenstgöring såsom Missionaire i Lappmarken. Stockholm, 1831. 8vo.

——— Fortsättning af Journalen öfver Missions-Resor i Lappmarken. Stockholm, 1833. 8vo.

Lampridius. Vita Elagabali in Script. Hist., August.

Landgrebe, Georg. Naturgeschichte der Vulcane. Gotha, 1855. 2 vols. 8vo.

Laurent, Ch. Mémoires sur le Sahara Oriental au point de vue des Puits Artésiens. Paris, 1859. 8vo. *pamphlet.* Also, in Mém de la Soc. des Ingénieurs Civils, and the Bulletin de la Soc. Géologique de France.

Laval. Mémoire sur les Dunes du Golfe de Gascogne; in Annales des Ponts et Chaussées, 1847, 2me sémestre, pp. 218–268.

Lavergne, M. L. de. Économie Rurale de la France, depuis 1789. 2me édition, Paris, 1861. 12mo.

Le Alpi che cingono l'Italia. Parte 1er, vol. 1er. Torino, 1845. 8vo.

Lefort. Notice sur les travaux de Fixation des Dunes; in Annales des Ponts et Chaussées, 1831, 2me sémestre, pp. 320–332.

Lenormant. Note relative à l'Execution d'un Puits Artésien en Egypte sous la XVIIIme Dynastie; Académie des Inscriptions et Belles-Lettres, 12 Novembre, 1852.

Liber Albus: The White Book of the City of London. London, 1861. 4to.

Loftus, W. K. Travels and Researches in Chaldæa and Susiana. New York, 1857. 8vo.

Lombardini. Cenni Idrografi sulla Lombardia; Intorno al Sistema Idraulico del Pô; epitomized by Baumgarten in Annales des Ponts et Chaussées, 1847, 1er sémestre, pp. 129, 199; and in Dumont, Des Travaux Publics, pp. 268, 335.

——— Sui progetti intesi ad estendere l'irrigazione della Pianura del Pô. Politecnico. Gennajo, 1863, pp. 5–50.

Lorentz. Cours Élémentaire de Culture des Bois, complété et publie par A. Parade, 4me edition. Paris et Nancy, 1860. 8vo.

Lyell, Sir Charles. The Geological Evidence of the Antiquity of Man. London, 1863. 8vo. Principles of Geology. New York, 1862. 8vo.

Mardigny, M. de. Mémoire sur les Inondations des Rivières de l'Ardèche. Paris, 1860. 8vo.

Marschand, A. Ueber die Entwaldung der Gebirge. Bern, 1849. 12mo. *pamphlet.*

Martineau. Endeavors after the Christian Life. Boston, 1858.

Martins. Revue des Deux Mondes, Avril, 1863.

Maury, M. F. The Physical Geography of the Sea. Tenth edition. London, 1861. 8vo.

Medlicott, Dr. Observations of, quoted from London Athenæum, 1863.

Meguscher, Francesco. Memorie sulla migliore maniera per rimettere i Boschi della Lombardia, etc. Milano, 1859. 8vo.

Mejdell, Th. Om Foranstaltninger til Behandling af Norges Skove. Christiania, 1858. 8vo.

Mella. Delle Inondazioni del Mella nella notte del 14 al 15 Agosto, 1850. Brescia, 1851. 8vo.

Meyer, J. Physik der Schweiz. Leipzig, 1854. 8vo.

Michelet, J. L'Insecte, 4me edition. Paris, 1860. 12mo.

——— L'Oiseau, 7me edition. Paris, 1861. 12mo.

Monestier-Savignat, A. Étude sur les Phénomènes, l'Aménagement et la Législation des Eaux au point de vue des Inondations. Paris, 1858. 8vo.

Montluisant. Note sur les Desséchements, les Endiguements et les Irrigations; in Annales des Ponts et Chaussées, 1833, 2me sémestre, pp. 281–294.

Morozzi, Ferdinando. Dello Stato Antico e Moderno del Fiume Arno. Firenze, 1762. 4to.

Müller, K. Das Buch der Pflanzenwelt. Leipzig, 1857. 2 vols. 12mo.

Nangis, Guillaume de. Extracts from, in Nouvelle Collection des Mémoires pour servir par Michaud et Poujoulat. Vol. i. Paris, 1836.

Nanquette, Henri. Cours d'Aménagement des Forêts. Paris et Nancy, 1860. 8vo.

Newberry, Dr. Report in Pacific Railroad Report, vol. vi.

Niebelunge-Lied, Der. Abdruck der Handschrift von Joseph von Lassberg. Leipzig, 1840. Folio.

Niel. L'Agriculture des États Sardes. Turin, 1857. 8vo.

Pacific Railroad Report.. Reports of Explorations and Surveys for a Railroad Route to the Pacific. Washington, various years. 12 vols. 4to.

Palissy, Bernard. Œuvres Complètes, avec des Notes, etc., par Paul-Antoine Cap. Paris, 1844. 12mo.

Parade, A. See *Lorentz.*

Paramelle, Abbé. Quellenkunde, Lehre von der Bildung und Auffindung der Quellen; mit einem Vorwort von B. Cotta. Leipzig, 1856. 12mo.

Parish, Dr. Life of Dr. Eleazer Wheelock. 8vo.

Parry, C. C. Report in United States and Mexican Boundary Survey, vol. i.

Parthey, G. Wanderungen durch Sicilien und die Levante. Berlin, 1834. 2 vols. 12mo.

Piper, R. U. The Trees of America. Boston, 1858, Nos. i–iv. 4to.

Plinii, Historia Naturalis, ed. Hardouin. Paris, 1723. 3 vols. folio.

Ponz, Antonio. Viage de España. Madrid, 1788, etc. 18 vols. 12mo.

Quatrefages, A. de. Souvenirs d'un Naturaliste. Paris, 1854. 2 vols. 12mo.

Reclus, Elisée. Le Littoral de la France; Revue des Deux Mondes, 15 Decembre, 1862.

Rentzsch, Hermann. Der Wald im Haushalt der Natur und der Volkswirthschaft. Leipzig, 1862. 8vo.

Ribbe, Charles de. La Provence au point de vue des Bois, des Torrents et des Inondations. Paris, 1857. 8vo.

Ridolfi, Cosimo. Lezioni Orali. Firenze, 1862. 2 vols. 8vo.

Ritter, Carl. Einleitung zur allgemeinen vergleichenden Geographie. Berlin, 1852. 8vo.

——— Die Erdkunde im Verhältniss zur Natur und zur Geschichte des Menschen. Berlin, various years. 19 vols. 8vo.

Rosa, G. Le Condizioni de' boschi, de' fiumi e de' torrenti nella provincia di Bergamo. Politecnico, Dicembre, 1861, pp. 606, 621.

——— Studii sui Boschi. Politecnico, Maggio, 1862, pp. 232, 238.

Rossmässler, C. A. Der Wald. Leipzig und Heidelberg, 1863. 8vo.

Roth, J. Der Vesuv und die Umgebung von Neapel. Berlin, 1857. 8vo.

Rozet, M. Moyens de forcer les Torrents des Montagnes de rendre une partie du sol qu'ils ravagent. Paris, 1856. 8vo. *pamphlet.*

Salvagnoli-Marchetti, Antonio. Memorie Economico-Statistiche sulle Maremme Toscane. Firenze, 1846. 8vo.

——— Raccolta di Documenti sul Bonificamento delle Maremme Toscane. Firenze, 1861. 8vo.

——— Rapporto sul Bonificamento delle Maremme Toscane. Firenze, 1859. 8vo.

——— Rapporto sulle Operazioni Idrauliche ed Economiche eseguite nel 1859–'60 nelle Maremme Toscane. Firenze, 1860. 8vo.

Sandys, George. A Relation of a Journey begun An. Dom. 1610. London, 1627. Folio.

Schacht, H. Les Arbres, Études sur leur Structure et leur Végétation, traduit par E. Morren. Bruxelles et Leipzig, 1862. 8vo.

Schleiden, M. J. Die Landenge von Suês. Leipzig, 1858. 8vo.

——— Die Pflanze und ihr Leben. Leipzig, 1848. 8vo.

Schubert, W. von. Resa genom Sverige, Norrige, Lappland, etc. Stockholm, 1823. 3 vols. 8vo.

Seneca, L. A. Opera Omnia quæ supersunt, ex rec. Ruhkopf. Aug. Taurinorum, 1831. 6 vols. 8vo.

Simonde, J. E. L. Tableau de l'Agriculture Toscane. Genève, 1801. 8vo.

Smith, Dr. William. A Dictionary of the Bible. London, 1860. 3 vols. 8vo.

——— A Dictionary of Greek and Roman Geography. London, 1854, 1857. 2 vols. 8vo.

Smith, John. Historie of Virginia. London, 1624. Folio.

Somerville, Mary. Physical Geography. Fifth edition. London, 1862. 12mo.

Springer, John S. Forest-Life and Forest-Trees. New York, 1851. 12mo.

Stanley, Dr. Lectures on the History of the Jewish Church. London, 1863. 8vo.

Staring, W. H. De Bodem van Nederland. Haarlem, 1856. 2 vols. 8vo.

——— Voormaals en Thans. Haarlem, 1858. 8vo.

Stevens, Gov. Report in Pacific Railroad Report, vol. xii.

Strain, Lieut. I. C. Darien Exploring Expedition, by J. T. Headley, in Harper's Magazine. New York, March, April, and May, 1855.

Streffleur, V. Ueber die Natur und die Wirkungen der Wildbäche. Sitz. Ber. der M. N. W. Classe der Kaiserl. Akad. der Wis. February, 1852, viii, p. 248.

Ström, Isr. Om Skogarnas Vård och Skötsel. Upsala, 1853. *Pamphlet.*

Surell, Alexandre. Étude sur les Torrents des Hautes Alpes. Paris, 1844. 4to.

Tartini, Ferdinando. Memorie sul Bonificamento delle Maremme Toscane. Firenze, 1838. Folio.

Thomas and Baldwin. Gazetteer. Philadelphia, 1855. 1 vol. 8vo.

Thompson, Z. History of Vermont, Natural, Civil, and Statistical. Burlington, 1842. 8vo.

——— Appendix to History of Vermont. Burlington, 1853. 8vo.

Titcomb, Timothy. Lessons in Life. New York, 1861. 12mo.

Treadwell, Dr. Observations of, quoted from Report of Commissioner of Patents.

Troy, Paul. Étude sur le Reboisement des Montagnes. Paris et Toulouse, 1861. 8vo. *pamphlet.*

Tschudi, Friedrich von. Ueber die Landwirthschaftliche Bedeutung der Vögel. St. Gallen, 1854. 12mo.

Tschudi, J. J. von. Travels in Peru. New York, 1848. 8vo.

Vallès, M. F. Études sur les Inondations, leurs causes et leurs effets. Paris, 1857. 8vo.

Valvasor, Johann Weichard. Die Ehre des Herzogthums Crain. Laybach, 1689. 4 vols. folio.

Van Lennep. Extracts from Journal of, in the Missionary Herald.

Vaupell, Chr. Bögens Indvandring i de Danske Skove. Kjöbenhavn, 1857. 8vo.

——— De Nordsjællandske Skovmoser. Kjöbenhavn, 1851. 4to. *pamphlet.*

Venema, G. A. Over het Dalen van de Noordelijke Kuststreken van ons Land. Groningen, 1854. 8vo.

Villa, Antonio Giovanni Batt. Necessità dei Boschi nella Lombardia. Milano, 1850. 4to.

Viollet, J. B. Théorie des Puits Artésiens. Paris, 1840. 8vo.

Waltershausen, W. Sartorius von. Ueber den Sicilianischen Ackerbau Göttingen, 1863.

Webster, Noah. A Collection of Papers on Political, Literary, and Moral Subjects. New York, 1843. 8vo.

Wessely, Joseph. Die Oesterreichischen Alpenländer und ihre Forste. Wien, 1853. 2 vols. 8vo.

Wetzstein, J. G. Reisebericht über Hauran und die Trachonen. Berlin, 1860. 8vo.

Wild, Albert. Die Niederlande. Leipzig, 1862. 2 vols. 8vo.

Wilhelm, Gustav. Der Boden und das Wasser. Wien, 1861. 8vo.

Williams, Dr. History of Vermont. 2 vols. 8vo.

Wittwer, W. C. Die Physikalische Geographie. Leipzig, 1855. 8vo.

Young, Arthur. Voyages en France, pendant les années 1787, 1788, 1789, précédée d'une introduction par Lavergne. Paris, 1860. 2 vols. 12mo.

——— Voyages en Italie et en Espagne, pendant les années 1787, 1789. Paris, 1860. 1 vol. 12mo.

TABLE OF CONTENTS.

CHAPTER I.

INTRODUCTORY.

CHAPTER II.

TRANSFER, MODIFICATION, AND EXTIRPATION OF VEGETABLE AND OF ANIMAL SPECIES.

CHAPTER III.

THE WOODS.

CHAPTER IV.

THE WATERS.

CHAPTER V.

THE SANDS.

CHAPTER VI.

PROJECTED OR POSSIBLE GEOGRAPHICAL CHANGES BY MAN.

CHAPTER I.

INTRODUCTORY.

NATURAL ADVANTAGES OF THE TERRITORY OF THE ROMAN EMPIRE—PHYSICAL DECAY OF THAT TERRITORY AND OF OTHER PARTS OF THE OLD WORLD—CAUSES OF THE DECAY—NEW SCHOOL OF GEOGRAPHERS—REACTION OF MAN UPON NATURE—OBSERVATION OF NATURE—COSMICAL AND GEOLOGICAL INFLUENCES—GEOGRAPHICAL INFLUENCE OF MAN—UNCERTAINTY OF OUR METEOROLOGICAL KNOWLEDGE—MECHANICAL EFFECTS PRODUCED BY MAN ON THE SURFACE OF THE EARTH—IMPORTANCE AND POSSIBILITY OF PHYSICAL RESTORATION—STABILITY OF NATURE—RESTORATION OF DISTURBED HARMONIES—DESTRUCTIVENESS OF MAN—PHYSICAL IMPROVEMENT—HUMAN AND BRUTE ACTION COMPARED—FORMS AND FORMATIONS MOST LIABLE TO PHYSICAL DEGRADATION—PHYSICAL DECAY OF NEW COUNTRIES—CORRUPT INFLUENCE OF PRIVATE CORPORATIONS, *note.*

Natural Advantages of the Territory of the Roman Empire.

THE Roman Empire, at the period of its greatest expansion, comprised the regions of the earth most distinguished by a happy combination of physical advantages. The provinces bordering on the principal and the secondary basins of the Mediterranean enjoyed a healthfulness and an equability of climate, a fertility of soil, a variety of vegetable and mineral products, and natural facilities for the transportation and distribution of exchangeable commodities, which have not been possessed in an equal degree by any territory of like extent in the Old World or the New. The abundance of the land and of the waters adequately supplied every material want, ministered liberally to every sensuous enjoyment. Gold and silver, indeed, were not found in the profusion which has proved so baneful to the industry of lands richer in veins of the precious metals; but mines and river beds yielded them in the spare measure most favorable to stability of value in the medium of exchange, and, consequently, to the regularity of commercial transactions. The ornaments of the barbaric pride of the East, the pearl, the ruby, the sapphire, and the diamond—though not unknown to the luxury of a people whose conquests and whose wealth commanded whatever the habitable world could contribute to augment the material splendor of their social life—were scarcely native to the territory of the empire; but the comparative rarity of these gems in Europe, at somewhat earlier periods, was, perhaps, the very circumstance that led the cunning artists of classic antiquity to enrich softer stones with engravings, which invest the common onyx and carnelian with a worth surpassing, in cultivated eyes, the lustre of the most brilliant oriental jewels.

Of these manifold blessings the temperature of the air, the distribution of the rains, the relative disposition of land and water, the plenty of the sea, the composition of the soil, and the raw material of some of the arts, were wholly gratuitous gifts. Yet the spontaneous nature of Europe, of Western Asia, of Libya, neither fed nor clothed the civilized inhabitants of those provinces. Every loaf was eaten in the sweat of the brow. All must be earned by toil. But toil was nowhere else rewarded by so generous wages; for nowhere would a given amount of intelligent labor produce so abundant, and, at the same time, so varied returns of the good things of material existence. The luxuriant harvests of cereals that waved on every field from the shores of the Rhine to the banks of the Nile, the vines that festooned the hillsides of Syria, of Italy and of Greece, the olives of Spain, the fruits of the gardens of the Hesperides, the domestic quadrupeds and fowls known in ancient rural husbandry—all these were original products of foreign climes, naturalized in new homes, and gradually ennobled by the art of man, while centuries of persevering labor were expelling the wild vegetation, and fitting the earth for the production of more generous growths

Only for the sense of landscape beauty did unaided nature

make provision. Indeed, the very commonness of this source of refined enjoyment seems to have deprived it of half its value; and it was only in the infancy of lands where all the earth was fair, that Greek and Roman humanity had sympathy enough with the inanimate world to be alive to the charms of rural and of mountain scenery. In later generations, when the glories of the landscape had been heightened by plantation, and decorative architecture, and other forms of picturesque improvement, the poets of Greece and Rome were blinded by excess of light, and became, at last, almost insensible to beauties that now, even in their degraded state, enchant every eye, except, too often, those which a lifelong familiarity has dulled to their attractions.

Physical Decay of the Territory of the Roman Empire, and of other parts of the Old World.

If we compare the present physical condition of the countries of which I am speaking, with the descriptions that ancient historians and geographers have given of their fertility and general capability of ministering to human uses, we shall find that more than one half of their whole extent—including the provinces most celebrated for the profusion and variety of their spontaneous and their cultivated products, and for the wealth and social advancement of their inhabitants—is either deserted by civilized man and surrendered to hopeless desolation, or at least greatly reduced in both productiveness and population. Vast forests have disappeared from mountain spurs and ridges; the vegetable earth accumulated beneath the trees by the decay of leaves and fallen trunks, the soil of the alpine pastures which skirted and indented the woods, and the mould of the upland fields, are washed away; meadows, once fertilized by irrigation, are waste and unproductive, because the cisterns and reservoirs that supplied the ancient canals are broken, or the springs that fed them dried up; rivers famous in history and song have shrunk to humble brooklets; the willows that ornamented and protected the banks of the lesser watercourses are gone, and the rivulets have ceased to exist as perennial currents, because the little water that finds its way into their old channels is evaporated by the droughts of summer, or absorbed by the parched earth, before it reaches the lowlands; the beds of the brooks have widened into broad expanses of pebbles and gravel, over which, though in the hot season passed dryshod, in winter sealike torrents thunder; the entrances of navigable streams are obstructed by sandbars, and harbors, once marts of an extensive commerce, are shoaled by the deposits of the rivers at whose mouths they lie; the elevation of the beds of estuaries, and the consequently diminished velocity of the streams which flow into them, have converted thousands of leagues of shallow sea and fertile lowland into unproductive and miasmatic morasses.

Besides the direct testimony of history to the ancient fertility of the regions to which I refer—Northern Africa, the greater Arabian peninsula, Syria, Mesopotamia, Armenia and many other provinces of Asia Minor, Greece, Sicily, and parts of even Italy and Spain—the multitude and extent of yet remaining architectural ruins, and of decayed works of internal improvement, show that at former epochs a dense population inhabited those now lonely districts. Such a population could have been sustained only by a productiveness of soil of which we at present discover but slender traces; and the abundance derived from that fertility serves to explain how large armies, like those of the ancient Persians, and of the Crusaders and the Tartars in later ages, could, without an organized commmissariat, secure adequate supplies in long marches through territories which, in our times, would scarcely afford forage for a single regiment.

It appears, then, that the fairest and fruitfulest provinces of the Roman Empire, precisely that portion of terrestrial surface, in short, which, about the commencement of the Christian era, was endowed with the greatest superiority of soil, climate, and position, which had been carried to the highest pitch of physical improvement, and which thus combined the natural and artificial conditions best fitting it for the habitation and enjoyment of a dense and highly refined and cultivated population, is now completely exhausted of its fertility, or so diminished in productiveness, as, with the exception of a few favored oases that have escaped the general ruin, to be no longer capable of affording sustenance to civilized man. If to this realm of desolation we add the now wasted and solitary soils of Persia and the remoter East, that once fed their millions with milk and honey, we shall see that a territory larger than all Europe, the abundance of which sustained in bygone centuries a population scarcely inferior to that of the whole Christian world at the present day, has been entirely withdrawn from human use, or, at best, is thinly inhabited by tribes too few in numbers, too poor in superfluous products, and too little advanced in culture and the social arts, to contribute anything to the general moral or material interests of the great commonwealth of man.

Causes of this Decay.

The decay of these once flourishing countries is partly due, no doubt, to that class of geological causes, whose action we can neither resist nor guide, and partly also to the direct violence of hostile human force; but it is, in a far greater proportion, either the result of man's ignorant disregard of the laws of nature, or an incidental consequence of war, and of civil and ecclesiastical tyranny and misrule. Next to ignorance of these laws, the primitive source, the *causa causarum*, of the acts and neglects which have blasted with sterility and physical decrepitude the noblest half of the empire of the Cæsars, is, first, the brutal and exhausting despotism which Rome herself exercised over her conquered kingdoms, and even over her Italian territory; then, the host of temporal and spiritual tyrannies which she left as her dying curse to all her wide dominion, and which, in some form of violence or of fraud, still brood over almost every soil subdued by the Roman legions.* Man can-

* In the Middle Ages, feudalism, and a nominal Christianity whose corruptions had converted the most beneficent of religions into the most

not struggle at once against human oppression and the destructive forces of inorganic nature. When both are combined against him, he succumbs after a shorter or a longer struggle, and the fields he has won from the primeval wood relapse into their original state of wild and luxuriant, but

baneful of superstitions, perpetuated every abuse of Roman tyranny, and added new oppressions and new methods of extortion to those invented by older despotisms. The burdens in question fell most heavily on the provinces that had been longest colonized by the Latin race, and these are the portions of Europe which have suffered the greatest physical degradation. "Feudalism," says Blanqui, "was a concentration of scourges. The peasant, stripped of the inheritance of his fathers, became the property of inflexible, ignorant, indolent masters; he was obliged to travel fifty leagues with their carts whenever they required it; he labored for them three days in the week, and surrendered to them half the product of his earnings during the other three; without their consent he could not change his residence, or marry. And why, indeed, should he wish to marry, when he could scarcely save enough to maintain himself? The Abbot Alcuin had twenty thousand slaves, called *serfs*, who were forever attached to the soil. This is the great cause of the rapid depopulation observed in the Middle Ages, and of the prodigious multitude of monasteries which sprang up on every side. It was doubtless a relief to such miserable men to find in the cloisters a retreat from oppression; but the human race never suffered a more cruel outrage, industry never received a wound better calculated to plunge the world again into the darkness of the rudest antiquity. It suffices to say that the prediction of the approaching end of the world, industriously spread by the rapacious monks at this time, was received without terror."—*Résumé de l'Histoire du Commerce*, p. 156.

The abbey of Saint-Germain-des-Prés, which, in the time of Charlemagne, had possessed a million of acres, was, down to the Revolution, still so wealthy, that the personal income of the abbot was 300,000 livres. The abbey of Saint-Denis was nearly as rich as that of Saint-Germain-des-Prés.—LAVERGNE, *Économie Rurale de la France*, p. 104.

Paul Louis Courier quotes from La Bruyère the following striking picture of the condition of the French peasantry in his time: "One sees certain dark, livid, naked, sunburnt, wild animals, male and female, scattered over the country and attached to the soil, which they root and turn over with indomitable perseverance. They have, as it were, an articulate voice, and when they rise to their feet, they show a human face. They are, in fact, men; they creep at night into dens, where they live on black bread, water, and roots. They spare other men the labor of ploughing,

unprofitable forest growth, or fall into that of a dry and barren wilderness.

Rome imposed on the products of agricultural labor in the rural districts taxes which the sale of the entire harvest would scarcely discharge; she drained them of their population by military conscription; she impoverished the peasantry by forced and unpaid labor on public works; she hampered industry and internal commerce by absurd restrictions and unwise regulations. Hence, large tracts of land were left uncultivated, or altogether deserted, and exposed to all the destructive forces which act with such energy on the surface of the earth when it is deprived of those protections by which nature originally guarded it, and for which, in well-ordered husbandry, human ingenuity has contrived more or less efficient substitutes.* Similar abuses have tended to perpetuate and extend these evils in later ages, and it is but recently that, even in the most populous parts of Europe, public attention

sowing, and harvesting, and therefore deserve some small share of the bread they have grown." "These are his own words," adds Courier; "he is speaking of the fortunate peasants, of those who had work and bread, and they were then the few."—*Pétition à la Chambre des Députés pour les Villageois que l'on empêche de danser.*

Arthur Young, who travelled in France from 1787 to 1789, gives, in the twenty-first chapter of his Travels, a frightful account of the burdens of the rural population even at that late period. Besides the regular governmental taxes, and a multitude of heavy fines imposed for trifling offences, he enumerates about thirty seignorial rights, the very origin and nature of some of which are now unknown, while those of some others, claimed and enforced by ecclesiastical as well as by temporal lords, are as repulsive to humanity and morality, as the worst abuses ever practised by heathen despotism. Most of these, indeed, had been commuted for money payments, and were levied on the peasantry as pecuniary imposts for the benefit of prelates and lay lords, who, by virtue of their nobility, were exempt from taxation. Who can wonder at the hostility of the French plebeian classes toward the aristocracy in the days of the Revolution?

* The temporary depopulation of an exhausted soil may be, in some cases, a physical, though, like fallows in agriculture, a dear-bought advantage. Under favorable circumstances, the withdrawal of man and his flocks allows the earth to clothe itself again with forests, and in a few generations to recover its ancient productiveness. In the Middle Ages,

has been half awakened to the necessity of restoring the disturbed harmonies of nature, whose well-balanced influences are so propitious to all her organic offspring, of repaying to our great mother the debt which the prodigality and the thriftlessness of former generations have imposed upon their successors—thus fulfilling the command of religion and of practical wisdom, to use this world as not abusing it.

New School of Geographers.

The labors of Humboldt, of Ritter, of Guyot and their followers, have given to the science of geography a more philosophical, and, at the same time, a more imaginative character than it had received from the hands of their predecessors. Perhaps the most interesting field of speculation, thrown open by the new school to the cultivators of this attractive study, is the inquiry: how far external physical conditions, and especially the configuration of the earth's surface, and the distribution, outline, and relative position of land and water, have influenced the social life and social progress of man.

Reaction of Man on Nature.

But, as we have seen, man has reacted upon organized and inorganic nature, and thereby modified, if not determined, the material structure of his earthly home. The measure of that reaction manifestly constitutes a very important element in the appreciation of the relations between mind and matter, as well as in the discussion of many purely physical problems. But though the subject has been incidentally touched upon by many geographers, and treated with much fulness of detail in regard to certain limited fields of human effort, and to certain specific effects of human action, it has not, as a whole, so far as I know, been made matter of special observation, or of his-

worn-out fields were depopulated, in many parts of the Continent, by civil and ecclesiastical tyrannies, which insisted on the surrender of the half of a loaf already too small to sustain its producer. Thus abandoned, these lands often relapsed into the forest state, and, some centuries later, were again brought under cultivation with renovated fertility.

torical research by any scientific inquirer.* Indeed, until the influence of physical geography upon human life was recognized as a distinct branch of philosophical investigation, there was no motive for the pursuit of such speculations; and it was desirable to inquire whether we have or can become the architects of our own abiding place, only when it was known how the mode of our physical, moral, and intellectual being is affected by the character of the home which Providence has appointed, and we have fashioned, for our material habitation.†

It is still too early to attempt scientific method in discussing this problem, nor is our present store of the necessary facts by any means complete enough to warrant me in promising any approach to fulness of statement respecting them. Systematic observation in relation to this subject has hardly yet begun,‡ and the scattered data which have chanced to be recorded have never been collected. It has now no place in the general scheme of physical science, and is matter of sug-

* The subject of climatic change, with and without reference to human action as a cause, has been much discussed by Moreau de Jonnes, Dureau de la Malle, Arago, Humboldt, Fuster, Gasparin, Becquerel, and many other writers in Europe, and by Noah Webster, Forry, Drake, and others in America. Fraas has endeavored to show, by the history of vegetation in Greece, not merely that clearing and cultivation have affected climate, but that change of climate has essentially modified the character of vegetable life. See his *Klima und Pflanzenwelt in der Zeit*.

† Gods Almagt wenkte van den troon,
En schiep elk volk een land ter woon:
Hier vestte Zij een grondgebied,
Dat Zij ons zelven scheppen liet.

‡ The udometric measurements of Belgrand, reported in the *Annales Forestières* for 1854, and discussed by Vallès in chap. vi of his *Études sur les Inondations*, constitute the earliest, and, in some respects, the most remarkable series known to me, of persevering and systematic observations bearing directly and exclusively upon the influence of human action on climate, or, to speak more accurately, on precipitation and natural drainage. The conclusions of Belgrand, however, and of Vallès, who adopts them, have not been generally accepted by the scientific world, and they seem to have been, in part at least, refuted by the arguments of Héricourt and the observations of Cantegril, Jeandel, and Belland. See chapter iii: *The Woods*.

gestion and speculation only, not of established and positive conclusion. At present, then, all that I can hope is to excite an interest in a topic of much economical importance, by pointing out the directions and illustrating the modes in which human action has been or may be most injurious or most beneficial in its influence upon the physical conditions of the earth we inhabit.

Observation of Nature.

In these pages, as in all I have ever written or propose to write, it is my aim to stimulate, not to satisfy, curiosity, and it is no part of my object to save my readers the labor of observation or of thought. For labor is life, and

Death lives where power lives unused.*

Self is the schoolmaster whose lessons are best worth his wages; and since the subject I am considering has not yet become a branch of formal instruction, those whom it may interest can, fortunately, have no pedagogue but themselves. To the natural philosopher, the descriptive poet, the painter, and the sculptor, as well as to the common observer, the power most important to cultivate, and, at the same time, hardest to acquire, is that of seeing what is before him. Sight is a faculty; seeing, an art. The eye is a physical, but not a self-acting apparatus, and in general it sees only what it seeks. Like a mirror, it reflects objects presented to it; but it may be as insensible as a mirror, and it does not necessarily perceive what it reflects.† It is disputed whether the purely material

* Verses addressed by G. C. to Sir Walter Raleigh.—HAKLUYT, i, p. 668.

† ——— I troer, at Synets Sands er lagt i Öiet,
Mens dette kun er Redskab. Synet strömmer
Fra Sjælens Dyb, og Öiets fine Nerver
Gaae ud fra Hjernens hemmelige Værksted.
HENRIK HERTZ, *Kong René's Datter*, sc. ii.

In the material eye, you think, sight lodgeth!
The *eye* is but an organ. *Seeing* streameth
From the soul's inmost depths. The fine perceptive
Nerve springeth from the brain's mysterious workshop.

sensibility of the eye is capable of improvement and cultivation. It has been maintained by high authority, that the natural acuteness of none of our sensuous faculties can be heightened by use, and hence that the minutest details of the image formed on the retina are as perfect in the most untrained, as in the most thoroughly disciplined organ. This may well be doubted, and it is agreed on all hands that the power of multifarious perception and rapid discrimination may be immensely increased by well-directed practice.* This exercise of the eye

* Skill in marksmanship, whether with firearms or with other projectile weapons, depends more upon the training of the eye than is generally supposed, and I have often found particularly good shots to possess an almost telescopic vision. In the ordinary use of the rifle, the barrel serves as a guide to the eye, but there are sportsmen who fire with the but of the gun at the hip. In this case, as in the use of the sling, the lasso, and the bolas, in hurling the knife (see BABINET, *Lectures*, vii, p. 84), in throwing the boomerang, the javelin, or a stone, and in the employment of the blow pipe and the bow, the movements of the hand and arm are guided by that mysterious sympathy which exists between the eye and the unseeing organs of the body.

In shooting the tortoises of the Amazon and its tributaries, the Indians use an arrow with a long twine and a float attached to it. Avé-Lallemant (*Die Benutzung der Palmen am Amazonenstrom*, p. 32) thus describes their mode of aiming: "As the arrow, if aimed directly at the floating tortoise, would strike it at a small angle, and glance from its flat and wet shell, the archers have a peculiar method of shooting. They are able to calculate exactly their own muscular effort, the velocity of the stream, the distance and size of the tortoise, and they shoot the arrow directly up into the air, so that it falls almost vertically upon the shell of the tortoise, and sticks in it." Analogous calculations—if such physico-mental operations can properly be so called—are made in the use of other missiles; for no projectile flies in a right line to its mark. But the exact training of the eye lies at the bottom of all of them, and marksmanship depends almost wholly upon the power of that organ, whose directions the blind muscles implicitly follow. It is perhaps not out of place to observe here that our English word *aim* comes from the Latin *æstimo*, I calculate or estimate. See WEDGWOOD's *Dictionary of English Etymology*, and the note to the American edition, under *Aim*.

Another proof of the control of the limbs by the eye has been observed in deaf-and-dumb schools, and others where pupils are first taught to write on large slates or blackboards. The writing is in large characters, the

I desire to promote, and, next to moral and religious doctrine, I know no more important practical lessons in this earthly life of ours—which, to the wise man, is a school from the cradle to the grave—than those relating to the employment of the sense of vision in the study of nature.

The pursuit of physical geography, embracing actual observation of terrestrial surface, affords to the eye the best general training that is accessible to all. The majority of even cultivated men have not the time and means of acquiring anything beyond a very superficial acquaintance with any branch of physical knowledge. Natural science has become so vastly extended, its recorded facts and its unanswered questions so immensely multiplied, that every strictly scientific man must be a specialist, and confine the researches of a whole life within a comparatively narrow circle. The study I am recommending, in the view I propose to take of it, is yet in that imperfectly developed state which allows its votaries to occupy themselves with such broad and general views as are attainable by every person of culture, and it does not now require a

small letters being an inch or more high. They are formed with chalk or a slate pencil firmly grasped in the fingers, and by appropriate motions of the wrist, elbow, and shoulder, not of the finger joints. Nevertheless, when a pen is put into the hand of a pupil thus taught, his handwriting, though produced by a totally different set of muscles and muscular movements, is identical in character with that which he has practised on the blackboard.

It has been much doubted whether the artists of the classic ages possessed a more perfect sight than those of modern times, or whether, in executing their minute mosaics and gem engravings, they used magnifiers. Glasses ground convex have been found at Pompeii, but they are too rudely fashioned and too imperfectly polished to have been of any practical use for optical purposes. But though the ancient artists may have had a microscopic vision, their astronomers cannot have had a telescopic power of sight; for they did not discover the satellites of Jupiter, which are often seen with the naked eye at Oormeeah, in Persia, and sometimes, as I can testify by personal observation, at Cairo.

For a very remarkable account of the restoration of vision impaired from age, by judicious training, see *Lessons in Life*, by TIMOTHY TITCOMB, **lesson xi.**

knowledge of special details which only years of application can master. It may be profitably pursued by all; and every traveller, every lover of rural scenery, every agriculturist, who will wisely use the gift of sight, may add valuable contributions to the common stock of knowledge on a subject which, as I hope to convince my readers, though long neglected, and now inartificially presented, is not only a very important, but a very interesting field of inquiry.

Cosmical and Geological Influences.

The revolutions of the seasons, with their alternations of temperature and of length of day and night, the climates of different zones, and the general condition and movements of the atmosphere and the seas, depend upon causes for the most part cosmical, and, of course, wholly beyond our control. The elevation, configuration, and composition of the great masses of terrestrial surface, and the relative extent and distribution of land and water, are determined by geological influences equally remote from our jurisdiction. It would hence seem that the physical adaptation of different portions of the earth to the use and enjoyment of man is a matter so strictly belonging to mightier than human powers, that we can only accept geographical nature as we find her, and be content with such soils and such skies as she spontaneously offers.

Geographical Influence of Man.

But it is certain that man has done much to mould the form of the earth's surface, though we cannot always distinguish between the results of his action and the effects of purely geological causes; that the destruction of the forests, the drainage of lakes and marshes, and the operations of rural husbandry and industrial art have tended to produce great changes in the hygrometric, thermometric, electric, and chemical condition of the atmosphere, though we are not yet able to measure the force of the different elements of disturbance, or to say how far they have been compensated by each other, or by still obscurer influences; and, finally, that the myriad forms of animal and vegetable life, which covered the earth when man first entered upon the theatre of a nature whose harmonies he was destined to derange, have been, through his action, greatly changed in numerical proportion, sometimes much modified in form and product, and sometimes entirely extirpated.

The physical revolutions thus wrought by man have not all been destructive to human interests. Soils to which no nutritious vegetable was indigenous, countries which once brought forth but the fewest products suited for the sustenance and comfort of man—while the severity of their climates created and stimulated the greatest number and the most imperious urgency of physical wants—surfaces the most rugged and intractable, and least blessed with natural facilities of communication, have been made in modern times to yield and distribute all that supplies the material necessities, all that contributes to the sensuous enjoyments and conveniences of civilized life. The Scythia, the Thule, the Britain, the Germany, and the Gaul which the Roman writers describe in such forbidding terms, have been brought almost to rival the native luxuriance and easily won plenty of Southern Italy; and, while the fountains of oil and wine that refreshed old Greece and Syria and Northern Africa have almost ceased to flow, and the soils of those fair lands are turned to thirsty and inhospitable deserts, the hyperborean regions of Europe have conquered, or rather compensated, the rigors of climate, and attained to a material wealth and variety of product that, with all their natural advantages, the granaries of the ancient world can hardly be said to have enjoyed.

These changes for evil and for good have not been caused by great natural revolutions of the globe, nor are they by any means attributable wholly to the moral and physical action or inaction of the peoples, or, in all cases, even of the races that now inhabit these respective regions. They are products of a complication of conflicting or coincident forces, acting through a long series of generations; here, improvidence, wastefulness, and wanton violence; there, foresight and wisely guided persevering industry. So far as they are purely the calculated and desired results of those simple and familiar operations of agriculture and of social life which are as universal as civilization—the removal of the forests which covered the soil required for the cultivation of edible fruits, the drying of here and there a few acres too moist for profitable husbandry, by draining off the surface waters, the substitution of domesticated and nutritious for wild and unprofitable vegetable growths, the construction of roads and canals and artificial harbors—they belong to the sphere of rural, commercial, and political economy more properly than to geography, and hence are but incidentally embraced within the range of our present inquiries, which concern physical, not financial balances. I propose to examine only the greater, more permanent, and more comprehensive mutations which man has produced, and is producing, in earth, sea, and sky, sometimes, indeed, with conscious purpose, but for the most part, as unforseen though natural consequences of acts performed for narrower and more immediate ends.

The exact measurement of the geographical changes hitherto thus effected is, as I have hinted, impracticable, and we possess, in relation to them, the means of only qualitative, not quantitative analysis. The fact of such revolutions is established partly by historical evidence, partly by analogical deduction from effects produced in our own time by operations similar in character to those which must have taken place in more or less remote ages of human action. Both sources of information are alike defective in precision; the latter, for general reasons too obvious to require specification; the former, because the facts to which it bears testimony occurred before the habit or the means of rigorously scientific observation upon any branch of physical research, and especially upon climatic changes, existed.

Uncertainty of our Meteorological Knowledge.

The invention of measures of heat, and of atmospheric moisture, pressure, and precipitation, is extremely recent. Hence, ancient physicists have left us no thermometric or barometric records, no tables of the fall, evaporation, and flow of waters, and even no accurate maps of coast lines and the course of rivers. Their notices of these phenomena are almost wholly confined to excessive and exceptional instances of high or of low temperatures, extraordinary falls of rain and snow, and unusual floods or droughts. Our knowledge of the meteorological condition of the earth, at any period more than two centuries before our own time, is derived from these imperfect details, from the vague statements of ancient historians and geographers in regard to the volume of rivers and the relative extent of forest and cultivated land, from the indications furnished by the history of the agriculture and rural economy of past generations, and from other almost purely casual sources of information.

Among these latter we must rank certain newly laid open fields of investigation, from which facts bearing on the point now under consideration have been gathered. I allude to the discovery of artificial objects in geological formations older than any hitherto recognized as exhibiting traces of the existence of man; to the ancient lacustrine habitations of Switzerland, containing the implements of the occupants, remains of their food, and other relics of human life; to the curious revelations of the Kjökkenmöddinger, or heaps of kitchen refuse, in Denmark, and of the peat mosses in the same and other northern countries; to the dwellings and other evidences of the industry of man in remote ages sometimes laid bare by the movement of sand dunes on the coasts of France and of the North Sea; and to the facts disclosed on the shores of the latter, by excavations in inhabited mounds which were, perhaps, raised before the period of the Roman Empire. These remains are memorials of races which have left no written records, because they perished before the historical period of the countries they occupied began. The plants and animals that furnished the relics found in the deposits were certainly contemporaneous with man; for they are associated with his works, and have evidently served his uses. In some cases, the animals belonged to species well ascertained to be now altogether extinct; in some others, both the animals and the vegetables, though extant elsewhere, have ceased to inhabit the regions where their remains are discovered. From the character of the artificial objects, as compared with others belonging to known dates, or at least to known periods of civilization, ingenious inferences have been drawn as to their age; and from the vegetation, remains of which accompany them, as to the climates of Central and Northern Europe at the time of their production.

There are, however, sources of error which have not always been sufficiently guarded against in making these estimates. When a boat, composed of several pieces of wood fastened together by pins of the same material, is dug out of a bog, it is inferred that the vessel, and the skeletons and implements found with it, belong to an age when the use of iron was not known to the builders. But this conclusion is not warranted by the simple fact that metals were not employed in its construction; for the Nubians at this day build boats large enough to carry half a dozen persons across the Nile, out of small pieces of acacia wood pinned together entirely with wooden bolts. Nor is the occurrence of flint arrow heads and knives, in conjunction with other evidences of human life, conclusive proof as to the antiquity of the latter. Lyell informs us that some Oriental tribes still continue to use the same stone implements as their ancestors, "after that mighty empires, where the use of metals in the arts was well known, had flourished for three thousand years in their neighborhood;" * and the North American Indians now manufacture and use weapons of stone, and even of glass, chipping them in the latter case out of the bottoms of thick bottles, with great facility.†

* *Antiquity of Man*, p. 377.

† "One of them [the Indians] seated himself near me, and made from

We may also be misled by our ignorance of the commercial relations existing between savage tribes. Extremely rude nations, in spite of their jealousies and their perpetual wars, sometimes contrive to exchange the products of provinces very widely separated from each other. The mounds of Ohio contain pearls, thought to be marine, which must have come from the Gulf of Mexico, or perhaps even from California, and the knives and pipes found in the same graves are often formed of far-fetched material, that was naturally paid for by some home product exported to the locality whence the material was derived. The art of preserving fish, flesh, and fowl by drying and smoking is widely diffused, and of great antiquity. The Indians of Long Island Sound are said to have carried on a trade in dried shell fish with tribes residing very far inland. From the earliest ages, the inhabitants of the Faroe and Orkney Islands, and of the opposite mainland coasts, have smoked wild fowl and other flesh. Hence it is possible that the animal and the vegetable food, the remains of which are found in the ancient deposits I am speaking of, may sometimes have been brought from climates remote from that where it was consumed.

The most important, as well as the most trustworthy con-

a fragment of quartz, with a simple piece of round bone, one end of which was hemispherical, with a small crease in it (as if worn by a thread) the sixteenth of an inch deep, an arrow head which was very sharp and piercing, and such as they use on all their arrows. The skill and rapidity with which it was made, without a blow, but by simply breaking the sharp edges with the creased bone by the strength of his hands—for the crease merely served to prevent the instrument from slipping, affording no leverage—was remarkable."—*Reports of Explorations and Surveys for Pacific Railroad*, vol. ii, 1855, *Lieut.* Beckwith's *Report*, p. 43.

It has been said that stone weapons are not found in Sicily, except in certain caves half filled with the skeletons of extinct animals. If they have not been found in that island in more easily accessible localities, I suspect it is because eyes familiar with such objects have not sought for them. In January, 1854, I picked up an arrow head of quartz in a little ravine or furrow just washed out by a heavy rain, in a field near the Simeto. It is rudely fashioned, but its artificial character and its special purpose are quite unequivocal.

clusions with respect to the climate of ancient Europe and Asia, are those drawn from the accounts given by the classical writers of the growth of cultivated plants; but these are by no means free from uncertainty, because we can seldom be sure of an identity of species, almost never of an identity of race or variety, between vegetables known to the agriculturists of Greece and Rome and those of modern times which are thought most nearly to resemble them. Besides this, there is always room for doubt whether the habits of plants long grown in different countries may not have been so changed by domestication that the conditions of temperature and humidity which they required twenty centuries ago were different from those at present demanded for their advantageous cultivation.*

* Probably no cultivated vegetable affords so good an opportunity of studying the laws of acclimation of plants as maize or Indian corn. Maize is grown from the tropics to at least lat. 47° in Northeastern America, and farther north in Europe. Every two or three degrees of latitude brings you to a new variety, with new climatic adaptations, and the capacity of the plant to accommodate itself to new conditions of temperature and season seems almost unlimited. We may easily suppose a variety of this grain, which had become acclimated in still higher latitudes, to have been lost, and in such case the failure to raise a crop from seed brought from some distance to the south would not prove that the climate had become colder.

Many persons now living remember that, when the common tomato was first introduced into Northern New England, it often failed to ripen; but, in the course of a very few years, it completely adapted itself to the climate, and now not only matures both its fruit and its seeds with as much certainty as any cultivated vegetable, but regularly propagates itself by self-sown seed. Meteorological observations, however, do not show any amelioration of the summer climate in those States within that period. See *Appendix*, No. 1.

Maize and the tomato, if not new to human use, have not been long known to civilization, and were, very probably, reclaimed and domesticated at a much more recent period than the plants which form the great staples of agricultural husbandry in Europe and Asia. Is the great power of accomodation to climate possessed by them due to this circumstance? There is some reason to suppose that the character of maize has been sensibly changed by cultivation in South America; for, according to Pöppig,

Even if we suppose an identity of species, of race, and of habit to be established between a given ancient and modern plant, the negative fact that the latter will not grow now where it flourished two thousand years ago does not in all cases prove a change of climate. The same result might follow from the exhaustion of the soil,* or from a change in the quantity of moisture it habitually contains. After a district of country has been completely or even partially cleared of its forest growth, and brought under cultivation, the drying of the soil, under favorable circumstances, goes on for generations, perhaps for ages.† In other cases, from injudicious

the ears of this grain found in old Peruvian tombs belong to varieties not now known in Peru.—*Travels in Peru*, chap. vii.

* The cultivation of madder is said to have been introduced into Europe by an Oriental in the year 1765, and it was first planted in the neighborhood of Avignon. Of course, it has been grown in that district for less than a century; but upon soils where it has been a frequent crop, it is already losing much of its coloring properties.—LAVERGNE, *Économie Rurale de la France*, pp. 259–291.

I believe there is no doubt that the cultivation of madder in the vicinity of Avignon is of recent introduction; but it appears from Fuller and other evidence, that this plant was grown in Europe before the middle of the seventeenth century. The madder brought to France from Persia may be of a different species, or, at least, variety. "Some two years since," says Fuller, "madder was sown by Sir Nicholas Crispe at Debtford, and I hope will have good success; first because it groweth in Zeland in the same (if not a more *northern*) *latitude*. Secondly, because *wild madder* grows here in abundance; and why may not *tame madder* if *cicurated* by art. Lastly, because as good as any grew some thirty years since at Barn-Elms, in Surrey, though it quit not cost through some error in the first planter thereof, which now we hope will be rectified."—FULLER, *Worthies of England*, ii, pp. 57, 58.

Perhaps the recent diseases of the olive, the vine, and the silkworm—the prevailing malady of which insect is supposed by some to be the effect of an incipient decay of the mulberry tree—may be, in part, due to changes produced in the character of the soil by exhaustion through long cultivation.

† In many parts of New England there are tracts, miles in extent, and presenting all varieties of surface and exposure, which were partially cleared sixty or seventy years ago, and where little or no change in the proportion of cultivated ground, pasturage, and woodland has taken place since. In

husbandry, or the diversion or choking up of natural water-courses, it may become more highly charged with humidity. An increase or diminution of the moisture of a soil almost necessarily supposes an elevation or a depression of its winter or its summer heat, and of its extreme, if not of its mean annual temperature, though such elevation or depression may be so slight as not sensibly to raise or lower the mercury in a thermometer exposed to the open air. Any of these causes, more or less humidity, or more or less warmth of soil, would affect the growth both of wild and of cultivated vegetation, and consequently, without any appreciable change in atmospheric temperature, precipitation, or evaporation, plants of a particular species might cease to be advantageously cultivated where they had once been easily reared.*

some cases, these tracts compose basins apparently scarcely at all exposed to any local influence in the way of percolation or infiltration of water toward or from neighboring valleys. But in such situations, apart from accidental disturbances, the ground is growing drier and drier, from year to year, springs are still disappearing, and rivulets still diminishing in their summer supply of water. A probable explanation of this is to be found in the rapid drainage of the surface of cleared ground, which prevents the subterranean natural reservoirs, whether cavities or merely strata of bibulous earth, from filling up. How long this process is to last before an equilibrium is reached, none can say. It may be, for years; it may be, for centuries.

Livingstone states facts which favor the supposition that a secular desiccation is still going on in central Africa. When the regions where the earth is growing drier were cleared of wood, or, indeed, whether forests ever grew there, we are unable to say, but the change appears to have been long in progress. There is reason to suspect a similar revolution in Arabia Petræa. In many of the wadis, and particularly in the gorges between Wadi Feiran and Wadi Esh Sheikh, there are water-worn banks showing that, at no very remote period, the winter floods must have risen fifty feet in channels where the growth of acacias and tamarisks and the testimony of the Arabs concur to prove that they have not risen six feet within the memory or tradition of the present inhabitants. There is little probability that any considerable part of the Sinaitic peninsula has been wooded since its first occupation by man, and we must seek the cause of its increasing dryness elsewhere than in the removal of the forest.

* The soil of newly subdued countries is generally in a high degree

We are very imperfectly acquainted with the present mean and extreme temperature, or the precipitation and the evaporation of any extensive region, even in countries most densely peopled and best supplied with instruments and observers. The progress of science is constantly detecting errors of method in older observations, and many laboriously constructed tables of meteorological phenomena are now thrown aside as fallacious, and therefore worse than useless, because some condition necessary to secure accuracy of result was neglected, in obtaining the data on which they were founded.

To take a familiar instance: it is but recently that attention has been drawn to the great influence of slight changes of

favorable to the growth of the fruits of the garden and the orchard, but usually becomes much less so in a very few years. Plums, of many varieties, were formerly grown, in great perfection and abundance, in many parts of New England where at present they can scarcely be reared at all; and the peach, which, a generation or two ago, succeeded admirably in the southern portion of the same States, has almost ceased to be cultivated there. The disappearance of these fruits is partly due to the ravages of insects, which have in later years attacked them; but this is evidently by no means the sole, or even the principal cause of their decay. In these cases, it is not to the exhaustion of the particular acres on which the fruit trees have grown that we are to ascribe their degeneracy, but to a general change in the condition of the soil or the air; for it is equally impossible to rear them successfully on absolutely new land in the neighborhood of grounds where, not long since, they bore the finest fruit.

I remember being told, many years ago, by one of the earliest settlers of the State of Ohio, a very intelligent and observing person, that the apple trees raised there from seed sown soon after the land was cleared, bore fruit in less than half the time required to bring to bearing those reared from seed sown when the ground had been twenty years under cultivation.

In the peat mosses of Denmark, Scotch firs and other trees not now growing in the same localities, are found in abundance. Every generation of trees leaves the soil in a different state from that in which it found it; every tree that springs up in a group of trees of another species than its own, grows under different influences of light and shade and atmosphere from its predecessors. Hence the succession of crops, which occurs in all natural forests, seems to be due rather to changes of condition than of climate. See chapter iii, *post.*

station upon the results of observations of temperature and precipitation. A thermometer removed but a few hundred yards from its first position differs not unfrequently five, sometimes even ten degrees in its readings; and when we are told that the annual fall of rain on the roof of the observatory at Paris is two inches less than on the ground by the side of it, we may see that the level of the rain-gauge is a point of much consequence in making estimates from its measurements. The data from which results have been deduced with respect to the hygrometrical and thermometrical conditions, the climate in short, of different countries, have very often been derived from observations at single points in cities or districts separated by considerable distances. The tendency of errors and accidents to balance each other authorizes us, indeed, to entertain greater confidence than we could otherwise feel in the conclusions drawn from such tables; but it is in the highest degree probable that they would be much modified by more numerous series of observations, at different stations within narrow limits.*

* The nomenclature of meteorology is vague and sometimes equivocal. Not long since, it was suspected that the observers reporting to a scientific institution did not agree in their understanding of the mode of expressing the direction of the wind prescribed by their instructions. It was found, upon inquiry, that very many of them used the names of the compass-points to indicate the quarter *from* which the wind blew, while others employed them to signify the quarter *toward* which the atmospheric currents were moving. In some instances, the observers were no longer within the reach of inquiry, and of course their tables of the wind were of no value.

"Winds," says Mrs. Somerville, "are named from the points whence they blow, currents exactly the reverse. An easterly wind comes from the east; whereas an easterly current comes from the west, and flows toward the east."—*Physical Geography*, p. 229.

There is no philological ground for this distinction, and it probably originated in a confusion of the terminations *-wardly* and *-erly*, both of which are modern. The root of the former ending implies the direction *to* or *to-ward* which motion is supposed. It corresponds to, and is probably allied with, the Latin *versus*. The termination *-erly* is a corruption or softening of *-ernly*, easterly for easternly, and many authors of the sev-

There is one branch of research which is of the utmost importance in reference to these questions, but which, from the great difficulty of direct observation upon it, has been less successfully studied than almost any other problem of physical science. I refer to the proportions between precipitation, superficial drainage, absorption, and evaporation. Precise actual measurement of these quantities upon even a single acre of ground is impossible; and in all cabinet experiments on the subject, the conditions of the surface observed are so different from those which occur in nature, that we cannot safely reason from one case to the other. In nature, the inclination of the ground, the degree of freedom or obstruction of the surface, the composition and density of the soil, upon which its permeability by water and its power of absorbing and retaining or transmitting moisture depend, its temperature, the dryness or saturation of the subsoil, vary at comparatively short distances; and though the precipitation upon and the superficial flow from very small geographical basins may be estimated with an

enteenth century so write it. In Hakluyt (i, p. 2), *easterly* is applied to place, "*easterly* bounds," and means *eastern*. In a passage in Drayton, "*easterly* winds" must mean winds *from* the east; but the same author, in speaking of nations, uses *northerly* for *northern*. Hakewell says: "The sonne cannot goe more *southernely* from vs, nor come more *northernely* towards vs." Holland, in his translation of Pliny, referring to the moon, has: "When shee is *northerly*," and "shee is gone *southerly*." Richardson, to whom I am indebted for the above citations, quotes a passage from Dampier where *westerly* is applied to the wind, but the context does not determine the direction. The only example of the termination in *-wardly* given by this lexicographer is from Donne, where it means *toward* the west.

Shakspeare, in *Hamlet* (v. ii), uses *northerly* wind for wind *from* the north. Milton does not employ either of these terminations, nor were they known to the Anglo-Saxons, who, however, had adjectives of direction in *-an* or *-en*, *ern* and *-weard*, the last always meaning the point *toward* which motion is supposed, the others that *from* which it proceeds.

We use an *east* wind, an *eastern* wind, and an *easterly* wind, to signify the same thing. The two former expressions are old, and constant in meaning; the last is recent, superfluous, and equivocal. See *Appendix*, No. 2.

approach to precision, yet even here we have no present means of knowing how much of the water absorbed by the earth is restored to the atmosphere by evaporation, and how much carried off by infiltration or other modes of underground discharge. When, therefore, we attempt to use the phenomena observed on a few square or cubic yards of earth, as a basis of reasoning upon the meteorology of a province, it is evident that our data must be insufficient to warrant positive general conclusions. In discussing the climatology of whole countries, or even of comparatively small local divisions, we may safely say that none can tell what percentage of the water they receive from the atmosphere is evaporated; what absorbed by the ground and conveyed off by subterranean conduits; what carried down to the sea by superficial channels; what drawn from the earth or the air by a given extent of forest, of short pasture vegetation, or of tall meadow-grass; what given out again by surfaces so covered, or by bare ground of various textures and composition, under different conditions of atmospheric temperature, pressure, and humidity; or what is the amount of evaporation from water, ice, or snow, under the varying exposures to which, in actual nature, they are constantly subjected. If, then, we are so ignorant of all these climatic phenomena in the best-known regions inhabited by man, it is evident that we can rely little upon theoretical deductions applied to the former more natural state of the same regions—less still to such as are adopted with respect to distant, strange, and primitive countries.

Mechanical Effects produced by Man on the Surface of the Earth more easily ascertainable.

In investigating the mechanical effects of human action on superficial geography, we are treading on safer ground, and dealing with much less subtile phenomena, less intractable elements. Great physical changes can, in some cases, be positively shown, in some almost certainly inferred, to have been produced by the operations of rural industry, and by the labors of man in other spheres of material effort; and hence, in this most important part of our subject, we can arrive at many positive generalizations, and obtain practical results of no small economical value.

Importance and Possibility of Physical Restoration.

Many circumstances conspire to invest with great present interest the questions: how far man can permanently modify and ameliorate those physical conditions of terrestrial surface and climate on which his material welfare depends; how far he can compensate, arrest, or retard the deterioration which many of his agricultural and industrial processes tend to produce; and how far he can restore fertility and salubrity to soils which his follies or his crimes have made barren or pestilential. Among these circumstances, the most prominent, perhaps, is the necessity of providing new homes for a European population which is increasing more rapidly than its means of subsistence, new physical comforts for classes of the people that have now become too much enlightened and have imbibed too much culture to submit to a longer deprivation of a share in the material enjoyments which the privileged ranks have hitherto monopolized.

To supply new hives for the emigrant swarms, there are, first, the vast unoccupied prairies and forests of America, of Australia, and of many other great oceanic islands, the sparsely inhabited and still unexhausted soils of Southern and even Central Africa, and, finally, the impoverished and half-depopulated shores of the Mediterranean, and the interior of Asia Minor and the farther East. To furnish to those who shall remain after emigration shall have conveniently reduced the too dense population of many European states, those means of sensuous and of intellectual well-being which are styled "artificial wants" when demanded by the humble and the poor, but are admitted to be "necessaries" when claimed by the noble and the rich, the soil must be stimulated to its highest powers of production, and man's utmost ingenuity and energy must be tasked to renovate a nature drained, by his improvidence, of fountains which a wise economy would have made plenteous and perennial sources of beauty, health, and wealth.

In those yet virgin lands which the progress of modern discovery in both hemispheres has brought and is still bringing to the knowledge and control of civilized man, not much improvement of great physical conditions is to be looked for. The proportion of forest is indeed to be considerably reduced, superfluous waters to be drawn off, and routes of internal communication to be constructed; but the primitive geographical and climatic features of these countries ought to be, as far as possible, retained.

Stability of Nature.

Nature, left undisturbed, so fashions her territory as to give it almost unchanging permanence of form, outline, and proportion, except when shattered by geologic convulsions; and in these comparatively rare cases of derangement, she sets herself at once to repair the superficial damage, and to restore, as nearly as practicable, the former aspect of her dominion. In new countries, the natural inclination of the ground, the self-formed slopes and levels, are generally such as best secure the stability of the soil. They have been graded and lowered or elevated by frost and chemical forces and gravitation and the flow of water and vegetable deposit and the action of the winds, until, by a general compensation of conflicting forces, a condition of equilibrium has been reached which, without the action of man, would remain, with little fluctuation, for countless ages.

We need not go far back to reach a period when, in all that portion of the North American continent which has been occupied by British colonization, the geographical elements very nearly balanced and compensated each other. At the commencement of the seventeenth century, the soil, with insignificant exceptions, was covered with forests;* and

* I do not here speak of the vast prairie region of the Mississippi val-

whenever the Indian, in consequence of war or the exhaustion of the beasts of the chase, abandoned the narrow fields he had planted and the woods he had burned over, they speedily returned, by a succession of herbaceous, arborescent, and arboreal growths, to their original state. Even a single generation sufficed to restore them almost to their primitive luxuriance of forest vegetation.* The unbroken forests had attained to their maximum density and strength of growth, and, as the older trees decayed and fell, they were succeeded by new shoots or seedlings, so that from century to century no perceptible change seems to have occurred in the wood, except the slow, spontaneous succession of crops. This succession involved no interruption of growth, and but little break in the "boundless contiguity of shade;" for, in the husbandry of nature, there are no fallows. Trees fall singly, not by square roods, and the tall pine is hardly prostrate, before the light and heat, admitted to the ground by the removal of the dense crown of foliage which had shut them out, stimulate the germination of the seeds of broad-leaved trees that had lain, waiting this kindly influence, perhaps for centuries. Two natural causes, destructive in character, were, indeed, in operation in the primitive American forests, though, in the Northern colonies, at least, there were sufficient compensations; for we do not discover that any considerable permanent change was produced by them. I refer to the action of

ley, which cannot properly be said ever to have been a field of British colonization; but of the original colonies, and their dependencies in the territory of the present United States, and in Canada. It is, however, equally true of the Western prairies as of the Eastern forest land, that they had arrived at a state of equilibrium, though under very different conditions.

* The great fire of Miramichi in 1825, probably the most extensive and terrific conflagration recorded in authentic history, spread its ravages over nearly six thousand square miles, chiefly of woodland, and was of such intensity that it seemed to consume the very soil itself. But so great are the recuperative powers of nature, that, in twenty-five years, the ground was thickly covered again with trees of fair dimensions, except where cultivation and pasturage kept down the forest growth.

beavers and of fallen trees in producing bogs,* and of smaller animals, insects, and birds, in destroying the woods. Bogs are less numerous and extensive in the Northern States of the American union, because the natural inclination of the surface favors drainage; but they are more frequent, and cover more ground, in the Southern States, for the opposite reason.†

* The English nomenclature of this geographical feature does not seem well settled. We have *bog*, *swamp*, *marsh*, *morass*, *moor*, *fen*, *turf moss*, *peat moss*, *quagmire*, all of which, though sometimes more or less accurately discriminated, are often used interchangeably, or are perhaps employed, each exclusively, in a particular district. In Sweden, where, especially in the Lappish provinces, this terr-aqueous formation is very extensive and important, the names of its different kinds are more specific in their application. The general designation of all soils permanently pervaded with water is *Kärr*. The elder Læstadius divides the *Kärr* into two genera: *Myror* (sing. *myra*), and *Mossar* (sing. *mosse*). "The former," he observes, "are grass-grown, and overflowed with water through almost the whole summer; the latter are covered with mosses and always moist, but very seldom overflowed." He enumerates the following species of *Myra*, the character of which will perhaps be sufficiently understood by the Latin terms into which he translates the vernacular names, for the benefit of strangers not altogether familiar with the language and the subject: 1. *Hömyror*, paludes graminosæ. 2. *Dy*, paludes profundæ. 3. *Flarkmyror*, or proper *kärr*, paludes limosæ. 4. *Fjällmyror*, paludes uliginosæ. 5. *Tufmyror*, paludes cæspitosæ. 6. *Rismyror*, paludes virgatæ. 7. *Starrängar*, prata irrigata, with their subdivisions, dry *starrängar* or *risängar*, wet *starrängar* and *fräkengropar*. 8. *Pölar*, lacunæ. 9. *Gölar*, fossæ inundatæ. The *Mossar*, paludes turfosæ, which are of great extent, have but two species: 1. *Torfmossar*, called also *Mossmyror* and *Snottermyror*, and, 2. *Björnmossar*.

The accumulations of stagnant or stagnating water originating in bogs are distinguished into *Träsk*, stagna, and *Tjernar* or *Tjärnar* (sing. *Tjern* or *Tjärn*), stagnatiles. *Träsk* are pools fed by bogs, or water emanating from them, and their bottoms are slimy; *Tjernar* are small *Träsk* situated within the limits of *Mossar*.—L. L. Læstadius, *om Möjligheten af Uppodlingar i Lappmarken*, pp. 23, 24.

† Although the quantity of bog land in New England is less than in many other regions of equal area, yet there is a considerable extent of this formation in some of the Northeastern States. Dana (*Manual of Geology*, p. 614) states that the quantity of peat in Massachusetts is estimated at 120,000,000 cords, or nearly 569,000,000 cubic yards, but he does not give

They generally originate in the checking of watercourses by the falling of timber, or of earth and rocks, across their channels. If the impediment thus created is sufficient to retain a permanent accumulation of water behind it, the trees whose roots are overflowed soon perish, and then by their fall increase the obstruction, and, of course, occasion a still wider spread of the stagnating stream. This process goes on until the water finds a new outlet, at a higher level, not liable to similar interruption. The fallen trees not completely covered by water are soon overgrown with mosses; aquatic and semi-aquatic plants propagate themselves, and spread until they more or less completely fill up the space occupied by the water, and the surface is gradually converted from a pond to a quaking morass.* The morass is slowly solidified by vegetable

either the area or the depth of the deposits. In any event, however, bogs cover but a small percentage of the territory in any of the Northern States, while it is said that one tenth of the whole surface of Ireland is composed of bogs, and there are still extensive tracts of undrained marsh in England.

Bogs, independently of their importance in geology as explaining the origin of some kinds of mineral coal, have a present value as repositories of fuel. Peat beds have sometimes a thickness of ten or twelve yards, or even more. A depth of ten yards would give 48,000 cubic yards to the acre. The greatest quantity of firewood yielded by the forests of New England to the acre is 100 cords solid measure, or 474 cubic yards; but this comprises only the trunks and larger branches. If we add the small branches and twigs, it is possible that 600 cubic yards might, in some cases, be cut on an acre. This is only one eightieth part of the quantity of peat sometimes found on the same area. It is true that a yard of peat and a yard of wood are not the equivalents of each other, but the fuel on an acre of deep peat is worth much more than that on an acre of the best woodland. Besides this, wood is perishable, and the quantity on an acre cannot be increased beyond the amount just stated; peat is indestructible, and the beds are always growing.

* "Aquatic plants have a utility in raising the level of marshy grounds, which renders them very valuable, and may well be called a geological function. * * *

"The engineer drains ponds at a great expense by lowering the surface of the water; nature attains the same end, gratuitously, by raising the level of the soil without depressing that of the water; but she proceeds more slowly. There are, in the Landes, marshes where this natural filling

production and deposit, then very often restored to the forest condition by the growth of black ashes, cedars, or, in southern latitudes, cypresses, and other trees suited to such a soil, and thus the interrupted harmony of nature is at last reëstablished.

I am disposed to think that more bogs in the Northern States owe their origin to beavers than to accidental obstructions of rivulets by wind-fallen or naturally decayed trees; for there are few swamps in those States, at the outlets of which we may not, by careful search, find the remains of a beaver dam. The beaver sometimes inhabits natural lakelets, but he prefers to owe his pond to his own ingenuity and toil. The reservoir once constructed, its inhabitants rapidly multiply and as its harvests of pond lilies, and other aquatic plants on which this quadruped feeds in winter, become too small for the growing population, the beaver metropolis sends out expeditions of discovery and colonization. The pond gradually fills up, by the operation of the same causes as when it owes its existence to an accidental obstruction, and when, at last, the original settlement is converted into a bog by the usual processes of vegetable life, the remaining inhabitants abandon it and build on some virgin brooklet a new city of the waters.

In countries somewhat further advanced in civilization than those occupied by the North American Indians, as in mediæval Ireland, the formation of bogs may be commenced by the neglect of man to remove, from the natural channels of superficial drainage, the tops and branches of trees felled

has a thickness of four mètres, and some of them, at first lower than the sea, have been thus raised and drained so as to grow summer crops, such, for example, as maize."—Boitel, *Mise en valeur des Terres pauvres*, p. 227.

The bogs of Denmark—the examination of which by Steenstrup and Vaupell has presented such curious results with respect to the natural succession of forest trees—appear to have gone through this gradual process of drying, and the birch, which grows freely in very wet soils, has contributed very effectually by its annual deposits to raise the surface above the water level, and thus to prepare the ground for the oak.—Vaupell, *Bögens Indvandring*, pp. 39, 40.

for the various purposes to which wood is applicable in his rude industry; and, when the flow of the water is thus checked, nature goes on with the processes I have already described. In such half-civilized regions, too, windfalls are more frequent than in those where the forest is unbroken, because, when openings have been made in it, for agricultural or other purposes, the entrance thus afforded to the wind occasions the sudden overthrow of hundreds of trees which might otherwise have stood for generations, and thus have fallen to the ground, only one by one, as natural decay brought them down.* Besides this, the flocks bred by man in the pastoral state, keep down the incipient growth of trees on the half-dried bogs, and prevent them from recovering their primitive condition.

Young trees in the native forest are sometimes girdled and killed by the smaller rodent quadrupeds, and their growth is checked by birds which feed on the terminal bud; but these animals, as we shall see, are generally found on the skirts of the wood only, not in its deeper recesses, and hence the mischief they do is not extensive. The insects which damage primitive forests by feeding upon products of trees essential to their growth, are not numerous, nor is their appearance, in destructive numbers, frequent; and those which perforate the stems and branches, to deposit and hatch their eggs, more commonly select dead trees for that purpose, though, unhappily, there are important exceptions to this latter remark.† I

* Careful examination of the peat mosses in North Sjælland—which are so abundant in fossil wood that, within thirty years, they have yielded above a million of trees—shows that the trees have generally fallen from age and not from wind. They are found in depressions on the declivities of which they grew, and they lie with the top lowest, always falling toward the bottom of the valley.—VAUPELL, *Bögens Indvandring i de Danske Skove*, pp. 10, 14.

† The locust insect, *Clitus pictus*, which deposits its eggs in the American locust, *Robinia pseudacacia*, is one of these, and its ravages have been and still are most destructive to that very valuable tree, so remarkable for combining rapidity of growth with strength and durability of wood. This insect, I believe, has not yet appeared in Europe, where, since the so gen-

do not know that we have any evidence of the destruction or serious injury of American forests by insects, before or even soon after the period of colonization; but since the white man has laid bare a vast proportion of the earth's surface, and thereby produced changes favorable, perhaps, to the multiplication of these pests, they have greatly increased in numbers, and, apparently, in voracity also. Not many years ago, the pines on thousands of acres of land in North Carolina, were destroyed by insects not known to have ever done serious injury to that tree before. In such cases as this and others of the like sort, there is good reason to believe that man is the indirect cause of an evil for which he pays so heavy a penalty. Insects increase whenever the birds which feed upon them disappear. Hence, in the wanton destruction of the robin and other insectivorous birds, the *bipes implumis*, the featherless biped, man, is not only exchanging the vocal orchestra which greets the rising sun for the drowsy beetle's evening drone, and depriving his groves and his fields of their fairest ornament, but he is waging a treacherous warfare on his natural allies.*

eral employment of the *Robinia* to clothe and protect embankments and the scarps of deep cuts on railroads, it would do incalculable mischief. As a traveller, however, I should find some compensation for this evil in the destruction of these acacia hedges, which as completely obstruct the view on hundreds of miles of French and Italian railways, as the garden walls of the same countries do on the ordinary roads. See *Appendix*, No. 4.

* In the artificial woods of Europe, insects are far more numerous and destructive to trees than in the primitive forests of America, and the same remark may be made of the smaller rodents, such as moles, mice, and squirrels. In the dense native wood, the ground and the air are too humid, the depth of shade too great for many tribes of these creatures, while near the natural meadows and other open grounds, where circumstances are otherwise more favorable for their existence and multiplication, their numbers are kept down by birds, serpents, foxes, and smaller predacious quadrupeds. In civilized countries, these natural enemies of the worm, the beetle and the mole, are persecuted, sometimes almost exterminated, by man, who also removes from his plantations the decayed or wind-fallen trees, the shrubs and underwood, which, in a state of nature, furnished food and shelter to the borer and the rodent, and often

In fine, in countries untrodden by man, the proportions and relative positions of land and water, the atmospheric precipitation and evaporation, the thermometric mean, and the distribution of vegetable and animal life, are subject to change only from geological influences so slow in their operation that the geographical conditions may be regarded as constant and immutable. These arrangements of nature it is, in most cases, highly desirable substantially to maintain, when such regions become the seat of organized commonwealths. It is, therefore, a matter of the first importance, that, in commencing the process of fitting them for permanent civilized occupation, the transforming operations should be so conducted as not unnecessarily to derange and destroy what, in too many cases, it is beyond the power of man to rectify or restore.

also to the animals that preyed upon them. Hence the insect and the gnawing quadruped are allowed to increase, from the expulsion of the police which, in the natural wood, prevent their excessive multiplication, and they become destructive to the forest because they are driven to the living tree for nutriment and cover. The forest of Fontainebleau is almost wholly without birds, and their absence is ascribed by some writers to the want of water, which, in the thirsty sands of that wood, does not gather into running brooks; but the want of undergrowth is perhaps an equally good reason for their scarcity. In a wood of spontaneous growth, ordered and governed by nature, the squirrel does not attack trees, or at least the injury he may do is too trifling to be perceptible, but he is a formidable enemy to the plantation. "The squirrels bite the cones of the pine and consume the seed which might serve to restock the wood; they do still more mischief by gnawing off, near the leading shoot, a strip of bark, and thus often completely girdling the tree. Trees so injured must be felled, as they would never acquire a vigorous growth. The squirrel is especially destructive to the pine in Sologne, where he gnaws the bark of trees twenty or twenty-five years old." But even here, nature sometimes provides a compensation, by making the appetite of this quadruped serve to prevent an excessive production of seed cones, which tends to obstruct the due growth of the leading shoot. "In some of the pineries of Brittany which produce cones so abundantly as to strangle the development of the leading shoot of the maritime pine, it has been observed that the pines are most vigorous where the squirrels are most numerous, a result attributed to the repression of the cones by this rodent."—BOITEL, *Mise en valeur des Terres pauvres*, p. 50. See *Appendix*, No. 5.

Restoration of Disturbed Harmonies.

In reclaiming and reoccupying lands laid waste by human improvidence or malice, and abandoned by man, or occupied only by a nomade or thinly scattered population, the task of the pioneer settler is of a very different character. He is to become a co-worker with nature in the reconstruction of the damaged fabric which the negligence or the wantonness of former lodgers has rendered untenantable. He must aid her in reclothing the mountain slopes with forests and vegetable mould, thereby restoring the fountains which she provided to water them; in checking the devastating fury of torrents, and bringing back the surface drainage to its primitive narrow channels; and in drying deadly morasses by opening the natural sluices which have been choked up, and cutting new canals for drawing off their stagnant waters. He must thus, on the one hand, create new reservoirs, and, on the other, remove mischievous accumulations of moisture, thereby equalizing and regulating the sources of atmospheric humidity and of flowing water, both which are so essential to all vegetable growth, and, of course, to human and lower animal life.

Destructiveness of Man.

Man has too long forgotten that the earth was given to him for usufruct alone, not for consumption, still less for profligate waste. Nature has provided against the absolute destruction of any of her elementary matter, the raw material of her works; the thunderbolt and the tornado, the most convulsive throes of even the volcano and the earthquake, being only phenomena of decomposition and recomposition. But she has left it within the power of man irreparably to derange the combinations of inorganic matter and of organic life, which through the night of æons she had been proportioning and balancing, to prepare the earth for his habitation, when, in the fulness of time, his Creator should call him forth to enter into its possession.

Apart from the hostile influence of man, the organic and

the inorganic world are, as I have remarked, bound together by such mutual relations and adaptations as secure, if not the absolute permanence and equilibrium of both, a long continuance of the established conditions of each at any given time and place, or at least, a very slow and gradual succession of changes in those conditions. But man is everywhere a disturbing agent. Wherever he plants his foot, the harmonies of nature are turned to discords. The proportions and accommodations which insured the stability of existing arrangements are overthrown. Indigenous vegetable and animal species are extirpated, and supplanted by others of foreign origin, spontaneous production is forbidden or restricted, and the face of the earth is either laid bare or covered with a new and reluctant growth of vegetable forms, and with alien tribes of animal life. These intentional changes and substitutions constitute, indeed, great revolutions; but vast as is their magnitude and importance, they are, as we shall see, insignificant in comparison with the contingent and unsought results which have flowed from them.

The fact that, of all organic beings, man alone is to be regarded as essentially a destructive power, and that he wields energies to resist which, nature—that Nature whom all material life and all inorganic substance obey—is wholly impotent, tends to prove that, though living in physical nature, he is not of her, that he is of more exalted parentage, and belongs to a higher order of existences than those born of her womb and submissive to her dictates.

There are, indeed, brute destroyers, beasts and birds and insects of prey—all animal life feeds upon, and, of course, destroys other life,—but this destruction is balanced by compensations. It is, in fact, the very means by which the existence of one tribe of animals or of vegetables is secured against being smothered by the encroachments of another; and the reproductive powers of species, which serve as the food of others, are always proportioned to the demand they are destined to supply. Man pursues his victims with reckless destructiveness; and, while the sacrifice of life by the lower animals is limited by the cravings of appetite, he unsparingly persecutes, even to extirpation, thousands of organic forms which he cannot consume.*

* The terrible destructiveness of man is remarkably exemplified in the chase of large mammalia and birds for single products, attended with the entire waste of enormous quantities of flesh, and of other parts of the animal, which are capable of valuable uses. The wild cattle of South America are slaughtered by millions for their hides and horns; the buffalo of North America for his skin or his tongue; the elephant, the walrus, and the narwhal for their tusks; the cetacea, and some other marine animals, for their oil and whalebone; the ostrich and other large birds, for their plumage. Within a few years, sheep have been killed in New England by whole flocks, for their pelts and suet alone, the flesh being thrown away; and it is even said that the bodies of the same quadrupeds have been used in Australia as fuel for limekilns. What a vast amount of human nutriment, of bone, and of other animal products valuable in the arts, is thus recklessly squandered! In nearly all these cases, the part which constitutes the motive for this wholesale destruction, and is alone saved, is essentially of insignificant value as compared with what is thrown away. The horns and hide of an ox are not economically worth a tenth part as much as the entire carcass.

One of the greatest benefits to be expected from the improvements of civilization is, that increased facilities of communication will render it possible to transport to places of consumption much valuable material that is now wasted because the price at the nearest market will not pay freight. The cattle slaughtered in South America for their hides would feed millions of the starving population of the Old World, if their flesh could be economically preserved and transported across the ocean.

We are beginning to learn a better economy in dealing with the inorganic world. The utilization—or, as the Germans more happily call it, the Verwerthung, the *beworthing*—of waste from metallurgical, chemical, and manufacturing establishments, is among the most important results of the application of science to industrial purposes. The incidental products from the laboratories of manufacturing chemists often become more valuable than those for the preparation of which they were erected. The slags from silver refineries, and even from smelting houses of the coarser metals, have not unfrequently yielded to a second operator a better return than the first had derived from dealing with the natural ore; and the saving of lead carried off in the smoke of furnaces has, of itself, given a large profit on the capital invested in the works. A few years ago, an officer of an American mint was charged with embezzling gold committed to him for coinage. He insisted, in his defence, that much of the metal was vola-

The earth was not, in its natural condition, completely adapted to the use of man, but only to the sustenance of wild animals and wild vegetation. These live, multiply their kind in just proportion, and attain their perfect measure of strength and beauty, without producing or requiring any change in the natural arrangements of surface, or in each other's spontaneous tendencies, except such mutual repression of excessive increase as may prevent the extirpation of one species by the encroachments of another. In short, without man, lower animal and spontaneous vegetable life would have been constant in type, distribution, and proportion, and the physical geography of the earth would have remained undisturbed for indefinite periods, and been subject to revolution only from possible, unknown cosmical causes, or from geological action.

But man, the domestic animals that serve him, the field and garden plants the products of which supply him with food and clothing, cannot subsist and rise to the full development of their higher properties, unless brute and unconscious nature be effectually combated, and, in a great degree, vanquished by human art. Hence, a certain measure of transformation of terrestrial surface, of suppression of natural, and stimulation of artificially modified productivity becomes necessary. This measure man has unfortunately exceeded. He has felled the forests whose network of fibrous roots bound the mould to the rocky skeleton of the earth; but had he allowed here and there a belt of woodland to reproduce itself by spontaneous propagation, most of the mischiefs which his reckless destruction of the natural protection of the soil has occasioned would have been averted. He has broken up the mountain reservoirs, the percolation of whose waters through unseen channels supplied the fountains that refreshed his cattle and fertilized his fields; but he has neglected to maintain the cisterns and the canals of irrigation which a wise antiquity

tilized and lost in refining and melting, and upon scraping the chimneys of the melting furnaces and the roofs of the adjacent houses, gold enough was found in the soot to account for no small part of the deficiency.

had constructed to neutralize the consequences of its own imprudence. While he has torn the thin glebe which confined the light earth of extensive plains, and has destroyed the fringe of semi-aquatic plants which skirted the coast and checked the drifting of the sea sand, he has failed to prevent the spreading of the dunes by clothing them with artificially propagated vegetation. He has ruthlessly warred on all the tribes of animated nature whose spoil he could convert to his own uses, and he has not protected the birds which prey on the insects most destructive to his own harvests.

Purely untutored humanity, it is true, interferes comparatively little with the arrangements of nature,* and the destruc-

* It is an interesting and not hitherto sufficiently noticed fact, that the domestication of the organic world, so far as it has yet been achieved, belongs, not indeed to the savage state, but to the earliest dawn of civilization, the conquest of inorganic nature almost as exclusively to the most advanced stages of artificial culture. It is familiarly known to all who have occupied themselves with the psychology and habits of the ruder races, and of persons with imperfectly developed intellects in civilized life, that although these humble tribes and individuals sacrifice, without scruple, the lives of the lower animals to the gratification of their appetites and the supply of their other physical wants, yet they nevertheless seem to cherish with brutes, and even with vegetable life, sympathies which are much more feebly felt by civilized men. The popular traditions of the simpler peoples recognize a certain community of nature between man, brute animals, and even plants; and this serves to explain why the apologue or fable, which ascribes the power of speech and the faculty of reason to birds, quadrupeds, insects, flowers, and trees, is one of the earliest forms of literary composition.

In almost every wild tribe, some particular quadruped or bird, though persecuted as a destroyer of more domestic beasts, or hunted for food, is regarded with peculiar respect, one might almost say, affection. Some of the North American aboriginal nations celebrate a propitiatory feast to the manes of the intended victim before they commence a bear hunt; and the Norwegian peasantry have not only retained an old proverb which ascribes to the same animal "*ti Mænds Styrke og tolv Mænds Vid*," ten men's strength and twelve men's cunning, but they still pay to him something of the reverence with which ancient superstition invested him. The student of Icelandic literature will find in the saga of *Finnbogi hinn rami* a curious illustration of this feeling, in an account of a dialogue between a

tive agency of man becomes more and more energetic and unsparing as he advances in civilization, until the impoverishment, with which his exhaustion of the natural resources of the soil is threatening him, at last awakens him to the neces-

Norwegian bear and an Icelandic champion—dumb show on the part of Bruin, and chivalric words on that of Finnbogi—followed by a duel, in which the latter, who had thrown away his arms and armor in order that the combatants might meet on equal terms, was victorious. Drummond Hay's very interesting work on Morocco contains many amusing notices of a similar feeling entertained by the Moors toward the redoubtable enemy of their flocks—the lion.

This sympathy helps us to understand how it is that most if not all the domestic animals—if indeed they ever existed in a wild state—were appropriated, reclaimed and trained before men had been gathered into organized and fixed communities, that almost every known esculent plant had acquired substantially its present artificial character, and that the properties of nearly all vegetable drugs and poisons were known at the remotest period to which historical records reach. Did nature bestow upon primitive man some instinct akin to that by which she teaches the brute to select the nutritious and to reject the noxious vegetables indiscriminately mixed in forest and pasture?

This instinct, it must be admitted, is far from infallible, and, as has been hundreds of times remarked by naturalists, it is in many cases not an original faculty but an acquired and transmitted habit. It is a fact familiar to persons engaged in sheep husbandry in New England—and I have seen it confirmed by personal observation—that sheep bred where the common laurel, as it is called, *Kalmia angustifolia*, abounds, almost always avoid browsing upon the leaves of that plant, while those brought from districts where laurel is unknown, and turned into pastures where it grows, very often feed upon it and are poisoned by it. A curious acquired and hereditary instinct, of a different character, may not improperly be noticed here. I refer to that by which horses bred in provinces where quicksands are common avoid their dangers or extricate themselves from them. See Brémontier, *Mémoire sur les Dunes*, *Annales des Ponts et Chaussées*, 1833: *premier sémestre*, pp. 155–157.

It is commonly said in New England, and I believe with reason, that the crows of this generation are wiser than their ancestors. Scarecrows which were effectual fifty years ago are no longer respected by the plunderers of the cornfield, and new terrors must from time to time be invented for its protection. See *Appendix*, No. 6.

Civilization has added little to the number of vegetable or animal species grown in our fields or bred in our folds, while, on the contrary,

sity of preserving what is left, if not of restoring what has been wantonly wasted. The wandering savage grows no cultivated vegetable, fells no forest, and extirpates no useful plant, no noxious weed. If his skill in the chase enables him to entrap numbers of the animals on which he feeds, he compensates this loss by destroying also the lion, the tiger, the wolf, the otter, the seal, and the eagle, thus indirectly protecting the feebler quadrupeds and fish and fowls, which would otherwise become the booty of beasts and birds of prey. But with stationary life, or rather with the pastoral state, man at once commences an almost indiscriminate warfare upon all the forms of animal and vegetable existence around him, and as he advances in civilization, he gradually eradicates or transforms every spontaneous product of the soil he occupies.*

Human and Brute Action Compared.

It has been maintained by authorities as high as any known to modern science, that the action of man upon nature, though greater in *degree*, does not differ in *kind*, from

the subjugation of the inorganic forces, and the consequent extension of man's sway over, not the annual products of the earth only, but her substance and her springs of action, is almost entirely the work of highly refined and cultivated ages. The employment of the elasticity of wood and of horn, as a projectile power in the bow, is nearly universal among the rudest savages. The application of compressed air to the same purpose, in the blowpipe, is more restricted, and the use of the mechanical powers, the inclined plane, the wheel and axle, and even the wedge and lever, seems almost unknown except to civilized man. I have myself seen European peasants to whom one of the simplest applications of this latter power was a revelation.

* The difference between the relations of savage life, and of incipient civilization, to nature, is well seen in that part of the valley of the Mississippi which was once occupied by the mound builders and afterward by the far less developed Indian tribes. When the tillers of the fields, which must have been cultivated to sustain the large population that once inhabited those regions perished, or were driven out, the soil fell back to the normal forest state, and the savages who succeeded the more advanced race interfered very little, if at all, with the ordinary course of spontaneous nature.

that of wild animals. It appears to me to differ in essential character, because, though it is often followed by unforeseen and undesired results, yet it is nevertheless guided by a self-conscious and intelligent will aiming as often at secondary and remote as at immediate objects. The wild animal, on the other hand, acts instinctively, and, so far as we are able to perceive, always with a view to single and direct purposes. The backwoodsman and the beaver alike fell trees; the man that he may convert the forest into an olive grove that will mature its fruit only for a succeeding generation, the beaver that he may feed upon their bark or use them in the construction of his habitation. Human differs from brute action, too, in its influence upon the material world, because it is not controlled by natural compensations and balances. Natural arrangements, once disturbed by man, are not restored until he retires from the field, and leaves free scope to spontaneous recuperative energies; the wounds he inflicts upon the material creation are not healed until he withdraws the arm that gave the blow. On the other hand, I am not aware of any evidence that wild animals have ever destroyed the smallest forest, extirpated any organic species or modified its natural character, occasioned any permanent change of terrestrial surface, or produced any disturbance of physical conditions which nature has not, of herself, repaired without the expulsion of the animal that had caused it.*

The form of geographical surface, and very probably the climate of a given country, depend much on the character of the vegetable life belonging to it. Man has, by domestication, greatly changed the habits and properties of the plants he rears; he has, by voluntary selection, immensely modified the forms and qualities of the animated creatures that serve him; and he has, at the same time, completely rooted out many forms of animal if not of vegetable being.† What is there, in

* There is a possible—but only a possible—exception in the case of the American bison. See note on that subject in chap. iii, *post*.

† Whatever may be thought of the modification of organic species by natural selection, there is certainly no evidence that animals have exerted

the influence of brute life, that corresponds to this? We have no reason to believe that in that portion of the American continent which, though peopled by many tribes of quadruped and fowl, remained uninhabited by man, or only thinly occupied by purely savage tribes, any sensible geographical change had occurred within twenty centuries before the epoch of discovery and colonization, while, during the same period, man had changed millions of square miles, in the fairest and most fertile regions of the Old World, into the barrenest deserts.

The ravages committed by man subvert the relations and destroy the balance which nature had established between her organized and her inorganic creations; and she avenges herself upon the intruder, by letting loose upon her defaced provinces destructive energies hitherto kept in check by organic forces destined to be his best auxiliaries, but which he has unwisely dispersed and driven from the field of action. When the forest is gone, the great reservoir of moisture stored up in its vegetable mould is evaporated, and returns only in deluges of rain to wash away the parched dust into which that mould has been converted. The well-wooded and humid hills are turned to ridges of dry rock, which encumbers the low grounds and chokes the watercourses with its debris, and—except in countries favored with an equable distribution of rain through the seasons, and a moderate and regular inclination of surface—the whole earth, unless rescued by human art from the physical degradation to which it tends, becomes an assemblage of bald mountains, of barren, turfless hills, and of swampy and malarious plains. There are parts of Asia Minor, of Northern Africa, of Greece, and even of Alpine Europe, where the operation of causes set in action by man has brought the face of the earth to a desolation almost as complete as that of the moon; and though, within that brief space

upon any form of life an influence analogous to that of domestication upon plants, quadrupeds, and birds reared artificially by man; and this is as true of unforeseen as of purposely effected improvements accomplished by voluntary selection of breeding animals.

of time which we call "the historical period," they are known to have been covered with luxuriant woods, verdant pastures, and fertile meadows, they are now too far deteriorated to be reclaimable by man, nor can they become again fitted for human use, except through great geological changes, or other mysterious influences or agencies of which we have no present knowledge, and over which we have no prospective control. The earth is fast becoming an unfit home for its noblest inhabitant, and another era of equal human crime and human improvidence, and of like duration with that through which traces of that crime and that improvidence extend, would reduce it to such a condition of impoverished productiveness, of shattered surface, of climatic excess, as to threaten the depravation, barbarism, and perhaps even extinction of the species.*

Physical Improvement.

True, there is a partial reverse to this picture. On narrow theatres, new forests have been planted; inundations of flowing streams restrained by heavy walls of masonry and other constructions; torrents compelled to aid, by depositing the slime with which they are charged, in filling up lowlands, and

* ——"And it may be remarked that, as the world has passed through these several stages of strife to produce a Christendom, so by relaxing in the enterprises it has learnt, does it tend downwards, through inverted steps, to wildness and the waste again. Let a people give up their contest with moral evil; disregard the injustice, the ignorance, the greediness, that may prevail among them, and part more and more with the Christian element of their civilization; and in declining this battle with sin, they will inevitably get embroiled with men. Threats of war and revolution punish their unfaithfulness; and if then, instead of retracing their steps, they yield again, and are driven before the storm, the very arts they had created, the structures they had raised, the usages they had established, are swept away; 'in that very day their thoughts perish.' The portion they had reclaimed from the young earth's ruggedness is lost; and failing to stand fast against man, they finally get embroiled with nature, and are thrust down beneath her ever-living hand."—MARTINEAU's *Sermon*, "*The Good Soldier of Jesus Christ.*"

raising the level of morasses which their own overflows had created; ground submerged by the encroachments of the ocean, or exposed to be covered by its tides, has been rescued from its dominion by diking;* swamps and even lakes have been drained, and their beds brought within the domain of agricultural industry; drifting coast dunes have been checked and made productive by plantation; seas and inland waters have been repeopled with fish, and even the sands of the Sahara have been fertilized by artesian fountains. These achievements are more glorious than the proudest triumphs of war, but, thus far, they give but faint hope that we shall yet make full atonement for our spendthrift waste of the bounties of nature.

It is, on the one hand, rash and unphilosophical to attempt to set limits to the ultimate power of man over inorganic nature, and it is unprofitable, on the other, to speculate on what may be accomplished by the discovery of now unknown and unimagined natural forces, or even by the invention of new arts and new processes. But since we have seen aerostation, the motive power of elastic vapors, the wonders of modern telegraphy, the destructive explosiveness of gunpowder, and even of a substance so harmless, unresisting, and inert as cotton, nothing in the way of mechanical achievement seems impossible, and it is hard to restrain the imagination from wandering forward a couple of generations to an epoch when our descendants shall have advanced as far beyond us in physical conquest, as we have marched beyond the trophies erected by our grandfathers.

I must therefore be understood to mean only, that no agencies now known to man and directed by him seem adequate to the reducing of great Alpine precipices to such

* The dependence of man upon the aid of spontaneous nature, in his most arduous material works, is curiously illustrated by the fact that one of the most serious difficulties to be encountered in executing the proposed gigantic scheme of draining the Zuiderzee in Holland, is that of procuring brushwood for the fascines to be employed in the embankments. See DIGGELEN's pamphlet, "*Groote Werken in Nederland.*"

slopes as would enable them to support a vegetable clothing, or to the covering of large extents of denuded rock with earth, and planting upon them a forest growth. But among the mysteries which science is yet to reveal, there may be still undiscovered methods of accomplishing even grander wonders than these. Mechanical philosophers have suggested the possibility of accumulating and treasuring up for human use some of the greater natural forces, which the action of the elements puts forth with such astonishing energy. Could we gather, and bind, and make subservient to our control, the power which a West Indian hurricane exerts through a small area in one continuous blast, or the momentum expended by the waves, in a tempestuous winter, upon the breakwater at Cherbourg,* or the lifting power of the tide, for a month, at the head of the Bay of Fundy, or the pressure of a square mile of sea water at the depth of five thousand fathoms, or a moment of the might of an earthquake or a volcano, our age—which moves no mountains and casts them into the sea by faith alone—might hope to scarp the rugged walls of the Alps and Pyrenees and Mount Taurus, robe them once more in a vegetation as rich as that of their pristine woods, and turn their wasting torrents into refreshing streams.†

* In heavy storms, the force of the waves as they strike against a sea wall is from one and a half to two tons to the square foot, and Stevenson, in one instance at Skerryvore, found this force equal to three tons per foot.

The seaward front of the breakwater at Cherbourg exposes a surface of about 2,500,000 square feet. In rough weather the waves beat against this whole face, though at the depth of twenty-two yards, which is the height of the breakwater, they exert a very much less violent motive force than at and near the surface of the sea, because this force diminishes in geometrical, as the distance below the surface increases in arithmetical proportion. The shock of the waves is received several thousand times in the course of twenty-four hours, and hence the sum of impulse which the breakwater resists in one stormy day amounts to many thousands of millions of tons. The breakwater is entirely an artificial construction. If then man could accumulate and control the forces which he is able effectually to resist, he might be said to be, physically speaking, omnipotent.

† Some well known experiments show that it is quite possible to accumulate the solar heat by a simple apparatus, and thus to obtain a temper-

Could this old world, which man has overthrown, be rebuilded, could human cunning rescue its wasted hillsides and its deserted plains from solitude or mere nomade occupation, from barrenness, from nakedness, and from insalubrity, and restore the ancient fertility and healthfulness of the Etruscan sea coast, the Campagna and the Pontine marshes, of Calabria, of Sicily, of the Peloponnesus and insular and continental Greece, of Asia Minor, of the slopes of Lebanon and Hermon, of Palestine, of the Syrian desert, of Mesopotamia and the delta of the Euphrates, of the Cyrenaica, of Africa proper, Numidia, and Mauritania, the thronging millions of Europe might still find room on the Eastern continent, and the main current of emigration be turned toward the rising instead of the setting sun.

But changes like these must await great political and moral revolutions in the governments and peoples by whom those regions are now possessed, a command of pecuniary and of mechanical means not at present enjoyed by those nations, and a more advanced and generally diffused knowledge of the processes by which the amelioration of soil and climate is possible, than now anywhere exists. Until such circumstances shall conspire to favor the work of geographical regeneration, the countries I have mentioned, with here and there a local exception, will continue to sink into yet deeper desolation, and

ature which might be economically important even in the climate of Switzerland. Saussure, by receiving the sun's rays in a nest of boxes blackened within and covered with glass, raised a thermometer enclosed in the inner box to the boiling point; and under the more powerful sun of the cape of Good Hope, Sir John Herschel cooked the materials for a family dinner by a similar process, using, however, but a single box, surrounded with dry sand and covered with two glasses. Why should not so easy a method of economizing fuel be resorted to in Italy, and even in more northerly climates?

The unfortunate John Davidson records in his journal that he saved fuel in Morocco by exposing his teakettle to the sun on the roof of his house, where the water rose to the temperature of one hundred and forty degrees, and, of course, needed little fire to bring it to boil. But this was the direct and simple, not the accumulated heat of the sun.

in the mean time, the American continent, Southern Africa, Australia, and the smaller oceanic islands, will be almost the only theatres where man is engaged, on a great scale, in transforming the face of nature.

Arrest of Physical Decay of New Countries.

Comparatively short as is the period through which the colonization of foreign lands by European emigrants extends, great, and, it is to be feared, sometimes irreparable, injury has been already done in the various processes by which man seeks to subjugate the virgin earth; and many provinces, first trodden by the *homo sapiens Europæ* within the last two centuries, begin to show signs of that melancholy dilapidation which is now driving so many of the peasantry of Europe from their native hearths. It is evidently a matter of great moment, not only to the population of the states where these symptoms are manifesting themselves, but to the general interests of humanity, that this decay should be arrested, and that the future operations of rural husbandry and of forest industry, in districts yet remaining substantially in their native condition, should be so conducted as to prevent the widespread mischiefs which have been elsewhere produced by thoughtless or wanton destruction of the natural safeguards of the soil. This can be done only by the diffusion of knowledge on this subject among the classes that, in earlier days, subdued and tilled ground in which they had no vested rights, but who, in our time, own their woods, their pastures, and their ploughlands as a perpetual possession for them and theirs, and have, therefore, a strong interest in the protection of their domain against deterioration.

Forms and Formations most liable to Physical Degradation.

The character and extent of the evils under consideration depend very much on climate and the natural forms and constitution of surface. If the precipitation, whether great or small in amount, be equally distributed through the seasons, so that there are neither torrential rains nor parching droughts, and if, further, the general inclination of ground be moderate, so that the superficial waters are carried off without destructive rapidity of flow, and without sudden accumulation in the channels of natural drainage, there is little danger of the degradation of the soil in consequence of the removal of forest or other vegetable covering, and the natural face of the earth may be considered as substantially permanent. These conditions are well exemplified in Ireland, in a great part of England, in extensive districts in Germany and France, and, fortunately, in an immense proportion of the valley of the Mississippi and the basin of the great American lakes, as well as in many parts of the continents of South America and of Africa.

Destructive changes are most frequent in countries of irregular and mountainous surface, and in climates where the precipitation is confined chiefly to a single season, and where the year is divided into a wet and a dry period, as is the case throughout a great part of the Ottoman empire, and, more or less strictly, the whole Mediterranean basin. It is partly, though by no means entirely, owing to topographical and climatic causes that the blight, which has smitten the fairest and most fertile provinces of Imperial Rome, has spared Britannia, Germania, Pannonia, and Mœsia, the comparatively inhospitable homes of barbarous races, who, in the days of the Cæsars, were too little advanced in civilized life to possess either the power or the will to wage that war against the order of nature which seems, hitherto, an almost inseparable condition precedent of high social culture, and of great progress in fine and mechanical art.*

* In the successive stages of social progress, the most destructive periods of human action upon nature are the pastoral condition, and that of incipient stationary civilization, or, in the newly discovered countries of modern geography, the colonial, which corresponds to the era of early civilization in older lands. In more advanced states of culture, conservative influences make themselves felt; and if highly civilized communities do not always restore the works of nature, they at least use a less wasteful expenditure than their predecessors in consuming them.

In mountainous countries, on the other hand, various causes combine to expose the soil to constant dangers. The rain and snow usually fall in greater quantity, and with much inequality of distribution; the snow on the summits accumulates for many months in succession, and then is not unfrequently almost wholly dissolved in a single thaw, so that the entire precipitation of months is in a few hours hurried down the flanks of the mountains, and through the ravines that furrow them; the natural inclination of the surface promotes the swiftness of the gathering currents of diluvial rain and of melting snow, which soon acquire an almost irresistible force, and power of removal and transportation; the soil itself is less compact and tenacious than that of the plains, and if the sheltering forest has been destroyed, it is confined by few of the threads and ligaments by which nature had bound it together, and attached it to the rocky groundwork. Hence every considerable shower lays bare its roods of rock, and the torrents sent down by the thaws of spring, and by occasional heavy discharges of the summer and autumnal rains, are seas of mud and rolling stones that sometimes lay waste, and bury beneath them acres, and even miles, of pasture and field and vineyard.*

Physical Decay of New Countries.

I have remarked that the effects of human action on the forms of the earth's surface could not always be distinguished from those resulting from geological causes, and there is also much uncertainty in respect to the precise influence of the

* The character of geological formation is an element of very great importance in determining the amount of erosion produced by running water, and, of course, in measuring the consequences of clearing off the forests. The soil of the French Alps yields very readily to the force of currents, and the declivities of the northern Apennines are covered with earth which becomes itself a fluid when saturated with water. Hence the erosion of such surfaces is vastly greater than on many other mountains of equal steepness of inclination. This point is fully considered by the authors referred to in chap. iii, *post.*

clearing and cultivating of the ground, and of other rural operations, upon climate. It is disputed whether either the mean or the extremes of temperature, the periods of the seasons, or the amount or distribution of precipitation and of evaporation, in any country whose annals are known, have undergone any change during the historical period. It is, indeed, impossible to doubt that many of the operations of the pioneer settler tend to produce great modifications in atmospheric humidity, temperature, and electricity; but we are at present unable to determine how far one set of effects is neutralized by another, or compensated by unknown agencies. This question scientific research is inadequate to solve, for want of the necessary data; but well conducted observation, in regions now first brought under the occupation of man, combined with such historical evidence as still exists, may be expected at no distant period to throw much light on this subject.

Australia is, perhaps, the country from which we have a right to expect the fullest elucidation of these difficult and disputable problems. Its colonization did not commence until the physical sciences had become matter of almost universal attention, and is, indeed, so recent that the memory of living men embraces the principal epochs of its history; the peculiarities of its fauna, its flora, and its geology are such as to have excited for it the liveliest interest of the votaries of natural science; its mines have given its people the necessary wealth for procuring the means of instrumental observation, and the leisure required for the pursuit of scientific research; and large tracts of virgin forest and natural meadow are rapidly passing under the control of civilized man. Here, then, exist greater facilities and stronger motives for the careful study of the topics in question than have ever been found combined in any other theatre of European colonization.

In North America, the change from the natural to the artificial condition of terrestrial surface began about the period when the most important instruments of meteorological observation were invented. The first settlers in the territory now constituting the United States and the British American provinces had other things to do than to tabulate barometrical and thermometrical readings, but there remain some interesting physical records from the early days of the colonies,* and there is still an immense extent of North American soil where the industry and the folly of man have as yet produced little appreciable change. Here, too, with the present increased facilities for scientific observation, the future effects, direct and contingent, of man's labors, can be measured, and such precautions taken in those rural processes which we call improvements, as to mitigate evils, perhaps, in some degree, inseparable from every attempt to control the action of natural laws.

In order to arrive at safe conclusions, we must first obtain a more exact knowledge of the topography, and of the present superficial and climatic condition of countries where the natural surface is as yet more or less unbroken. This can only be accomplished by accurate surveys, and by a great multiplication of the points of meteorological registry,† already so

* The Travels of Dr. Dwight, president of Yale College, which embody the results of his personal observations, and of his inquiries among the early settlers, in his vacation excursions in the Northern States of the American Union, though presenting few instrumental measurements or tabulated results, are of value for the powers of observation they exhibit, and for the sound common sense with which many natural phenomena, such for instance as the formation of the river meadows, called "intervales," in New England, are explained. They present a true and interesting picture of physical conditions, many of which have long ceased to exist in the theatre of his researches, and of which few other records are extant.

† The general law of temperature is that it decreases as we ascend. But, in hilly regions, the law is reversed in cold, still weather, the cold air descending, by reason of its greater gravity, into the valleys. If there be wind enough, however, to produce a disturbance and intermixture of higher and lower atmospheric strata, this exception to the general law does not take place. These facts have long been familiar to the common people of Switzerland and of New England, but their importance has not been sufficiently taken into account in the discussion of meteorological observations. The descent of the cold air and the rise of the warm affect

numerous; and as, moreover, considerable changes in the proportion of forest and of cultivated land, or of dry and wholly or partially submerged surface, will often take place within brief periods, it is highly desirable that the attention of observers, in whose neighborhood the clearing of the soil, or the drainage of lakes and swamps, or other great works of rural improvement, are going on or meditated, should be especially drawn not only to revolutions in atmospheric temperature and precipitation, but to the more easily ascertained and perhaps more important local changes produced by these operations in the temperature and the hygrometric state of the superficial strata of the earth, and in its spontaneous vegetable and animal products.

The rapid extension of railroads, which now everywhere keeps pace with, and sometimes even precedes, the occupation of new soil for agricultural purposes, furnishes great facilities for enlarging our knowledge of the topography of the territory they traverse, because their cuttings reveal the composition and general structure of surface, and the inclination and elevation of their lines constitute known hypsometrical sections, which give numerous points of departure for the measurement of higher and lower stations, and of course for determining the relief and depression of surface, the slope of the beds of watercourses, and many other not less important questions.*

the relative temperatures of hills and valleys to a much greater extent than has been usually supposed. A gentleman well known to me kept a thermometrical record for nearly half a century, in a New England country town, at an elevation of at least 1,500 feet above the sea. During these years his thermometer never fell lower than 26° Fahrenheit, while at the shire town of the county, situated in a basin one thousand feet lower, and ten miles distant, as well as at other points in similar positions, the mercury froze several times in the same period.

* Railroad surveys must be received with great caution where any motive exists for *cooking* them. Capitalists are shy of investments in roads with steep grades, and of course it is important to make a fair show of facilities in obtaining funds for new routes. Joint-stock companies have no souls; their managers, in general, no consciences. Cases can be cited

The geological, hydrographical, and topographical surveys, which almost every general and even local government of the civilized world is carrying on, are making yet more important contributions to our stock of geographical and general physical knowledge, and, within a comparatively short space, there will

where engineers and directors of railroads, with long grades above one hundred feet to the mile, have regularly sworn in their annual reports, for years in succession, that there were no grades upon their routes exceeding half that elevation. In fact, every person conversant with the history of these enterprises knows that in their public statements falsehood is the rule, truth the exception.

What I am about to remark is not exactly relevant to my subject; but it is hard to "get the floor" in the world's great debating society, and when a speaker who has anything to say once finds access to the public ear, he must make the most of his opportunity, without inquiring too nicely whether his observations are "in order." I shall harm no honest man by endeavoring, as I have often done elsewhere, to excite the attention of thinking and conscientious men to the dangers which threaten the great moral and even political interests of Christendom, from the unscrupulousness of the private associations that now control the monetary affairs, and regulate the transit of persons and property, in almost every civilized country. More than one American State is literally governed by unprincipled corporations, which not only defy the legislative power, but have, too often, corrupted even the administration of justice. Similar evils have become almost equally rife in England, and on the Continent; and I believe the decay of commercial morality, and indeed of the sense of all higher obligations than those of a pecuniary nature, on both sides of the Atlantic, is to be ascribed more to the influence of joint-stock banks and manufacturing and railway companies, to the workings, in short, of what is called the principle of "associate action," than to any other one cause of demoralization.

The apophthegm, "the world is governed too much," though unhappily too truly spoken of many countries—and perhaps, in some aspects, true of all—has done much mischief whenever it has been too unconditionally accepted as a political axiom. The popular apprehension of being over-governed, and, I am afraid, more emphatically the fear of being over-taxed, has had much to do with the general abandonment of certain governmental duties by the ruling powers of most modern states. It is theoretically the duty of government to provide all those public facilities of intercommunication and commerce, which are essential to the prosperity of civilized commonwealths, but which individual means are inadequate to furnish, and for the due administration of which individual guar-

be an accumulation of well established constant and historical facts, from which we can safely reason upon all the relations of action and reaction between man and external nature.

But we are, even now, breaking up the floor and wainscoting and doors and window frames of our dwelling, for fuel to warm our bodies and seethe our pottage, and the world cannot afford to wait till the slow and sure progress of exact science has taught it a better economy. Many practical lessons have been learned by the common observation of unschooled men; and the teachings of simple experience, on topics where natural philosophy has scarcely yet spoken, are not to be despised.

In these humble pages, which do not in the least aspire to rank among scientific expositions of the laws of nature, I shall

anties are insufficient. Hence public roads, canals, railroads, postal communications, the circulating medium of exchange, whether metallic or representative, armies, navies, being all matters in which the nation at large has a vastly deeper interest than any private association can have, ought legitimately to be constructed and provided only by that which is the visible personification and embodiment of the nation, namely, its legislative head. No doubt the organization and management of these institutions by government are liable, as are all things human, to great abuses. The multiplication of public placeholders, which they imply, is a serious evil. But the corruption thus engendered, foul as it is, does not strike so deep as the rottenness of private corporations; and official rank, position, and duty have, in practice, proved better securities for fidelity and pecuniary integrity in the conduct of the interests in question, than the suretyships of private corporate agents, whose bondsmen so often fail or abscond before their principal is detected.

Many theoretical statesmen have thought that voluntary associations for strictly pecuniary and industrial purposes, and for the construction and control of public works, might furnish, in democratic countries, a compensation for the small and doubtful advantages, and at the same time secure an exemption from the great and certain evils, of aristocratic institutions. The example of the American States shows that private corporations—whose rule of action is the interest of the association, not the conscience of the individual—though composed of ultra-democratic elements, may become most dangerous enemies to rational liberty, to the moral interests of the commonwealth, to the purity of legislation and of judicial action, and to the sacredness of private rights.

attempt to give the most important practical conclusions suggested by the history of man's efforts to replenish the earth and subdue it; and I shall aim to support those conclusions by such facts and illustrations only as address themselves to the understanding of every intelligent reader, and as are to be found recorded in works capable of profitable perusal, or at least consultation, by persons who have not enjoyed a special scientific training.

CHAPTER II.

TRANSFER, MODIFICATION, AND EXTIRPATION OF VEGETABLE AND OF ANIMAL SPECIES.

MODERN GEOGRAPHY EMBRACES ORGANIC LIFE—TRANSFER OF VEGETABLE LIFE—FOREIGN PLANTS GROWN IN THE UNITED STATES—AMERICAN PLANTS GROWN IN EUROPE—MODES OF INTRODUCTION OF FOREIGN PLANTS—VEGETABLES, HOW AFFECTED BY TRANSFER TO FOREIGN SOILS—EXTIRPATION OF VEGETABLES—ORIGIN OF DOMESTIC PLANTS—ORGANIC LIFE AS A GEOLOGICAL AND GEOGRAPHICAL AGENCY—ORIGIN AND TRANSFER OF DOMESTIC ANIMALS —EXTIRPATION OF ANIMALS—NUMBERS OF BIRDS IN THE UNITED STATES—BIRDS AS SOWERS AND CONSUMERS OF SEEDS, AND AS DESTROYERS OF INSECTS—DIMINUTION AND EXTIRPATION OF BIRDS—INTRODUCTION OF BIRDS—UTILITY OF INSECTS AND WORMS—INTRODUCTION OF INSECTS—DESTRUCTION OF INSECTS—REPTILES—DESTRUCTION OF FISH—INTRODUCTION AND BREEDING OF FISH—EXTIRPATION OF AQUATIC ANIMALS—MINUTE ORGANISMS.

Modern Geography embraces Organic Life.

IT was a narrow view of geography which confined that science to delineation of terrestrial surface and outline, and to description of the relative position and magnitude of land and water. In its improved form, it embraces not only the globe itself, but the living things which vegetate or move upon it, the varied influences they exert upon each other, the reciprocal action and reaction between them and the earth they inhabit. Even if the end of geographical studies were only to obtain a knowledge of the external forms of the mineral and fluid masses which constitute the globe, it would still be necessary to take into account the element of life; for every plant, every animal, is a geographical agency, man a destructive, vegetables, and even wild beasts, restorative powers. The rushing waters sweep down earth from the uplands; in the first moment of repose, vegetation seeks to reëstablish itself on the bared surface, and, by the slow deposit of its decaying products, to raise again the soil which the torrent had lowered. So important an element of reconstruction is this, that it has been seriously questioned whether, upon the whole, vegetation does not contribute as much to elevate, as the waters to depress, the level of the surface.

Whenever man has transported a plant from its native habitat to a new soil, he has introduced a new geographical force to act upon it, and this generally at the expense of some indigenous growth which the foreign vegetable has supplanted. The new and the old plants are rarely the equivalents of each other, and the substitution of an exotic for a native tree, shrub, or grass, increases or diminishes the relative importance of the vegetable element in the geography of the country to which it is removed. Further, man sows that he may reap. The products of agricultural industry are not suffered to rot upon the ground, and thus raise it by an annual stratum of new mould. They are gathered, transported to greater or less distances, and after they have served their uses in human economy, they enter, on the final decomposition of their elements, into new combinations, and are only in small proportion returned to the soil on which they grew. The roots of the grasses, and of many other cultivated plants, however, usually remain and decay in the earth, and contribute to raise its surface, though certainly not in the same degree as the forest.

The vegetables, which have taken the place of trees, unquestionably perform many of the same functions. They radiate heat, they condense the humidity of the atmosphere, they act upon the chemical constitution of the air, their roots penetrate the earth to greater depths than is commonly supposed, and form an inextricable labyrinth of filaments which bind the soil together and prevent its erosion by water. The broad-leaved annuals and perennials, too, shade the ground, and prevent the evaporation of moisture from its surface by wind and sun.* At a certain stage of growth, grass land is probably a more energetic radiator and condenser than even the forest, but this powerful action is exerted, in its full intensity, for a few days only, while trees continue such functions, with unabated vigor, for many months in succession. Upon the whole, it seems quite certain, that no cultivated ground is as efficient in tempering climatic extremes, or in conservation of geographical surface and outline, as is the soil which nature herself has planted.

Transfer of Vegetable Life.

It belongs to vegetable and animal geography, which are almost sciences of themselves, to point out in detail what man has done to change the distribution of plants and of animated life and to revolutionize the aspect of organic nature; but some of the more important facts bearing on this subject may pertinently be introduced here. Most of the fruit trees grown

* It is impossible to say how far the abstraction of water from the earth by broad-leaved field and garden plants—such as maize, the gourd family, the cabbage, &c.—is compensated by the condensation of dew, which sometimes pours from them in a stream, by the exhalation of aqueous vapor from their leaves, which is directly absorbed by the ground, and by the shelter they afford the soil from sun and wind, thus preventing evaporation. American farmers often say that after the leaves of Indian corn are large enough to "shade the ground," there is little danger that the plants will suffer from drought; but it is probable that the comparative security of the fields from this evil is in part due to the fact that, at this period of growth, the roots penetrate down to a permanently humid stratum of soil, and draw from it the moisture they require. Stirring the ground between the rows of maize with a light harrow or cultivator, in very dry seasons, is often recommended as a preventive of injury by drought. It would seem, indeed, that loosening and turning over the surface earth might aggravate the evil by promoting the evaporation of the little remaining moisture; but the practice is founded partly on the belief that the hygroscopicity of the soil is increased by it to such a degree that it gains more by absorption than it loses by evaporation, and partly on the doctrine that to admit air to the rootlets, or at least to the earth near them, is to supply directly elements of vegetable growth.

in Europe and the United States are believed, and—if the testimony of Pliny and other ancient naturalists is to be depended upon—many of them are historically known, to have originated in the temperate climates of Asia. The wine grape has been thought to be truly indigenous only in the regions bordering on the eastern end of the Black Sea, where it now, particularly on the banks of the Rion, the ancient Phasis, propagates itself spontaneously, and grows with unexampled luxuriance.* But some species of the vine seem native to Europe, and many varieties of grape have been too long known as common to every part of the United States to admit of the supposition that they were all introduced by European colonists.†

It is an interesting fact that the commerce—or at least the maritime carrying trade—and the agricultural and mechanical industry of the world are, in very large proportion, dependent on vegetable and animal products little or not at all known to ancient Greek, Roman, and Jewish civilization. In many instances, the chief supply of these articles comes from countries to which they are probably indigenous, and where they are still almost exclusively grown; but in many others, the plants or animals from which they are derived have been

* The vine-wood planks of the ancient great door of the cathedral at Ravenna, which measured thirteen feet in length by a foot and a quarter in width, are traditionally said to have been brought from the Black Sea, by way of Constantinople, about the eleventh or twelfth century. No vines of such dimensions are now found in any other part of the East, and, though I have taken some pains on the subject, I never found in Syria or in Turkey a vine stock exceeding six inches in diameter, bark excluded.

† The Northmen who—as I think it has been indisputably established by Professor Rafn of Copenhagen—visited the coast of Massachusetts about the year 1000, found grapes growing there in profusion, and the vine still flourishes in great variety and abundance in the southeastern counties of that State. The townships in the vicinity of the Dighton rock, supposed by many—with whom, however, I am sorry I cannot agree—to bear a Scandinavian inscription, abound in wild vines, and I have never seen a region which produced them so freely. I have no doubt that the cultivation of the grape will become, at no distant day, one of the most important branches of rural industry in that district.

introduced by man into the regions now remarkable for their most successful cultivation, and that, too, in comparatively recent times, or, in other words, within two or three centuries.

Foreign Plants grown in the United States.

According to Bigelow, the United States had, on the first of June, 1860, in round numbers, 163,000,000 acres of improved land, the quantity having been increased by 50,000,000 acres within the ten years next preceding.* Not to mention less important crops, this land produced, in the year ending on the day last mentioned, in round numbers, 171,000,000 bushels of wheat, 21,000,000 bushels of rye, 172,000,000 bushels of oats, 15,000,000 bushels of pease and beans, 16,000,000 bushels of barley, orchard fruits to the value of $20,000,000, 900,000 bushels of cloverseed, 900,000 bushels of other grass seed, 104,000 tons of hemp, 4,000,000 pounds of flax, and 600,000 pounds of flaxseed. These vegetable growths were familiar to ancient European agriculture, but they were all introduced into North America after the close of the sixteenth century.

Of the fruits of agricultural industry unknown to the Greeks and Romans, or too little employed by them to be of any commercial importance, the United States produced, in the same year, 187,000,000 pounds of rice, 18,000,000 bushels of buckwheat, 2,075,000,000 pounds of ginned cotton,†

* *Les États Unis d'Amérique en* 1863, p. 360. By "improved" land, in the reports on the census of the United States, is meant "cleared land used for grazing, grass, or tillage, or which is now fallow, connected with or belonging to a farm."—*Instructions to Marshals and Assistants, Census of* 1850, schedule 4, §§ 2, 3.

† Cotton, though cultivated in Asia and Africa from the remotest antiquity, and known as a rare and costly product to the Latins and the Greeks, was not used by them to any considerable extent, nor did it enter into their commerce as a regular article of importation. The early voyagers found it in common use in the West Indies and in the provinces first colonized by the Spaniards; but it was introduced into the territory of the United States by European settlers, and did not become of any importance

302,000,000 pounds of cane sugar, 16,000,000 gallons of cane molasses, 7,000,000 gallons of sorghum molasses, all yielded by vegetables introduced into that country within two hundred years, and—with the exception of buckwheat, the origin of which is uncertain, and of cotton—all, directly or indirectly, from the East Indies; besides, from indigenous plants unknown to ancient agriculture, 830,000,000 bushels of Indian corn or maize, 429,000,000 pounds of tobacco, 110,000,000 bushels of potatoes, 42,000,000 bushels of sweet potatoes, 39,000,000 pounds of maple sugar, and 2,000,000 gallons of maple molasses. To all this we are to add 19,000,000 tons of hay, produced partly by new, partly by long known, partly by exotic, partly by native herbs and grasses, an incalculable quantity of garden vegetables, chiefly of European or Asiatic origin, and many minor agricultural products.

The weight of this harvest of a year would be not less than 60,000,000 tons—which is eleven times the tonnage of all the shipping of the United States at the close of the year 1861—and, with the exception of the maple sugar, the maple molasses, and the products of the Western prairie lands and some small Indian clearings, it was all grown upon lands wrested from the forest by the European race within little more than two hundred years. The wants of Europe have introduced into the colonies of tropical America the sugar cane, the coffee plant, the orange and the lemon,* all of Oriental origin, have

until after the Revolution. Cotton seed was sown in Virginia as early as 1621, but was not cultivated with a view to profit for more than a century afterward. Sea-island cotton was first grown on the coast of Georgia in 1786, the seed having been brought from the Bahamas, where it had been introduced from Anguilla.—Bigelow, *Les États Unis en* 1863, p. 370.

* The sugar cane was introduced by the Arabs into Sicily and Spain as early as the ninth century, and though it is now scarcely grown in those localities, I am not aware of any reason to doubt that its cultivation might be revived with advantage. From Spain it was carried to the West Indies, though different varieties have since been introduced into those islands from other sources. Tea is now cultivated with a certain success in Brazil, and promises to become an important crop in the Southern States of the American Union. The lemon is, I think, readily recognizable, by Pliny's

immensely stimulated the cultivation of the former two in the countries of which they are natives, and, of course, promoted agricultural operations which must have affected the geography of those regions to an extent proportionate to the scale on which they have been pursued.

American Plants grown in Europe.

America has partially repaid her debt to the Eastern continent. Maize and the potato are very valuable additions to the field agriculture of Europe and the East, and the tomato is no mean gift to the kitchen gardens of the Old World, though certainly not an adequate return for the multitude of esculent roots and leguminous plants which the European colonists carried with them.* I wish I could believe, with some, that America is not alone responsible for the introduction of the filthy weed, tobacco, the use of which is the most vulgar and pernicious habit engrafted by the semi-barbarism of modern civilization upon the less multifarious sensualism of ancient life; † but the alleged occurrence of pipe-like objects in Scla-

description, as known to the ancients, but it does not satisfactorily appear that they were acquainted with the orange.

* John Smith mentions, in his *Historie of Virginia*, 1624, pease and beans as having been cultivated by the natives before the arrival of the whites, and there is no doubt, I believe, that the pumpkin and several other cucurbitaceous plants are of American origin; but most, if not all the varieties of pease, beans, and other pod fruits now grown in American gardens, are from European and other foreign seed. See *Appendix*, No, 8.

† There are some usages of polite society which are inherently low in themselves, and debasing in their influence and tendency, and which no custom or fashion can make respectable or fit to be followed by self-respecting persons. It is essentially vulgar to smoke or chew tobacco, and especially to take snuff; it is unbecoming a gentleman to perform the duties of his coachman; it is indelicate in a lady to wear in the street skirts so long that she cannot walk without grossly soiling them. Not that all these things are not practised by persons justly regarded as gentlemen and ladies; but the same individuals would be, and feel themselves to be, much more emphatically gentlemen and ladies, if they abstained from them.

vonic, and, it has been said, in Hungarian sepulchres, is hardly sufficient evidence to convict those races of complicity in this grave offence against the temperance and the refinement of modern society.

Modes of Introduction of Foreign Plants.

Besides the vegetables I have mentioned, we know that many plants of smaller economical value have been the subjects of international exchange in very recent times. Busbequius, Austrian ambassador at Constantinople about the middle of the sixteenth century—whose letters contain one of the best accounts of Turkish life which have appeared down to the present day—brought home from the Ottoman capital the lilac and the tulip. The Belgian Clusius about the same time introduced from the East the horse chestnut, which has since wandered to America. The weeping willows of Europe and the United States are said to have sprung from a slip received from Smyrna by the poet Pope, and planted by him in an English garden; and the Portuguese declare that the progenitor of all the European and American oranges was an Oriental tree transplanted to Lisbon, and still living in the last generation.* The present favorite flowers of the parterres of

* The name *portogallo*, so generally applied to the orange in Italy, seems to favor this claim. The orange, however, was known in Europe before the discovery of the Cape of Good Hope, and, therefore, before the establishment of direct relations between Portugal and the East.

A correspondent of the *Athenæum*, in describing the newly excavated villa, which has been named Livia's Villa, near the Porta del Popolo at Rome, states that: "The walls of one of the rooms are, singularly enough, decorated with landscape paintings, a grove of palm and *orange* trees, with fruits and birds on the branches—the colors all as fresh and lively as if painted yesterday." The writer remarks on the character of this decoration as something very unusual in Roman architecture; and if the trees in question are really orange, and not lemon trees, this circumstance may throw some doubt on the antiquity of the painting. If, on the other hand, it proves really ancient, it shows that the orange was known to the Roman painters, if not gardeners. The landscape may perhaps represent Oriental,

Europe have been imported from America, Japan and other remote Oriental countries, within a century and a half, and, in fine, there are few vegetables of any agricultural importance, few ornamental trees or decorative plants, which are not now common to the three civilized continents.

The statistics of vegetable emigration exhibit numerical results quite surprising to those not familiar with the subject. The lonely island of St. Helena is described as producing, at the time of its discovery in the year 1501, about sixty vegetable species, including some three or four known to grow elsewhere also. At the present time its flora numbers seven hundred and fifty species. Humboldt and Bonpland found, among the unquestionably indigenous plants of tropical America, monocotyledons only, all the dicotyledons of those extensive regions having been probably introduced after the colonization of the New World by Spain.

The faculty of spontaneous reproduction and perpetuation necessarily supposes a greater power of accommodation, within a certain range, than we find in most domesticated plants, for it would rarely happen that the seed of a wild plant would fall into ground as nearly similar, in composition and condition, to that where its parent grew, as the soils of different fields artificially prepared for growing a particular vegetable are to each other. Accordingly, though every wild species affects a habitat of a particular character, it is found that, if accidentally or designedly sown elsewhere, it will grow under conditions extremely unlike those of its birthplace.* Cooper says: "We

not European scenery. The accessories of the picture would probably determine that question.—*Athenæum*, No. 1859, June 13, 1863.

Müller, *Das Buch der Pflanzenwelt*, p. 86, asserts that in 1802 the ancestor of all the mulberries in France, planted in 1500, was still standing in a garden in the village of Allan-Montélimart.

* The vegetables which, so far as we know their history, seem to have been longest the objects of human care, can, by painstaking industry, be made to grow under a great variety of circumstances, and some of them—the vine for instance—prosper nearly equally well, when planted and tended, on soils of almost any geological character; but their seeds vegetate only in artificially prepared ground, they have little self-sustaining

cannot say positively that *any* plant is *uncultivable* anywhere until it has been tried;" and this seems to be even more true of wild than of domesticated vegetation.

The seven hundred new species which have found their way to St. Helena within three centuries and a half, were certainly not all, or even in the largest proportion, designedly planted there by human art, and if we were well acquainted with vegetable emigration, we should probably be able to show that man has intentionally transferred fewer plants than he has accidentally introduced into countries foreign to them. After the wheat, follow the tares that infest it. The weeds that grow among the cereal grains, the pests of the kitchen garden, are the same in America as in Europe.* The overturning of a wagon, or any of the thousand accidents which befall the emigrant in his journey across the Western plains, may scatter upon the ground the seeds he designed for his garden, and the herbs which fill so important a place in the rustic materia medica of the Eastern States, spring up along the prairie paths but just opened by the caravan of the settler.†

power, and they soon perish when the nursing hand of man is withdrawn from them. In range of climate, wild plants are much more limited than domestic, but much less so with regard to the state of the soil in which they germinate and grow. See *Appendix*, No. 9.

Dr. Dwight remarks that the seeds of American forest trees will not vegetate when dropped on grassland. This is one of the very few errors of personal observation to be found in that author's writings. There are seasons, indeed, when few tree seeds germinate in the meadows and the pastures, and years favorable to one species are not always propitious to another; but there is no American forest tree known to me which does not readily propagate itself by seed in the thickest greensward, if its germs are not disturbed by man or animals.

* Some years ago I made a collection of weeds in the wheatfields of Upper Egypt, and another in the gardens on the Bosphorus. Nearly all the plants were identical with those which grow under the same conditions in New England. I do not remember to have seen in America the scarlet wild poppy so common in European grainfields. I have heard, however, that it has lately crossed the Atlantic, and I am not sorry for it. With our abundant harvests of wheat, we can well afford to pay now and then a loaf of bread for the cheerful radiance of this brilliant flower.

† Josselyn, who wrote about fifty years after the foundation of the first

The hortus siccus of a botanist may accidentally sow seeds from the foot of the Himalayas on the plains that skirt the Alps; and it is a fact of very familiar observation, that exotics, transplanted to foreign climates suited to their growth, often escape from the flower garden and naturalize themselves among the spontaneous vegetation of the pastures. When the cases containing the artistic treasures of Thorvaldsen were opened in the court of the museum where they are deposited, the straw and grass employed in packing them were scattered upon the ground, and the next season there sprang up from the seeds no less than twenty-five species of plants belonging to the Roman campagna, some of which were preserved and cultivated as a new tribute to the memory of the great Scandinavian sculptor, and at least four are said to have spontaneously naturalized themselves about Copenhagen.* In the campaign of 1814, the Russian troops brought, in the stuffing of their saddles and by other accidental means, seeds from the banks of the Dnieper to the valley of the Rhine, and even introduced the plants of the steppes into the environs of Paris. The Turkish armies, in their incursions into Europe, brought Eastern vegetables in their train, and left the seeds of Oriental wall plants to grow upon the ramparts of Buda and Vienna.†

British colony in New England, says that the settlers at Plymouth had observed more than twenty English plants springing up spontaneously near their improvements.

Every country has many plants not now, if ever, made use of by man, and therefore not designedly propagated by him, but which cluster around his dwelling, and continue to grow luxuriantly on the ruins of his rural habitation after he has abandoned it. The site of a cottage, the very foundation stones of which have been carried off, may often be recognized, years afterward, by the rank weeds which cover it, though no others of the same species are found for miles.

"Mediæval Catholicism," says Vaupell, "brought us the red horsehoof—whose reddish-brown flower buds shoot up from the ground when the snow melts, and are followed by the large leaves—*lægekulsukker* and snake-root, which grow only where there were convents and other dwellings in the Middle Ages."—*Bögens Indvandring i de Danske Skove*, pp. 1, 2.

* Vaupell, *Bögens Invandring i de Danske Skove*, p. 2.

† It is, I believe, nearly certain that the Turks inflicted tobacco upon

The Canada thistle, *Erigeron Canadense*, is said to have sprung up in Europe, two hundred years ago, from a seed which dropped out of the stuffed skin of a bird.*

Vegetables, how affected by Transfer to Foreign Soils.

Vegetables, naturalized abroad either by accident or design, sometimes exhibit a greatly increased luxuriance of growth. The European cardoon, an esculent thistle, has broken out from the gardens of the Spanish colonies on the La Plata, acquired a gigantic stature, and propagated itself, in impenetrable thickets, over hundreds of leagues of the Pampas; and the *Anacharis alsinastrum*, a water plant not much inclined to spread in its native American habitat, has found its way into English rivers, and extended itself to such a degree as to form a serious obstruction to the flow of the current, and even to navigation.

Not only do many wild plants exhibit a remarkable facility of accommodation, but their seeds usually possess great tenacity of life, and their germinating power resists very severe trials. Hence, while the seeds of very many cultivated vegetables lose their vitality in two or three years, and can be transported safely to distant countries only with great precautions, the weeds that infest those vegetables, though not cared for by man, continue to accompany him in his migrations, and find a new home on every soil he colonizes. Nature fights in

Hungary, and probable that they in some measure compensated the injury by introducing maize also, which, as well as tobacco, has been claimed as Hungarian by patriotic Magyars.

* Accidents sometimes limit, as well as promote, the propagation of foreign vegetables in countries new to them. The Lombardy poplar is a diœcious tree, and is very easily grown from cuttings. In most of the countries into which it has been introduced the cuttings have been taken from the male, and as, consequently, males only have grown from them, the poplar does not produce seed in those regions. This is a fortunate circumstance, for otherwise this most worthless and least ornamental of trees would spread with a rapidity that would make it an annoyance to the agriculturist. See *Appendix*, No. 10.

defence of her free children, but wars upon them when they have deserted her banners and tamely submitted to the dominion of man.*

Not only is the wild plant much hardier than the domesticated vegetable, but the same law prevails in animated brute and even human life. The beasts of the chase are more capable of endurance and privation and more tenacious of life, than the domesticated animals which most nearly resemble them. The savage fights on, after he has received half a dozen mortal wounds, the least of which would have instantly paralyzed the strength of his civilized enemy, and, like the wild boar,† he has been known to press forward along the shaft of the spear which was transpiercing his vitals, and to deal a deathblow on the soldier who wielded it.

True, domesticated plants can be gradually acclimatized to bear a degree of heat or of cold, which, in their wild state, they would not have supported; the trained English racer outstrips the swiftest horse of the pampas or prairies, perhaps even the less systematically educated courser of the Arab; the strength of the European, as tested by the dynamometer, is greater than that of the New Zealander. But all these are instances of excessive development of particular capacities and faculties at the expense of general vital power. Expose untamed and domesticated forms of life, together, to an entire set of physical conditions equally alien to the former habits of both, so that every power of resistance and accommodation shall be called into action, and the wild plant or animal will live, while the domesticated will perish.

The saline atmosphere of the sea is specially injurious both to seeds and to very many young plants, and it is only recently

* Tempests, violent enough to destroy all cultivated plants, often spare those of spontaneous growth. During the present summer, I have seen in Northern Italy, vineyards, maize fields, mulberry and fruit trees completely stripped of their foliage by hail, while the forest trees scattered through the meadows, and the shrubs and brambles which sprang up by the wayside, passed through the ordeal with scarcely the loss of a leaflet.

† The boar spear is provided with a short crossbar, to enable the hunter to keep the infuriated animal at bay after he has transfixed him.

that the transportation of some very important vegetables across the ocean has been made practicable, through the invention of Ward's airtight glass cases. It is by this means that large numbers of the trees which produce the Jesuit's bark have been successfully transplanted from America to the British possessions in the East, where it is hoped they will become fully naturalized.

Extirpation of Vegetables.

Lamentable as are the evils produced by the too general felling of the woods in the Old World, I believe it does not satisfactorily appear that any species of native forest tree has yet been extirpated by man on the Eastern continent. The roots, stumps, trunks, and foliage found in bogs are recognized as belonging to still extant species. Except in some few cases where there is historical evidence that foreign material was employed, the timber of the oldest European buildings, and even of the lacustrine habitations of Switzerland, is evidently the product of trees still common in or near the countries where such architectural remains are found; nor have the Egyptian catacombs themselves revealed to us the former existence of any woods not now familiar to us as the growth of still living trees.* It is, however, said that the yew tree, *Taxus baccata*, formerly very common in England, Germany, and—as we are authorized to infer from Theophrastus—in Greece, has almost wholly disappeared from the latter country, and seems to be dying out in Germany. The wood of the yew surpasses that of any other European tree in closeness and fineness of grain, and it is well known for the elasticity which of old made it so great a favorite with the English

* Some botanists think that a species of water lily represented in many Egyptian tombs has become extinct, and the papyrus, which must have once been abundant in Egypt, is now found only in a very few localities near the mouth of the Nile. It grows very well and ripens its seeds in the waters of the Anapus near Syracuse, and I have seen it in garden ponds at Messina and in Malta. There is no apparent reason for believing that it could not be easily cultivated in Egypt, to any extent, if there were any special motive for encouraging its growth.

archer. It is much in request among wood carvers and turners, and the demand for it explains, in part, its increasing scarcity. It is also worth remarking that no insect depends upon it for food or shelter, or aids in its fructification, no bird feeds upon its berries—the latter a circumstance of some importance, because the tree hence wants one means of propagation or diffusion common to so many other plants. But it is alleged that the reproductive power of the yew is exhausted, and that it can no longer be readily propagated by the natural sowing of its seeds, or by artificial methods. If further investigation and careful experiment should establish this fact, it will go far to show that a climatic change, of a character unfavorable to the growth of the yew, has really taken place in Germany, though not yet proved by instrumental observation, and the most probable cause of such change would be found in the diminution of the area covered by the forests.

The industry of man is said to have been so successful in the local extirpation of noxious or useless vegetables in China, that, with the exception of a few water plants in the rice grounds, it is sometimes impossible to find a single weed in an extensive district; and the late eminent agriculturist, Mr. Coke, is reported to have offered in vain a considerable reward for the detection of a weed in a large wheatfield on his estate in England. In these cases, however, there is no reason to suppose that diligent husbandry has done more than to eradicate the pests of agriculture within a comparatively limited area, and the cockle and the darnel will probably remain to plague the slovenly cultivator as long as the cereal grains continue to bless him.*

* Although it is not known that man has extirpated any vegetable, the mysterious diseases which have, for the last twenty years, so injuriously affected the potato, the vine, the orange, the olive, and silk husbandry—whether in this case the malady resides in the mulberry or in the insect—are ascribed by some to a climatic deterioration produced by excessive destruction of the woods. As will be seen in the next chapter, a retardation in the period of spring has been observed in numerous localities in Southern Europe, as well as in the United States. This change has been thought to favor the multiplication of the obscure parasites which cause the injury to the vegetables just mentioned.

Babinet supposes the parasites which attack the grape and the potato

Origin of Domestic Plants.

One of the most important, and, at the same time, most difficult questions connected with our subject is: how far we are to regard our cereal grains, our esculent bulbs and roots, and the multiplied tree fruits of our gardens, as artificially

to be animal, not vegetable, and he ascribes their multiplication to excessive manuring and stimulation of the growth of the plants on which they live. They are now generally, if not universally, regarded as vegetable, and if they are so, Babinet's theory would be even more plausible than on his own supposition.—*Études et Lectures*, ii, p. 269.

It is a fact of some interest in agricultural economy, that the öidium, which is so destructive to the grape, has produced no pecuniary loss to the proprietors of the vineyards in France. "The price of wine," says Lavergne, "has quintupled, and as the product of the vintage has not diminished in the same proportion, the crisis has been, on the whole, rather advantageous than detrimental to the country."—*Économie Rurale de la France*, pp. 263, 264.

France produces a considerable surplus of wines for exportation, and the sales to foreign consumers are the principal source of profit to French vinegrowers. In Northern Italy, on the contrary, which exports little wine, there has been no such increase in the price of wine as to compensate the great diminution in the yield of the vines, and the loss of this harvest is severely felt. In Sicily, however, which exports much wine, prices have risen as rapidly as in France. Waltershausen informs us that in the years 1838–'42, the red wine of Mount Etna sold at the rate of one kreuzer and a half, or one cent the bottle, and sometimes even at but two thirds that price, but that at present it commands five or six times as much.

The grape disease has operated severely on small cultivators whose vineyards only furnished a supply for domestic use, but Sicily has received a compensation in the immense increase which it has occasioned in both the product and the profits of the sulphur mines. Flour of sulphur is applied to the vine as a remedy against the disease, and the operation is repeated from two to three or four—and even, it is said, eight or ten times—in a season. Hence there is a great demand for sulphur in all the vinegrowing countries of Europe, and Waltershausen estimates the annual consumption of that mineral for this single purpose at 850,000 *centner*, or more than forty thousand tons. The price of sulphur has risen in about the same proportion as that of wine.—WALTERSHAUSEN, *Ueber den Sicilianischen Ackerbau*, pp. 19, 20.

modified and improved forms of wild, self-propagating vegetation. The narratives of botanical travellers have often announced the discovery of the original form and habitat of domesticated plants, and scientific journals have described the experiments by which the identity of particular wild and cultivated vegetables has been thought to be established. It is confidently affirmed that maize and the potato—which we must suppose to have been first cultivated at a much later period than the breadstuffs and most other esculent vegetables of Europe and the East—are found wild and self-propagating in Spanish America, though in forms not recognizable by the common observer as identical with the familiar corn and tuber of modern agriculture. It was lately asserted, upon what seemed very strong evidence, that the *Ægilops ovata*, a plant growing wild in Southern France, had been actually converted into common wheat; but, upon a repetition of the experiments, later observers have declared that the apparent change was only a case of temporary hybridation or fecundation by the pollen of true wheat, and that the grass alleged to be transformed into wheat could not be perpetuated as such from its own seed.

The very great modifications which cultivated plants are constantly undergoing under our eyes, and the numerous varieties and races which spring up among them, certainly countenance the doctrine, that every domesticated vegetable, however dependent upon human care for growth and propagation in its present form, may have been really derived, by a long succession of changes, from some wild plant not now much resembling it. But it is, in every case, a question of evidence. The only satisfactory proof that a given wild plant is identical with a given garden or field vegetable, is the test of experiment, the actual growing of the one from the seed of the other, or the conversion of the one into the other by transplantation and change of conditions. It is hardly contended that any of the cereals or other plants important as human aliment, or as objects of agricultural industry, exist and propagate themselves uncultivated in the same form and with the same properties as when sown and reared by human art.* fact, the cases are rare where the identity of a wild with a domesticated plant is considered by the best authorities as conclusively established, and we are warranted in affirming of but few of the latter, as a historically known or experimentally proved fact, that they ever did exist, or could exist, independently of man.†

* Some recent observations of the learned traveller Wetzstein are worthy of special notice. "The soil of the Haurân," he remarks, "produces, in its primitive condition, much wild rye, which is not known as a cultivated plant in Syria, and much wild barley and oats. These cereals precisely resemble the corresponding cultivated plants in leaf, ear, size, and height of straw, but their grains are sensibly flatter and poorer in flour."—*Reisebericht über Haurân und die Trachonen*, p. 40.

† This remark is much less applicable to fruit trees than to garden vegetables and the cerealia. The wild orange of Florida, though once considered indigenous, is now generally thought by botanists to be descended from the European orange introduced by the early colonists. The fig and the olive are found growing wild in every country where those trees are cultivated. The wild fig differs from the domesticated in its habits, its season of fructification, and its insect population, but is, I believe, not specifically distinguishable from the garden fig, though I do not know that it is reclaimable by cultivation. The wild olive, which is so abundant in the Tuscan Maremma, produces good fruit without further care, when thinned out and freed from the shade of other trees, and is particularly suited for grafting. See SALVAGNOLI, *Memorie sulle Maremme*, pp. 63–73. See *Appendix*, No. 12.

FRAAS, *Klima und Pflanzenwelt in der Zeit*, pp. 35–38, gives, upon the authority of Link and other botanical writers, a list of the native habitats of most cereals and of many fruits, or at least of localities where these plants are said to be now found wild; but the data do not appear to rest, in general, upon very trustworthy evidence. Theoretically, there can be little doubt that all our cultivated plants are modified forms of spontaneous vegetation, but the connection is not historically shown, nor are we able to say that the originals of some domesticated vegetables may not be now extinct and unrepresented in the existing wild flora. See, on this subject, HUMBOLDT, *Ansichten der Natur*, i, pp. 208, 209. The following are interesting incidents: "A negro slave of the great Cortez was the first who sowed wheat in New Spain. He found three grains of it among the rice which had been brought from Spain as food for the soldiers. In the Franciscan monastery at Quito, I saw the earthen pot which contained the first

Organic Life as a Geological and Geographical Agency.

The quantitative value of organic life, as a geological agency, seems to be inversely as the volume of the individual organism; for nature supplies by numbers what is wanting in the bulk of the plant or animal out of whose remains or structures she forms strata covering whole provinces, and builds up from the depths of the sea large islands, if not continents. There are, it is true, near the mouths of the great Siberian rivers which empty themselves into the Polar Sea, drift islands composed, in an incredibly large proportion, of the bones and tusks of elephants, mastodons, and other huge pachyderms, and many extensive caves in various parts of the world are half filled with the skeletons of quadrupeds, sometimes lying loose in the earth, sometimes cemented together into an osseous breccia by a calcareous deposit or other binding material. These remains of large animals, though found in comparatively late formations, generally belong to extinct species, and their modern congeners or representatives do not exist in sufficient numbers to be of sensible importance in geology or in geography by the mere mass of their skeletons.* But the vegetable

wheat sown there by Friar Jodoco Rixi, of Ghent. It was preserved as a relic."

The Adams of modern botany and zoology have been put to hard shifts in finding names for the multiplied organisms which the Creator has brought before them, "to see what they would call them;" and naturalists and philosophers have shown much moral courage in setting at naught the laws of philology in the coinage of uncouth words to express scientific ideas. It is much to be wished that some bold neologist would devise English technical equivalents for the German *verwildert*, run-wild, and *veredelt*, improved by cultivation.

* Could the bones and other relics of the domestic quadrupeds destroyed by disease or slaughtered for human use in civilized countries be collected into large deposits, as obscure causes have gathered together those of extinct animals, they would soon form aggregations which might almost be called mountains. There were in the United States, in 1860, as we shall see hereafter, nearly one hundred and two millions of horses, black cattle, sheep, and swine. There are great numbers of all the same animals in the British American Provinces and in Mexico, and there are large herds of

products found with them, and, in rare cases, in the stomachs of some of them, are those of yet extant plants; and besides this evidence, the recent discovery of works of human art,

wild horses on the plains, and of tamed among the independent Indian tribes of North America. It would perhaps not be extravagant to suppose that all these cattle may amount to two thirds as many as those of the United States, and thus we have in North America a total of 170,000,000 domestic quadrupeds belonging to species introduced by European colonization, besides dogs, cats, and other four-footed household pets and pests, also of foreign origin.

If we allow half a solid foot to the skeleton and other slowly destructible parts of each animal, the remains of these herds would form a cubical mass measuring not much short of four hundred and fifty feet to the side, or a pyramid equal in dimensions to that of Cheops, and as the average life of these animals does not exceed six or seven years, the accumulations of their bones, horns, hoofs, and other durable remains would amount to at least fifteen times as great a volume in a single century. It is true that the actual mass of solid matter, left by the decay of dead domestic quadrupeds and permanently added to the crust of the earth, is not so great as this calculation makes it. The greatest proportion of the soft parts of domestic animals, and even of the bones, is soon decomposed, through direct consumption by man and other carnivora, industrial use, and employment as manure, and enters into new combinations in which its animal origin is scarcely traceable; there is, nevertheless, a large annual residuum, which, like decayed vegetable matter, becomes a part of the superficial mould; and in any event, brute life immensely changes the form and character of the superficial strata, if it does not sensibly augment the quantity of the matter composing them.

The remains of man, too, add to the earthy coating that covers the face of the globe. The human bodies deposited in the catacombs during the long, long ages of Egyptian history, would perhaps build as large a pile as one generation of the quadrupeds of the United States. In the barbarous days of old Moslem warfare, the conquerors erected large pyramids of human skulls. The soil of cemeteries in the great cities of Europe has sometimes been raised several feet by the deposit of the dead during a few generations. In the East, Turks and Christians alike bury bodies but a couple of feet beneath the surface. The grave is respected as long as the tombstone remains, but the sepultures of the ignoble poor, and of those whose monuments time or accident has removed, are opened again and again to receive fresh occupants. Hence the ground in Oriental cemeteries is pervaded with relics of humanity, if not wholly composed of them; and an examination of the soil of the lower part of the *Petit Champ des Morts* deposited in juxtaposition with fossil bones, and evidently at the same time and by the same agency which buried these latter—not to speak of alleged human bones found in the same strata—proves that the animals whose former existence they testify were contemporaneous with man, and possibly even extirpated by him.* I do not propose to enter upon the thorny question, whether the existing races of man are genealogically connected with these ancient types of humanity, and I advert to these facts only for the sake of the suggestion, that man, in his earliest known stages of existence, was probably a destructive power upon the earth, though perhaps not so emphatically as his present representatives.

The larger wild animals are not now numerous enough in any one region to form extensive deposits by their remains; but they have, nevertheless, a certain geographical importance.

at Pera, by the naked eye alone, shows the observer that it consists almost exclusively of the comminuted bones of his fellow man.

* It is asserted that the bones of mammoths and mastodons, in many instances, appear to have been grazed or cut by flint arrow-heads or other stone weapons. These accounts have often been discredited, because it has been assumed that the extinction of these animals was more ancient than the existence of man. Recent discoveries render it highly probable, if not certain, that this conclusion has been too hastily adopted. Lyell observes: "These stories * * must in future be more carefully inquired into, for we can scarcely doubt that the mastodon in North America lived down to a period when the mammoth coexisted with man in Europe."—*Antiquity of Man*, p. 354.

On page 143 of the volume just quoted, the same very distinguished writer remarks that man "no doubt played his part in hastening the era of the extinction" of the large pachyderms and beasts of prey; but, as contemporaneous species of other animals, which man cannot be supposed to have extirpated, have also become extinct, he argues that the disappearance of the quadrupeds in question cannot be ascribed to human action alone.

On this point it may be observed that, as we cannot know what precise physical conditions were necessary to the existence of a given extinct organism, we cannot say how far such conditions may have been modified by the action of man, and he may therefore have influenced the life of such organisms in ways, and to an extent, of which we can form no just idea.

If the myriads of large browsing and grazing quadrupeds which wander over the plains of Southern Africa—and the slaughter of which by thousands is the source of a ferocious pleasure and a brutal triumph to professedly civilized hunters —if the herds of the American bison, which are numbered by hundreds of thousands, do not produce visible changes in the forms of terrestrial surface, they have at least an immense influence on the growth and distribution of vegetable life, and, of course, indirectly upon all the physical conditions of soil and climate between which and vegetation a mutual interdependence exists.

The influence of wild quadrupeds upon vegetable life has been little studied, and not many facts bearing upon it have been recorded, but, so far as it is known, it appears to be conservative rather than pernicious.* Few if any of them depend for their subsistence on vegetable products obtainable only by the destruction of the plant, and they seem to confine their consumption almost exclusively to the annual harvest of leaf or twig, or at least of parts of the vegetable easily reproduced. If there are exceptions to this rule, they are in cases where the numbers of the animal are so proportioned to the abundance of the vegetable, that there is no danger of the extermination of the plant from the voracity of the quadruped, or of the extinction of the quadruped from the scarcity of the plant. In diet and natural wants the bison resembles the ox, the ibex

* Evelyn thought the depasturing of grass by cattle serviceable to its growth. "The biting of cattle," he remarks, "gives a gentle loosening to the roots of the herbage, and makes it to grow fine and sweet, and their very breath and treading as well as soil, and the comfort of their warm bodies, is wholesome and marvellously cherishing."—*Terra, or Philosophical Discourse of Earth*, p. 36.

In a note upon this passage, Hunter observes: "Nice farmers consider the lying of a beast upon the ground, for one night only, as a sufficient tilth for the year. The breath of graminivorous quadrupeds does certainly enrich the roots of grass; a circumstance worthy of the attention of the philosophical farmer."—*Terra*, same page.

The "philosophical farmer" of the present day will not adopt these opinions without some qualification.

and the chamois assimilate themselves to the goat and the sheep; but while the wild animal does not appear to be a destructive agency in the garden of nature, his domestic congeners are eminently so. This is partly from the change of habits resulting from domestication and association with man, partly from the fact that the number of reclaimed animals is not determined by the natural relation of demand and spontaneous supply which regulates the multiplication of wild creatures, but by the convenience of man, who is, in comparatively few things, amenable to the control of the merely physical arrangements of nature. When the domesticated animal escapes from human jurisdiction, as in the case of the ox, the horse, the goat, and perhaps the ass—which, so far as I know, are the only well-authenticated instances of the complete emancipation of household quadrupeds—he becomes again an unresisting subject of nature, and all his economy is governed by the same laws as that of his fellows which have never been enslaved by man; but, so long as he obeys a human lord, he is an auxiliary in the warfare his master is ever waging against all existences except those which he can tame to a willing servitude.

Number of Quadrupeds in the United States.

Civilization is so intimately associated with, if not dependent upon, certain inferior forms of animal life, that cultivated man has never failed to accompany himself, in all his migrations, with some of these humble attendants. The ox, the horse, the sheep, and even the comparatively useless dog and cat, as well as several species of poultry, are voluntarily transported by every emigrant colony, and they soon multiply to numbers very far exceeding those of the wild genera most nearly corresponding to them.* According to the census of the United

* The rat and the mouse, though not voluntarily transported, are passengers by every ship that sails from Europe to a foreign port, and several species of these quadrupeds have, consequently, much extended their range and increased their numbers in modern times. From a story of Heliogabalus related by Lampridius, *Hist. Aug. Scriptores*, ed. Casaubon,

States for 1860,* the total number of horses in all the States of the American Union, was, in round numbers, 7,300,000; of asses and mules, 1,300,000; of the ox tribe, 29,000,000;† of sheep, 25,000,000; and of swine, 39,000,000. The only North

1690, p. 110, it would seem that mice at least were not very common in ancient Rome. Among the capricious freaks of that emperor, it is said that he undertook to investigate the statistics of the arachnoid population of the capital, and that 10,000 pounds of spiders (or spiders' webs—for *aranea* is equivocal) were readily collected; but when he got up a mouse show, he thought ten thousand mice a very fair number. I believe as many might almost be found in a single palace in modern Rome. Rats are not less numerous in all great cities, and in Paris, where their skins are used for gloves, and their flesh, it is whispered, in some very complex and equivocal dishes, they are caught by legions. I have read of a manufacturer who contracted to buy of the rat catchers, at a high price, all the rat skins they could furnish before a certain date, and failed, within a week, for want of capital, when the stock of peltry had run up to 600,000.

* BIGELOW, *Les États Unis en* 1863, pp. 379, 380. In the same paragraph this volume states the number of animals slaughtered in the United States by butchers, in 1859, at 212,871,653. This is an error of the press. Number is confounded with value. A reference to the tables of the census shows that the animals slaughtered that year were estimated at 212,871,653 *dollars;* the number of head is not given. The wild horses and horned cattle of the prairies and the horses of the Indians are not included in the returns.

† Of this total number, 2,240,000, or nearly nine per cent., are reported as working oxen. This would strike European, and especially English agriculturists, as a large proportion; but it is explained by the difference between a new country and an old, in the conditions which determine the employment of animal labor. Oxen are very generally used in the United States and Canada for hauling timber and firewood through and from the forests; for ploughing in ground still full of rocks, stumps, and roots; for breaking up the new soil of the prairies with its strong matting of native grasses, and for the transportation of heavy loads over the rough roads of the interior. In all these cases, the frequent obstructions to the passage of the timber, the plough, and the sled or cart, are a source of constant danger to the animals, the vehicles, and the harness, and the slow and steady step of the ox is attended with much less risk than the swift and sudden movements of the impatient horse. It is surprising to see the sagacity with which the dull and clumsy ox—hampered as he is by the rigid yoke, the most absurd implement of draught ever contrived by man—picks his way, when once trained to forest work, among rocks and roots,

American quadruped sufficiently gregarious in habits, and sufficiently multiplied in numbers, to form really large herds, is the bison, or, as he is commonly called in America, the buffalo; and this animal is confined to the prairie region of the Mississippi basin and Northern Mexico. The engineers sent out to survey railroad routes to the Pacific estimated the number of a single herd of bisons seen within the last ten years on the great plains near the Upper Missouri, at not less than 200,000, and yet the range occupied by this animal is now very much smaller in area than it was when the whites first established themselves on the prairies.* But it must be remarked that the American buffalo is a migratory animal, and that, at the season of his annual journeys, the whole stock of a vast extent of pasture ground is collected into a single army, which is seen at or very near any one point only for a few days during the entire season. Hence there is risk of great error in estimating the numbers of the bison in a given district from the magnitude of the herds seen at or about the same time at a single place of observation; and, upon the whole, it is neither proved nor probable that the bison was ever, at any one time, as numerous in North America as the domestic bovine species is at present. The elk, the moose, the musk ox, the caribou, and the smaller quadrupeds popularly embraced under the general name of deer,† though sufficient for the wants of a sparse savage popu-

and even climbs over fallen trees, not only moving safely, but drawing timber over ground wholly impracticable for the light and agile horse.

Cows, so constantly employed for draught in Italy, are never yoked or otherwise used for labor in America, except in the Slave States.

* "About five miles from camp we ascended to the top of a high hill, and for a great distance ahead every square mile seemed to have a herd of buffalo upon it. Their number was variously estimated by the members of the party; by some as high as half a million. I do not think it any exaggeration to set it down at 200,000."—STEVENS's *Narrative and Final Report. Reports of Explorations and Surveys for Railroad to Pacific*, vol. xii, book i, 1860.

The next day, the party fell in with a "buffalo trail," where at least 100,000 were thought to have crossed a slough.

† The most zealous and successful New England hunter of whom I have

lation, were never numerically very abundant, and the carnivora which fed upon them were still less so. It is almost needless to add that the Rocky Mountain sheep and goat must always have been very rare.

Summing up the whole, then, it is evident that the wild quadrupeds of North America, even when most numerous, were few compared with their domestic successors, that they required a much less supply of vegetable food, and consequently were far less important as geographical elements than the many millions of hoofed and horned cattle now fed by civilized man on the same continent.

Origin and Transfer of Domestic Quadrupeds.

Of the origin of our domestic animals, we know historically nothing, because their domestication belongs to the ages which preceded written history; but though they cannot all be specifically identified with now extant wild animals, it is presumable that they have been reclaimed from an originally wild state. Ancient annalists have preserved to us fewer data respecting the introduction of domestic animals into new countries than respecting the transplantation of domestic vegetables. Ritter, in his learned essay on the camel, has shown that this animal was not employed by the Egyptians until a comparatively late period in their history; that he was unknown to the Carthaginians until after the downfall of their commonwealth; and that his first appearance in Western Africa is more recent still. The Bactrian camel was certainly brought

any personal knowledge, and who continued to indulge his favorite passion much beyond the age which generally terminates exploits in woodcraft, lamented on his deathbed that he had not lived long enough to carry up the record of his slaughtered deer to the number of one thousand, which he had fixed as the limit of his ambition. He was able to handle the rifle, for sixty years, at a period when the game was still nearly as abundant as ever, but had killed only nine hundred and sixty of these quadrupeds, of all species. The exploits of this Nimrod have been far exceeded by prairie hunters, but I doubt whether, in the originally wooded territory of the Union, any single marksman has brought down a larger number.

from Asia Minor to the Northern shores of the Black Sea, by the Goths, in the third or fourth century.* The Arabian single-humped camel, or dromedary, has been carried to the Canary Islands, partially introduced into Australia, Greece, Spain, and even Tuscany, experimented upon to little purpose in Venezuela, and finally imported by the American Government into Texas and New Mexico, where it finds the climate and the vegetable products best suited to its wants, and promises to become a very useful agent in the promotion of the special civilization for which those regions are adapted. America had no domestic quadruped but a species of dog, the lama tribe, and, to a certain extent, the bison or buffalo.† Of course, it owes the horse, the ass, the ox, the sheep, the goat, and the swine, as does also Australia, to European colonization. Modern Europe has, thus far, not accomplished much in the way of importation of new animals, though some interesting essays have been made. The reindeer was successfully introduced into Iceland about a century ago, while similar attempts failed, about the same time, in Scotland. The Cashmere or Thibet goat was brought to France a generation since, and succeeds well. The same or an allied species and the Asiatic buffalo were carried to South Carolina about the year 1850, and the former, at least, is thought likely to prove of permanent value in the United States. The yak, or Tartary ox, seems to thrive in France, and success has attended the recent efforts to introduce the South American alpaca into Europe.

* *Erdkunde*, viii, *Asien*, 1*ste Abtheilung*, pp. 660, 758.

† See chapter iii, *post;* also HUMBOLDT, *Ansichten der Natur*, i, p. 71. From the anatomical character of the bones of the urus, or auerochs, found among the relics of the lacustrine population of ancient Switzerland, and from other circumstances, it is inferred that this animal had been domesticated by that people; and it is stated, I know not upon what authority, in *Le Alpi che cingono l'Italia*, that it had been tamed by the Veneti also. See LYELL, *Antiquity of Man*, pp. 24, 25, and the last-named work, p. 489. This is a fact of much interest, because it is, I believe, the only known instance of the extinction of a domestic quadruped, and the extreme improbability of such an event gives some countenance to the theory of the identity of the domestic ox with, and its descent from, the urus.

Extirpation of Quadrupeds.

Although man never fails greatly to diminish, and is perhaps destined ultimately to exterminate, such of the larger wild quadrupeds as he cannot profitably domesticate, yet their numbers often fluctuate, and even after they seem almost extinct, they sometimes suddenly increase, without any intentional steps to promote such a result on his part. During the wars which followed the French Revolution, the wolf multiplied in many parts of Europe, partly because the hunters were withdrawn from the woods to chase a nobler game, and partly because the bodies of slain men and horses supplied this voracious quadruped with more abundant food. The same animal became again more numerous in Poland after the general disarming of the rural population by the Russian Government. On the other hand, when the hunters pursue the wolf, the graminivorous wild quadrupeds increase, and thus in turn promote the multiplication of their great four-footed destroyer by augmenting the supply of his nourishment. So long as the fur of the beaver was extensively employed as a material for fine hats, it bore a very high price, and the chase of this quadruped was so keen that naturalists feared its speedy extinction. When a Parisian manufacturer invented the silk hat, which soon came into almost universal use, the demand for beavers' fur fell off, and this animal—whose habits, as we have seen, are an important agency in the formation of bogs and other modifications of forest nature—immediately began to increase, reappeared in haunts which he had long abandoned, and can no longer be regarded as rare enough to be in immediate danger of extirpation. Thus the convenience or the caprice of Parisian fashion has unconsciously exercised an influence which may sensibly affect the physical geography of a distant continent.

Since the invention of gunpowder, some quadrupeds have completely disappeared from many European and Asiatic countries where they were formerly numerous. The last wolf was killed in Great Britain two hundred years ago, and the bear was extirpated from that island still earlier. The British wild ox exists only in a few English and Scottish parks, while in Irish bogs, of no great apparent antiquity, are found antlers which testify to the former existence of a stag much larger than any extant European species. The lion is believed to have inhabited Asia Minor and Syria, and probably Greece and Sicily also, long after the commencement of the historical period, and he is even said to have been not yet extinct in the first-named two of these countries at the time of the first Crusades.* Two large graminivorous or browsing quadrupeds, the ur and the schelk, once common in Germany, are utterly extinct, the eland and the auerochs nearly so. The Nibelungen-Lied, which, in the oldest form preserved to us, dates from about the year 1,200, though its original composition no doubt belongs to an earlier period, thus sings:

> Then slowe the dowghtie Sigfrid a wisent and an elk,
> He smote four stonte uroxen and a grim and sturdie schelk.†

Modern naturalists identify the elk with the eland, the wisent with the auerochs. The period when the ur and the schelk became extinct is not known. The auerochs survived in Prussia until the middle of the last century, but unless it is identical with a similar quadruped said to be found on the Caucasus, it now exists only in the Russian imperial forest of

* In maintaining the recent existence of the lion in the countries named in the text, naturalists have, perhaps, laid too much weight on the frequent occurrence of representations of this animal in sculptures apparently of a historical character. It will not do to argue, twenty centuries hence, that the lion and the unicorn were common in Great Britain in Queen Victoria's time, because they are often seen "fighting for the crown" in the carvings and paintings of that period.

> † Dar nach slûger schiere. einen wisent unt elch.
> Starcher ûre viere. unt einen grimmen schelch.
>
> *XVI Auentiure.*

The testimony of the *Nibelungen-Lied* is not conclusive evidence that these quadrupeds existed in Germany at the time of the composition of that poem. It proves too much; for, a few lines above those just quoted, Sigfrid is said to have killed a lion, an animal which the most patriotic Teuton will hardly claim as a denizen of mediæval Germany.

Bialowitz, where about a thousand are still preserved, and in some great menageries, as for example that at Schönbrunn, near Vienna, which, in 1852, had four specimens. The eland, which is closely allied to the American wapiti, if not specifically the same animal, is still kept in the royal preserves of Prussia, to the number of four or five hundred individuals. The chamois is becoming rare, and the ibex or steinbock, once common in all the high Alps, is now believed to be confined to the Cogne mountains in Piedmont, between the valleys of the Dora Baltea and the Orco.

Number of Birds in the United States.

The tame fowls play a much less conspicuous part in rural life than the quadrupeds, and, in their relations to the economy of nature, they are of very much less moment than four-footed animals, or than the undomesticated birds. The domestic turkey * is probably more numerous in the territory of the United States than the wild bird of the same species ever was, and the grouse cannot, at the period of their greatest abundance, have counted as many as we now number of the common hen. The dove, however, must fall greatly short of the wild pigeon in multitude, and it is hardly probable that the flocks of domestic geese and ducks are as numerous as once were those of their wild congeners. The pigeon, indeed, seems to have multiplied immensely, for some years after the first clearings in the woods, because the settlers warred unsparingly upon the hawk, while the crops of grain and other vegetable growths increased the supply of food within the reach of the young birds, at the age when their power of flight is not

* The wild turkey takes readily to the water, and is able to cross rivers of very considerable width by swimming. By way of giving me an idea of the former abundance of this bird, an old and highly respectable gentleman who was among the early white settlers of the West, told me that he once counted, in walking down the northern bank of the Ohio River, within a distance of four miles, eighty-four turkeys as they landed singly, or at most in pairs, after swimming over from the Kentucky side.

yet great enough to enable them to seek it over a wide area.* The pigeon is not described by the earliest white inhabitants of the American States as filling the air with such clouds of winged life as astonish naturalists in the descriptions of Audubon, and, at the present day, the net and the gun have so reduced its abundance, that its appearance in large numbers is recorded only at long intervals, and it is never seen in the great flocks remembered by many still living observers as formerly very common.

Birds as Sowers and Consumers of Seeds, and as Destroyers of Insects.

Wild birds form of themselves a very conspicuous and interesting feature in the *staffage*, as painters call it, of the natural landscape, and they are important elements in the view we are taking of geography, whether we consider their immediate or their incidental influence. Birds affect vegetation directly by sowing seeds and by consuming them; they affect it indirectly by destroying insects injurious, or, in some cases, beneficial to vegetable life. Hence, when we kill a seed-sowing bird, we check the dissemination of a plant; when we kill a bird which digests the seed it swallows, we promote the increase of a vegetable. Nature protects the seeds of wild, much more effectually than those of domesticated plants. The cereal grains are completely digested when consumed by birds, but the germ of the smaller stone fruits and of very many other wild vegetables is uninjured, perhaps even stimulated to more vigorous growth, by the natural chemistry of the bird's stomach. The power of flight and the restless habits of the bird enable it to transport heavy seeds to far greater distances than

* The wood pigeon has been observed to increase in numbers in Europe also, when pains have been taken to exterminate the hawk. The pigeons, which migrated in flocks so numerous that they were whole days in passing a given point, were no doubt injurious to the grain, but probably less so than is generally supposed; for they did not confine themselves exclusively to the harvests for their nourishment.

they could be carried by the wind. A swift-winged bird may drop cherry stones a thousand miles from the tree they grow on; a hawk, in tearing a pigeon, may scatter from its crop the still fresh rice it had swallowed at a distance of ten degrees of latitude,* and thus the occurrence of isolated plants in situations where their presence cannot otherwise well be explained, is easily accounted for. There is a large class of seeds apparently specially fitted by nature for dissemination by animals. I refer to those which attach themselves, by means of hooks, or by viscous juices, to the coats of quadrupeds and the feathers of birds, and are thus transported wherever their living vehicles may chance to wander. Some birds, too, deliberately bury seeds, not indeed with a foresight aiming directly at the propagation of the plant, but from apparently purposeless secretiveness, or as a mode of preserving food for future use.

An unfortunate popular error greatly magnifies the injury done to the crops of grain and leguminous vegetables by wild birds. Very many of those generally supposed to consume large quantities of the seeds of cultivated plants really feed almost exclusively upon insects, and frequent the wheatfields, not for the sake of the grain, but for the eggs, larvæ, and fly of the multiplied tribes of insect life which are so destructive to the harvests. This fact has been so well established by the examination of the stomachs of great numbers of birds in Europe and New England, at different seasons of the year, that it is no longer open to doubt, and it appears highly probable that even the species which consume more or less grain generally make amends, by destroying insects whose ravages would have been still more injurious.† On this subject, we

* Pigeons were shot near Albany, in New York, a few years ago, with green rice in their crops, which it was thought must have been growing, a very few hours before, at the distance of seven or eight hundred miles.

† Professor Treadwell, of Massachusetts, found that a half-grown American robin in confinement ate in one day sixty-eight earthworms, weighing together nearly once and a half as much as the bird himself, and another had previously starved upon a daily allowance of eight or ten worms, or about twenty per cent. of his own weight. The largest of these

have much other evidence besides that derived from dissection. Direct observation has shown, in many instances, that the destruction of wild birds has been followed by a great multiplication of noxious insects, and, on the other hand, that these latter have been much reduced in numbers by the protection and increase of the birds that devour them. Many interesting facts of this nature have been collected by professed naturalists, but I shall content myself with a few taken from familiar and generally accessible sources. The following extract is from Michelet, *L'Oiseau* pp. 169, 170:

"The *stingy* farmer—an epithet justly and feelingly bestowed by Virgil. Avaricious, blind, indeed, who proscribes the birds—those destroyers of insects, those defenders of his harvests. Not a grain for the creature which, during the rains of winter, hunts the future insect, finds out the nests of the

numbers appeared, so far as could be judged by watching parent birds of the same species, as they brought food to their young, to be much greater than that supplied to them when fed in the nest; for the old birds did not return with worms or insects oftener than once in ten minutes on an average. If we suppose the parents to hunt for food twelve hours in a day, and a nest to contain four young, we should have seventy-two worms, or eighteen each, as the daily supply of the brood. It is probable enough that some of the food collected by the parents may be more nutritious than the earthworms, and consequently that a smaller quantity sufficed for the young in the nest than when reared under artificial conditions.

The supply required by growing birds is not the measure of their wants after they have arrived at maturity, and it is not by any means certain that great muscular exertion always increases the demand for nourishment, either in the lower animals or in man. The members of the English Alpine Club are not distinguished for appetites which would make them unwelcome guests to Swiss landlords, and I think every man who has had the personal charge of field or railway hands, must have observed that laborers who spare their strength the least are not the most valiant trencher champions. During the period when imprisonment for debt was permitted in New England, persons confined in country jails had no specific allowance, and they were commonly fed without stint. I have often inquired concerning their diet, and been assured by the jailers that their prisoners, who were not provided with work or other means of exercise, consumed a considerably larger supply of food than common out-door laborers.

larvæ, examines, turns over every leaf, and destroys, every day, thousands of incipient caterpillars. But sacks of corn for the mature insect, whole fields for the grasshoppers, which the bird would have made war upon. With eyes fixed upon his furrow, upon the present moment only, without seeing and without foreseeing, blind to the great harmony which is never broken with impunity, he has everywhere demanded or approved laws for the extermination of that necessary ally of his toil—the insectivorous bird. And the insect has well avenged the bird. It has become necessary to revoke in haste the proscription. In the Isle of Bourbon, for instance, a price was set on the head of the martin; it disappeared, and the grasshoppers took possession of the island, devouring, withering, scorching with a biting drought all that they did not consume. In North America it has been the same with the starling, the protector of Indian corn.* Even the sparrow, which really does attack grain, but which protects it still more, the pilferer, the outlaw, loaded with abuse and smitten with curses—it has been found in Hungary that they were likely to perish without him, that he alone could sustain the mighty war against the beetles and the thousand winged enemies that swarm in the lowlands; they have revoked the decree of banishment, recalled in haste this valiant militia, which, though deficient in discipline, is nevertheless the salvation of the country.†

* I hope Michelet has good authority for this statement, but I am unable to confirm it.

† Apropos of the sparrow—a single pair of which, according to Michelet, p. 315, carries to the nest four thousand and three hundred caterpillars or coleoptera in a week—I take from the *Record*, an English religious newspaper, of December 15, 1862, the following article communicated to a country paper by a person who signs himself "A real friend to the farmer:"

"*Crawley Sparrow Club.*—The annual dinner took place at the George Inn on Wednesday last. The first prize was awarded to Mr. I. Redford, Worth, having destroyed within the last year 1,467. Mr. Heayman took the second with 1,448 destroyed. Mr. Stone, third, with 982 affixed. Total destroyed, 11,944. Old birds, 8,663; young ditto, 722; eggs, 2,556."

This trio of valiant fowlers, and their less fortunate—or rather less un-

"Not long since, in the neighborhood of Rouen and in the valley of Monville, the blackbird was for some time proscribed. The beetles profited well by this proscription; their larvæ, infinitely multiplied, carried on their subterranean labors with such success, that a meadow was shown me, the surface of which was completely dried up, every herbaceous root was consumed, and the whole grassy mantle, easily loosened, might have been rolled up and carried away like a carpet."

Diminution and Extirpation of Birds.

The general hostility of the European populace to the smaller birds is, in part, the remote effect of the reaction created by the game laws. When the restrictions imposed upon the chase by those laws were suddenly removed in France, the whole people at once commenced a destructive campaign against every species of wild animal. Arthur Young, writing in Provence, on the 30th of August, 1789, soon after the National Assembly had declared the chase free, thus complains of the annoyance he experienced from the use made by the peasantry of their newly won liberty. "One would think that every rusty firelock in all Provence was at work in the indiscriminate destruction of all the birds. The wadding buzzed by my ears, or fell into my carriage, five or six times in the course of the day." * * "The declaration of the Assembly that every man is free to hunt on his own land

fortunate, but not therefore less guilty—associates, have rescued by their prowess, it may be, a score of pecks of grain from being devoured by the voracious sparrow, but every one of the twelve thousand hatched and unhatched birds, thus sacrificed to puerile vanity and ignorant prejudice, would have saved his bushel of wheat by preying upon insects that destroy the grain. Mr. Redford, Mr. Heayman, and Mr. Stone ought to contribute the value of the bread they have wasted to the fund for the benefit of the Lancashire weavers; and it is to be hoped that the next Byron will satirize the sparrowcide as severely as the first did the prince of anglers, Walton. in the well known lines:

> "The quaint, old, cruel coxcomb in his gullet
> Should have a hook, and a small trout to pull it."

* * has filled all France with an intolerable cloud of sportsmen. * * The declaration speaks of compensations and indemnities [to the *seigneurs*], but the ungovernable populace takes advantage of the abolition of the game laws and laughs at the obligation imposed by the decree."

The French Revolution removed similar restrictions, with similar results, in other countries. The habits then formed have become hereditary on the Continent, and though game laws still exist in England, there is little doubt that the blind prejudices of the ignorant and half-educated classes in that country against birds are, in some degree, at least, due to a legislation, which, by restricting the chase of all game worth killing, drives the unprivileged sportsman to indemnify himself by slaughtering all wild life which is not reserved for the amusement of his betters. Hence the lord of the manor buys his partridges and his hares by sacrificing the bread of his tenants, and so long as the farmers of Crawley are forbidden to follow higher game, they will suicidally revenge themselves by destroying the sparrows which protect their wheatfields.

On the Continent, and especially in Italy, the comparative scarcity and dearness of animal food combine with the feeling I have just mentioned to stimulate still further the destructive passions of the fowler. In the Tuscan province of Grosseto, containing less than 2,000 square miles, nearly 300,000 thrushes and other small birds are annually brought to market.*

* SALVAGNOLI, *Memorie sulle Maremme Toscane*, p. 143. The country about Naples is filled with slender towers fifteen or twenty feet high, which are a standing puzzle to strangers. They are the stations of the fowlers who watch from them the flocks of small birds and drive them down in to the nets by throwing stones over them. See *Appendix*, No. 14.

Tschudi has collected in his little work, *Ueber die Landwirthschaftliche Bedeutung der Vögel*, many interesting facts respecting the utility of birds, and the wanton destruction of them in Italy and elsewhere. Not only the owl, but many other birds more familiarly known as predacious in their habits, are useful by destroying great numbers of mice and moles. The importance of this last service becomes strikingly apparent when it is known that the burrows of the mole are among the most frequent causes of rupture in the dikes of the Po, and, consequently, of inundations which

Birds are less hardy in constitution, they possess less facility of accommodation,* and they are more severely affected by climatic excess than quadrupeds. Besides, they generally want the means of shelter against the inclemency of the weather and against pursuit by their enemies, which holes and dens afford to burrowing animals and to some larger beasts of prey. The egg is exposed to many dangers before hatching, and the young bird is especially tender, defenceless, and helpless. Every cold rain, every violent wind, every hailstorm during the breeding season, destroys hundreds of nestlings, and the parent often perishes with her progeny while brooding over it in the vain effort to protect it.† The great proportional numbers of birds, their migratory habits, and the ease with which they may escape most dangers that beset them, would seem to secure them from extirpation, and even from very great numerical reduction. But experience shows that when not pro-

lay many square miles under water.—*Annales des Ponts et Chaussées*, 1847, 1re sémestre, p. 150. See also VOGT, *Nützliche u. schädliche Thiere.*

* Wild birds are very tenacious in their habits. The extension of particular branches of agriculture introduces new birds; but unless in the case of such changes in physical conditions, particular species seem indissolubly attached to particular localities. The migrating tribes follow almost undeviatingly the same precise line of flight in their annual journeys, and establish themselves in the same breeding places from year to year. The stork is a strong-winged bird and roves far for food, but very rarely establishes new colonies. He is common in Holland, but unknown in England. Not above five or six pairs of storks commonly breed in the suburbs of Constantinople along the European shore of the narrow Bosphorus, while —much to the satisfaction of the Moslems, who are justly proud of the marked partiality of so orthodox a bird—dozens of chimneys of the true believers on the Asiatic side are crowned with his nests. See *App.* No. 15.

† It is not the unfledged and the nursing bird alone that are exposed to destruction by severe weather. Whole flocks of adult and strong-winged tribes are killed by hail. Severe winters are usually followed by a sensible diminution in the numbers of the non-migrating birds, and a cold storm in summer often proves fatal to the more delicate species. On the 10th of June, 184–, five or six inches of snow fell in Northern Vermont. The next morning I found a humming bird killed by the cold, and hanging by its claws just below a loose clapboard on the wall of a small wooden building where it had sought shelter.

tected by law, by popular favor or superstition, or by other special circumstances, they yield very readily to the hostile influences of civilization, and, though the first operations of the settler are favorable to the increase of many species, the great extension of rural and of mechanical industry is, in a variety of ways, destructive even to tribes not directly warred upon by man.*

* LYELL, *Antiquity of Man*, p. 409, observes: "Of birds it is estimated that the number of those which die every year equals the aggregate number by which the species to which they respectively belong is, on the average, permanently represented."

A remarkable instance of the influence of new circumstances upon birds was observed upon the establishment of a lighthouse on Cape Cod some years since. The morning after the lamps were lighted for the first time, more than a hundred dead birds of several different species, chiefly water fowl, were found at the foot of the tower. They had been killed in the course of the night by flying against the thick glass or grating of the lantern. See *Appendix*, No. 16.

Migrating birds, whether for greater security from eagles, hawks, and other enemies, or for some unknown reason, perform a great part of their annual journeys by night; and it is observed in the Alps that they follow the high roads in their passage across the mountains. This is partly because the food in search of which they must sometimes descend is principally found near the roads. It is, however, not altogether for the sake of consorting with man, or of profiting by his labors, that their line of flight conforms to the paths he has traced, but rather because the great roads are carried through the natural depressions in the chain, and hence the birds can cross the summit by these routes without rising to a height where at the seasons of migration the cold would be excessive.

The instinct which guides migratory birds in their course is not in all cases infallible, and it seems to be confounded by changes in the condition of the surface. I am familiar with a village in New England, at the junction of two valleys, each drained by a mill stream, where the flocks of wild geese which formerly passed, every spring and autumn, were very frequently lost, as it was popularly phrased, and I have often heard their screams in the night as they flew wildly about in perplexity as to the proper course. Perhaps the village lights embarrassed them, or perhaps the constant changes in the face of the country, from the clearings then going on, introduced into the landscape features not according with the ideal map handed down in the anserine family, and thus deranged its traditional geography.

Nature sets bounds to the disproportionate increase of birds, while at the same time, by the multitude of their resources, she secures them from extinction through her own spontaneous agencies. Man both preys upon them and wantonly destroys them. The delicious flavor of game birds, and the skill implied in the various arts of the sportsman who devotes himself to fowling, make them favorite objects of the chase, while the beauty of their plumage, as a military and feminine decoration, threatens to involve the sacrifice of the last survivor of many once numerous species. Thus far, but few birds described by ancient or modern naturalists are known to have become absolutely extinct, though there are some cases in which they are ascertained to have utterly disappeared from the face of the earth in very recent times. The most familiar instances are those of the dodo, a large bird peculiar to the Mauritius or Isle of France, exterminated about the year 1690, and now known only by two or three fragments of skeletons, and the solitary, which inhabited the islands of Bourbon and Rodriguez, but has not been seen for more than a century. A parrot and some other birds of the Norfolk Island group are said to have lately become extinct. The wingless auk, *Alca impennis*, a bird remarkable for its excessive fatness, was very abundant two or three hundred years ago in the Faroe Islands, and on the whole Scandinavian seaboard. The early voyagers found either the same or a closely allied species, in immense numbers, on all the coasts and islands of Newfoundland. The value of its flesh and its oil made it one of the most important resources of the inhabitants of those sterile regions, and it was naturally an object of keen pursuit. It is supposed to be now completely extinct, and few museums can show even its skeleton.

There seems to be strong reason to believe that our boasted modern civilization is guiltless of one or two sins of extermination which have been committed in recent ages. New Zealand formerly possessed three species of dinornis, one of which, called *moa* by the islanders, was much larger than the ostrich. The condition in which the bones of these birds have been found and the traditions of the natives concur to prove that, though the aborigines had probably extirpated them before the discovery of New Zealand by the whites, they still existed at a comparatively late period. The same remarks apply to a winged giant the eggs of which have been brought from Madagascar. This bird must have much exceeded the dimensions of the moa, at least so far as we can judge from the egg, which is eight times as large as the average size of the ostrich egg, or about one hundred and fifty times that of the hen.

But though we have no evidence that man has exterminated many species of birds, we know that his persecutions have caused their disappearance from many localities where they once were common, and greatly diminished their numbers in others. The cappercailzie, *Tetrao urogallus*, the finest of the grouse family, formerly abundant in Scotland, had become extinct in Great Britain, but has been reintroduced from Sweden.* The ostrich is mentioned by all the old trav-

* The cappercailzie, or tjäder, as he is called in Sweden, is a bird of singular habits, and seems to want some of the protective instincts which secure most other wild birds from destruction. The younger Læstadius frequently notices the tjäder, in his very remarkable account of the Swedish Laplanders—a work wholly unsurpassed as a genial picture of semi-barbarian life, and not inferior in minuteness of detail to Schlatter's description of the manners of the Nogai Tartars, or even to Lane's admirable and exhaustive work on the Modern Egyptians. The tjäder, though not a bird of passage, is migratory, or rather wandering in domicile, and appears to undertake very purposeless and absurd journeys. "When he flits," says Læstadius, "he follows a straight course, and sometimes pursues it quite out of the country. It is said that, in foggy weather, he sometimes flies out to sea, and, when tired, falls into the water and is drowned. It is accordingly observed that, when he flies westwardly, toward the mountains, he soon comes back again; but when he takes an eastwardly course, he returns no more, and for a long time is very scarce in Lapland. From this it would seem that he turns back from the bald mountains, when he discovers that he has strayed from his proper home, the wood; but when he finds himself over the Baltic, where he cannot alight to rest and collect himself, he flies on until he is exhausted and falls into the sea."—PETRUS LÆSTADIUS, *Journal af första året, etc.*, p. 325.

ellers, as common on the Isthmus of Suez down to the middle of the seventeenth century. It appears to have frequented Syria and even Asia Minor at earlier periods, but is now found only in the seclusion of remoter deserts.

The modern increased facilities of transportation have brought distant markets within reach of the professional hunter, and thereby given a new impulse to his destructive propensities. Not only do all Great Britain and Ireland contribute to the supply of game for the British capital, but the canvas-back duck of the Potomac, and even the prairie hen from the basin of the Mississippi, may be found at the stalls of the London poulterer. Kohl * informs us that on the coasts of the North Sea, twenty thousand wild ducks are usually taken in the course of the season in a single decoy, and sent to the large maritime towns for sale. The statistics of the great European cities show a prodigious consumption of game birds, but the official returns fall far below the truth, because they do not include the rural districts, and because neither the poacher nor his customers report the number of his victims. Reproduction, in cultivated countries, cannot keep pace with this excessive destruction, and there is no doubt that all the wild birds which are chased for their flesh or their plumage are diminishing with a rapidity which justifies the fear that the last of them will soon follow the dodo and the wingless auk.

Fortunately the larger birds which are pursued for their flesh or for their feathers, and those the eggs of which are used as food, are, so far as we know the functions appointed to them by nature, not otherwise specially useful to man, and, therefore, their wholesale destruction is an economical evil only in the same sense in which all waste of productive capital is an evil. If it were possible to confine the consumption of game fowl to a number equal to the annual increase, the world would be a gainer, but not to the same extent as it would be by checking the wanton sacrifice of millions of the smaller

* *Die Herzogthümer Schleswig und Holstein*, i, p. 208.

birds, which are of no real value as food, but which, as we have seen, render a most important service by battling, in our behalf, as well as in their own, against the countless legions of humming and of creeping things, with which the prolific powers of insect life would otherwise cover the earth.

Introduction of Birds.

Man has undesignedly introduced into new districts perhaps fewer species of birds than of quadrupeds; but the distribution of birds is very much influenced by the character of his industry, and the transplantation of every object of agricultural production is, at a longer or shorter interval, followed by that of the birds which feed upon its seeds, or more frequently upon the insects it harbors. The vulture, the crow, and other winged scavengers, follow the march of armies as regularly as the wolf. Birds accompany ships on long voyages, for the sake of the offal which is thrown overboard, and, in such cases, it might often happen that they would breed and become naturalized in countries where they had been unknown before.* There is a familiar story of an English bird which built its nest in an unused block in the rigging of a ship, and made one or two short voyages with the vessel while hatching its eggs. Had the young become fledged while lying in a foreign harbor, they would of course have claimed the rights of citizenship in the country where they first took to the wing.†

* Gulls hover about ships in port, and often far out at sea, diligently watching for the waste of the caboose. While the four great fleets, English, French, Turkish, and Egyptian, were lying in the Bosphorus, in the summer and autumn of 1853, a young lady of my family called my attention to the fact that the gulls were far more numerous about the ships of one of the fleets than about the others. This was verified by repeated observation, and the difference was owing no doubt to the greater abundance of the refuse from the cookrooms of the naval squadron most frequented by the birds. Persons acquainted with the economy of the navies of the states in question, will be able to conjecture which fleet was most favored with these delicate attentions.

† Birds do not often voluntarily take passage on board ships bound for

Some enthusiastic entomologist will, perhaps, by and by discover that insects and worms are as essential as the larger organisms to the proper working of the great terraqueous machine, and we shall have as eloquent pleas in defence of the mosquito, and perhaps even of the tzetze fly, as Toussenel and Michelet have framed in behalf of the bird.* The silk-worm and the bee need no apologist; a gallnut produced by the puncture of an insect on a Syrian oak is a necessary ingredient in the ink I am writing with, and from my windows I recognize the grain of the kermes and the cochineal in the gay habiliments of the holiday groups beneath them. But agriculture, too, is indebted to the insect and the worm. The ancients, according to Pliny, were accustomed to hang branches

foreign countries, but I can testify to one such case. A stork, which had nested near one of the palaces on the Bosphorus, had, by some accident, injured a wing, and was unable to join his fellows when they commenced their winter migration to the banks of the Nile. Before he was able to fly again, he was caught, and the flag of the nation to which the palace belonged was tied to his leg, so that he was easily identified at a considerable distance. As his wing grew stronger, he made several unsatisfactory experiments at flight, and at last, by a vigorous effort, succeeded in reaching a passing ship bound southward, and perched himself on a topsail yard. I happened to witness this movement, and observed him quietly maintaining his position as long as I could discern him with a spy-glass. I suppose he finished the voyage, for he certainly did not return to the palace.

* The enthusiasm of naturalists is not always proportioned to the magnitude or importance of the organisms they concern themselves with. It is not recorded that Adams, who found the colossal antediluvian pachyderm in a thick-ribbed mountain of Siberian ice, ran wild over his *trouvaille;* but Schmidl, in describing the natural history of the caves of the Karst, speaks of an eminent entomologist as "*der glückliche Entdecker*," the *happy* discoverer of a new coleopteron, in one of those dim caverns. How various are the sources of happiness! Think of a learned German professor, the bare enumeration of whose Rath-ships and scientific Mitglied-ships fills a page, made famous in the annals of science, immortal, happy, by the discovery of a beetle! Had that imperial *ennuyé*, who offered a premium for the invention of a new pleasure, but read Schmidl's *Höhlen des Karstes*, what splendid rewards would he not have heaped upon Kirby and Spence!

of the wild fig upon the domestic tree, in order that the insects which frequented the former might hasten the ripening of the cultivated fig by their punctures—or, as others suppose, might fructify it by transporting to it the pollen of the wild fruit—and this process, called caprification, is not yet entirely obsolete. The earthworms long ago made good their title to the respect and gratitude of the farmer as well as of the angler. The utility of the earthworms has been pointed out in many scientific as well as in many agricultural treatises. The following extract, cut from a newspaper, will answer my present purpose:

"Mr. Josiah Parkes, the consulting engineer of the Royal Agricultural Society of England, says that worms are great assistants to the drainer, and valuable aids to the farmer in keeping up the fertility of the soil. He says they love moist, but not wet soils; they will bore down to, but not into water; they multiply rapidly on land after drainage, and prefer a deeply dried soil. On examining with Mr. Thomas Hammond, of Penhurst, Kent, part of a field which he had deeply drained, after long-previous shallow drainage, he found that the worms had greatly increased in number, and that their bores descended quite to the level of the pipes. Many worm bores were large enough to receive the little finger. Mr. Henry Handley had informed him of a piece of land near the sea in Lincolnshire, over which the sea had broken and killed all the worms—the field remained sterile until the worms again inhabited it. He also showed him a piece of pasture land near to his house, in which worms were in such numbers that he thought their casts interfered too much with its produce, which induced him to have it rolled at night in order to destroy the worms. The result was, that the fertility of the field greatly declined, nor was it restored until they had recruited their numbers, which was aided by collecting and transporting multitudes of worms from the fields.

"The great depth into which worms will bore, and from which they push up fine fertile soil, and cast it on the surface, has been admirably traced by Mr. C. Darwin, of Down, Kent, who has shown that in a few years they have actually elevated the surface of fields by a large layer of rich mould, several inches thick—thus affording nourishment to the roots of grasses, and increasing the productiveness of the soil."

It should be added that the writer quoted, and others who have discussed the subject, have overlooked one very important element in the fertilization produced by earthworms. I refer to the enrichment of the soil by their excreta during life, and by the decomposition of their remains when they die. The manure thus furnished is as valuable as the like amount of similar animal products derived from higher organisms, and when we consider the prodigious numbers of these worms found on a single square yard of some soils, we may easily see that they furnish no insignificant contribution to the nutritive material required for the growth of plants.*

The perforations of the earthworm mechanically affect the texture of the soil and its permeability by water, and they therefore have a certain influence on the form and character of surface. But the geographical importance of insects proper, as well as of worms, depends principally on their connection

* I believe there is no foundation for the supposition that earthworms attack the tuber of the potato. Some of them, especially one or two species employed by anglers as bait, if natives of the woods, are at least rare in shaded grounds, but multiply very rapidly after the soil is brought under cultivation. Forty or fifty years ago they were so scarce in the newer parts of New England, that the rustic fishermen of every village kept secret the few places where they were to be found in their neighborhood, as a professional mystery, but at present one can hardly turn over a shovelful of rich moist soil anywhere, without unearthing several of them. A very intelligent lady, born in the woods of Northern New England, told me that, in her childhood, these worms were almost unknown in that region, though anxiously sought for by the anglers, but that they increased as the country was cleared, and at last became so numerous in some places, that the water of springs, and even of shallow wells, which had formerly been excellent, was rendered undrinkable by the quantity of dead worms that fell into them. The increase of the robin and other small birds which follow the settler when he has prepared a suitable home for them, at last checked the excessive multiplication of the worms, and abated the nuisance.

with vegetable life as agents of its fecundation, and of its destruction.* I am acquainted with no single fact so strikingly illustrative of this importance, as the following statement which I take from a notice of Darwin's volume, On Various Contrivances by which British and Foreign Orchids are Fertilized by Insects, in the *Saturday Review*, of October 18, 1862: "The net result is, that some six thousand species of orchids are absolutely dependent upon the agency of insects for their fertilization. That is to say, were those plants unvisited by insects, they would all rapidly disappear." What is true of the orchids is more or less true of many other vegetable families. We do not know the limits of this agency, and many of the insects habitually regarded as unqualified pests, may directly or indirectly perform functions as important to the most valuable plants as the services rendered by certain tribes to the orchids. I say directly or indirectly, because, besides the other arrangements of nature for checking the undue multiplication of particular species, she has established a police among insects themselves, by which some of them keep down or promote the increase of others; for there are insects, as well as birds and beasts, of prey. The existence of an insect which fertilizes a useful vegetable may depend on that of

* I have already remarked that the remains of extant animals are rarely, if ever, gathered in sufficient quantities to possess any geographical importance by their mere mass; but the decayed exuviæ of even the smaller and humbler forms of life are sometimes abundant enough to exercise a perceptible influence on soil and atmosphere. "The plain of Cumana," says Humboldt, "presents a remarkable phenomenon, after heavy rains. The moistened earth, when heated by the rays of the sun, diffuses the musky odor common in the torrid zone to animals of very different classes, to the jaguar, the small species of tiger cat, the cabiaï, the gallinazo vulture, the crocodile, the viper, and the rattlesnake. The gaseous emanations, the vehicles of this aroma, appear to be disengaged in proportion as the soil, which contains the remains of an innumerable multitude of reptiles, worms, and insects, begins to be impregnated with water. Wherever we stir the earth, we are struck with the mass of organic substances which in turn are developed and become transformed or decomposed. Nature in these climes seems more active, more prolific, and so to speak, more prodigal of life."

another, which constitutes his food in some stage of his life, and this other again may be as injurious to some plant as his destroyer is beneficial to another. The equation of animal and vegetable life is too complicated a problem for human intelligence to solve, and we can never know how wide a circle of disturbance we produce in the harmonies of nature when we throw the smallest pebble into the ocean of organic life.

This much, however, we seem authorized to conclude: as often as we destroy the balance by deranging the original proportions between different orders of spontaneous life, the law of self-preservation requires us to restore the equilibrium, by either directly returning the weight abstracted from one scale, or removing a corresponding quantity from the other. In other words, destruction must be either repaired by reproduction, or compensated by new destruction in an opposite quarter.

The parlor aquarium has taught even those to whom it is but an amusing toy, that the balance of animal and vegetable life must be preserved, and that the excess of either is fatal to the other, in the artificial tank as well as in natural waters. A few years ago, the water of the Cochituate aqueduct at Boston became so offensive in smell and taste as to be quite unfit for use. Scientific investigation found the cause in the too scrupulous care with which aquatic vegetation had been excluded from the reservoir, and the consequent death and decay of the animalculæ which could not be shut out, nor live in the water without the vegetable element.*

* It is remarkable that Palissy, to whose great merits as an acute observer I am happy to have frequent occasion to bear testimony, had noticed that vegetation was necessary to maintain the purity of water in artificial reservoirs, though he mistook the rationale of its influence, which he ascribed to the elemental "salt" supposed by him to play an important part in all the operations of nature. In his treatise upon Waters and Fountains, p. 174, of the reprint of 1844, he says: "And in special, thou shalt note one point, the which is understood of few: that is to say, that the leaves of the trees which fall upon the parterre, and the herbs growing beneath, and singularly the fruits, if any there be upon the trees, being decayed, the waters of the parterre shall draw unto them the salt of the

Introduction of Insects.

The general tendency of man's encroachments upon spontaneous nature has been to increase insect life at the expense of vegetation and of the smaller quadrupeds and birds. Doubtless there are insects in all woods, but in temperate climates they are comparatively few and harmless, and the most numerous tribes which breed in the forest, or rather in its waters, and indeed in all solitudes, are those which little injure vegetation, such as mosquitoes, gnats, and the like. With the cultivated plants of man come the myriad tribes which feed or breed upon them, and agriculture not only introduces new species, but so multiplies the number of individuals as to defy calculation. Newly introduced vegetables frequently escape for years the insect plagues which had infested them in their native habitat; but the importation of other varieties of the plant, the exchange of seed, or some mere accident, is sure in the long run to carry the egg, the larva, or the chrysalis to the most distant shores where the plant assigned to it by nature as its possession has preceded it. For many years after the colonization of the United States, few or none of the insects which attack wheat in its different stages of growth, were known in America. During the Revolutionary war, the Hessian fly, *Cecidomyia destructor*, made its appearance, and it was so called because it was first observed in the year when the Hessian troops were brought over, and was popularly supposed to have been accidentally imported by those unwelcome strangers. Other destroyers of cereal grains have since found their way across the Atlantic, and a noxious European aphis has first attacked the American wheatfields within the last four or five years. Unhappily, in these cases of migration, the natural corrective of excessive multiplication, the parasitic or voracious enemy of the noxious insect, does not always accompany the wanderings of its prey, and

said fruits, leaves, and herbs, the which shall greatly better the water of thy fountains, and hinder the putrefaction thereof."

the bane long precedes the antidote. Hence, in the United States, the ravages of imported insects injurious to cultivated crops, not being checked by the counteracting influences which nature had provided to limit their devastations in the Old World, are much more destructive than in Europe. It is not known that the wheat midge is preyed upon in America by any other insect, and in seasons favorable to it, it multiplies to a degree which would prove almost fatal to the entire harvest, were it not that, in the great territorial extent of the United States, there is room for such differences of soil and climate as, in a given year, to present in one State all the conditions favorable to the increase of a particular insect, while in another, the natural influences are hostile to it. The only apparent remedy for this evil is, to balance the disproportionate development of noxious foreign species by bringing from their native country the tribes which prey upon them. This, it seems, has been attempted. The United States' Census Report for 1860, p. 82, states that the New York Agricultural Society "has introduced into this country from abroad certain parasites which Providence has created to counteract the destructive powers of some of these depredators."

This is, however, not the only purpose for which man has designedly introduced foreign forms of insect life. The eggs of the silkworm are known to have been brought from the farther East to Europe in the sixth century, and new silk spinners which feed on the castor oil bean and the ailanthus, have recently been reared in France and in South America with promising success. The cochineal, long regularly bred in aboriginal America, has been transplanted to Spain, and both the kermes insect and the cantharides have been transferred to other climates than their own. The honey bee must be ranked next to the silkworm in economical importance.* This useful

* Between the years 1851 and 1858, both inclusive, the United States exported 2,665,857 pounds of beeswax, besides a considerable quantity employed in the manufacture of candles for exportation. This is an average of more than 330,000 pounds per year. The census of 1850 gave the total production of wax and honey for that year at 14,853,128 pounds. In

creature was carried to the United States by European colonists, in the latter part of the seventeenth century; it did not cross the Mississippi till the close of the eighteenth, and it is only within the last five or six years that it has been transported to California, where it was previously unknown. The Italian stingless bee has very lately been introduced into the United States.

The insects and worms intentionally transplanted by man bear but a small proportion to those accidentally introduced by him. Plants and animals often carry their parasites with them, and the traffic of commercial countries, which exchange their products with every zone and every stage of social existence, cannot fail to transfer in both directions the minute organisms that are, in one way or another, associated with almost every object important to the material interests of man.*

The tenacity of life possessed by many insects, their prodigious fecundity, the length of time they often remain in the different phases of their existence,† the security of the retreats

1860, it amounted to 26,370,813 pounds, the increase being partly due to the introduction of improved races of bees from Italy and Switzerland.—Bigelow, *Les États Unis en* 1863, p. 376.

* A few years ago, a laborer, employed at a North American port in discharging a cargo of hides from the opposite extremity of the continent, was fatally poisoned by the bite or the sting of an unknown insect, which ran out from a hide he was handling.

† In many insects, some of the stages of life regularly continue for several years, and they may, under peculiar circumstances, be almost indefinitely prolonged. Dr. Dwight mentions the following remarkable case of this sort, which may be new to many readers: "While I was here [at Williamstown, Mass.], Dr. Fitch showed me an insect, about an inch in length, of a brown color tinged with orange, with two antennæ, not unlike a rosebug. This insect came out of a tea table, made of the boards of an apple tree." Dr. Dwight examined the table, and found the "cavity whence the insect had emerged into the light," to be "about two inches in length, nearly horizontal, and inclining upward very little, except at the mouth. Between the hole, and the outside of the leaf of the table, there were forty grains of the wood." It was supposed that the sawyer and the cabinet maker must have removed at least thirteen grains more, and the table had been in the possession of its proprietor for twenty years.

into which their small dimensions enable them to retire, are all circumstances very favorable not only to the perpetuity of their species, but to their transportation to distant climates and their multiplication in their new homes. The teredo, so destructive to shipping, has been carried by the vessels whose wooden walls it mines to almost every part of the globe. The termite, or white ant, is said to have been brought to Rochefort by the commerce of that port a hundred years ago.* This creature is more injurious to wooden structures and implements than any other known insect. It eats out almost the entire substance of the wood, leaving only thin partitions between the galleries it excavates in it; but as it never gnaws through the surface to the air, a stick of timber may be almost wholly consumed without showing any external sign of the damage it has sustained. The termite is found also in other parts of France, and particularly at Rochelle, where, thus far, its ravages are confined to a single quarter of the city. A borer, of similar habits, is not uncommon in Italy, and you may see in that country, handsome chairs and other furniture which have been reduced by this insect to a framework of powder of post, covered, and apparently held together, by nothing but the varnish.

The carnivorous, and often the herbivorous insects render an important service to man by consuming dead and decaying animal and vegetable matter, the decomposition of which would otherwise fill the air with effluvia noxious to health. Some of them, the grave-digger beetle, for instance, bury the small animals in which they lay their eggs, and thereby prevent the escape of the gases disengaged by putrefaction. The prodigious rapidity of development in insect life, the great numbers of the individuals in many species, and the voracity of most of them while in the larva state, justify the appellation of nature's scavengers which has been bestowed upon them, and there is very little doubt that, in warm countries,

* It does not appear to be quite settled whether the termites of France are indigenous or imported. See QUATREFAGES, *Souvenirs d'un Naturaliste*, ii, pp. 400, 542, 543.

they consume a much larger quantity of putrescent organic material than the quadrupeds and the birds which feed upon such aliment.

Destruction of Insects.

It is well known to naturalists, but less familiarly to common observers, that the aquatic larvæ of some insects constitute, at certain seasons, a large part of the food of fresh-water fish, while other larvæ, in their turn, prey upon the spawn and even the young of their persecutors.* The larvæ of the mosquito and the gnat are the favorite food of the trout in the wooded regions where those insects abound.† Earlier in the year the trout feeds on the larvæ of the May fly, which is itself very destructive to the spawn of the salmon, and hence, by a sort of house-that-Jack-built, the destruction of the mos-

* I have seen the larva of the dragon fly in an aquarium, bite off the head of a young fish as long as itself.

† Insects and fish—which prey upon and feed each other—are the only forms of animal life that are numerous in the native woods, and their range is, of course, limited by the extent of the waters. The great abundance of the trout, and of other more or less allied genera in the lakes of Lapland, seems to be due to the supply of food provided for them by the swarms of insects which in the larva state inhabit the waters, or, in other stages of their life, are accidentally swept into them. All travellers in the north of Europe speak of the gnat and the mosquito as very serious drawbacks upon the enjoyments of the summer tourist, who visits the head of the Gulf of Bothnia to see the midnight sun, and the brothers Læstadius regard them as one of the great plagues of sub-Arctic life. "The persecutions of these insects," says Lars Levi Læstadius [*Culex pipiens*, *Culex reptans*, and *Culex pulicaris*], "leave not a moment's peace, by day or night, to any living creature. Not only man, but cattle, and even birds and wild beasts, suffer intolerably from their bite." He adds in a note, "I will not affirm that they have ever devoured a living man, but many young cattle, such as lambs and calves, have been worried out of their lives by them. All the people of Lapland declare that young birds are killed by them, and this is not improbable, for birds are scarce after seasons when the midge, the gnat, and the mosquito are numerous."—*Om Uppodlingar i Lappmarken*, p. 50.

Petrus Læstadius makes similar statements in his *Journal för första arct*, p. 285.

quito, that feeds the trout that preys on the May fly that destroys the eggs that hatch the salmon that pampers the epicure, may occasion a scarcity of this latter fish in waters where he would otherwise be abundant. Thus all nature is linked together by invisible bonds, and every organic creature, however low, however feeble, however dependent, is necessary to the well-being of some other among the myriad forms of life with which the Creator has peopled the earth.

I have said that man has promoted the increase of the insect and the worm, by destroying the bird and the fish which feed upon them. Many insects, in the four different stages of their growth, inhabit in succession the earth, the water, and the air. In each of these elements they have their special enemies, and, deep and dark as are the minute recesses in which they hide themselves, they are pursued to the remotest, obscurest corners by the executioners that nature has appointed to punish their delinquencies, and furnished with cunning contrivances for ferreting out the offenders and dragging them into the light of day. One tribe of birds, the woodpeckers, seems to depend for subsistence almost wholly on those insects which breed in dead or dying trees, and it is, perhaps, needless to say that the injury these birds do the forest is imaginary. They do not cut holes in the trunk of the tree to prepare a lodgment for a future colony of boring larvæ, but to extract the worm which has already begun his mining labors. Hence these birds are not found where the forester removes trees as fast as they become fit habitations for such insects. In clearing new lands in the United States, dead trees, especially of the spike-leaved kinds, too much decayed to serve for timber, and which, in that state, are worth little for fuel, are often allowed to stand until they fall of themselves. Such *stubs*, as they are popularly called, are filled with borers, and often deeply cut by the woodpeckers, whose strong bills enable them to penetrate to the very heart of the tree and drag out the lurking larvæ. After a few years, the stubs fall, or, as wood becomes valuable, are cut and carried off for firewood, and, at the same time, the farmer selects for felling, in the forest he has reserved as a permanent source of supply of fuel and timber, the decaying trees which, like the dead stems in the fields, serve as a home for both the worm and his pursuer. We thus gradually extirpate this tribe of insects, and, with them, the species of birds which subsist principally upon them. Thus the fine, large, red-headed woodpecker, *Picus erythrocephalus*, formerly very common in New England, has almost entirely disappeared from those States, since the dead trees are gone, and the apples, his favorite vegetable food, are less abundant.

There are even large quadrupeds which feed almost exclusively upon insects. The ant bear is strong enough to pull down the clay houses built by the species of termites that constitute his ordinary diet, and the curious ai-ai, a climbing quadruped of Madagascar—of which I believe only a single specimen, secured by Mr. Sandwith, has yet reached Europe—is provided with a very slender, hook-nailed finger, long enough to reach far into a hole in the trunk of a tree, and extract the worm which bored it.

Reptiles.

But perhaps the most formidable foes of the insect, and even of the small rodents, are the reptiles. The chameleon approaches the insect perched upon the twig of a tree, with an almost imperceptible slowness of motion, until, at the distance of a foot, he shoots out his long, slimy tongue, and rarely fails to secure the victim. Even the slow toad catches the swift and wary housefly in the same manner; and in the warm countries of Europe, the numerous lizards contribute very essentially to the reduction of the insect population, which they both surprise in the winged state upon walls and trees, and consume as egg, worm, and chrysalis, in their earlier metamorphoses. The serpents feed much upon insects, as well as upon mice, moles, and small reptiles, including also other snakes. The disgust and fear with which the serpent is so universally regarded expose him to constant persecution by man, and perhaps no other animal is so relentlessly sacrificed by him. In temperate climates, snakes are consumed by scarcely any beast or bird of prey except the stork, and they have few dangerous enemies but man, though in the tropics other animals prey upon them.* It is doubtful whether any species of serpent has been exterminated within the human period, and even the dense population of China has not been able completely to rid itself of the viper. They have, however, almost entirely disappeared from particular localities. The rattlesnake is now wholly unknown in many large districts where it was extremely common half a century ago, and Palestine has long been, if not absolutely free from venomous serpents, at least very nearly so.†

Destruction of Fish.

The inhabitants of the waters seem comparatively secure from human pursuit or interference by the inaccessibility of their retreats, and by our ignorance of their habits—a natural

* It is very questionable whether there is any foundation for the popular belief in the hostility of swine and of deer to the rattlesnake, and careful experiments as to the former quadruped seem to show that the supposed enmity is wholly imaginary. Observing that the starlings, *stornelli*, which bred in an old tower in Piedmont, carried something from their nests and dropped it upon the ground, about as often as they brought food to their young, I watched their proceedings, and found every day lying near the tower numbers of dead or dying slowworms, and, in a few cases, small lizards, which had, in every instance, lost about two inches of the tail. This part I believe the starlings gave to their nestlings, and threw away the remainder.

† Russell denies the existence of poisonous snakes in Northern Syria, and states that the last instance of death known to have occurred from the bite of a serpent near Aleppo took place a hundred years before his time. In Palestine, the climate, the thinness of population, the multitude of insects and of lizards, all circumstances, in fact, seem very favorable to the multiplication of serpents, but the venomous species, at least, are extremely rare, if at all known, in that country. I have, however, been assured by persons very familiar with Mount Lebanon, that cases of poisoning from the bite of snakes had occurred within a few years, near Hasbeiyeh, and at other places on the southern declivities of Lebanon and Hermon. In

result of the difficulty of observing the ways of creatures living in a medium in which we cannot exist. Human agency has, nevertheless, both directly and incidentally, produced great changes in the population of the sea, the lakes, and the rivers, and if the effects of such revolutions in aquatic life are apparently of small importance in general geography, they are still not wholly inappreciable. The great diminution in the abundance of the larger fish employed for food or pursued for products useful in the arts is familiar, and when we consider how the vegetable and animal life on which they feed must be affected by the reduction of their numbers, it is easy to see that their destruction may involve considerable modifications in many of the material arrangements of nature. The whale does not appear to have been an object of pursuit by the ancients, for any purpose, nor do we know when the whale fishery first commenced.* It was, however, very actively prosecuted in the Middle Ages, and the Biscayans seem to have been particularly successful in this as indeed in other branches of nautical industry.† Five hundred years ago, whales abounded in every

Egypt, on the other hand, the cobra, the asp, and the cerastes are as numerous as ever, and are much dreaded by all the natives, except the professional snake charmers. See *Appendix*, No. 17.

* I use *whale* not in a technical sense, but as a generic term for all the large inhabitants of the sea popularly grouped under that name.

† From the narrative of Ohther, introduced by King Alfred into his translation of Orosius, it is clear that the Northmen pursued the whale fishery in the ninth century, and it appears, both from the poem called The Whale, in the Codex Exoniensis, and from the dialogue with the fisherman in the Colloquies of Aelfric, that the Anglo-Saxons followed this dangerous chase at a period not much later. I am not aware of any evidence to show that any of the Latin nations engaged in this fishery until a century or two afterward, though it may not be easy to disprove their earlier participation in it. In mediæval literature, Latin and Romance, very frequent mention is made of a species of vessel called in Latin, *balenaria*, *balenerium*, *balenerius*, *balaneria*, etc.; in Catalan, *balener;* in French, *balenier;* all of which words occur in many other forms. The most obvious etymology of these words would suggest the meaning, *whaler*, *baleinier;* but some have supposed that the name was descriptive of the great size of the ships, and others have referred it to a different root. From the

sea. They long since became so rare in the Mediterranean as not to afford encouragement for the fishery as a regular occupation; and the great demand for oil and whalebone for mechanical and manufacturing purposes, in the present century, has stimulated the pursuit of the "hugest of living creatures" to such activity, that he has now almost wholly disappeared from many favorite fishing grounds, and in others is greatly diminished in numbers.

What special functions, besides his uses to man, are assigned to the whale in the economy of nature, we do not know; but some considerations, suggested by the character of the food upon which certain species subsist, deserve to be specially noticed. None of the great mammals grouped under the general name of whale are rapacious. They all live upon small organisms, and the most numerous species feed almost wholly upon the soft gelatinous mollusks in which the sea abounds in all latitudes. We cannot calculate even approximately the number of the whales, or the quantity of organic nutriment consumed by an individual, and of course we can form no estimate of the total amount of animal matter withdrawn by them, in a given period, from the waters of the sea. It is certain, however, that it must have been enormous when they were more abundant, and that it is still very considerable. A very few years since, the United States had more than six hundred whaling ships constantly employed in the Pacific, and the product of the American whale fishery for the year ending June 1st, 1860, was seven millions and a half of dollars.* The mere bulk of the whales destroyed in a single year

fourteenth century, the word occurs oftener, perhaps, in old Catalan, than in any other language; but Capmany does not notice the whale fishery as one of the maritime pursuits of the very enterprising Catalan people, nor do I find any of the products of the whale mentioned in the old Catalan tariffs. The *whalebone* of the mediæval writers, which is described as very white, is doubtless the ivory of the walrus or of the narwhale.

* In consequence of the great scarcity of the whale, the use of coal gas for illumination, the substitution of other fatty and oleaginous substances, such as lard, palm oil, and petroleum, for right-whale oil and spermaceti, the whale fishery has rapidly fallen off within a few years. The great

8

by the American and the European vessels engaged in this fishery would form an island of no inconsiderable dimensions, and each one of those taken must have consumed, in the course of his growth, many times his own weight of mollusks. The destruction of the whales must have been followed by a proportional increase of the organisms they feed upon, and if we had the means of comparing the statistics of these humble forms of life, for even so short a period as that between the years 1760 and 1860, we should find a difference sufficient, possibly, to suggest an explanation of some phenomena at present unaccounted for.

For instance, as I have observed in another work,* the phosphorescence of the sea was unknown to ancient writers, or at least scarcely noticed by them, and even Homer—who, blind as tradition makes him when he composed his epics, had seen, and marked, in earlier life, all that the glorious nature of the Mediterranean and its coasts discloses to unscientific observation—nowhere alludes to this most beautiful and striking of maritime wonders. In the passage just referred to, I have endeavored to explain the silence of ancient writers with respect to this as well as other remarkable phenomena on psychological grounds; but is it not possible that, in modern times, the animalculæ which produce it may have immensely multiplied, from the destruction of their natural enemies by man, and hence that the gleam shot forth by their decomposition, or by their living processes, is both more frequent and more brilliant than in the days of classic antiquity?

Although the whale does not prey upon smaller creatures resembling himself in form and habits, yet true fishes are extremely voracious, and almost every tribe devours unspar-

supply of petroleum, which is much used for lubricating machinery as well as for numerous other purposes, has produced a more perceptible effect on the whale fishery than any other single circumstance. According to Bigelow, *Les États Unis en* 1863, p. 346, the American whaling fleet was diminished by 29 in 1858, 57 in 1860, 94 in 1861, and 65 in 1862. The present number of American ships employed in that fishery is 353.

* The Origin and History of the English Language, &c., pp. 423, 424.

ingly the feebler species, and even the spawn and young of its own. The enormous destruction of the pike, the trout family, and other ravenous fish, as well as of the fishing birds, the seal, and the otter, by man, would naturally have occasioned a great increase in the weaker and more defenceless fish on which they feed, had he not been as hostile to them also as to their persecutors. We have little evidence that any fish employed as human food has naturally multiplied in modern times, while all the more valuable tribes have been immensely reduced in numbers.* This reduction must have affected the more voracious species not used as food by man, and accordingly the shark, and other fish of similar habits, though not objects of systematic pursuit, are now comparatively rare in many waters where they formerly abounded. The result is, that man has greatly reduced the numbers of all larger marine animals, and consequently indirectly favored the multiplication of the smaller aquatic organisms which entered into their nutriment. This change in the relations of the organic and inorganic matter of the sea must have exercised an influence on the latter. What that influence has been, we cannot say, still less

* Among the unexpected results of human action, the destruction or multiplication of fish, as well as of other animals, is a not unfrequent occurrence. I shall have occasion to mention on a following page the extermination of the fish in a Swedish river by a flood occasioned by the sudden discharge of the waters of a pond. Williams, in his *History of Vermont*, i, p. 149, quoted in Thompson's *Natural History of Vermont*, p. 142, records a case of the increase of trout from an opposite cause. In a pond formed by damming a small stream to obtain water power for a sawmill, and covering one thousand acres of primitive forest, the increased supply of food brought within reach of the fish multiplied them to that degree, that, at the head of the pond, where, in the spring, they crowded together in the brook which supplied it, they were taken by the hands at pleasure, and swine caught them without difficulty. A single sweep of a small scoopnet would bring up half a bushel, carts were filled with them as fast as if picked up on dry land, and in the fishing season they were commonly sold at a shilling (eightpence halfpenny, or about seventeen cents) a bushel. The increase in the size of the trout was as remarkable as the multiplication of their numbers.

can we predict what it will be hereafter; but its action is not for that reason the less certain.

Introduction and Breeding of Fish.

The introduction and successful breeding of fish of foreign species appears to have been long practised in China and was not unknown to the Greeks and Romans. This art has been revived in modern times, but thus far without any important results, economical or physical, though there seems to be good reason to believe it may be employed with advantage on an extended scale. As in the case of plants, man has sometimes undesignedly introduced new species of aquatic animals into countries distant from their birthplace. The accidental escape of the Chinese goldfish from ponds where they were bred as a garden ornament, has peopled some European, and it is said American streams with this species. Canals of navigation and irrigation interchange the fish of lakes and rivers widely separated by natural barriers, as well as the plants which drop their seeds into the waters. The Erie Canal, as measured by its own channel, has a length of about three hundred and sixty miles, and it has ascending and descending locks in both directions. By this route, the fresh-water fish of the Hudson and the Upper Lakes, and some of the indigenous vegetables of these respective basins, have intermixed, and the fauna and flora of the two regions have now more species common to both than before the canal was opened. Some accidental attraction not unfrequently induces fish to follow a vessel for days in succession, and they may thus be enticed into zones very distant from their native habitat. Several years ago, I was told at Constantinople, upon good authority, that a couple of fish, of a species wholly unknown to the natives, had just been taken in the Bosphorus. They were alleged to have followed an English ship from the Thames, and to have been frequently observed by the crew during the passage, but I was unable to learn their specific character.

Many of the fish which pass the greater part of the year in

salt water spawn in fresh, and some fresh-water species, the common brook trout of New England for instance, which, under ordinary circumstances, never visit the sea, will, if transferred to brooks emptying directly into the ocean, go down into the salt water after spawning time, and return again the next season. Sea fish, the smelt among others, are said to have been naturalized in fresh water, and some naturalists have argued from the character of the fish of Lake Baikal, and especially from the existence of the seal in that locality, that all its inhabitants were originally marine species, and have changed their habits with the gradual conversion of the saline waters of the lake—once, as is assumed, a maritime bay—into fresh.* The presence of the seal is hardly conclusive on this point, for it is sometimes seen in Lake Champlain at the distance of some hundreds of miles from even brackish water. One of these animals was killed on the ice in that lake in February, 1810, another in February, 1846,† and remains of the seal have been found at other times in the same waters.

The remains of the higher orders of aquatic animals are generally so perishable that, even where most abundant, they do not appear to be now forming permanent deposits of any considerable magnitude; but it is quite otherwise with shell fish, and, as we shall see hereafter, with many of the minute limeworkers of the sea. There are, on the southern coast of the United States, beds of shells so extensive that they were formerly supposed to have been naturally accumulated, and were appealed to as proofs of an elevation of the coast by geological causes; but they are now ascertained to have been derived from oysters, consumed in the course of long ages by

* Babinet, *Études et Lectures*, ii, pp. 108, 110.

† Thompson, *Natural History of Vermont*, p. 38, and Appendix, p. 13. There is no reason to believe that the seal breeds in Lake Champlain, but the individual last taken there must have been some weeks, at least, in its waters. It was killed on the ice in the widest part of the lake, on the 23d of February, thirteen days after the surface was entirely frozen, except the usual small cracks, and a month or two after the ice closed at all points north of the place where the seal was found.

the inhabitants of Indian towns. The planting of a bed of oysters in a new locality might, very probably, lead, in time, to the formation of a bank, which, in connection with other deposits, might perceptibly affect the line of a coast, or, by changing the course of marine currents, or the outlet of a river, produce geographical changes of no small importance. The transplantation of oysters to artificial ponds has long been common, and it appears to have recently succeeded well on a large scale in the open sea on the French coast. A great extension of this fishery is hoped for, and it is now proposed to introduce upon the same coast the American soft clam, which is so abundant in the tide-washed beach sands of Long Isand Sound as to form an important article in the diet of the neighboring population.

The intentional naturalization of foreign fish, as I have said, has not thus far yielded important fruits ; but though this particular branch of what is called, not very happily, *pisciculture*, has not yet established its claims to the attention of the physical geographer or the political economist, the artificial breeding of domestic fish has already produced very valuable results, and is apparently destined to occupy an extremely conspicuous place in the history of man's efforts to compensate his prodigal waste of the gifts of nature. The restoration of the primitive abundance of salt and fresh water fish, is one of the greatest material benefits that, with our present physical resources, governments can hope to confer upon their subjects. The rivers, lakes, and seacoasts once restocked, and protected by law from exhaustion by taking fish at improper seasons, by destructive methods, and in extravagant quantities, would continue indefinitely to furnish a very large supply of most healthful food, which, unlike all domestic and agricultural products, would spontaneously renew itself and cost nothing but the taking. There are many sterile or wornout soils in Europe so situated that they might, at no very formidable cost, be converted into permanent lakes, which would serve not only as reservoirs to retain the water of winter rains and snow, and give it out in the dry season for irrigation, but as breeding ponds for fish, and would thus, without further cost, yield a larger supply of human food than can at present be obtained from them even at a great expenditure of capital and labor in agricultural operations. The additions which might be made to the nutriment of the civilized world by a judicious administration of the resources of the waters, would allow some restriction of the amount of soil at present employed for agricultural purposes, and a corresponding extension of the area of the forest, and would thus facilitate a return to primitive geographical arrangements which it is important partially to restore.

Extirpation of Aquatic Animals.

It does not seem probable that man, with all his rapacity and all his enginery, will succeed in totally extirpating any salt-water fish, but he has already exterminated at least one marine warm-blooded animal—Steller's sea cow—and the walrus, the sea lion, and other large amphibia, as well as the principal fishing quadrupeds, are in imminent danger of extinction. Steller's sea cow, *Rhytina Stelleri*, was first seen by Europeans in the year 1741, on Bering's Island. It was a huge amphibious mammal, weighing not less than eight thousand pounds, and appears to have been confined exclusively to the islands and coasts in the neighborhood of Bering's Strait. Its flesh was very palatable, and the localities it frequented were easily accessible from the Russian establishments in Kamtschatka. As soon as its existence and character, and the abundance of fur animals in the same waters, were made known to the occupants of those posts by the return of the survivors of Bering's expedition, so active a chase was commenced against the amphibia of that region, that, in the course of twenty-seven years, the sea cow, described by Steller as extremely numerous in 1741, is believed to have been completely extirpated, not a single individual having been seen since the year 1768. The various tribes of seals in the Northern and Southern Pacific, the walrus and the sea otter, are already so reduced in numbers that they seem destined soon to follow the sea cow, unless protected by legislation stringent enough, and a police energetic enough, to repress the ardent cupidity of their pursuers.

The seals, the otter tribe, and many other amphibia which feed almost exclusively upon fish, are extremely voracious, and of course their destruction or numerical reduction must have favored the multiplication of the species of fish principally preyed upon by them. I have been assured by the keeper of several tamed seals that, if supplied at frequent intervals, each seal would devour not less than fourteen pounds of fish, or about a quarter of his own weight, in a day.* A very intelligent and observing hunter, who has passed a great part of his life in the forest, after carefully watching the habits of the fresh-water otter of the Northern American States, estimates their consumption of fish at about four pounds per day.

Man has promoted the multiplication of fish by making war on their brute enemies, but he has by no means thereby compensated his own greater destructiveness.† The bird and beast of prey, whether on land or in the water, hunt only as long as they feel the stimulus of hunger, their ravages are limited by the demands of present appetite, and they do not wastefully destroy what they cannot consume. Man, on the

* See page 89, note, *ante*.

† According to Hartwig, the United Provinces of Holland had, in 1618, three thousand herring busses and nine thousand vessels engaged in the transport of these fish to market. The whole number of persons employed in the Dutch herring fishery was computed at 200,000.

In the latter part of the eighteenth century, this fishery was most successfully prosecuted by the Swedes, and in 1781, the town of Gottenburg alone exported 136,649 barrels, each containing 1,200 herrings, making a total of about 164,000,000; but so rapid was the exhaustion of the fish, from this keen pursuit, that in 1799 it was found necessary to prohibit the exportation of them altogether.—*Das Leben des Meeres*, p. 182.

In 1855, the British fisheries produced 900,000 barrels, or enough to supply a fish to every human inhabitant of the globe.

On the shores of Long Island Sound, the white fish, a species of herring too bony to be easily eaten, is used as manure in very great quantities. Ten thousand are employed as a dressing for an acre, and a single net has sometimes taken 200,000 in a day.—Dwight's *Travels*, ii, pp. 512, 515.

contrary, angles to-day that he may dine to-morrow; he takes and dries millions of fish on the banks of Newfoundland, that the fervent Catholic of the shores of the Mediterranean may have wherewithal to satisfy the cravings of the stomach during next year's Lent, without imperilling his soul by violating the discipline of the papal church; and all the arrangements of his fisheries are so organized as to involve the destruction of many more fish than are secured for human use, and the loss of a large proportion of the annual harvest of the sea in the process of curing, or in transportation to the places of its consumption.*

Fish are more affected than quadrupeds by slight and even imperceptible differences in their breeding places and feeding grounds. Every river, every brook, every lake stamps a special character upon its salmon, its shad, and its trout, which is at once recognized by those who deal in or consume them. No skill can give the fish fattened by food selected and prepared by man the flavor of those which are nourished at the table of nature, and the trout of the artificial ponds in Germany and Switzerland are so inferior to the brook fish of the same species and climate, that it is hard to believe them identical. The superior sapidity of the American trout to the

* The indiscriminate hostility of man to inferior forms of animated life is little creditable to modern civilization, and it is painful to reflect that it becomes keener and more unsparing in proportion to the refinement of the race. The savage slays no animal, not even the rattlesnake, wantonly; and the Turk, whom we call a barbarian, treats the dumb beast as gently as a child. One cannot live many weeks in Turkey without witnessing touching instances of the kindness of the people to the lower animals, and I have found it very difficult to induce even the boys to catch lizards and other reptiles for preservation as specimens. See *Appendix*, No. 18.

The fearless confidence in man, so generally manifested by wild animals in newly discovered islands, ought to have inspired a gentler treatment of them; but a very few years of the relentless pursuit, to which they are immediately subjected, suffice to make them as timid as the wildest inhabitants of the European forest. This timidity, however, may easily be overcome. The squirrels introduced by Mayor Smith into the public parks of Boston are so tame as to feed from the hands of passengers, and they not unfrequently enter the neighboring houses.

European species, which is familiar to every one acquainted with both continents, is probably due less to specific difference than to the fact that, even in the parts of the New World which have been longest cultivated, wild nature is not yet tamed down to the character it has assumed in the Old, and which it will acquire in America also when her civilization shall be as ancient as is now that of Europe.

Man has hitherto hardly anywhere produced such climatic or other changes as would suffice of themselves totally to banish the wild inhabitants of the dry land, and the disappearance of the native birds and quadrupeds from particular localities is to be ascribed quite as much to his direct persecutions as to the want of forest shelter, of appropriate food, or of other conditions indispensable to their existence. But almost all the processes of agriculture, and of mechanical and chemical industry, are fatally destructive to aquatic animals within reach of their influence. When, in consequence of clearing the woods, the changes already described as thereby produced in the beds and currents of rivers, are in progress, the spawning grounds of fish are exposed from year to year to a succession of mechanical disturbances; the temperature of the water is higher in summer, colder in winter, than when it was shaded and protected by wood; the smaller organisms, which formed the sustenance of the young fry, disappear or are reduced in numbers, and new enemies are added to the old foes that preyed upon them; the increased turbidness of the water in the annual inundations chokes the fish; and, finally, the quickened velocity of its current sweeps them down into the larger rivers or into the sea, before they are yet strong enough to support so great a change of circumstances.* Industrial oper-

* A fact mentioned by Schubert—and which in its causes and many of its results corresponds almost precisely with those connected with the escape of Barton Pond in Vermont, so well known to geological students—is important, as showing that the diminution of the fish in rivers exposed to inundations is chiefly to be ascribed to the mechanical action of the current, and not mainly, as some have supposed, to changes of temperature occasioned by clearing. Our author states that, in 1796, a terrible inun-

ations are not less destructive to fish which live or spawn in fresh water. Milldams impede their migrations, if they do not absolutely prevent them, the sawdust from lumber mills clogs their gills, and the thousand deleterious mineral substances, discharged into rivers from metallurgical, chemical, and manufacturing establishments, poison them by shoals.

Minute Organisms.

Besides the larger creatures of the land and of the sea, the quadrupeds, the reptiles, the birds, the amphibia, the crustacea, the fish, the insects, and the worms, there are other countless forms of vital being. Earth, water, the ducts and fluids of vegetable and of animal life, the very air we breathe, are peopled by minute organisms which perform most important functions in both the living and the inanimate kingdoms of nature. Of the offices assigned to these creatures, the most familiar to common observation is the extraction of lime, and more rarely, of silex, from the waters inhabited by them, and the deposit of these minerals in a solid form, either as the material of their habitations or as the exuviæ of their bodies. The microscope and other means of scientific observation assure us that the chalk beds of England and of France, the coral reefs of marine waters in warm climates, vast calcareous and silicious deposits in the sea and in many fresh-water ponds, the common polishing earths and slates, and many species of apparently dense and solid rock, are the work of the humble organisms of which I speak, often, indeed, of animalculæ so small as to become visible only by the aid of lenses magnifying a hundred times the linear measures. It is pop-

dation was produced in the Indalself, which rises in the Storsjö in Jemtland, by drawing off into it the waters of another lake near Ragunda. The flood destroyed houses and fields; much earth was swept into the channel, and the water made turbid and muddy; the salmon and the smaller fish forsook the river altogether, and never returned. The banks of the river have never regained their former solidity, and portions of their soil are still continually falling into the water.—*Resa genom Sverge*, ii, p. 51.

ularly supposed that animalculæ, or what are commonly embraced under the vague name of infusoria, inhabit the water alone, but the atmospheric dust transported by every wind and deposited by every calm is full of microscopic life or of its relics. The soil on which the city of Berlin stands, contains at the depth of ten or fifteen feet below the surface, living elaborators of silex; * and a microscopic examination of a handful of earth connected with the material evidences of guilt has enabled the naturalist to point out the very spot where a crime was committed. It has been computed that one sixth part of the solid matter let fall by great rivers at their outlets consists of still recognizable infusory shells and shields, and, as the friction of rolling water must reduce much of these fragile structures to a state of comminution which even the microscope cannot resolve into distinct particles and identify as relics of animal or of vegetable life, we must conclude that a considerably larger proportion of river deposits is really the product of animalcules.†

It is evident that the chemical, and in many cases the mechanical character of a great number of the objects important in the material economy of human life, must be affected by the presence of so large an organic element in their substance, and it is equally obvious that all agricultural and all industrial operations tend to disturb the natural arrangements of this element, to increase or to diminish the special adaptation of every medium in which it lives to the particular orders of

* WITTWER, *Physikalische Geographie*, p. 142.

† To vary the phrase, I make occasional use of *animalcule*, which, as a popular designation, embraces all microscopic organisms. The name is founded on the now exploded supposition that all of them are animated, which was the general belief of naturalists when attention was first drawn to them. It was soon discovered that many of them were unquestionably vegetable, and there are numerous genera the true classification of which is matter of dispute among the ablest observers. There are cases in which objects formerly taken for living animalcules turn out to be products of the decomposition of matter once animated, and it is admitted that neither spontaneous motion nor even apparent irritability are sure signs of animal life.

being inhabited by it. The conversion of woodland into pasturage, of pasture into plough land, of swamp or of shallow sea into dry ground, the rotations of cultivated crops, must prove fatal to millions of living things upon every rood of surface thus deranged by man, and must, at the same time, more or less fully compensate this destruction of life by promoting the growth and multiplication of other tribes equally minute in dimensions.

I do not know that man has yet endeavored to avail himself, by artificial contrivances, of the agency of these wonderful architects and manufacturers. We are hardly well enough acquainted with their natural economy to devise means to turn their industry to profitable account, and they are in very many cases too slow in producing visible results for an age so impatient as ours. The over-civilization of the nineteenth century cannot wait for wealth to be amassed by infinitesimal gains, and we are in haste to *speculate* upon the powers of nature, as we do upon objects of bargain and sale in our trafficking one with another. But there are still some cases where the little we know of a life, whose workings are invisible to the naked eye, suggests the possibility of advantageously directing the efforts of troops of artisans that we cannot see. Upon coasts occupied by the corallines, the reef-building animalcule does not work near the mouth of rivers. Hence the change of the outlet of a stream, often a very easy matter, may promote the construction of a barrier to coast navigation at one point, and check the formation of a reef at another, by diverting a current of fresh water from the former and pouring it into the sea at the latter. Cases may probably be found in tropical seas, where rivers have prevented the working of the coral animalcules in straits separating islands from each other or from the mainland. The diversion of such streams might remove this obstacle, and reefs consequently be formed which should convert an archipelago into a single large island, and finally join that to the neighboring continent.

Quatrefages proposed to destroy the teredo in harbors by impregnating the water with a mineral solution fatal to them.

Perhaps the labors of the coralline animals might be arrested over a considerable extent of sea coast by similar means. The reef builders are leisurely architects, but the precious coral is formed so rapidly that the beds may be refished advantageously as often as once in ten years.* It does not seem impossible that this coral might be transplanted to the American coast, where the Gulf stream would furnish a suitable temperature beyond the climatic limits that otherwise confine its growth; and thus a new source of profit might perhaps be added to the scanty returns of the hardy fisherman.

In certain geological formations, the diatomaceæ deposit, at the bottom of fresh-water ponds, beds of silicious shields, valuable as a material for a species of very light firebrick, in the manufacture of water glass and of hydraulic cement, and ultimately, doubtless, in many yet undiscoverd industrial processes. An attentive study of the conditions favorable to the propagation of the diatomaceæ might perhaps help us to profit directly by the productivity of this organism, and, at the same time, disclose secrets of nature capable of being turned to valuable account in dealing with silicious rocks, and the metal which is the base of them. Our acquaintance with the obscure and infinitesimal life of which I have now been treating is very recent, and still very imperfect. We know that it is of vast importance in the economy of nature, but we are so ambitious to grasp the great, so little accustomed to occupy ourselves with the minute, that we are not yet prepared to enter seriously upon the question how far we can control and direct the operations, not of unembodied physical forces, but of beings, in popular apprehension, almost as immaterial as they.

Nature has no unit of magnitude by which she measures her works. Man takes his standards of dimension from himself. The hair's breadth was his minimum until the microscope told him that there are animated creatures to which one

* See an interesting report on the coral fishery, by Sant' Agabio, Italian Consul-General at Algiers, in the *Bollettino Consolare*, published by the Department of Foreign Affairs, 1862, pp. 139, 151, and in the *Annali di Agricoltura, Industria e Commercio*, No. ii, pp. 360, 373.

of the hairs of his head is a larger cylinder than is the trunk of the giant California redwood to him. He borrows his inch from the breadth of his thumb, his palm and span from the width of his hand and the spread of his fingers, his foot from the length of the organ so named; his cubit is the distance from the tip of his middle finger to his elbow, and his fathom is the space he can measure with his outstretched arms. To a being who instinctively finds the standard of all magnitudes in his own material frame, all objects exceeding his own dimensions are absolutely great, all falling short of them absolutely small. Hence we habitually regard the whale and the elephant as essentially large and therefore important creatures, the animalcule as an essentially small and therefore unimportant organism. But no geological formation owes its origin to the labors or the remains of the huge mammal, while the animalcule composes, or has furnished, the substance of strata thousands of feet in thickness, and extending, in unbroken beds, over many degrees of terrestrial surface. If man is destined to inhabit the earth much longer, and to advance in natural knowledge with the rapidity which has marked his progress in physical science for the last two or three centuries, he will learn to put a wiser estimate on the works of creation, and will derive not only great instruction from studying the ways of nature in her obscurest, humblest walks, but great material advantage from stimulating her productive energies in provinces of her empire hitherto regarded as forever inaccessible, utterly barren.*

* The fermentation of liquids, and in many cases the decomposition of semi-solids, formerly supposed to be owing purely to chemical action, are now ascertained to be due to vital processes of living minute organisms both vegetable and animal, and consequently to physiological, as well as to chemical forces. Even alcohol is stated to be an animal product. See an interesting article by Auguste Laugel on the recent researches of Pasteur, in the *Revue des Deux Mondes*, for September 15th, 1863.

CHAPTER III.

THE WOODS.

THE HABITABLE EARTH ORIGINALLY WOODED—THE FOREST DOES NOT FURNISH FOOD FOR MAN—FIRST REMOVAL OF THE WOODS—EFFECTS OF FIRE ON FOREST SOIL—EFFECTS OF THE DESTRUCTION OF THE FOREST—ELECTRICAL INFLUENCE OF TREES—CHEMICAL INFLUENCE OF THE FOREST.

INFLUENCE OF THE FOREST, CONSIDERED AS INORGANIC MATTER, ON TEMPERATURE: *a*, ABSORBING AND EMITTING SURFACE; *b*, TREES AS CONDUCTORS OF HEAT; *c*, TREES IN SUMMER AND IN WINTER; *d*, DEAD PRODUCTS OF TREES; *e*, TREES AS A SHELTER TO GROUNDS TO THE LEEWARD OF THEM; *f*, TREES AS A PROTECTION AGAINST MALARIA—THE FOREST, AS INORGANIC MATTER, TENDS TO MITIGATE EXTREMES.

TREES AS ORGANISMS: SPECIFIC TEMPERATURE—TOTAL INFLUENCE OF THE FOREST ON TEMPERATURE.

INFLUENCE OF FORESTS ON THE HUMIDITY OF THE AIR AND THE EARTH: *a*, AS INORGANIC MATTER; *b*, AS ORGANIC—WOOD MOSSES AND FUNGI—FLOW OF SAP—ABSORPTION AND EXHALATION OF MOISTURE BY TREES—BALANCE OF CONFLICTING INFLUENCES—INFLUENCE OF THE FOREST ON TEMPERATURE AND PRECIPITATION—INFLUENCE OF THE FOREST ON THE HUMIDITY OF THE SOIL—ITS INFLUENCE ON THE FLOW OF SPRINGS—GENERAL CONSEQUENCES OF THE DESTRUCTION OF THE WOODS—LITERATURE AND CONDITION OF THE FOREST IN DIFFERENT COUNTRIES—THE INFLUENCE OF THE FOREST ON INUNDATIONS—DESTRUCTIVE ACTION OF TORRENTS—THE PO AND ITS DEPOSITS—MOUNTAIN SLIDES—PROTECTION AGAINST THE FALL OF ROCKS AND AVALANCHES BY TREES—PRINCIPAL CAUSES OF THE DESTRUCTION OF THE FOREST—AMERICAN FOREST TREES—SPECIAL CAUSES OF THE DESTRUCTION OF EUROPEAN WOODS—ROYAL FORESTS AND GAME LAWS—SMALL FOREST PLANTS, VITALITY OF SEEDS—UTILITY OF THE FOREST—THE FORESTS OF EUROPE—FORESTS OF THE UNITED STATES AND CANADA—THE ECONOMY OF THE FOREST—EUROPEAN AND AMERICAN TREES COMPARED—SYLVICULTURE—INSTABILITY OF AMERICAN LIFE.

The Habitable Earth Originally Wooded.

THERE is good reason to believe that the surface of the habitable earth, in all the climates and regions which have been the abodes of dense and civilized populations, was, with few exceptions, already covered with a forest growth when it first became the home of man. This we infer from the extensive vegetable remains—trunks, branches, roots, fruits, seeds, and leaves of trees—so often found in conjunction with works of primitive art, in the boggy soil of districts where no forests appear to have existed within the eras through which written annals reach; from ancient historical records, which prove that large provinces, where the earth has long been wholly bare of trees, were clothed with vast and almost unbroken woods when first made known to Greek and Roman civilization;* and from the state of much of North and of South America when they were discovered and colonized by the European race.†

These evidences are strengthened by observation of the natural economy of our own time; for, whenever a tract of country, once inhabited and cultivated by man, is abandoned by him and by domestic animals,‡ and surrendered to the

* The recorded evidence in support of the proposition in the text has been collected by L. F. Alfred Maury, in his *Histoire des grandes Forêts de la Gaule et de l'ancienne France*, and by Becquerel, in his important work, *Des climats et de l'Influence qu'exercent les Sols boisés et non boisés*, livre ii, chap. i to iv.

We may rank among historical evidences on this point, if not technically among historical records, old geographical names and terminations etymologically indicating forest or grove, which are so common in many parts of the Eastern Continent now entirely stripped of woods—such as, in Southern Europe, Breuil, Broglio, Brolio, Brolo; in Northern, Brühl, -wald, -wold, -wood, -shaw, -skog, and -skov.

† The island of Madeira, whose noble forests were devastated by fire not long after its colonization by European settlers, derives its name from the Portuguese word for wood.

‡ Browsing animals, and most of all the goat, are considered by foresters as more injurious to the growth of young trees, and, therefore, to the reproduction of the forest, than almost any other destructive cause. "According to Beatson's *Saint Helena*, introductory chapter, and Darwin's *Journal of Researches in Geology and Natural History*, pp. 582, 583," says Emsmann, in the notes to his translation of Foissac, p. 654, "it was the goats which destroyed the beautiful forests that, three hundred and fifty years ago, covered a continuous surface of not less than two thousand acres in the

undisturbed influences of spontaneous nature, its soil sooner or later clothes itself with herbaceous and arborescent plants, and, at no long interval, with a dense forest growth. Indeed, upon surfaces of a certain stability, and not absolutely precipitous inclination, the special conditions required for the spontaneous

interior of the island [of St. Helena], not to mention scattered groups of trees. Darwin observes: 'During our stay at Valparaiso, I was most positively assured that sandal wood formerly grew in abundance on the island of Juan Fernandez, but that this tree had now become entirely extinct there, having been extirpated by the goats which early navigators had introduced. The neighboring islands, to which goats have not been carried, still abound in sandal wood.' "

In the winter, the deer tribe, especially the great American moose deer, subsists much on the buds and young sprouts of trees; yet—though from the destruction of the wolves or from some not easily explained cause, these latter animals have recently multiplied so rapidly in some parts of North America, that, not long since, four hundred of them are said to have been killed, in one season, on a territory in Maine not comprising more than one hundred and fifty square miles—the wild browsing quadrupeds are rarely, if ever, numerous enough in regions uninhabited by man to produce any sensible effect on the condition of the forest. A reason why they are less injurious than the goat to young trees may be that they resort to this nutriment only in the winter, when the grasses and shrubs are leafless or covered with snow, whereas the goat feeds upon buds and young shoots principally in the season of growth. However this may be, the natural law of consumption and supply keeps the forest growth, and the wild animals which live on its products, in such a state of equilibrium as to insure the indefinite continuance of both, and the perpetuity of neither is endangered until man, who is above natural law, interferes and destroys the balance.

When, however, deer are bred and protected in parks, they multiply like domestic cattle, and become equally injurious to trees. "A few years ago," says Clavé, "there were not less than two thousand deer of different ages in the forest of Fontainebleau. For want of grass, they are driven to the trees, and they do not spare them. * * It is calculated that the browsing of these animals, and the consequent retardation of the growth of the wood, diminishes the annual product of the forest to the amount of two hundred thousand cubic feet per year, * * and besides this, the trees thus mutilated are soon exhausted and die. The deer attack the pines, too, tearing off the bark in long strips, or rubbing their heads against them when shedding their horns; and sometimes, in groves of

propagation of trees may all be negatively expressed and reduced to these three: exemption from defect or excess of moisture, from perpetual frost, and from the depredations of man and browsing quadrupeds. Where these requisites are secured, the hardest rock is as certain to be overgrown with wood as the most fertile plain, though, for obvious reasons, the process is slower in the former than in the latter case. Lichens and mosses first prepare the way for a more highly organized vegetation. They retain the moisture of rains and dews, and bring it to act, in combination with the gases evolved by their organic processes, in decomposing the surface of the rocks they cover; they arrest and confine the dust which the wind scatters over them, and their final decay adds new material to the soil already half formed beneath and upon them. A very thin stratum of mould is sufficient for the germination of seeds of the hardy evergreens and birches, the roots of which are often found in immediate contact with the rock, supplying their trees with nourishment from a soil derived from the decomposition of their own foliage, or sending out long rootlets into the surrounding earth in search of juices to feed them.

The eruptive matter of volcanoes, forbidding as is its aspect, does not refuse nutriment to the woods. The refractory lava of Etna, it is true, remains long barren, and that of the great eruption of 1669 is still almost wholly devoid of vegetation.* But the cactus is making inroads even here, while the volcanic sand and molten rock thrown out by Vesuvius soon

more than a hundred hectares, not one pine is found uninjured by them."—*Revue des Deux Mondes*, Mai, 1863, p. 157. See also *Appendix*, No. 21.

Beckstein computes that a park of 2,500 acres, containing 250 acres of marsh, 250 of fields and meadows, and the remaining 2,000 of wood, may keep 364 deer of different species, 47 wild boars, 200 hares, 100 rabbits, and an indefinite number of pheasants. These animals would require, in winter, 123,000 pounds of hay, and 22,000 pounds of potatoes, besides what they would pick up themselves. The natural forest most thickly peopled with wild animals would not, in temperate climates, contain, upon the average, one tenth of these numbers to the same extent of surface.

* Even the volcanic dust of Etna remains very long unproductive. Near Nicolosi is a great extent of coarse black sand, thrown out in 1669,

becomes productive. George Sandys, who visited this latter mountain in 1611, after it had reposed for several centuries, found the throat of the volcano at the bottom of the crater "almost choked with broken rocks and *trees* that are falne therein." "Next to this," he continues, "the matter thrown up is ruddy, light, and soft: more removed, blacke and ponderous: the uttermost brow, that declineth like the seates in a theater, flourishing with trees and excellent pasturage. The midst of the hill is shaded with chestnut trees, and others bearing sundry fruits." *

I am convinced that forests would soon cover many parts of the Arabian and African deserts, if man and domestic animals, especially the goat and the camel, were banished from them. The hard palate and tongue and strong teeth and jaws of this latter quadruped enable him to break off and masticate tough and thorny branches as large as the finger. He is particularly fond of the smaller twigs, leaves, and seedpods of the *sont* and other acacias, which, like the American Robinia,

which, for almost two centuries, lay entirely bare, and can be made to grow plants only by artificial mixtures and much labor.

The increase in the price of wines, in consequence of the diminution of the product from the grape disease, however, has brought even these ashes under cultivation. "I found," says Waltershausen, referring to the years 1861–'62, "plains of volcanic sand and half-subdued lava streams, which twenty years ago lay utterly waste, now covered with fine vineyards. The ashfield of ten square miles above Nicolosi, created by the eruption of 1669, which was entirely barren in 1835, is now planted with vines almost to the summits of Monte Rosso, at a height of three thousand feet."—*Ueber den Sicilianischen Ackerbau*, p. 19.

* *A Relation of a Journey Begun An. Dom.* 1610, lib. 4, p. 260, edition of 1627. The testimony of Sandys on this point is confirmed by that of Pighio, Braccini, Magliocco, Salimbeni, and Nicola di Rubeo, all cited by Roth, *Der Vesuv.*, p. 9. There is some uncertainty about the date of the last eruption previous to the great one of 1631. Ashes, though not lava, appear to have been thrown out about the year 1500, and some chroniclers have recorded an eruption in the year 1306; but this seems to be an error for 1036, when a great quantity of lava was ejected. In 1139, ashes were thrown out for many days. I take these dates from the work of Roth just cited.

thrive well on dry and sandy soils, and he spares no tree the branches of which are within his reach, except, if I remember right, the tamarisk that produces manna. Young trees sprout plentifully around the springs and along the winter water-courses of the desert, and these are just the halting stations of the caravans and their routes of travel. In the shade of these trees, annual grasses and perennial shrubs shoot up, but are mown down by the hungry cattle of the Bedouin, as fast as they grow. A few years of undisturbed vegetation would suffice to cover such points with groves, and these would gradually extend themselves over soils where now scarcely any green thing but the bitter colocynth and the poisonous foxglove is ever seen.

The Forest does not Furnish Food for Man.

In a region absolutely covered with trees, human life could not long be sustained, for want of animal and vegetable food. The depths of the forest seldom furnish either bulb or fruit suited to the nourishment of man; and the fowls and beasts on which he feeds are scarcely seen except upon the margin of the wood, for here only grow the shrubs and grasses, and here only are found the seeds and insects, which form the sustenance of the non-carnivorous birds and quadrupeds.*

* Except upon the banks of rivers or of lakes, the woods of the interior of North America, far from the habitations of man, are almost destitute of animal life. Dr. Newberry, describing the vast forests of the yellow pine of the West, *Pinus ponderosa*, remarks: "In the arid and desert regions of the interior basin, we made whole days' marches in forests of yellow pine, of which neither the monotony was broken by other forms of vegetation, nor its stillness by the flutter of a bird or the hum of an insect."—*Pacific Railroad Report*, vol. vi, 1857. Dr. NEWBERRY's *Report on Botany*, p. 37.

The wild fruit and nut trees, the Canada plum, the cherries, the many species of walnut, the butternut, the hazel, yield very little, frequently nothing, so long as they grow in the woods; and it is only when the trees around them are cut down, or when they grow in pastures, that they become productive. The berries, too—the strawberry, the blackberry, the

First Removal of the Forest.

As soon as multiplying man had filled the open grounds along the margin of the rivers, the lakes, and the sea, and sufficiently peopled the natural meadows and savannas of the interior, where such existed,* he could find room for expansion

raspberry, the whortleberry, scarcely bear fruit at all except in cleared ground.

The North American Indians did not inhabit the interior of the forests. Their settlements were upon the shores of rivers and lakes, and their weapons and other relics are found only in the narrow open grounds which they had burned over and cultivated, or in the margin of the woods around their villages.

The rank forests of the tropics are as unproductive of human aliment as the less luxuriant woods of the temperate zone. In Strain's unfortunate expedition across the great American isthmus, where the journey lay principally through thick woods, several of the party died of starvation, and for many days the survivors were forced to subsist on the scantiest supplies of unnutritious vegetables perhaps never before employed for food by man. See the interesting account of that expedition in *Harper's Magazine* for March, April, and May, 1855.

Clavé, as well as many earlier writers, supposes that primitive man derived his nutriment from the spontaneous productions of the wood. "It is to the forests," says he, "that man was first indebted for the means of subsistence. Exposed alone, without defence, to the rigor of the seasons, as well as to the attacks of animals stronger and swifter than himself, he found in them his first shelter, drew from them his first weapons. In the first period of humanity, they provided for all his wants: they furnished him wood for warmth, fruits for food, garments to cover his nakedness, arms for his defence."—*Études sur l'Économie Forestière*, p. 13.

But the history of savage life, as far as it is known to us, presents man in that condition as inhabiting only the borders of the forest and the open grounds that skirt the waters and the woods, and as finding only there the aliments which make up his daily bread.

* The origin of the great natural meadows, or prairies as they are called, of the valley of the Mississippi, is obscure. There is, of course, no historical evidence on the subject, and I believe that remains of forest vegetation are seldom or never found beneath the surface, even in the *sloughs*, where the perpetual moisture would preserve such remains indefinitely. The want of trees upon them has been ascribed to the occasional long-continued droughts of summer, and the excessive humidity of the soil

and further growth, only by the removal of a portion of the forest that hemmed him in. The destruction of the woods, then, was man's first geographical conquest, his first violation of the harmonies of inanimate nature.

Primitive man had little occasion to fell trees for fuel, or,

in winter; but it is, in very many instances, certain that, by whatever means the growth of forests upon them was first prevented or destroyed, the trees have been since kept out of them only by the annual burning of the grass, by grazing animals, or by cultivation. The groves and belts of trees which are found upon the prairies, though their seedlings are occasionally killed by drought, or by excess of moisture, extend themselves rapidly over them when the seeds and shoots are protected against fire, cattle, and the plough. The prairies, though of vast extent, must be considered as a local, and, so far as our present knowledge extends, abnormal exception to the law which clothes all suitable surfaces with forest; for there are many parts of the United States—Ohio, for example—where the physical conditions appear to be nearly identical with those of the States lying farther west, but where there were comparatively few natural meadows. The prairies were the proper feeding grounds of the bison, and the vast number of those animals is connected, as cause or consequence, with the existence of these vast pastures. The bison, indeed, could not convert the forest into a pasture, but he would do much to prevent the pasture from becoming a forest.

There is positive evidence that some of the American tribes possessed large herds of domesticated bisons. See HUMBOLDT, *Ansichten der Natur*, i, pp. 71–73. What authorizes us to affirm that this was simply the wild bison reclaimed, and why may we not, with equal probability, believe that the migratory prairie buffalo is the progeny of the domestic animal run wild?

There are, both on the prairies, as in Wisconsin, and in deep forests, as in Ohio, extensive remains of a primitive people, who must have been more numerous and more advanced in art than the present Indian tribes. There can be no doubt that the woods where such earthworks are found in Ohio were cleared by them, and that the vicinity of these fortresses or temples was inhabited by a large population. Nothing forbids the supposition that the prairies were cleared by the same or a similar people, and that the growth of trees upon them has been prevented by fires and grazing, while the restoration of the woods in Ohio may be due to the abandonment of that region by its original inhabitants. The climatic conditions unfavorable to the spontaneous growth of trees on the prairies may be an effect of too extensive clearings, rather than a cause of the want of woods. See *Appendix*, No. 22.

for the construction of dwellings, boats, and the implements of his rude agriculture and handicrafts. Windfalls would furnish a thin population with a sufficient supply of such material, and if occasionally a growing tree was cut, the injury to the forest would be too insignificant to be at all appreciable.

The accidental escape and spread of fire, or, possibly, the combustion of forests by lightning, must have first suggested the advantages to be derived from the removal of too abundant and extensive woods, and, at the same time, have pointed out a means by which a large tract of surface could readily be cleared of much of this natural incumbrance. As soon as agriculture had commenced at all, it would be observed that the growth of cultivated plants, as well as of many species of wild vegetation, was particularly rapid and luxuriant on soils which had been burned over, and thus a new stimulus would be given to the practice of destroying the woods by fire, as a means of both extending the open grounds, and making the acquisition of a yet more productive soil. After a few harvests had exhausted the first rank fertility of the virgin mould, or when weeds and briers and the sprouting roots of the trees had begun to choke the crops of the half-subdued soil, the ground would be abandoned for new fields won from the forest by the same means, and the deserted plain or hillock would soon clothe itself anew with shrubs and trees, to be again subjected to the same destructive process, and again surrendered to the restorative powers of vegetable nature.* This

* In many parts of the North American States, the first white settlers found extensive tracts of thin woods, of a very park-like character, called "oak openings," from the predominance of different species of that tree upon them. These were the semi-artificial pasture grounds of the Indians, brought into that state, and so kept, by partial clearing, and by the annual burning of the grass. The object of this operation was to attract the deer to the fresh herbage which sprang up after the fire. The oaks bore the annual scorching, at least for a certain time; but if it had been indefinitely continued, they would very probably have been destroyed at last. The soil would have then been much in the prairie condition, and would have needed nothing but grazing for a long succession of years to make the resemblance perfect. That the annual fires alone occasioned the peculiar

rude economy would be continued for generations, and wasteful as it is, is still largely pursued in Northern Sweden, Swedish Lapland, and sometimes even in France and the United States.*

character of the oak openings, is proved by the fact, that as soon as the Indians had left the country, young trees of many species sprang up and grew luxuriantly upon them. See a very interesting account of the oak openings in DWIGHT's *Travels*, iv, pp. 58–63.

* The practice of burning over woodland, at once to clear and manure the ground, is called in Swedish *svedjande*, a participial noun from the verb *att svedja*, to burn over. Though used in Sweden as a preparation for crops of rye or other grain, it is employed in Lapland more frequently to secure an abundant growth of pasturage, which follows in two or three years after the fire; and it is sometimes resorted to as a mode of driving the Laplanders and their reindeer from the vicinity of the Swedish backwoodsman's grass grounds and haystacks, to which they are dangerous neighbors. The forest, indeed, rapidly recovers itself, but it is a generation or more before the reindeer moss grows again. When the forest consists of pine, *tall*, the ground, instead of being rendered fertile by this process, becomes hopelessly barren, and for a long time afterward produces nothing but weeds and briers.—LÆSTADIUS, *Om Uppodlingar i Lappmarken*, p. 15. See also SCHUBERT, *Resa i Sverge*, ii, p. 375.

In some parts of France this practice is so general that Clavé says: "In the department of Ardennes it (*le sartage*) is the basis of agriculture. The northern part of the department, comprising the arrondissements of Rocroi and Mézières, is covered by steep wooded mountains with an argillaceous, compact, moist and cold soil; it is furrowed by three valleys, or rather three deep ravines, at the bottom of which roll the waters of the Meuse, the Semoy, and the Sormonne, and villages show themselves wherever the walls of the valleys retreat sufficiently from the rivers to give room to establish them. Deprived of arable soil, since the nature of the ground permits neither regular clearing nor cultivation, the peasant of the Ardennes, by means of burning, obtains from the forest a subsistence which, without this resource, would fail him. After the removal of the disposable wood, he spreads over the soil the branches, twigs, briars, and heath, sets fire to them in the dry weather of July and August, and sows in September a crop of rye, which he covers by a light ploughing. Thus prepared, the ground yields from seventeen to twenty bushels an acre, besides a ton and a half or two tons of straw of the best quality for the manufacture of straw hats."—CLAVÉ, *Études sur l'Économie Forestière*, p. 21.

Clavé does not expressly condemn the *sartage*, which indeed seems the only practicable method of obtaining crops from the soil he describes, but as we shall see hereafter, it is regarded by most writers as a highly nicious practice.

Effects of Fire on Forest Soil.

Aside from the mechanical and chemical effects of the disturbance of the soil by agricultural operations, and of the freer admission of sun, rain, and air to the ground, the fire of itself exerts an important influence on its texture and condition. It consumes a portion of the half-decayed vegetable mould which served to hold its mineral particles together and to retain the water of precipitation, and thus loosens, pulverizes, and dries the earth; it destroys reptiles, insects, and worms, with their eggs, and the seeds of trees and of smaller plants; it supplies, in the ashes which it deposits on the surface, important elements for the growth of a new forest clothing, as well as of the usual objects of agricultural industry; and by the changes thus produced, it fits the ground for the reception of a vegetation different in character from that which had spontaneously covered it. These new conditions help to explain the natural succession of forest crops, so generally observed in all woods cleared by fire and then abandoned. There is no doubt, however, that other influences contribute to the same result, because effects more or less analogous follow when the trees are destroyed by other causes, as by high winds, by the woodman's axe, and even by natural decay.*

* The remarkable mounds and other earthworks constructed in the valley of the Ohio and elsewhere in the territory of the United States, by a people apparently more advanced in culture than the modern Indian, were overgrown with a dense clothing of forest when first discovered by the whites. But though the ground where they were erected must have been occupied by a large population for a considerable length of time, and therefore entirely cleared, the trees which grew upon the ancient fortresses and the adjacent lands were not distinguishable in species, or even in dimensions and character of growth, from the neighboring forests, where the soil seemed never to have been disturbed. This apparent exception to the law of change of crop in natural forest growth was ingeniously explained by General Harrison's suggestion, that the lapse of time since the era of the mound builders was so great as to have embraced several successive generations of 'rees, and occasioned, by their rotation, a return to the original vegetation.

he successive changes in the spontaneous growth of the forest, as

Effects of Destruction of the Forest.

The physico-geographical effects of the destruction of the forests may be divided into two great classes, each having an important influence on vegetable and on animal life in all their manifestations, as well as on every branch of rural economy and productive industry, and, therefore, on all the material interests of man. The first respects the meteorology of the countries exposed to the action of these influences; the second, their superficial geography, or, in other words, configuration, consistence, and clothing of surface.

For reasons assigned in the first chapter, the meteorological or climatic branch of the subject is the most obscure, and the conclusions of physicists respecting it are, in a great degree, inferential only, not founded on experiment or direct observation. They are, as might be expected, somewhat discordant, though certain general results are almost universally accepted, and seem indeed too well supported to admit of serious question.

proved by the character of the wood found in bogs, is not unfrequently such as to suggest the theory of a considerable change of climate during the human period. But the laws which govern the germination and growth of forest trees must be further studied, and the primitive local conditions of the sites where ancient woods lie buried must be better ascertained, before this theory can be admitted upon the evidence in question. In fact, the order of succession—for a rotation or alternation is not yet proved—may move in opposite directions in different countries with the same climate and at the same time. Thus in Denmark and in Holland the spike-leaved firs have given place to the broad-leaved beech, while in Northern Germany the process has been reversed, and evergreens have supplanted the oaks and birches of deciduous foliage. The principal determining cause seems to be the influence of light upon the germination of the seeds and the growth of the young tree. In a forest of firs, for instance, the distribution of the light and shade, to the influence of which seeds and shoots are exposed, is by no means the same as in a wood of beeches or of oaks, and hence the growth of different species will be stimulated in the two forests. See Berg, *Das Verdrängen der Laubwälder im Nördlichen Deutschland*, 1844. Heyer, *Das Verhalten der Waldbäume gegen Licht und Schatten*, 1852. Staring, *De Bodem van Nederland*, 1856, i, pp. 120–200. Vaupell, *Om Bögens Indvandring i de Danske Skove*, 1857. Knorr, *Studien über die Buchen-Wirthschaft*, 1863.

Electrical Influence of Trees.

The properties of trees, singly and in groups, as exciters or conductors of electricity, and their consequent influence upon the electrical state of the atmosphere, do not appear to have been much investigated; and the conditions of the forest itself are so variable and so complicated, that the solution of any general problem respecting its electrical influence would be a matter of extreme difficulty. It is, indeed, impossible to suppose that a dense cloud, a sea of vapor, can pass over miles of surface bristling with good conductors, without undergoing some change of electrical condition. Hypothetical cases may be put in which the character of the change could be deduced from the known laws of electrical action. But in actual nature, the elements are too numerous for us to seize. The true electrical condition of neither cloud nor forest could be known, and it could seldom be predicted whether the vapors would be dissolved as they floated over the wood, or discharged upon it in a deluge of rain. With regard to possible electrical influences of the forest, wider still in their range of action, the uncertainty is even greater. The data which alone could lead to certain, or even probable, conclusions are wanting, and we should, therefore, only embarrass our argument by any attempt to discuss this meteorological element, important as it may be, in its relations of cause and effect to more familiar and better understood meteoric phenomena. It may, however, be observed that hail storms—which were once generally supposed, and are still held by many, to be produced by a specific electrical action, and which, at least, are always accompanied by electrical disturbances—are believed, in all countries particularly exposed to that scourge, to have become more frequent and destructive in proportion as the forests have been cleared. Caimi observes: "When the chains of the Alps and the Apennines had not yet been stripped of their magnificent crown of woods, the May hail, which now desolates the fertile plains of Lombardy, was much less frequent; but since the general prostration of the forest, these tempests are laying waste even the mountain soils whose older inhabitants scarcely knew this plague.* The *paragrandini*,† which the learned curate of Rivolta advised to erect, with sheaves of straw set up vertically, over a great extent of cultivated country, are but a Lilliputian image of the vast paragrandini, pines, larches, firs, which nature had planted by millions on the crests and ridges of the Alps and the Apennines." ‡ "Electrical action being diminished," says Meguscher, "and the rapid congelation of vapors by the abstraction of heat being impeded by the influence of the woods, it is rare that hail or waterspouts are produced within the precincts of a large forest when it is assailed by the tempest." § Arthur Young was told that since the forests which covered the mountains between the Riviera and the county of Montferrat had disappeared, hail had become more destructive in the district of Acqui, ‖ and it appears

* There are, in Northern Italy and in Switzerland, joint-stock companies which insure against damage by hail, as well as by fire and lightning. Between the years 1854 and 1861, a single one of these companies, La Riunione Adriatica, paid, for damage by hail in Piedmont, Venetian Lombardy, and the Duchy of Parma, above 6,500,000 francs, or nearly $200,000 per year.

† The *paragrandine*, or, as it is called in French, the *paragrêle*, is a species of conductor by which it has been hoped to protect the harvests in countries particularly exposed to damage by hail. It was at first proposed to employ for this purpose poles supporting sheaves of straw connected with the ground by the same material; but the experiment was afterward tried in Lombardy on a large scale, with more perfect electrical conductors, consisting of poles secured to the top of tall trees and provided with a pointed wire entering the ground and reaching above the top of the pole. It was at first thought that this apparatus, erected at numerous points over an extent of several miles, was of some service as a protection against hail, but this opinion was soon disputed, and does not appear to be supported by well-ascertained facts. The question of a repetition of the experiment over a wide area has been again agitated within a very few years in Lombardy; but the doubts expressed by very able physicists as to its efficacy, and as to the point whether hail is an electrical phenomenon, have discouraged its advocates from attempting it.

‡ *Cenni sulla Importanza e Coltura dei Boschi*, p. 6.

§ *Memoria sui Boschi, etc.*, p. 44.

‖ *Travels in Italy*, chap. iii.

upon good authority, that a similar increase in the frequency and violence of hail storms in the neighborhood of Saluzzo and Mondovì, the lower part of the Valtelline, and the territory of Verona and Vicenza, is probably to be ascribed to a similar cause.*

Chemical Influence of the Forest.

We know that the air in a close apartment is appreciably affected through the inspiration and expiration of gases by plants growing in it. The same operations are performed on a gigantic scale by the forest, and it has even been supposed that the absorption of carbon, by the rank vegetation of earlier geological periods, occasioned a permanent change in the constitution of the terrestrial atmosphere.† To the effects thus produced, are to be added those of the ultimate gaseous decomposition of the vast vegetable mass annually shed by trees, and of their trunks and branches when they fall a prey to time. But the quantity of gases thus abstracted from and restored to the atmosphere is inconsiderable—infinitesimal, one might almost say—in comparison with the ocean of air from which they are drawn and to which they return; and though the exhalations from bogs, and other low grounds covered with decaying vegetable matter, are highly deleterious to human health, yet, in general, the air of the forest is hardly chemically distinguishable from that of the sand plains, and we can as little trace the influence of the woods in the analysis of the atmosphere, as we can prove that the mineral ingredients of

* *Le Alpi che cingono l'Italia*, i, p. 377.

† "Long before the appearance of man, * * * they [the forests] had robbed the atmosphere of the enormous quantity of carbonic acid it contained, and thereby transformed it into respirable air. Trees heaped upon trees had already filled up the ponds and marshes, and buried with them in the bowels of the earth—to restore it to us after thousands of ages in the form of bituminous coal and of anthracite—the carbon which was destined to become, by this wonderful condensation, a precious store of future wealth."—CLAVÉ, *Études sur l'Économie Forestière*, p. 13.

This opinion of the modification of the atmosphere by vegetation is contested.

land springs sensibly affect the chemistry of the sea. I may, then, properly dismiss the chemical, as I have done the electrical influences of the forest, and treat them both alike, if not as unimportant agencies, at least as quantities of unknown value in our meteorological equation.* Our inquiries upon this branch of the subject will accordingly be limited to the thermometrical and hygrometrical influences of the woods.

Influence of the Forest, considered as Inorganic Matter, on Temperature.

The evaporation of fluids, and the condensation and expansion of vapors and gases, are attended with changes of temperature; and the quantity of moisture which the air is capable of containing, and, of course, the evaporation, rise and fall with the thermometer. The hygroscopical and the thermoscopical conditions of the atmosphere are, therefore, inseparably connected as reciprocally dependent quantities, and neither can be fully discussed without taking notice of the other. But the forest, regarded purely as inorganic matter, and without reference to its living processes of absorption and exhalation of water and gases, has, as an absorbent, a radiator and a conductor of heat, and as a mere covering of the ground, an influence on the temperature of the air and the earth, which may be considered by itself.

* Schacht ascribes to the forest a specific, if not a measurable, influence upon the constitution of the atmosphere. "Plants imbibe from the air carbonic acid and other gaseous or volatile products exhaled by animals or developed by the natural phenomena of decomposition. On the other hand, the vegetable pours into the atmosphere oxygen, which is taken up by animals and appropriated by them. The tree, by means of its leaves and its young herbaceous twigs, presents a considerable surface for absorption and evaporation; it abstracts the carbon of carbonic acid, and solidifies it in wood, fecula, and a multitude of other compounds. The result is that a forest withdraws from the air, by its great absorbent surface, much more gas than meadows or cultivated fields, and exhales proportionally a considerably greater quantity of oxygen. The influence of the forests on the chemical composition of the atmosphere is, in a word, of the highest importance."—*Les Arbres*, p. 111. See *Appendix*, No. 23.

a. *Absorbing and Emitting Surface.*

A given area of ground, as estimated by the every-day rule of measurement in yards or acres, presents always the same apparent quantity of absorbing, radiating, and reflecting surface; but the real extent of that surface is very variable, depending, as it does, upon its configuration, and the bulk and form of the adventitious objects it bears upon it; and, besides, the true superficies remaining the same, its power of absorption, radiation, reflection, and conduction of heat will be much affected by its consistence, its greater or less humidity, and its color, as well as by its inclination of plane and exposure.*

* Composition, texture and color of soil are important elements to be considered in estimating the effects of the removal of the forest upon its thermoscopic action. "Experience has proved," says Becquerel, "that when the soil is bared, it becomes more or less heated [by the rays of the sun] according to the nature and the color of the particles which compose it, and according to its humidity, and that, in the refrigeration resulting from radiation, we must take into the account the conducting power of those particles also. Other things being equal, silicious and calcareous sands, compared in equal volumes with different argillaceous earths, with calcareous powder or dust, with humus, with arable and with garden earth, are the soils which least conduct heat. It is for this reason that sandy ground, in summer, maintains a high temperature even during the night. We may hence conclude that when a sandy soil is stripped of wood, the local temperature will be raised. After the sands follow successively argillaceous, arable, and garden ground, then humus, which occupies the lowest rank. If we represent the power of calcareous sand to retain heat by 100, we have, according to Schubler,

For [silicious ?] sand	95.6
" arable calcareous soil	74.3
" argillaceous earth	68.4
" garden earth	64.8
" humus	49.0

"The retentive power of humus, then, is but half as great as that of calcareous sand. We will add that the power of retaining heat is proportional to the density. It has also a relation to the magnitude of the particles. It is for this reason that ground covered with silicious pebbles cools more slowly than silicious sand, and that pebbly soils are best suited to the cultivation of the vine, because they advance the ripening of the grape more rapidly than chalky and clayey earths, which cool quickly.

An acre of chalk, rolled hard and smooth, would have great reflecting power, but its radiation would be much increased by breaking it up into clods, because the actually exposed surface would be greater, though the outline of the field remained the same. The area of a triangle being equal to its base multiplied by half the length of a perpendicular let fall from its apex, it follows that the entire superficies of the triangular faces of a quadrangular pyramid, the perpendicular of whose sides should be twice the length of the base, would be four times the area of the ground it covered, and would add to the field on which it stood so much surface capable of receiving and emitting heat, though, in consequence of obliquity and direction of plane, its actual absorption and emission of heat might not be so great as that of an additional quantity of level ground containing four times the area of its base. The lesser inequalities which always occur in the surface of ordinary earth affect in the same way its quantity of superficies acting upon the temperature of the atmosphere, and acted on by it, though the amount of this action and reaction is not susceptible of measurement.

Analogous effects are produced by other objects, of whatever form or character, standing or lying upon the earth, and no solid can be placed upon a flat piece of ground, without itself exposing a greater surface than it covers. This applies, of course, to forest trees and their leaves, and indeed to all vegetables, as well as to other prominent bodies. If we suppose forty trees to be planted on an acre, one being situated in the centre of every square of two rods the side, and to grow until their branches and leaves everywhere meet, it is evident that, when in full foliage, the trunks, branches, and leaves would present an amount of thermoscopic surface much greater than that of an acre of bare earth; and besides this, the fallen leaves lying scattered on the ground, would some-

Hence we see that in examining the calorific effects of clearing forests, it is important to take into account the properties of the soil laid bare."—BECQUEREL, *Des Climats et des Sols boisés*, p. 137.

what augment the sum total.* On the other hand, the growing leaves of trees generally form a succession of stages, or, loosely speaking, layers, corresponding to the annual growth of the branches, and more or less overlying each other. This disposition of the foliage interferes with that free communication between sun and sky above, and leaf surface below, on which the amount of radiation and absorption of heat depends. From all these considerations, it appears that though the effective thermoscopic surface of a forest in full leaf does not exceed that of bare ground in the same proportion as does its measured superficies, yet the actual quantity of area capable of receiving and emitting heat must be greater in the former than in the latter case.†

It must further be remembered that the form and texture of a given surface are important elements in determining its thermoscopic character. Leaves are porous, and admit air and light more or less freely into their substance; they are generally smooth and even glazed on one surface; they are usually covered on one or both sides with spiculæ, and they very commonly present one or more acuminated points in their outline—all circumstances which tend to augment their power of emitting heat by reflection or radiation. Direct experiment on growing trees is very difficult, nor is it in any case practicable to distinguish how far a reduction of temperature produced by vegetation is due to radiation, and how far to exhalation of the fluids of the plant in a gaseous form; for both processes usually go on together. But the frigorific effect of leafy structure is well observed in the deposit of dew and the occurrence of hoarfrost on the foliage of grasses and other small vegetables, and on other objects of similar form and con-

* "The Washington elm at Cambridge—a tree of no extraordinary size—was some years ago estimated to produce a crop of seven millions of leaves, exposing a surface of two hundred thousand square feet, or about five acres of foliage."—GRAY, *First Lessons in Botany and Vegetable Physiology*, as quoted by COULTAS, *What may be learned from a Tree*, p. 34.

† See, on this particular point, and on the general influence of the forest on temperature, HUMBOLDT, *Ansichten der Natur*, i, 158.

sistence, when the temperature of the air a few yards above has not been brought down to the dew point, still less to 32°, the degree of cold required to congeal dew to frost.*

b. *Trees as Conductors of Heat.*

We are also to take into account the action of the forest as a conductor of heat between the atmosphere and the earth. In the most important countries of America and Europe, and especially in those which have suffered most from the destruction of the woods, the superficial strata of the earth are colder in winter, and warmer in summer than those a few inches lower, and their shifting temperature approximates to the atmospheric mean of the respective seasons. The roots of large trees penetrate beneath the superficial strata, and reach earth of a nearly constant temperature, corresponding to the mean for the entire year. As conductors, they convey the heat of the atmosphere to the earth when the earth is colder than the air, and transmit it in the contrary direction when the temperature of the earth is higher than that of the atmosphere. Of course, then, as conductors, they tend to equalize the temperature of the earth and the air.

c. *Trees in Summer and Winter.*

In countries where the questions I am considering have the greatest practical importance, a very large proportion, if not a majority, of the trees are of deciduous foliage, and their radiating as well as their shading surface is very much greater in summer than in winter. In the latter season, they little obstruct the reception of heat by the ground or the radiation from it; whereas, in the former, they often interpose a complete

* The radiating and refrigerating power of objects by no means depends on their form alone. Melloni cut sheets of metal into the shape of leaves and grasses, and found that they produced little cooling effect, and were not moistened under atmospheric conditions which determined a plentiful deposit of dew on the leaves of vegetables

canopy between the ground and the sky, and materially interfere with both processes.

d. *Dead Products of Trees.*

Besides this various action of standing trees considered as inorganic matter, the forest exercises, by the annual moulting of its foliage, still another influence on the temperature of the earth, and, consequently, of the atmosphere which rests upon it. If you examine the constitution of the superficial soil in a primitive or an old and undisturbed artificially planted wood, you find, first, a deposit of undecayed leaves, twigs, and seeds, lying in loose layers on the surface; then, more compact beds of the same materials in incipient, and, as you descend, more and more advanced stages of decomposition; then, a mass of black mould, in which traces of organic structure are hardly discoverable except by microscopic examination; then, a stratum of mineral soil, more or less mixed with vegetable matter carried down into it by water, or resulting from the decay of roots; and, finally, the inorganic earth or rock itself. Without this deposit of the dead products of trees, this latter would be the superficial stratum, and as its powers of absorption, radiation, and conduction of heat would differ essentially from those of the layers with which it has been covered by the droppings of the forest, it would act upon the temperature of the atmosphere, and be acted on by it, in a very different way from the leaves and mould which rest upon it. Leaves, still entire, or partially decayed, are very indifferent conductors of heat, and, therefore, though they diminish the warming influence of the summer sun on the soil below them, they, on the other hand, prevent the escape of heat from that soil in winter, and, consequently, in cold climates, even when the ground is not covered by a protecting mantle of snow, the earth does not freeze to as great a depth in the wood as in the open field.

e. *Trees as a Shelter to Ground to the Leeward.*

The action of the forest, considered merely as a mechanical shelter to grounds lying to the leeward of it, would seem to be an influence of too restricted a character to deserve much notice; but many facts concur to show that it is an important element in local climate, and that it is often a valuable means of defence against the spread of miasmatic effluvia, though, in this last case, it may exercise a chemical as well as a mechanical agency. In the report of a committee appointed in 1836 to examine an article of the forest code of France, Arago observes: "If a curtain of forest on the coasts of Normandy and of Brittany were destroyed, these two provinces would become accessible to the winds from the west, to the mild breezes of the sea. Hence a decrease of the cold of winter. If a similar forest were to be cleared on the eastern border of France, the glacial east wind would prevail with greater strength, and the winters would become more severe. Thus the removal of a belt of wood would produce opposite effects in the two regions." *

This opinion receives confirmation from an observation of Dr. Dwight, who remarks, in reference to the woods of New England: "Another effect of removing the forest will be the free passage of the winds, and among them of the southern winds, over the surface. This, I think, has been an increasing fact within my own remembrance. As the cultivation of the country has extended farther to the north, the winds from the south have reached distances more remote from the ocean, and imparted their warmth frequently, and in such degrees as, forty years since, were in the same places very little known. This fact, also, contributes to lengthen the summer, and to shorten the winter-half of the year." †

It is thought in Italy that the clearing of the Apennines has very materially affected the climate of the valley of the Po. It is asserted in Le Alpi che cingono l'Italia that: "In

* Becquerel, *Des Climats, etc., Discours Prélim.* vi. † *Travels*, i, p. 61.

consequence of the felling of the woods on the Apennines, the sirocco prevails greatly on the right bank of the Po, in the Parmesan territory, and in a part of Lombardy; it injures the harvests and the vineyards, and sometimes ruins the crops of the season. To the same cause many ascribe the meteorological changes in the precincts of Modena and of Reggio. In the communes of these districts, where formerly straw roofs resisted the force of the winds, tiles are now hardly sufficient; in others, where tiles answered for roofs, large slabs of stone are now ineffectual; and in many neighboring communes the grapes and the grain are swept off by the blasts of the south and southwest winds."

On the other hand, according to the same authority, the pinery of Porto, near Ravenna—which is 33 kilometres long, and is one of the oldest pine woods in Italy—having been replanted with resinous trees after it was unfortunately cut, has relieved the city from the sirocco to which it had become exposed, and in a great degree restored its ancient climate.*

The felling of the woods on the Atlantic coast of Jutland has exposed the soil not only to drifting sands, but to sharp sea winds, that have exerted a sensible deteriorating effect on the climate of that peninsula, which has no mountains to serve at once as a barrier to the force of the winds, and as a storehouse of moisture received by precipitation or condensed from atmospheric vapors.†

It is evident that the effect of the forest, as a mechanical impediment to the passage of the wind, would extend to a very considerable distance above its own height, and hence protect while standing, or lay open when felled, a much larger surface than might at first thought be supposed. The atmosphere, movable as are its particles, and light and elastic as are its masses, is nevertheless held together as a continuous whole by the gravitation of its atoms and their consequent pressure on each other, if not by attraction between them, and, therefore, an obstruction which mechanically impedes the movement of a given

* *Le Alpi che cingono l'Italia*, pp. 370, 371.

† Bergsöe, *Reventlovs Virksomhed*, ii, p. 125.

stratum of air, will retard the passage of the strata above and below it. To this effect may often be added that of an ascending current from the forest itself, which must always exist when the atmosphere within the wood is warmer than the stratum of air above it, and must be of almost constant occurrence in the case of cold winds, from whatever quarter, because the still air in the forest is slow in taking up the temperature of the moving columns and currents around and above it. Experience, in fact, has shown that mere rows of trees, and even much lower obstructions, are of essential service in defending vegetation against the action of the wind. Hardy proposes planting, in Algeria, belts of trees at the distance of one hundred mètres from each other, as a shelter which experience had proved to be useful in France.* "In the valley of the Rhone," says Becquerel, "a simple hedge, two mètres in height, is a sufficient protection for a distance of twenty-two mètres." † The mechanical shelter acts, no doubt, chiefly as a defence against the mechanical force of the wind, but its uses are by no means limited to this effect. If the current of air which it resists moves horizontally, it would prevent the access of cold or parching blasts to the ground for a great distance; and did the wind even descend at a large angle with the surface, still a considerable extent of ground would be protected by a forest to the windward of it. If we suppose the trees of a wood to have a mean height of only twenty yards, they would often beneficially affect the temperature or the moisture of a belt of land two or three hundred yards in width, and thus perhaps rescue valuable crops from destruction.‡

* BECQUEREL, *Des Climats, etc.*, p. 179. † Ibid., p. 116.

‡ The following well-attested instance of a local change of climate is probably to be referred to the influence of the forest as a shelter against cold winds. To supply the extraordinary demand for Italian iron occasioned by the exclusion of English iron in the time of Napoleon I, the furnaces of the valleys of Bergamo were stimulated to great activity. "The ordinary production of charcoal not sufficing to feed the furnaces and the forges, the woods were felled, the copses cut before their time, and the whole economy of the forest was deranged. At Piazzatorre there was

The local retardation of spring so much complained of in Italy, France, and Switzerland, and the increased frequency of late frosts at that season, appear to be ascribable to the admission of cold blasts to the surface, by the felling of the forests which formerly both screened it as by a wall, and communicated the warmth of their soil to the air and earth to the leeward. Caimi states that since the cutting down of the woods of the Apennines, the cold winds destroy or stunt the vegetation, and that, in consequence of "the usurpation of winter on the domain of spring," the district of Mugello has lost all its mulberries, except the few which find in the lee of buildings a protection like that once furnished by the forest.*

"It is proved," says Clavé, "Études," p. 44, that the department of Ardèche, which now contains not a single considerable wood, has experienced within thirty years a climatic disturbance, of which the late frosts, formerly unknown in the country, are one of the most melancholy effects. Similar results have been observed in the plain of Alsace, in consequence of the denudation of several of the crests of the Vosges."

such a devastation of the woods, and consequently such an increased severity of climate, that maize no longer ripened. An association, formed for the purpose, effected the restoration of the forest, and maize flourishes again in the fields of Piazzatorre."—Report by G. Rosa, in *Il Politecnico*, Dicembre, 1861, p. 614.

Similar ameliorations have been produced by plantations in Belgium. In an interesting series of articles by Baude, entitled "Les Cotes de la Manche," in the *Revue des Deux Mondes*, I find this statement: "A spectator placed on the famous bell tower of the cathedral of Antwerp, saw, not long since, on the opposite side of the Schelde only a vast desert plain; now he sees a forest, the limits of which are confounded with the horizon. Let him enter within its shade. The supposed forest is but a system of regular rows of trees, the oldest of which is not forty years of age. These plantations have ameliorated the climate which had doomed to sterility the soil where they are planted. While the tempest is violently agitating their tops, the air a little below is still, and sands far more barren than the plateau of La Hague have been transformed, under their protection, into fertile fields."—*Revue des Deux Mondes*, January, 1859, p. 277.

* *Cenni sulla Importanza e Coltura dei Boschi*, p. 31.

Dussard, as quoted by Ribbe,* maintains that even the *mistral*, or northwest wind, whose chilling blasts are so fatal to tender vegetation in the spring, "is the child of man, the result of his devastations." "Under the reign of Augustus," continues he, "the forests which protected the Cévennes were felled, or destroyed by fire, in mass. A vast country, before covered with impenetrable woods—powerful obstacles to the movement and even to the formation of hurricanes—was suddenly denuded, swept bare, stripped, and soon after, a scourge hitherto unknown struck terror over the land from Avignon to the Bouches du Rhone, thence to Marseilles, and then extended its ravages, diminished indeed by a long career which had partially exhausted its force, over the whole maritime frontier. The people thought this wind a curse sent of God. They raised altars to it and offered sacrifices to appease its rage." It seems, however, that this plague was less destructive than at present, until the close of the sixteenth century, when further clearings had removed most of the remaining barriers to its course. Up to that time, the northwest wind appears not to have attained to the maximum of specific effect which now characterizes it as a local phenomenon. Extensive districts, from which the rigor of the seasons has now banished valuable crops, were not then exposed to the loss of their harvests by tempests, cold, or drought. The deterioration was rapid in its progress. Under the Consulate, the clearings had exerted so injurious an effect upon the climate, that the cultivation of the olive had retreated several leagues, and since the winters and springs of 1820 and 1836, this branch of rural industry has been abandoned in a great number of localities where it was advantageously pursued before. The orange now flourishes only at a few sheltered points of the coast, and it is threatened even at Hyères, where the clearing of the hills near the town has proved very prejudicial to this valuable tree.

Marchand informs us that, since the felling of the woods, late spring frosts are more frequent in many localities north

* *La Provence au point de vue des Torrents et des Inondations*, p. 19.

of the Alps; that fruit trees thrive well no longer, and that it is difficult to raise young trees.*

f. *Trees as a Protection against Malaria.*

The influence of forests in preventing the diffusion of miasmatic vapors is a matter of less familiar observation, and perhaps does not come strictly within the sphere of the present inquiry, but its importance will justify me in devoting some space to the subject. "It has been observed" (I quote again from Becquerel) "that humid air, charged with miasmata, is deprived of them in passing through the forest. Rigaud de Lille observed localities in Italy where the interposition of a screen of trees preserved everything beyond it, while the unprotected grounds were subject to fevers."† Few European countries present better opportunities for observation on this point than Italy, because in that kingdom the localities exposed to miasmatic exhalations are numerous, and belts of trees, if not forests, are of so frequent occurrence that their efficacy in this respect can be easily tested. The belief that rows of trees afford an important protection against malarious influences is very general among Italians best qualified by intelligence and professional experience to judge upon the subject. The commissioners appointed to report on the measures to be adopted for the improvement of the Tuscan Maremme advised the planting of three or four rows of poplars, *Populus alba*, in such directions as to obstruct the currents of air from malarious localities, and thus intercept a great proportion of the pernicious exhalations."‡ Lieutenant Maury even believed that a few rows of sunflowers, planted between the Washington Observatory and the marshy banks of the Potomac, had saved the inmates of that establishment from the intermittent fevers to which they had been formerly liable.

* *Ueber die Entwaldung der Gebirge*, p. 28.

† BECQUEREL, *Des Climats, etc.*, p. 9.

‡ SALVAGNOLI, *Rapporto sul Bonificamento delle Maremme Toscane*, pp. xli, 124.

Maury's experiments have been repeated in Italy. Large plantations of sunflowers have been made upon the alluvial deposits of the Oglio, above its entrance into the Lake of Iseo near Pisogne, and it is said with favorable results to the health of the neighborhood.* In fact, the generally beneficial effects of a forest wall or other vegetable screen, as a protection against noxious exhalations from marshes or other sources of disease situated to the windward of them, are very commonly admitted.

It is argued that, in these cases, the foliage of trees and of other vegetables exercises a chemical as well as a mechanical effect upon the atmosphere, and some, who allow that forests may intercept the circulation of the miasmatic effluvia of swampy soils, or even render them harmless by decomposing them, contend, nevertheless, that they are themselves active causes of the production of malaria. The subject has been a good deal discussed in Italy, and there is some reason to think that under special circumstances the influence of the forest in this respect may be prejudicial rather than salutary, though this does not appear to be generally the case.† It is, at all events, well known that the great swamps of Virginia and the Carolinas, in climates nearly similar to that of Italy, are healthy even to the white man, so long as the forests in and around them remain, but become very insalubrious when the woods are felled.‡

The Forest, as Inorganic Matter, tends to mitigate Extremes.

The surface which trees and leaves present augments the general superficies of the earth exposed to the absorption of

* *Il Politecnico, Milano, Aprile e Maggio*, 1863, p. 35.

† SALVAGNOLI, *Memorie sulle Maremme Toscane*, pp. 213, 214.

‡ Except in the seething marshes of the tropics, where vegetable decay is extremely rapid, the uniformity of temperature and of atmospheric humidity renders all forests eminently healthful. See HOHENSTEIN's observations on this subject, *Der Wald*, p. 41.

There is no question that open squares and parks conduce to the salubrity of cities, and many observers are of opinion that the trees and other vegetables with which such grounds are planted contribute essentially to their beneficial influence. See an article in *Aus. der Natur*, xxii, p. 813.

heat, and increases the radiating and reflecting area in the same proportion. It is impossible to measure the relative value of these two elements—increase of absorbing and increase of emitting surface—as thermometrical influences, because they exert themselves under infinitely varied conditions; and it is equally impossible to make a quantitative estimate of any partial, still more of the total effect of the forest, considered as dead matter, on the temperature of the atmosphere, and of the portion of the earth's surface acted on by it. But it seems probable that its greatest influence in this respect is due to its character of a screen, or mechanical obstacle to the transmission of heat between the earth and the air; and this is equally true of the standing tree and of the dead foliage which it deposits in successive layers at its foot.

The complicated action of trees and their products, as dead absorbents, radiators, reflectors, and conductors of heat, and as intercepters of its transmission, is so intimately connected with their effects upon the humidity of the air and the earth, and with all their living processes, that it is difficult to separate the former from the latter class of influences; but upon the whole, the forest must thus far be regarded as tending to mitigate extremes, and, therefore, as an equalizer of temperature.

TREES AS ORGANISMS.

Specific Heat.

Trees, considered as organisms, produce in themselves, or in the air, a certain amount of heat, by absorbing and condensing atmospheric vapor, and they exert an opposite influence by absorbing water and exhaling it in the form of vapor; but there is still another mode by which their living processes may warm the air around them, independently of the thermometric effects of condensation and evaporation. The vital heat of a dozen persons raises the temperature of a room. If trees possess a specific temperature of their own, an organic power of generating heat, like that with which the warm-blooded animals are gifted, though by a different process, a certain amount of weight is to be ascribed to this element, in estimating the action of the forest upon atmospheric temperature.

"Observation shows," says Meguscher, "that the wood of a living tree maintains a temperature of + 12° or 13° Cent. [= 54°, 56° Fahr.] when the temperature of the air stands at 3°, 7°, and 8° [= 37°, 46°, 47° F.] above zero, and that the internal warmth of the tree does not rise and fall in proportion to that of the atmosphere. So long as the latter is below 18° [= 67° Fahr.], that of the tree is always the highest; but if the temperature of the air rises to 18°, that of the vegetable growth is the lowest. Since, then, trees maintain at all seasons a constant mean temperature of 12° [= 54° Fahr.], it is easy to see why the air in contact with the forest must be warmer in winter, cooler in summer, than in situations where it is deprived of that influence." *

Boussingault remarks: "In many flowers there has been observed a very considerable evolution of heat, at the approach of fecundation. In certain *arums* the temperature rises to 40° or 50° Cent. [= 104° or 122° Fahr.]. It is very probable that this phenomenon is general, and varies only in the intensity with which it is manifested." †

If we suppose the fecundation of the flowers of forest trees to be attended with a tenth only of this calorific power, they could not fail to exert an important influence on the warmth of the atmospheric strata in contact with them.

In a paper on Meteorology by Professor Henry, published in the United States Patent Office Report for 1857, p. 504, that distinguished physicist observes: "As a general deduction from chemical and mechanical principles, we think no change of temperature is ever produced where the actions belonging to one or both of these principles are not present. Hence, in midwinter, when all vegetable functions are dormant, we do not believe that any heat is developed by a tree, or that its interior differs in temperature from its exterior

* *Memoria sui Boschi di Lombardia*, p. 45. † *Économie Rurale*, i, p. 22.

further than it is protected from the external air. The experiments which have been made on this point, we think, have been directed by a false analogy. During the active circulation of the sap and the production of new tissue, variations of temperature belonging exclusively to the plant may be observed; but it is inconsistent with general principles that heat should be generated where no change is taking place."

There can be no doubt that moisture is given out by trees and evaporated in extremely cold winter-weather, and unless new fluid were supplied from the roots, the tree would be exhausted of its juices before winter was over. But this is not observed to be the fact, and, though the point is disputed, respectable authorities declare that "wood felled in the depth of winter is the heaviest and fullest of sap." * Warm weather in winter, of too short continuance to affect the temperature of the ground sensibly, stimulates a free flow of sap in the maple. Thus, in the last week of December, 1862, and the first week of January, 1863, sugar was made from that tree, in various parts of New England. "A single branch of a tree, admitted into a warm room in winter through an aperture in a window, opened its buds and developed its leaves while the rest of the tree in the external air remained in its winter sleep." † The roots of forest trees in temperate climates, remain, for the most part, in a moist soil, of a temperature not much below the annual mean, through the whole winter; and we cannot account for the uninterrupted moisture of the tree, unless we suppose that the roots furnish a constant supply of water.

Atkinson describes a ravine in a valley in Siberia, which was filled with ice to the depth of twenty-five feet. Poplars were growing in this ice, which was thawed to the distance of some inches from the stem. But the surface of the soil beneath it must have remained still frozen, for the holes around the trees were full of water resulting from its melting, and this

* Rossmässler, *Der Wald*, p. 158. † Ibid., p. 160.

would have escaped below if the ground had been thawed. In this case, although the roots had not thawed the thick covering of earth above them, the trunks must have melted the ice in contact with them. The trees, when observed by Atkinson, were in full leaf, but it does not appear at what period the ice around their stems had melted.

From these facts, and others of the like sort, it would seem that "all vegetable functions are" not absolutely "dormant" in winter, and, therefore, that trees may give out *some* heat at that season. But, however this may be, the "circulation of the sap" commences at a very early period in the spring, and the temperature of the air in contact with trees may then be sufficiently affected by heat evolved in the vital processes of vegetation, to raise the thermometric mean of wooded countries for that season, and, of course, for the year.*

Total Influence of the Forest on Temperature.

It has not yet been found practicable to measure, sum up, and equate the total influence of the forest, its processes and its products, dead and living, upon temperature, and investigators differ much in their conclusions on this subject. It seems

* The low temperature of air and soil at which, in the frigid zone, as well as in warmer latitudes under special circumstances, the processes of vegetation go on, seems to necessitate the supposition that all the manifestations of vegetable life are attended with an evolution of heat. In the United States, it is common to protect ice, in icehouses, by a covering of straw, which naturally sometimes contains kernels of grain. These often sprout, and even throw out roots and leaves to a considerable length, in a temperature very little above the freezing point. Three or four years since, I saw a lump of very clear and apparently solid ice, about eight inches long by six thick, on which a kernel of grain had sprouted in an icehouse, and sent half a dozen or more very slender roots into the pores of the ice and through the whole length of the lump. The young plant must have thrown out a considerable quantity of heat; for though the ice was, as I have said, otherwise solid, the pores through which the roots passed were enlarged to perhaps double the diameter of the fibres, but still not so much as to prevent the retention of water in them by capillary attraction. See *App.* 24.

probable that in every particular case the result is, if not determined, at least so much modified by local conditions which are infinitely varied, that no general formula is applicable to the question.

In the report to which I referred on page 149, Gay-Lussac says: "In my opinion we have not yet any positive proof that the forest has, in itself, any real influence on the climate of a great country, or of a particular locality. By closely examining the effects of clearing off the woods, we should perhaps find that, far from being an evil, it is an advantage; but these questions are so complicated when they are examined in a climatological point of view, that the solution of them is very difficult, not to say impossible."

Becquerel, on the other hand, considers it certain that in tropical climates, the destruction of the forests is accompanied with an elevation of the mean temperature, and he thinks it highly probable that it has the same effect in the temperate zones. The following is the substance of his remarks on this subject:—

"Forests act as frigorific causes in three ways:

"1. They shelter the ground against solar irradiation and maintain a greater humidity.

"2. They produce a cutaneous transpiration by the leaves.

"3. They multiply, by the expansion of their branches, the surfaces which are cooled by radiation.

"These three causes acting with greater or less force, we must, in the study of the climatology of a country, take into account the proportion between the area of the forests and the surface which is bared of trees and covered with herbs and grasses.

"We should be inclined to believe *à priori*, according to the foregoing considerations, that the clearing of the woods, by raising the temperature and increasing the dryness of the air, ought to react on climate. There is no doubt that, if the vast desert of the Sahara were to become wooded in the course of ages, the sands would cease to be heated as much as at the present epoch, when the mean temperature is twenty-nine degrees [centigrade, = 85° Fahr.]. In that case, the ascending currents of warm air would cease, or be less warm, and would not contribute, by descending in our latitudes, to soften the climate of Western Europe. Thus the clearing of a great country may react on the climates of regions more or less remote from it.

"The observations by Boussingault leave no doubt on this point. This writer determined the mean temperature of wooded and of cleared points, under the same latitude, and at the same elevation above the sea, in localities comprised between the eleventh degree of north and the fifth degree of south latitude, that is to say, in the portion of the tropics nearest to the equator, and where radiation tends powerfully during the night to lower the temperature under a sky without clouds." *

The result of these observations, which has been pretty generally adopted by physicists, is that the mean temperature of cleared land in the tropics appears to be about one degree centigrade, or a little less than two degrees of Fahrenheit, above that of the forest. On page 147 of the volume just cited, Becquerel argues that, inasmuch as the same and sometimes a greater difference is found in favor of the open ground, at points within the tropics so elevated as to have a temperate or even a polar climate, we must conclude that the forests in Northern America exert a refrigerating influence equally powerful. But the conditions of the soil are so different in the two regions compared, that I think we cannot, with entire confidence, reason from the one to the other, and it is much to be desired that observations be made on the summer and winter temperature of both the air and the ground in the depths of the North American forests, before it is too late.†

* BECQUEREL, *Des Climats, etc.*, pp. 139—141.

† Dr. Williams made some observations on this subject in 1789, and in 1791, but they generally belonged to the warmer months, and I do not know that any extensive series of comparisons between the temperature of the ground in the woods and the fields has been attempted in America.

INFLUENCE OF FORESTS ON THE HUMIDITY OF THE AIR AND THE EARTH.

a. *As Inorganic Matter.*

The most important influence of the forest on climate is, no doubt, that which it exercises on the humidity of the air and the earth, and this climatic action it exerts partly as dead, partly as living matter. By its interposition as a curtain between the sky and the ground, it intercepts a large proportion of the dew and the lighter showers, which would otherwise

Dr. Williams's thermometer was sunk to the depth of ten inches, and gave the following results:

Time.	Temperature of ground in pasture.	Temperature of ground in woods.	Difference.
May 23......	52	46	6
" 28......	57	48	9
June 15......	64	51	13
" 27......	62	51	11
July 16......	62	51	11
" 30......	65½	55½	10
Aug. 15......	68	58	10
" 31......	59½	55	4½
Sept. 15......	59½	55	4½
Oct. 1......	59½	55	4½
" 15......	49	49	0
Nov. 1......	43	43	0
" 16......	43½	43½	0

On the 14th of January, 1791, in a winter remarkable for its extreme severity, he found the ground, on a plain open field where the snow had been blown away, frozen to the depth of three feet and five inches; in the woods where the snow was three feet deep, and where the soil had frozen to the depth of six inches before the snow fell, the thermometer, at six inches below the surface of the ground, stood at 39°. In consequence of the covering of the snow, therefore, the previously frozen ground had been thawed and raised to seven degrees above the freezing point.—Williams's *Vermont*, i, p. 74.

Bodies of fresh water, so large as not to be sensibly affected by local influences of narrow reach or short duration, would afford climatic indications well worthy of special observation. Lake Champlain, which forms the boundary between the States of New York and Vermont, presents very favorable conditions for this purpose. This lake, which drains a basin of about 6,000 square miles, covers an area, excluding its islands, of about 500

moisten the surface of the soil, and restores it to the atmosphere by evaporation; while in heavier rains, the large drops which fall upon the leaves and branches are broken into smaller ones, and consequently strike the ground with less mechanical force, or are perhaps even dispersed into vapor without reaching it.* As a screen, it prevents the access of

square miles. It extends from lat. 43° 30′ to 45° 20′, in very nearly a meridian line, has a mean width of four and a half miles, with an extreme breadth, excluding bays almost land-locked, of thirteen miles. Its mean depth is not well known. It is, however, 400 feet deep in some places, and from 100 to 200 in many, and has few shoals or flats. The climate is of such severity that it rarely fails to freeze completely over, and to be safely crossed upon the ice, with heavy teams, for several weeks every winter. THOMPSON (*Vermont*, p. 14, and Appendix, p. 9) gives the following table of the times of the complete closing and opening of the ice, opposite Burlington, about the centre of the lake, and where it is ten miles wide.

Year.	Closing.	Opening.	Days closed.	Year.	Closing.	Opening.	Days closed.
1816	February 9			1836	January 27	April 21	85
1817	January 29	April 16	78	1837	January 15	April 26	101
1818	February 2	April 15	72	1838	February 2	April 13	70
1819	March 4	April 17	44	1839	January 25	April 6	71
1820	February 3	February	4	1840	January 25	February 20	26
	March 8	March 12		1841	February 18	April 19	61
1821	January 15	April 21	95	1842	not closed		
1822	January 24	March 30	75	1843	February 16	April 22	65
1823	February 7	April 5	57	1844	January 25	April 11	77
1824	January 22	February 11	20	1845	February 3	March 26	51
1825	February 9			1846	February 10	March 26	44
1826	February 1	March 24	51	1847	February 15	April 23	68
1827	January 21	March 31	68	1848	February 13	February 26	13
1828	not closed			1849	February 7	March 23	44
1829	January 31	April		1850	not closed		
1832	February 6	April 17	70	1851	February 1	March 12	39
1833	February 2	April 6	63	1852	January 18	April 19	92
1834	February 13	February 20	7				
1835	January 10	January 23	13				
	February 7	April 12	64				

In 1847, although, at the point indicated, the ice broke up on the 23d of April, it remained frozen much later at the North, and steamers were not able to traverse the whole length of the lake until May 6th.

* We are not, indeed, to suppose that condensation of vapor and evaporation of water are going on in the same stratum of air at the same time, or, in other words, that vapor is condensed into raindrops, and raindrops evaporated, under the same conditions; but rain formed in one stratum, may fall through another, where vapor would not be condensed. Two saturated strata of different temperatures may be brought into con-

the sun's rays to the earth, and, of course, an elevation of temperature which would occasion a great increase of evaporation. As a mechanical obstruction, it impedes the passage of air currents over the ground, which, as is well known, is one of the most efficient agents in promoting evaporation and the refrigeration resulting from it.* In the forest, the air is almost quiescent, and moves only as local changes of temperature affect the specific gravity of its particles. Hence there is often a dead calm in the woods when a furious blast is raging in the open country at a few yards' distance. The denser the forest —as for example, where it consists of spike-leaved trees, or is thickly intermixed with them—the more obvious is its effect, and no one can have passed from the field to the wood in cold, windy weather, without having remarked it.†

tact in the higher regions, and discharge large raindrops, which, if not divided by some obstruction, will reach the ground, though passing all the time through strata which would vaporize them if they were in a state of more minute division.

* It is perhaps too much to say that the influence of trees upon the wind is strictly limited to the mechanical resistance of their trunks, branches, and foliage. So far as the forest, by dead or by living action, raises or lowers the temperature of the air within it, so far it creates upward or downward currents in the atmosphere above it, and, consequently, a flow of air toward or from itself. These air streams have a certain, though doubtless a very small influence on the force and direction of greater atmospheric movements.

† As a familiar illustration of the influence of the forest in checking the movement of winds, I may mention the well-known fact, that the sensible cold is never extreme in thick woods, where the motion of the air is little felt. The lumbermen in Canada and the Northern United States labor in the woods, without inconvenience, when the mercury stands many degrees below the zero of Fahrenheit, while in the open grounds, with only a moderate breeze, the same temperature is almost insupportable. The engineers and firemen of locomotives, employed on railways running through forests of any considerable extent, observe that, in very cold weather, it is much easier to keep up the steam while the engine is passing through the woods than in the open ground. As soon as the train emerges from the shelter of the trees the steam gauge falls, and the stoker is obliged to throw in a liberal supply of fuel to bring it up again.

Another less frequently noticed fact, due, no doubt, in a great measure

The vegetable mould, resulting from the decomposition of leaves and of wood, carpets the ground with a spongy covering which obstructs the evaporation from the mineral earth below, drinks up the rains and melting snows that would otherwise

to the immobility of the air, is, that sounds are transmitted to incredible distances in the unbroken forest. Many instances of this have fallen under my own observation, and others, yet more striking, have been related to me by credible and competent witnesses familiar with a more primitive condition of the Anglo-American world. An acute observer of natural phenomena, whose childhood and youth were spent in the interior of one of the newer New England States, has often told me that when he established his home in the forest, he always distinctly heard, in still weather, the plash of horses' feet, when they forded a small brook nearly seven-eighths of a mile from his house, though a portion of the wood that intervened consisted of a ridge seventy or eighty feet higher than either the house or the ford.

I have no doubt that, in such cases, the stillness of the air is the most important element in the extraordinary transmissibility of sound; but it must be admitted that the absence of the multiplied and confused noises, which accompany human industry in countries thickly peopled by man, contributes to the same result. We become, by habit, almost insensible to the familiar and never-resting voices of civilization in cities and towns; but the indistinguishable drone, which sometimes escapes even the ear of him who listens for it, deadens and often quite obstructs the transmission of sounds which would otherwise be clearly audible. An observer, who wishes to appreciate that hum of civic life which he cannot analyze, will find an excellent opportunity by placing himself on the hill of Capo di Monte at Naples, in the line of prolongation of the street called Spaccanapoli.

It is probably to the stillness of which I have spoken, that we are to ascribe the transmission of sound to great distances at sea in calm weather. In June, 1853, I and my family were passengers on board a ship of war bound up the Ægean. On the evening of the 27th of that month, as we were discussing, at the tea table, some observations of Humboldt on this subject, the captain of the ship told us that he had once heard a single gun at sea at the distance of ninety nautical miles. The next morning, though a light breeze had sprung up from the north, the sea was of glassy smoothness when we went on deck. As we came up, an officer told us that he had heard a gun at sunrise, and the conversation of the previous evening suggested the inquiry whether it could have been fired from the combined French and English fleet then lying at Beshika Bay. Upon examination of our position we were found to have been, at sunrise, ninety sea miles

flow rapidly over the surface and perhaps be conveyed to the distant sea, and then slowly gives out, by evaporation, infiltration, and percolation, the moisture thus imbibed. The roots, too, penetrate far below the superficial soil, conduct the water along their surface to the lower depths to which they reach, and thus serve to drain the superior strata and remove the moisture out of the reach of evaporation.

b. *The Forest as Organic.*

These are the principal modes in which the humidity of the atmosphere is affected by the forest regarded as lifeless matter. Let us inquire how its organic processes act upon this meteorological element.

The commonest observation shows that the wood and bark of living trees are always more or less pervaded with watery and other fluids, one of which, the sap, is very abundant in trees of deciduous foliage when the buds begin to swell and the leaves to develop themselves in the spring. The outer bark of most trees is of a corky character, not admitting the absorption of much moisture from the atmosphere through its pores, and we can hardly suppose that the buds are able to

from that point. We continued beating up northward, and between sunrise and twelve o'clock meridian of the 28th, we had made twelve miles northing, reducing our distance from Beshika Bay to seventy-eight sea miles. At noon we heard several guns so distinctly that we were able to count the number. On the 29th we came up with the fleet, and learned from an officer who came on board that a royal salute had been fired at noon on the 28th, in honor of the day as the anniversary of the Queen of England's coronation. The report at sunrise was evidently the morning gun, those at noon the salute.

Such cases are rare, because the sea is seldom still, and the *κυμάτων ἀνήριϑμον γέλασμα* rarely silent, over so great a space as ninety or even seventy-eight nautical miles. I apply the epithet *silent* to *γέλασμα* advisedly. I am convinced that Æschylus meant the audible laugh of the waves, which is indeed of *countless* multiplicity, not the visible smile of the sea, which, belonging to the great expanse as one impersonation, is single, though, like the human smile, made up of the play of many features.

extract from the air a much larger supply. The obvious conclusion as to the source from which the extraordinary quantity of sap at this season is derived, is that to which scientific investigation leads us, namely, that it is absorbed from the earth by the roots, and thence distributed to all parts of the plant. Popular opinion, indeed, supposes that all the vegetable fluids, during the entire period of growth, are thus drawn from the bosom of the earth, and that the wood and other products of the tree are wholly formed from matter held in solution in the water abstracted by the roots from the ground. This is an error, for, not only is the solid matter of the tree, in a certain proportion not important to our present inquiry, received from the atmosphere in a gaseous form, through the pores of the leaves and of the young shoots, but water in the state of vapor is absorbed and contributed to the circulation, by the same organs.* The amount of water taken up by the roots, however, is vastly greater than that imbibed through the leaves, especially at the season when the juices are most abun-

* "The presence of watery vapor in the air is general. * * * Vegetable surfaces are endowed with the power of absorbing gases, vapors, and also, no doubt, the various soluble bodies which are presented to them. The inhalation of humidity is carried on by the leaves upon a large scale; the dew of a cold summer night revives the groves and the meadows, and a single shower of rain suffices to refresh the verdure of a forest which a long drought had parched."—SCHACHT, *Les Arbres*, ix, p. 340.

The absorption of the vapor of water by leaves is disputed. "The absorption of watery vapor by the leaves of plants is, according to Unger's experiments, inadmissible."—WILHELM, *Der Boden und das Wasser*, p. 19. If this latter view is correct, the apparently refreshing effects of atmospheric humidity upon vegetation must be ascribed to moisture absorbed by the ground from the air and supplied to the roots. In some recent experiments by Dr. Sachs, a porous flower-pot, with a plant growing in it, was left unwatered until the earth was dry, and the plant began to languish. The pot was then placed in a glass case containing air, which was kept always saturated with humidity, but no water was supplied, and the leaves of the plant were exposed to the open atmosphere. The soil in the flower pot absorbed from the air moisture enough to revive the foliage, and keep it a long time green, but not enough to promote development of new leaves. —Id., ibid., p. 18.

dant, and when, as we have seen, the leaves are yet in embryo. The quantity of water thus received from the air and the earth, in a single year, by a wood of even a hundred acres, is very great, though experiments are wanting to furnish the data for even an approximate estimate of its measure; for only the vaguest conclusions can be drawn from the observations which have been made on the imbibition and exhalation of water by trees and other plants reared in artificial conditions diverse from those of the natural forest.*

Wood Mosses and Fungi.

Besides the water drawn by the roots from the earth and the vapor absorbed by the leaves from the air, the wood mosses and fungi, which abound in all dense forests, take up a great quantity of moisture from the atmosphere when it is charged with humidity, and exhale it again when the air is dry. These humble organizations, which play a more important part in regulating the humidity of the air than writers on the forest have usually assigned to them, perish with the trees they grow on; but, in many situations, nature provides a compensation for the tree mosses in ground species, which, on cold soils, especially those with a northern exposure, spring up abundantly both before the woods are felled, and when the land is cleared and employed for pasturage, or deserted. These mosses discharge a portion of the functions appropriated to the wood, and while they render the soil of improved lands much less fit for agricultural use, they, at the same time, prepare it for the growth of a new harvest of trees, when the infertility they produce shall have driven man to abandon it and suffer it to relapse into the hands of nature.†

* The experiments of Hales and others, on the absorption and exhalation of water by vegetables, are of the highest physiological interest; but observations on sunflowers, cabbages, hops, and single branches of isolated trees, growing in artificially prepared soils and under artificial conditions, furnish no trustworthy data for computing the quantity of water received and given off by the natural wood.

† In the primitive forest, except where the soil is too wet for the dense

Flow of Sap.

The amount of sap which can be withdrawn from living trees furnishes, not indeed a measure of the quantity of water sucked up by their roots from the ground—for we cannot extract from a tree its whole moisture—but numerical data which may aid the imagination to form a general notion of the powerful action of the forest as an absorbent of humidity from the earth.

The only forest tree known to Europe and North America, the sap of which is largely enough applied to economical uses to have made the amount of its flow a matter of practical importance and popular observation, is the sugar maple, *Acer saccharinum*, of the Anglo-American Provinces and States. In the course of a single "sugar season," which lasts ordinarily from twenty-five to thirty days, a sugar maple two feet in diameter will yield not less than twenty gallons of sap, and sometimes much more.* This, however, is but a trifling pro-

growth of trees, the ground is generally too thickly covered with leaves to allow much room for ground mosses. In the more open woods of Europe, this form of vegetation is more frequent—as, indeed, are many other small plants of a more inviting character—than in the native American forest. See, on the cryptogams and wood plants, ROSSMÄSSLER, *Der Wald*, pp. 33 *et seqq.*

* Emerson (*Trees of Massachusetts*, p. 493) mentions a maple six feet in diameter, as having yielded a barrel, or thirty-one and a half gallons of sap in twenty-four hours, and another, the dimensions of which are not stated, as having yielded one hundred and seventy-five gallons in the course of the season. The *Cultivator*, an American agricultural journal, for June, 1842, states that twenty gallons of sap were drawn in eighteen hours from a single maple, two and a half feet in diameter, in the town of Warner, New Hampshire, and the truth of this account has been verified by personal inquiry made in my behalf. This tree was of the original forest growth, and had been left standing when the ground around it was cleared. It was tapped only every other year, and then with six or eight incisions. Dr. Williams (*History of Vermont*, i, p. 91) says: "A man much employed in making maple sugar, found that, for twenty-one days together, a maple tree discharged seven and a half gallons per day."

An intelligent correspondent, of much experience in the manufacture

portion of the water abstracted from the earth by the roots during this season, when the yet undeveloped leaves can hardly absorb an appreciable quantity of vapor from the atmosphere; * for all this fluid runs from two or three incisions or auger holes, so narrow as to intercept the current of comparatively few sap vessels, and besides, experience shows that large as is the quantity withdrawn from the circulation, it is relatively too small to affect very sensibly the growth of the tree.† The number of large maple trees on an acre is frequently not less than fifty, ‡ and of course the quantity of moisture ab-

of maple sugar, writes me that a second-growth maple, of about two feet in diameter, standing in open ground, tapped with four incisions, has, for several seasons, generally run eight gallons per day in fair weather. He speaks of a very large tree, from which sixty gallons were drawn in the course of a season, and of another, something more than three feet through, which made forty-two pounds of wet sugar, and must have yielded not less than one hundred and fifty gallons.

* "The buds of the maple," says the same correspondent, "do not start till toward the close of the sugar season. As soon as they begin to swell, the sap seems less sweet, and the sugar made from it is of a darker color, and with less of the distinctive maple flavor."

† "In this region, maples are usually tapped with a three-quarter inch bit, boring to the depth of one and a half or two inches. In the smaller trees, one incision only is made, two in those of eighteen inches in diameter, and four in trees of larger size. Two $\frac{3}{4}$-inch holes in a tree twenty-two inches in diameter $= \frac{1}{46}$ of the circumference, and $\frac{1}{169}$ of the area of section."

"Tapping does not check the growth, but does injure the quality of the wood of maples. The wood of trees often tapped is lighter and less dense than that of trees which have not been tapped, and gives less heat in burning. No difference has been observed in the starting of the buds of tapped and untapped trees."—*Same correspondent.*

‡ Dr. Rush, in a letter to Jefferson, states the number of maples fit for tapping on an acre at from thirty to fifty. "This," observes my correspondent, "is correct with regard to the original growth, which is always more or less intermixed with other trees; but in second growth, composed of maples alone, the number greatly exceeds this. I have had the maples on a quarter of an acre, which I thought about an average of second-growth 'maple orchards,' counted. The number was found to be fifty-two, of which thirty-two were ten inches or more in diameter, and, of course, large enough to tap. This gives two hundred and eight trees

stracted from the soil by this tree alone is measured by thousands of gallons to the acre. The sugar orchards, as they are called, contain also many young maples too small for tapping, and numerous other trees—two of which, at least, the black birch, *Betula lenta*, and yellow birch, *Betula excelsa*, both very common in the same climate, are far more abundant in sap than the maple *—are scattered among the sugar trees; for the North American native forests are remarkable for the mixture of their crops.

The sap of the maple, and of other trees with deciduous leaves which grow in the same climate, flows most freely in the early spring, and especially in clear weather, when the nights are frosty and the days warm; for it is then that the melting snows supply the earth with moisture in the justest proportion, and that the absorbent power of the roots is stimulated to its highest activity.†

to the acre, one hundred and twenty-eight of which were of proper size for tapping."

According to the census returns, the quantity of maple sugar made in the United States in 1850 was 34,253,436 pounds; in 1860, it was 38,863,884 pounds, besides 1,944,594 gallons of molasses. The cane sugar made in 1850 amounted to 237,133,000 pounds; in 1859, to 302,205,000.—*Preliminary Report on the Eighth Census*, p. 88.

According to Bigelow, *Les États Unis d'Amérique en* 1863, chap. iv, the sugar product of Louisiana alone for 1862 is estimated at 528,321,500 pounds.

* The correspondent already referred to informs me that a black birch, tapped about noon with two incisions, was found the next morning to have yielded sixteen gallons. Dr. Williams (*History of Vermont*, i, p. 91) says: "A large birch, tapped in the spring, ran at the rate of five gallons an hour when first tapped. Eight or nine days after, it was found to run at the rate of about two and a half gallons an hour, and at the end of fifteen days the discharge continued in nearly the same quantity. The sap continued to flow for four or five weeks, and it was the opinion of the observers that it must have yielded as much as sixty barrels [1,890 gallons]."

† "The best state of weather for a good run," says my correspondent, "is clear days, thawing fast in the daytime and freezing well at night, with a gentle west or northwest wind; though we sometimes have clear, fine, thawing days followed by frosty nights, without a good run of sap.

When the buds are ready to burst, and the green leaves begin to show themselves beneath their scaly covering, the ground has become drier, the thirst of the roots is quenched, and the flow of sap from them to the stem is greatly diminished.*

Absorption and Exhalation of Moisture.

The leaves now commence the process of absorption, and imbibe both uncombined gases and an unascertained but perhaps considerable quantity of watery vapor from the humid atmosphere of spring which bathes them.

The organic action of the tree, as thus far described, tends to the desiccation of air and earth; but when we consider what volumes of water are daily absorbed by a large tree, and how small a proportion of the weight of this fluid consists of matter which enters into new combinations, and becomes a

I have thought it probable that the irregular flow of sap on different days in the same season is connected with the variation in atmospheric pressure; for the atmospheric conditions above mentioned as those most favorable to a free flow of sap are also those in which the barometer usually indicates pressure considerably above the mean. With a south or southeast wind, and in lowering weather, which causes a fall in the barometer, the flow generally ceases, though the sap sometimes runs till after the beginning of the storm. With a *gentle* wind, south of west, maples sometimes run all night. When this occurs, it is oftenest shortly before a storm. Last spring, the sap of a sugar orchard in a neighboring town flowed the greater part of the time for two days and two nights successively, and did not cease till after the commencement of a rain storm."

The cessation of the flow of sap at night is perhaps in part to be ascribed to the nocturnal frost, which checks the melting of the snow, of course diminishing the supply of moisture in the ground, and sometimes congeals the strata from which the rootlets suck in water. From the facts already mentioned, however, and from other well-known circumstances—such, for example, as the more liberal flow of sap from incisions on the south side of the trunk—it is evident that the withdrawal of the stimulating influences of the sun's light and heat is the principal cause of the suspension of the circulation in the night.

* "The flow ceases altogether soon after the buds begin to swell."—*Letter before quoted.*

part of the solid framework of the vegetable, or a component of its deciduous products, it is evident that the superfluous moisture must somehow be carried off almost as rapidly as it flows into the tree.* At the very commencement of vegeta-

* We might obtain a contribution to an approximate estimate of the quantity of moisture abstracted by forest vegetation from the earth and the air, by ascertaining, as nearly as possible, the quantity of wood on a given area, the proportion of assimilable matter contained in the fluids of the tree at different seasons of the year, the ages of the trees respectively, and the quantity of leaf and seed annually shed by them. The results would, indeed, be very vague, but they might serve to check or confirm estimates arrived at by other processes. The following facts are items too loose perhaps to be employed as elements in such a computation.

Dr. Williams, who wrote when the woods of Northern New England were generally in their primitive condition, states the number of trees growing on an acre at from one hundred and fifty to six hundred and fifty, according to their size and the quality of the soil; the quantity of wood, at from fifty to two hundred cords, or from 238 to 952 cubic yards, but adds that on land covered with pines, the quantity of wood would be much greater. Whether he means to give the entire solid contents of the tree, or, as is usual in ordinary estimates in New England, the marketable wood only, the trunks and larger branches, does not appear. Next to the pine, the maple would probably yield a larger amount to a given area than any of the other trees mentioned by Dr. Williams, but mixed wood, in general, measures most. In a good deal of observation on this subject, the largest quantity of marketable wood I have ever known cut on an acre of virgin forest was one hundred and four cords, or 493 cubic yards, and half that amount is considered a very fair yield. The smaller trees, branches, and twigs would not increase the quantity more than twenty-five per cent., and if we add as much more for the roots, we should have a total of about 750 cubic yards. I think Dr. Williams's estimate too large, though it would fall much below the product of the great trees of the Mississippi Valley, of Oregon, and of California. It should be observed that these measurements are those of the wood as it lies when 'corded' or piled up for market, and exceed the real solid contents by not less than fifteen per cent.

"In a soil of medium quality," says Clavé, quoting the estimates of Pfeil, for the climate of Prussia, "thé volume of a hectare of pines twenty years old, would exceed 80 cubic mètres [42½ cubic yards to the acre]; it would amount to but 24 in a meagre soil. This tree attains its maximum of mean growth at the age of seventy-five years. At that age, in the sandy earth of Prussia, it produces annually about 5 cubic mètres, with a

tion in spring, some of this fluid certainly escapes through the buds, the nascent foliage, and the pores of the bark, and vegetable physiology tells us that there is a current of sap toward the roots as well as from them.* I do not know that the exudation of water into the earth, through the bark or at the extremities of these latter organs, has been directly proved, but the other known modes of carrying off the surplus do not seem adequate to dispose of it at the almost leafless period when it is most abundantly received, and it is therefore difficult to believe that the roots do not, to some extent, drain as well as flood the watercourses of their stem. Later in the season the roots absorb less, and the now developed leaves exhale a vastly increased quantity of moisture into the air. In any event, all the water derived by the growing tree from the atmosphere and the ground is returned again by transpiration

total volume of 311 cubic mètres per hectare [166 cubic yards per acre]. After this age the volume increases, but the mean rate of growth diminishes. At eighty years, for instance, the volume is 335 cubic mètres, the annual production 4 only. The beech reaches its maximum of annual growth at one hundred and twenty years. It then has a total volume of 633 cubic mètres to the hectare [335 cubic yards to the acre], and produces 5 cubic mètres per year."—CLAVÉ, *Études*, p. 151.

These measures, I believe, include the entire ligneous product of the tree, exclusive of the roots, and express the actual solid contents. The specific gravity of maple wood is stated to be 75. Maple sap yields sugar at the rate of about one pound *wet* sugar to three gallons of sap, and wet sugar is to dry sugar in about the proportion of nineteen to sixteen. Besides the sugar, there is a small residuum of "sand," composed of phosphate of lime, with a little silex, and it is certain that by the ordinary hasty process of manufacture, a good deal of sugar is lost; for the drops, condensed from the vapor of the boilers on the rafters of the rude sheds where the sap is boiled, have a decidedly sweet taste.

* "The elaborated sap, passing out of the leaves, is received into the inner bark, * * * and a part of what descends finds its way even to the ends of the roots, and is all along diffused laterally into the stem, where it meets and mingles with the ascending crude sap or raw material. So there is no separate circulation of the two kinds of sap; and no crude sap exists separately in any part of the plant. Even in the root, where it enters, this mingles at once with some elaborated sap already there."—GRAY, *How Plants Grow*, § 273.

or exudation, after having surrendered to the plant the small proportion of matter required for vegetable growth which it held in solution or suspension.* The hygrometrical equilibrium is then restored, so far as this: the tree yields up again the moisture it had drawn from the earth and the air, though it does not return it each to each; for the vapor carried off by transpiration greatly exceeds the quantity of water absorbed by the foliage from the atmosphere, and the amount, if any, carried back to the ground by the roots.

The evaporation of the juices of the plant, by whatever process effected, takes up atmospheric heat and produces refrigeration. This effect is not less real, though much less sensible, in the forest than in meadow or pasture land, and it cannot be doubted that the local temperature is considerably affected by it. But the evaporation that cools the air diffuses through it, at the same time, a medium which powerfully resists the escape of heat from the earth by radiation. Visible vapors or clouds, it is well known, prevent frosts by obstruct-

* Ward's tight glazed cases for raising, and especially for transporting plants, go far to prove that water only circulates through vegetables, and is again and again absorbed and transpired by organs appropriated to these functions. Seeds, growing grasses, shrubs, or trees planted in proper earth, moderately watered and covered with a glass bell or close frame of glass, live for months and even years, with only the original store of air and water. In one of Ward's early experiments, a spire of grass and a fern, which sprang up in a corked bottle containing a little moist earth introduced as a bed for a snail, lived and flourished for eighteen years without a new supply of either fluid. In these boxes the plants grow till the enclosed air is exhausted of the gaseous constituents of vegetation, and till the water has yielded up the assimilable matter it held in solution, and dissolved and supplied to the roots the nutriment contained in the earth in which they are planted. After this, they continue for a long time in a state of vegetable sleep, but if fresh air and water be introduced into the cases, or the plants be transplanted into open ground, they rouse themselves to renewed life, and grow vigorously, without appearing to have suffered from their long imprisonment. The water transpired by the leaves is partly absorbed by the earth directly from the air, partly condensed on the glass, along which it trickles down to the earth, enters the roots again, and thus continually repeats the circuit. See *Aus der Natur*, 21, B. S. 537.

ing radiation, or rather by reflecting back again the heat radiated by the earth, just as any mechanical screen would do. On the other hand, clouds intercept the rays of the sun also, and hinder its heat from reaching the earth. The invisible vapors given out by leaves impede the passage of heat reflected and radiated by the earth and by all terrestrial objects, but oppose much less resistance to the transmission of direct solar heat, and indeed the beams of the sun seem more scorching when received through clear air charged with uncondensed moisture than after passing through a dry atmosphere. Hence the reduction of temperature by the evaporation of moisture from vegetation, though sensible, is less than it would be if water in the gaseous state were as impervious to heat given out by the sun as to that emitted by terrestrial objects.

The hygroscopicity of vegetable mould is much greater than that of any mineral earth, and therefore the soil of the forest absorbs more atmospheric moisture than the open ground. The condensation of the vapor by absorption disengages heat, and consequently raises the temperature of the soil which absorbs it. Von Babo found the temperature of sandy earth thus elevated from 20° to 27° centigrade, making a difference of nearly thirteen degrees of Fahrenheit, and that of soil rich in humus from 20° to 31° centigrade, a difference of almost twenty degrees of Fahrenheit.*

Balance of Conflicting Influences.

We have shown that the forest, considered as dead matter, tends to diminish the moisture of the air, by preventing the sun's rays from reaching the ground and evaporating the

* WILHELM, *Der Boden und das Wasser*, p. 18. It is not ascertained in what proportions the dew is evaporated, and in what it is absorbed by the earth, in actual nature, but there can be no doubt that the amount of water taken up by the ground, both from vapor suspended in the air and from dew, is large. The annual fall of dew in England is estimated at five inches, but this quantity is much exceeded in many countries with a clearer sky. "In many of our Algerian campaigns," says Babinet, "when

water that falls upon the surface, and also by spreading over the earth a spongy mantle which sucks up and retains the humidity it receives from the atmosphere, while, at the same time, this covering acts in the contrary direction by accumulating, in a reservoir not wholly inaccessible to vaporizing influences, the water of precipitation which might otherwise suddenly sink deep into the bowels of the earth, or flow by superficial channels to other climatic regions. We now see that, as a living organism, it tends, on the one hand, to diminish the humidity of the air by absorbing moisture from it, and, on the other, to increase that humidity by pouring out into the atmosphere, in a vaporous form, the water it draws up through its roots. This last operation, at the same time, lowers the temperature of the air in contact with or proximity to the wood, by the same law as in other cases of the conversion of water into vapor.

As I have repeatedly said, we cannot measure the value of any one of these elements of climatic disturbance, raising or lowering of temperature, increase or diminution of humidity, nor can we say that in any one season, any one year, or any one fixed cycle, however long or short, they balance and compensate each other. They are sometimes, but certainly not always, contemporaneous in their action, whether their tendency is in the same or in opposite directions, and, therefore, their influence is sometimes cumulative, sometimes conflicting; but, upon the whole, their general effect seems to be to mitigate extremes of atmospheric heat and cold, moisture and drought. They serve as equalizers of temperature and humidity, and it is highly probable that, in analogy with most other works and workings of nature, they, at certain or uncertain periods, restore the equilibrium which, whether as lifeless masses or as living organisms, they may have temporarily disturbed.

it was wished to punish the brigandage of the unsubdued tribes, it was impossible to set their grain fields on fire until a late hour of the day; for the plants were so wet with the night dew that it was necessary to wait until the sun had dried them."—*Études et Lectures*, ii, p. 212.

When, therefore, man destroyed these natural harmonizers of climatic discords, he sacrificed an important conservative power, though it is far from certain that he has thereby affected the mean, however much he may have exaggerated the extremes of atmospheric temperature and humidity, or, in other words, may have increased the range and lengthened the scale of thermometric and hygrometric variation.

Influence of the Forest on Temperature and Precipitation.

Aside from the question of compensation, it does not seem probable that the forests sensibly affect the total quantity of precipitation, or the general mean of atmospheric temperature of the globe, or even that they had this influence when their extent was vastly greater than at present. The waters cover about three fourths of the face of the earth,* and if we deduct the frozen zones, the peaks and crests of lofty mountains and

* "It has been concluded that the dry land occupies about 49,800,000 square statute miles. This does not include the recently discovered tracts of land in the vicinity of the poles, and allowing for yet undiscovered land (which, however, can only exist in small quantity), if we assign 51,000,000 to the land, there will remain about 146,000,000 of square miles for the extent of surface occupied by the ocean."—Sir J. F. W. HERSCHEL, *Physical Geography*, 1861, p. 19.

It does not appear to which category Herschel assigns the inland seas and the fresh-water lakes and rivers of the earth; and Mrs. Somerville, who states that the "dry land occupies an area of 38,000,000 of square miles," and that "the ocean covers nearly three fourths of the surface of the globe," is equally silent on this point.—*Physical Geography*, fifth edition, p. 30. On the following page, Mrs. Somerville, in a note, cites Mr. Gardner as her authority, and says that, "according to his computation, the extent of land is about 37,673,000 square British miles, independently of Victoria Continent; and the sea occupies 110,849,000. Hence the land is to the sea as 1 to 4 nearly." Sir John F. W. Herschel makes the area of dry land and ocean together 197,000,000 square miles; Mrs. Somerville, or rather Mr. Gardner, 148,522,000. I suppose Sir John Herschel includes the islands in his aggregate of the "dry land," and the inland waters under the general designation of the "ocean," and that Mrs. Somerville excludes both.

their craggy slopes, the Sahara and other great African and Asiatic deserts, and all such other portions of the solid surface as are permanently unfit for the growth of wood, we shall find that probably not one tenth of the total superficies of our planet was ever, at any one time in the present geological epoch, covered with forests. Besides this, the distribution of forest land, of desert, and of water, is such as to reduce the possible influence of the former to a low expression; for the forests are, in large proportion, situated in cold or temperate climates, where the action of the sun is comparatively feeble both in elevating temperature and in promoting evaporation; while, in the torrid zone, the desert and the sea—the latter of which always presents an evaporable surface—enormously preponderate. It is, upon the whole, not probable that so small an extent of forest, so situated, could produce an appreciable influence on the *general* climate of the globe, though it might appreciably affect the local action of all climatic elements. The total annual amount of solar heat absorbed and radiated by the earth, and the sum of terrestrial evaporation and atmospheric precipitation must be supposed constant; but the distribution of heat and of humidity is exposed to disturbance in both time and place, by a multitude of local causes, among which the presence or absence of the forest is doubtless one.

So far as we are able to sum up the general results, it would appear that, in countries in the temperate zone still chiefly covered with wood, the summers would be cooler, moister, shorter, the winters milder, drier, longer, than in the same regions after the removal of the forest. The slender historical evidence we possess seems to point to the same conclusion, though there is some conflict of testimony and of opinion on this point, and some apparently well-established exceptions to particular branches of what appears to be the general law.

One of these occurs both in climates where the cold of winter is severe enough to freeze the ground to a considerable depth, as in Sweden and the Northern States of the American Union, and in milder zones, where the face of the earth is exposed to cold mountain winds, as in some parts of Italy and of France; for there, as we have seen, the winter is believed to extend itself into the months which belong to the spring, later than at periods when the forest covered the greater part of the ground.* More causes than one doubtless contribute to this result; but in the case of Sweden and the United States, the most obvious explanation of the fact is to be found in the loss of the shelter afforded to the ground by the thick coating of leaves which the forest sheds upon it, and the snow which the woods protect from blowing away, or from melting in the brief thaws of winter. I have already remarked that bare ground freezes much deeper than that which is covered by beds of leaves, and when the earth is thickly coated with snow, the strata frozen before it fell begin to thaw. It is not uncommon to find the ground in the woods, where the snow lies two or three feet deep, entirely free from frost, when the atmospheric temperature has been for several weeks below the freezing point, and for some days even below the zero of Fahrenheit. When the ground is cleared and brought under cultivation, the leaves are ploughed into the soil and decomposed, and the snow, especially upon knolls and eminences, is blown

* It has been observed in Sweden that the spring, in many districts where the forests have been cleared off, now comes on a fortnight later than in the last century.—ASBJÖRNSEN, *Om Skovene i Norge*, p. 101.

The conclusion arrived at by Noah Webster, in his very learned and able paper on the supposed change in the temperature of winter, read before the Connecticut Academy of Arts and Sciences in 1799, was as follows: "From a careful comparison of these facts, it appears that the weather, in modern winters, in the United States, is more inconstant than when the earth was covered with woods, at the first settlement of Europeans in the country; that the warm weather of autumn extends further into the winter months, and the cold weather of winter and spring encroaches upon the summer; that, the wind being more variable, snow is less permanent, and perhaps the same remark may be applicable to the ice of the rivers. These effects seem to result necessarily from the greater quantity of heat accumulated in the earth in summer since the ground has been cleared of wood and exposed to the rays of the sun, and to the greater depth of frost in the earth in winter by the exposure of its uncovered surface to the cold atmosphere."—*Collection of Papers by* NOAH WEBSTER, p. 162.

off, or perhaps half thawed, several times during the winter. The water from the melting snow runs into the depressions, and when, after a day or two of warm sunshine or tepid rain, the cold returns, it is consolidated to ice, and the bared ridges and swells of earth are deeply frozen.* It requires many days of mild weather to raise the temperature of soil in this condition, and of the air in contact with it, to that of the earth in the forests of the same climatic region. Flora is already plaiting her sylvan wreath before the corn flowers which are to deck the garland of Ceres have waked from their winter's sleep; and it is not a popular error to believe that, where man has substituted his artificial crops for the spontaneous harvest of nature, spring delays her coming.

In many cases, the apparent change in the period of the seasons is a purely local phenomenon, which is probably compensated by a higher temperature in other months, without any real disturbance of the average thermometrical equilibrium. We may easily suppose that there are analogous partial deviations from the general law of precipitation; and, without insisting that the removal of the forest has diminished the sum total of snow and rain, we may well admit that it has lessened the quantity which annually falls within particular limits. Various theoretical considerations make this probable, the most obvious argument, perhaps, being that drawn from the generally admitted fact, that the summer and even the mean temperature of the forest is below that of the open country in the same latitude. If the air in a wood is cooler than that around it, it must reduce the temperature of the atmospheric stratum immediately above it, and, of course, whenever a saturated current sweeps over it, it must produce precipitation which would fall upon or near it.

But the subject is so exceedingly complex and difficult,

* I have seen, in Northern New England, the surface of the open ground frozen to the depth of twenty-two inches, in the month of November, when in the forest earth no frost was discoverable; and later in the winter, I have known an exposed sand knoll to remain frozen six feet deep, after the ground in the woods was completely thawed.

that it is safer to regard it as a historical problem, or at least as what lawyers call a mixed question of law and fact, than to attempt to decide it upon *à priori* grounds. Unfortunately the evidence is conflicting in tendency, and sometimes equivocal in interpretation, but I believe that a majority of the foresters and physicists who have studied the question are of opinion that in many, if not in all cases, the destruction of the woods has been followed by a diminution in the annual quantity of rain and dew. Indeed, it has long been a popularly settled belief that vegetation and the condensation and fall of atmospheric moisture are reciprocally necessary to each other, and even the poets sing of

> Afric's barren sand,
> Where nought can grow, because it raineth not,
> And where no rain can fall to bless the land,
> Because nought grows there.*

Before stating the evidence on the general question and citing the judgments of the learned upon it, however, it is well to remark that the comparative variety or frequency of inundations in earlier and later centuries is not necessarily, in most cases not probably, entitled to any weight whatever, as a proof that more or less rain fell formerly than now; because the accumulation of water in the channel of a river depends far less upon the quantity of precipitation in its valley, than upon the rapidity with which it is conducted, on or under the surface of the ground, to the central artery that drains the basin. But this point will be more fully discussed in a subsequent chapter.

There is another important observation which may properly be introduced here. It is not universally, or even generally true, that the atmosphere returns its humidity to the local

> * ——Det golde Strög i Afrika,
> Der Intet voxe kan, da ei det regner,
> Og, omvendt, ingen Regn kan falde, da
> Der Intet voxer.
>
> PALUDAN-MÜLLER, *Adam Homo*, ii, 408.

source from which it receives it. The air is constantly in motion,

> ——howling tempests scour amain
> From sea to land, from land to sea; *

and, therefore, it is always probable that the evaporation drawn up by the atmosphere from a given river, or sea, or forest, or meadow, will be discharged by precipitation, not at or near the point where it rose, but at a distance of miles, leagues, or even degrees. The currents of the upper air are invisible, and they leave behind them no landmark to record their track. We know not whence they come, or whither they go. We have a certain rapidly increasing acquaintance with the laws of general atmospheric motion, but of the origin and limits, the beginning and end of that motion, as it manifests itself at any particular time and place, we know nothing. We cannot say where or when the vapor, exhaled to-day from the lake on which we float, will be condensed and fall; whether it will waste itself on a barren desert, refresh upland pastures, descend in snow on Alpine heights, or contribute to swell a distant torrent which shall lay waste square miles of fertile corn land; nor do we know whether the rain which feeds our brooklets is due to the transpiration from a neighboring forest, or to the evaporation from a far-off sea. If, therefore, it were proved that the annual quantity of rain and dew is now as great on the plains of Castile, for example, as it was when they were covered with the native forest, it would by no means follow that those woods did not augment the amount of precipitation elsewhere.

But I return to the question. Beginning with the latest authorities, I cite a passage from Clavé.† After arguing that we cannot reason from the climatic effects of the forest in tropical and sub-tropical countries as to its influence in temperate

* Und Stürme brausen um die Wette
Vom Meer aufs Land, vom Land aufs Meer.
GOETHE, *Faust*, *Song of the Archangels*.

† *Études sur l'Économie Forestière*, pp. 45, 46.

latitudes, the author proceeds: "The action of the forests on rain, a consequence of that which they exercise on temperature, is difficult to estimate in our climate, but is very pronounced in hot countries, and is established by numerous examples. M. Boussingault states that in the region comprised between the Bay of Cupica and the Gulf of Guayaquil, which is covered with immense forests, the rains are almost continual, and that the mean temperature of this humid country rises hardly to twenty-six degrees (= 80° Fahr.). M. Blanqui, in his 'Travels in Bulgaria,' informs us that at Malta rain has become so rare, since the woods were cleared to make room for the growth of cotton, that at the time of his visit in October, 1841, not a drop of rain had fallen for three years.* The terrible droughts which desolate the Cape Verd Islands must also be attributed to the destruction of the forests. In the Island of St. Helena, where the wooded surface has considerably extended within a few years, it has been observed that the rain has increased in the same proportion. It is now in quantity double what it was during the residence of Napoleon. In Egypt, recent plantations have caused rains, which hitherto were almost unknown."

Schacht observes: "In wooded countries, the atmosphere is generally humid, and rain and dew fertilize the soil. As the lightning rod abstracts the electric fluid from the stormy sky, so the forest attracts to itself the rain from the clouds, which, in falling, refreshes not it alone, but extends its benefits

* I am not aware of any evidence to show that Malta had any woods of importance at any time since the cultivation of cotton was introduced there; and if it is true, as has been often asserted, that its present soil was imported from Sicily, it can certainly have possessed no forests since a very remote period. In Sandys's time, 1611, there were no woods in the island, and it produced little cotton. He describes it as "a country altogether champion, being no other than a rocke couered ouer with earth, but two feete deepe where the deepest; hauing but few trees but such as beare fruite. * * * So that their wood they haue from Sicilia." They have "an indifferent quantity of cotton wooll, but that the best of all other."—SANDYS, *Travels*, p. 228.

† SCHACHT, *Les Arbres*, p. 412.

to the neighboring fields. * * The forest, presenting a considerable surface for evaporation, gives to its own soil and to all the adjacent ground an abundant and enlivening dew. There falls, it is true, less dew on a tall and thick wood than on the surrounding meadows, which, being more highly heated during the day by the influence of insolation, cool with greater rapidity by radiation. But it must be remarked, that this increased deposition of dew on the neighboring fields is partly due to the forests themselves; for the dense, saturated strata of air which hover over the woods descend in cool, calm evenings, like clouds, to the valley, and in the morning, beads of dew sparkle on the leaves of the grass and the flowers of the field. Forests, in a word, exert, in the interior of continents, an influence like that of the sea on the climate of islands and of coasts: both water the soil and thereby insure its fertility." In a note upon this passage, quoting as authority the *Historia de la Conquista de las siete islas de Gran Canaria, de Juan de Abreu Galindo*, 1632, p. 47, he adds: "Old historians relate that a celebrated laurel in Ferro formerly furnished drinkable water to the inhabitants of the island. The water flowed from its foliage, uninterruptedly, drop by drop, and was collected in cisterns. Every morning the sea breeze drove a cloud toward the wonderful tree, which attracted it to its huge top," where it was condensed to a liquid form.

In a number of the *Missionary Herald*, published at Boston, the date of which I have mislaid, the Rev. Mr. Van Lennep, well known as a competent observer, gives the following remarkable account of a similar fact witnessed by him in an excursion to the east of Tocat in Asia Minor:

"In this region, some 3,000 feet above the sea, the trees are mostly oak, and attain a large size. I noticed an illustration of the influence of trees in general in collecting moisture. Despite the fog, of a week's duration, the ground was everywhere perfectly dry. The dry oak leaves, however, had gathered the water, and the branches and trunks of the trees were more or less wet. In many cases the water had run down the trunk and moistened the soil around the roots of the tree. In two places, several trees had each furnished a small stream of water, and these, uniting, had run upon the road, so that travellers had to pass through the mud; although, as I said, everywhere else the ground was perfectly dry. Moreover, the collected moisture was not sufficient to drop directly from the leaves, but in every case it ran down the branches and trunk to the ground. Farther on we found a grove, and at the foot of each tree, on the north side, was a lump of ice, the water having frozen as it reached the ground. This is a most striking illustration of the acknowledged influence of trees in collecting moisture; and one cannot for a moment doubt, that the parched regions which commence at Sivas, and extend in one direction to the Persian Gulf, and in another to the Red Sea, were once a fertile garden, teeming with a prosperous population, before the forests which covered the hillsides were cut down—before the cedar and the fir tree were rooted up from the sides of Lebanon.

"As we now descended the northern side of the watershed, we passed through the grove of walnut, oak, and black mulberry trees, which shade the village of Oktab, whose houses, cattle, and ruddy children were indicative of prosperity."

Coultas thus argues: "The ocean, winds, and woods may be regarded as the several parts of a grand distillatory apparatus. The sea is the boiler in which vapor is raised by the solar heat, the winds are the guiding tubes which carry the vapor with them to the forests where a lower temperature prevails. This naturally condenses the vapor, and showers of rain are thus distilled from the cloud masses which float in the atmosphere, by the woods beneath them." *

Sir John F. W. Herschel enumerates among "the influences unfavorable to rain," "absence of vegetation in warm climates, and especially of trees. This is, no doubt," continues he, "one of the reasons of the extreme aridity of Spain. The hatred of a Spaniard toward a tree is proverbial. Many districts in France have been materially injured by denudation (Earl of

* *What may be learned from a Tree*, p. 117.

Lovelace on Climate, etc.), and, on the other hand, rain has become more frequent in Egypt since the more vigorous cultivation of the palm tree."

Hohenstein remarks: "With respect to the temperature in the forest, I have already observed that, at certain times of the day and of the year, it is less than in the open field. Hence the woods may, in the daytime, in summer and toward the end of winter, tend to increase the fall of rain; but it is otherwise in summer nights and at the beginning of winter, when there is a higher temperature in the forest, which is not favorable to that effect. * * * The wood is, further, like the mountain, a mechanical obstruction to the motion of rain clouds, and, as it checks them in their course, it gives them occasion to deposit their water. These considerations render it probable that the forest increases the quantity of rain; but they do not establish the certainty of this conclusion, because we have no positive numerical data to produce on the depression of temperature, and the humidity of the air in the woods." *

Barth presents the following view of the subject: "The ground in the forest, as well as the atmospheric stratum over it, continues humid after the woodless districts have lost their moisture; and the air, charged with the humidity drawn from them, is usually carried away by the winds before it has deposited itself in a condensed form on the earth. Trees constantly transpire through their leaves a great quantity of moisture, which they partly absorb again by the same organs, while the greatest part of their supply is pumped up through their widely ramifying roots from considerable depths in the ground. Thus a constant evaporation is produced, which keeps the forest atmosphere moist even in long droughts, when all other sources of humidity in the forest itself are dried up. * * * Little is required to compel the stratum of air resting upon a wood to give up its moisture, which thus, as rain, fog, or dew, is returned to the forest. * * * The warm, moist currents

* *Der Wald*, p. 13.

of air which come from other regions are cooled as they approach the wood by its less heated atmosphere, and obliged to let fall the humidity with which they are charged. The woods contribute to the same effect by mechanically impeding the motion of fog and rain cloud, whose particles are thus accumulated and condensed to rain. The forest thus has a greater power than the open ground to retain within its own limits already existing humidity, and to preserve it, and it attracts and collects that which the wind brings it from elsewhere, and forces it to deposit itself as rain or other precipitation. * * * In consequence of these relations of the forest to humidity, it follows that wooded districts have both more frequent and more abundant rain, and in general are more humid, than woodless regions; for what is true of the woods themselves, in this respect, is true also of their treeless neighborhood, which, in consequence of the ready mobility of the air and its constant changes, receives a share of the characteristics of the forest atmosphere, coolness and moisture. * * * When the districts stripped of trees have long been deprived of rain and dew, * * * and the grass and the fruits of the field are ready to wither, the grounds which are surrounded by woods are green and flourishing. By night they are refreshed with dew, which is never wanting in the moist air of the forest, and in due season they are watered by a beneficent shower, or a mist which rolls slowly over them." *

Asbjörnsen, after adducing the familiar theoretical arguments on this point, adds: "The rainless territories in Peru and North Africa establish this conclusion, and numerous other examples show that woods exert an influence in producing rain, and that rain fails where they are wanting; for many countries have, by the destruction of the forests, been deprived of rain, moisture, springs, and watercourses, which are necessary for vegetable growth. * * * The narratives of travellers show the deplorable consequences of felling the woods in the Island of Trinidad, Martinique, San Do-

* *Om Skovene og deres Forhold til Nationalœconomien*, pp. 131–133.

mingo, and indeed, in almost the entire West Indian group * * * In Palestine and many other parts of Asia and Northern Africa, which in ancient times were the granaries of Europe, fertile and populous, similar consequences have been experienced. These lands are now deserts, and it is the destruction of the forests alone which has produced this desolation. * * * In Southern France, many districts have, from the same cause, become barren wastes of stone, and the cultivation of the vine and the olive has suffered severely since the baring of the neighboring mountains. Since the extensive clearings between the Spree and the Oder, the inhabitants complain that the clover crop is much less productive than before. On the other hand, examples of the beneficial influence of planting and restoring the woods are not wanting. In Scotland, where many miles square have been planted with trees, this effect has been manifest, and similar observations have been made in several places in Southern France. In Lower Egypt, both at Cairo and near Alexandria, rain rarely fell in considerable quantity—for example, during the French occupation of Egypt, about 1798, it did not rain for sixteen months—but since Mehemet Aali and Ibrahim Pacha executed their vast plantations (the former alone having planted more than twenty millions of olive and fig trees, cottonwood, oranges, acacias, planes, &c.), there now falls a good deal of rain, especially along the coast, in the months of November, December, and January; and even at Cairo it rains both oftener and more abundantly, so that real showers are no rarity." *

Babinet, in one of his lectures,† cites the supposed fact of the increase of rain in Egypt in consequence of the planting of trees, and thus remarks upon it: "A few years ago it never rained in Lower Egypt. The constant north winds, which almost exclusively prevail there, passed without obstruction over a surface bare of vegetation. Grain was kept on

* *Om Skovene og om et ordnet Skovbrug i Norge*, p. 106.

† *Études et Lectures*, iv. p. 114.

the roofs in Alexandria, without being covered or otherwise protected from injury by the atmosphere; but since the making of plantations, an obstacle has been created which retards the current of air from the north. The air thus checked, accumulates, dilates, cools, and yields rain.* The forests of the

* The supposed increase in the frequency and quantity of rain in Lower Egypt is by no means established. I have heard it disputed on the spot by intelligent Franks, whose residence in that country began before the plantations of Mehemet Aali and Ibrahim Pacha, and I have been assured by them that meteorological observations, made at Alexandria about the beginning of this century, show an annual fall of rain as great as is usual at this day. The mere fact, that it did not rain during the French occupation, is not conclusive. Having experienced a gentle shower of nearly twenty-four hours' duration in Upper Egypt, I inquired of the local governor in relation to the frequency of this phenomenon, and was told by him that not a drop of rain had fallen at that point for more than two years previous.

The belief in the increase of rain in Egypt rests almost entirely on the observations of Marshal Marmont, and the evidence collected by him in 1836. His conclusions have been disputed, if not confuted, by Jomard and others, and are probably erroneous. See, FOISSAC, *Météorologie*, German translation, pp. 634–639.

It certainly sometimes rains briskly at Cairo, but evaporation is exceedingly rapid in Egypt—as any one, who ever saw a Fellah woman wash a napkin in the Nile, and dry it by shaking it a few moments in the air, can testify; and a heap of grain, wet a few inches below the surface, would probably dry again without injury. At any rate, the Egyptian Government often has vast quantities of wheat stored at Boulak, in uncovered yards through the winter, though it must be admitted that the slovenliness and want of foresight in Oriental life, public and private, are such that we cannot infer the safety of any practice followed in the East, merely from its long continuance.

Grain, however, may be long kept in the open air in climates much less dry than that of Egypt, without injury, except to the superficial layers; for moisture does not penetrate to a great depth in a heap of grain once well dried, and kept well aired. When Louis IX was making his preparations for his campaign in the East, he had large quantities of wine and grain purchased in the Island of Cyprus, and stored up, for two years, to await his arrival. "When we were come to Cyprus," says Joinville, *Histoire de Saint Louis*, §§ 72, 73, "we found there greate foison of the Kynge's purveyance. * * The wheate and the barley they had piled

Vosges and Ardennes produce the same effects in the north east of France, and send us a great river, the Meuse, which is as remarkable for its volume as for the small extent of its basin. With respect to the retardation of the atmospheric currents, and the effects of that retardation, one of my illustrious colleagues, M. Mignet, who is not less a profound thinker than an eloquent writer, suggested to me that, to produce rain, a forest was as good as a mountain, and this is literally true."

Monestier-Savignat arrives at this conclusion: "Forests on the one hand diminish evaporation; on the other, they act on the atmosphere as refrigerating causes. The second scale of the balance predominates over the other, for it is established that in wooded countries it rains oftener, and that, the quantity of rain being equal, they are more humid." *

Boussingault—whose observations on the drying up of lakes and springs, from the destruction of the woods, in tropical America, have often been cited as a conclusive proof that the quantity of rain was thereby diminished—after examining the question with much care, remarks: "In my judgment it is settled that very large clearings must diminish the annual fall of rain in a country;" and on a subsequent page, he concludes that, "arguing from meteorological facts collected in the equinoctial regions, there is reason to presume that clearings diminish the annual fall of rain." †

The same eminent author proposes series of observations on the level of natural lakes, especially on those without outlet, as a means of determining the increase or diminution of precipitation in their basins, and, of course, of measuring the

up in greate heapes in the feeldes, and to looke vpon, they were like vnto mountaynes; for the raine, the whyche hadde beaten vpon the wheate now a longe whyle, had made it to sproute on the toppe, so that it seemed as greene grasse. And whanne they were mynded to carrie it to Egypte, they brake that sod of greene herbe, and dyd finde under the same the wheate and the barley, as freshe as yf menne hadde but nowe thrashed it."

* *Étude sur les Eaux au point de vue des Inondations*, p. 91.

† *Économie Rurale*, ii, chap. xx, § 4, pp. 756–759. See also p. 733.

effect of clearing when such operations take place within those basins. But it must be observed that lakes without a visible outlet are of very rare occurrence, and besides, where no superficial conduit for the discharge of lacustrine waters exists, we can seldom or never be sure that nature has not provided subterranean channels for their escape. Indeed, when we consider that most earths, and even some rocks under great hydrostatic pressure, are freely permeable by water, and that fissures are frequent in almost all rocky strata, it is evident that we cannot know in what proportion the depression of the level of a lake is to be ascribed to infiltration, to percolation, or to evaporation.* Further, we are, in general, as little able to affirm that a given lake derives all its water from the fall of rain within its geographical basin, or that it receives all the water that falls in that basin except what evaporates from the ground, as we are to show that all its superfluous water is carried off by visible channels and by evaporation.

Suppose the strata of the mountains on two sides of a lake, east and west, to be tilted in the same direction, and that those of the hill on the east side incline toward the lake, those of that on the west side from it. In this case a large proportion of the rain which falls on the eastern slope of the eastern hill may find its way between the strata to the lake, and an equally large proportion of the precipitation upon the eastern slope of the western ridge may escape out of the basin by similar channels. In such case the clearing of the *outer* slopes of either or both mountains, while the forests of the *inner* declivities remained intact, might affect the quantity of water received by the lake, and it would always be impossible to know to what territorial extent influences thus affecting the level of a lake

* Jacini, speaking of the great Italian lakes, says: "A large proportion of the water of the lakes, instead of discharging itself by the Ticino, the Adda, the Oglio, the Mincio, filters through the silicious strata which underlie the hills, and follows subterranean channels to the plain, where it collects in the *fontanili*, and being thence conducted into the canals of irrigation, becomes a source of great fertility."—*La Proprietà Fondiaria, etc.*, p. 144.

might reach. Boussingault admits that extensive clearing *below* an alpine lake, even at a considerable distance, might affect the level of its waters. How it would produce this influence he does not inform us, but, as he says nothing of the natural subterranean drainage of surface waters, it is to be presumed that he refers to the supposed diminution of the quantity of rain from the removal of the forest, which might manifest itself at a point more elevated than the cause which occasioned it. The elevation or depression of the level of natural lakes, then, cannot be relied upon as a proof, still less as a measure of an increase or diminution in the fall of rain within their geographical basins, resulting from the felling of the woods which covered them; though such phenomena afford very strong presumptive evidence that the supply of water is somehow augmented or lessened. The supply is, in most cases, derived much less from the precipitation which falls directly upon the surface of lakes, than from waters which flow above or under the ground around them, and which, in the latter case, often come from districts not comprised within what superficial geography would regard as belonging to the lake basins.

It is, upon the whole, evident that the question can hardly be determined except by the comparison of pluviometrical observations made at a given station before and after the destruction of the woods. Such observations, unhappily, are scarcely to be found, and the opportunity for making them is rapidly passing away, except so far as a converse series might be collected in countries—France, for example—where forest plantation is now going on upon a large scale. The Smithsonian Institution at Washington is well situated for directing the attention of observers in the newer territory of the United States to this subject, and it is to be hoped that it will not fail to avail itself of its facilities for this purpose.

Numerous other authorities might be cited in support of the proposition that forests tend, at least in certain latitudes and at certain seasons, to produce rain; but though the arguments of the advocates of this doctrine are very plausible, not to say convincing, their opinions are rather *à priori* conclusions from general meteorological laws, than deductions from facts of observation, and it is remarkable that there is so little direct evidence on the subject.

On the other hand, Foissac expresses the opinion that forests have no influence on precipitation, beyond that of promoting the deposit of dew in their vicinity, and he states, as a fact of experience, that the planting of large vegetables, and especially of trees, is a very efficient means of drying morasses, because the plants draw from the earth a quantity of water larger than the average annual fall of rain.* Klöden, admitting that the rivers Oder and Elbe have diminished in quantity of water, the former since 1778, the latter since 1828, denies that the diminution of volume is to be ascribed to a decrease of precipitation in consequence of the felling of the forests, and states, what other physicists confirm, that, during the same period, meteorological records in various parts of Europe show rather an augmentation than a reduction of rain.†

The observations of Belgrand tend to show, contrary to the general opinion, that less rain falls in wooded than in denuded districts. He compared the precipitation for the year 1852, at Vezelay in the valley of the Bouchat, and at Avallon in the valley of the Grenetière. At the first of these places it was 881 millimètres, at the latter 581 millimètres. The two cities are not more than eight miles apart. They are at the same altitude, and it is stated that the only difference in their geographical conditions consists in the different proportions of forest and cultivated country around them, the basin of the Bouchat being entirely bare, while that of the Grenetière is well wooded.‡ Observations in the same valleys, considered

* *Météorologie*, German translation by EMSMANN, p. 605.

† *Handbuch der Physischen Geographie*, p. 658.

‡ *Annales des Ponts et Chaussées*, 1854, 1st sémestre, pp. 21 *et seqq.* See the comments of VALLÈS on these observations, in his *Études sur les Inondations*, pp. 441 *et seqq.*

with reference to the seasons, show the following pluviometric results:

FOR LA GRENETIÈRE.

February,	1852,	42.2	millimètres precipitation.
November,	"	28.8	" "
January,	1853,	35.4	" "
Total,		106.4	in three cold months.
September,	1851,	27.1	millimètres precipitation.
May,	1852,	20.9	" "
June,	"	56.3	" "
July,	"	22.8	" "
September,	"	22.8	" "
Total,		149.9	in five warm months.

FOR LE BOUCHAT.

February,	1852,	51.3	millimètres precipitation.
November,	"	36.6	" "
January,	1853,	92.0	" "
Total,		179.9	in three cold months.
September,	1851,	43.8	millimètres precipitation.
May,	1852,	13.2	" "
June,	"	55.5	" "
July,	"	19.5	" "
September,	"	26.5	" "
Total,		158.5	in five warm months.

These observations, so far as they go, seem to show that more rain falls in cleared than in wooded countries, but this result is so contrary to what has been generally accepted as a theoretical conclusion, that further experiment is required to determine the question.

Becquerel—whose treatise on the climatic effects of the destruction of the forest is the fullest general discussion of that subject known to me—does not examine this particular point, and as, in the summary of the results of his investigations, he does not ascribe to the forest any influence upon precipitation, the presumption is that he rejects the doctrine of its importance as an agent in producing the fall of rain.

The effect of the forest on precipitation, then, is not entirely free from doubt, and we cannot positively affirm that the total annual quantity of rain is diminished or increased by the destruction of the woods, though both theoretical considerations and the balance of testimony strongly favor the opinion that more rain falls in wooded than in open countries. One important conclusion, at least, upon the meteorological influence of forests is certain and undisputed: the proposition, namely, that, within their own limits, and near their own borders, they maintain a more uniform degree of humidity in the atmosphere than is observed in cleared grounds. Scarcely less can it be questioned that they promote the frequency of showers, and, if they do not augment the amount of precipitation, they equalize its distribution through the different seasons.

Influence of the Forest on the Humidity of the Soil.

I have hitherto confined myself to the influence of the forest on meteorological conditions, a subject, as has been seen, full of difficulty and uncertainty. Its comparative effects on the temperature, the humidity, the texture and consistence, the configuration and distribution of the mould or arable soil, and, very often, of the mineral strata below, and on the permanence and regularity of springs and greater superficial watercourses, are much less disputable as well as more easily estimated, and much more important, than its possible value as a cause of strictly climatic equilibrium or disturbance.

The action of the forest on the earth is chiefly mechanical, but the organic process of abstraction of water by its roots affects the quantity of that fluid contained in the vegetable mould, and in the mineral strata near the surface, and, consequently, the consistency of the soil. In treating of the effects of trees on the moisture of the atmosphere, I have said that the forest, by interposing a canopy between the sky and the ground, and by covering the surface with a thick mantle of fallen leaves, at once obstructed insolation and prevented the radiation of heat from the earth. These influences go far to balance each other; but familiar observation shows that, in summer, the forest soil is not raised to so high a temperature as open grounds exposed to irradiation. For this reason, and in consequence of the mechanical resistance opposed by the bed of dead leaves to the escape of moisture, we should expect that, except after recent rains, the superficial strata of woodland soil would be more humid than that of cleared land. This agrees with experience. The soil of the forest is always moist, except in the extremest droughts, and it is exceedingly rare that a primitive wood suffers from want of humidity. How far this accumulation of water affects the condition of neighboring grounds by lateral infiltration, we do not know, but we shall see, in a subsequent chapter, that water is conveyed to great distances by this process, and we may hence infer that the influence in question is an important one.

Influence of the Forest on the Flow of Springs.

It is well established that the protection afforded by the forest against the escape of moisture from its soil, insures the permanence and regularity of natural springs, not only within the limits of the wood, but at some distance beyond its borders, and thus contributes to the supply of an element essential to both vegetable and animal life. As the forests are destroyed, the springs which flowed from the woods, and, consequently, the greater watercourses fed by them, diminish both in number and in volume. This fact is so familiar throughout the American States and the British Provinces, that there are few old residents of the interior of those districts who are not able to testify to its truth as a matter of personal observation. My own recollection suggests to me many instances of this sort, and I remember one case where a small mountain spring, which disappeared soon after the clearing of the ground where it rose, was recovered about ten or twelve years ago, by simply allowing the bushes and young trees to grow up on a rocky knoll, not more than half an acre in extent, immediately above it, and has since continued to flow uninterruptedly. The uplands in the Atlantic States formerly abounded in sources and rills, but in many parts of those States which have been cleared for above a generation or two, the hill pastures now suffer severely from drought, and in dry seasons no longer afford either water or herbage for cattle.

Foissac, indeed, quotes from the elder Pliny (*Nat. Hist.*, xxxi, c. 30) a passage affirming that the felling of the woods gives rise to springs which did not exist before because the water of the soil was absorbed by the trees; and the same meteorologist declares, as I observed in treating of the effect of the forest on atmospheric humidity, that the planting of trees tends to drain marshy ground, because the roots absorb more water than falls from the air. But Pliny's statement rests on very doubtful authority, and Foissac cites no evidence in support of his own proposition.* In the American States, it is always observed that clearing the ground not only causes running springs to disappear, but dries up the stagnant pools and the spongy soils of the low grounds. The first roads in those States ran along the ridges, when practicable, because there only was the earth dry enough to allow of their construction, and, for the same reason, the cabins of the first settlers were perched upon the hills. As the forests have been from time to time removed, and the face of the earth laid open to the air and sun, the moisture has been evaporated, and the removal of the highways and of human habitations from the bleak hills to the sheltered valleys, is one of the most agree-

* The passage in Pliny is as follows: "Nascuntur fontes, decisis plerumque silvis, quos arborum alimenta consumebant, sicut in Hæmo, obsidente Gallos Cassandro, quum valli gratia cecidissent. Plerumque vero damnosi torrentes corrivantur, detracta collibus silva continere nimbos ac digerere consueta."—*Nat. Hist.*, xxxi, 30.

Seneca cites this case, and another similar one said to have been observed at Magnesia, from a passage in Theophrastus, not to be found in the extant works of that author; but he adds that the stories are incredible, because shaded grounds abound most in water: ferè aquosissima sunt quæcumque umbrosissima.—*Quæst. Nat.*, iii, 11. See *Appendix*, No. 25.

able among the many improvements which later generations have witnessed in the interior of New England and the other Northern States.

Almost every treatise on the economy of the forest adduces numerous facts in support of the doctrine that the clearing of the woods tends to diminish the flow of springs and the humidity of the soil, and it might seem unnecessary to bring forward further evidence on this point.* But the subject is of too much practical importance and of too great philosophical interest to be summarily disposed of; and it ought particularly to be noticed that there is at least one case—that of some loose soils which, when bared of wood, very rapidly absorb and transmit to lower strata the water they receive from the atmosphere, as argued by Vallès †—where the removal of the forest may increase the flow of springs at levels below it, by exposing to the rain and melted snow a surface more bibulous, and at the same time less retentive, than its original covering. Under such circumstances, the water of precipitation, which had formerly flowed off without penetrating through the superficial layers of leaves upon the ground—as, in very heavy showers, it sometimes does—or been absorbed by the vegetable mould and retained until it was evaporated, might descend through porous earth until it meets an impermeable stratum, and then be conducted along it, until, finally, at the outcrop-

* "Why go so far for the proof of a phenomenon that is repeated every day under our own eyes, and of which every Parisian may convince himself, without venturing beyond the Bois de Boulogne or the forest of Meudon? Let him, after a few rainy days, pass along the Chevreuse road, which is bordered on the right by the wood, on the left by cultivated fields. The fall of water and the continuance of the rain have been the same on both sides; but the ditch on the side of the forest will remain filled with water proceeding from the infiltration through the wooded soil, long after the other, contiguous to the open ground, has performed its office of drainage and become dry. The ditch on the left will have discharged in a few hours a quantity of water, which the ditch on the right requires several days to receive and carry down to the valley."—CLAVÉ, *Études, etc.*, pp. 53, 54.

† VALLÈS, *Études sur les Inondations*, p. 472.

ping of this stratum, it bursts from a hillside as a running spring. But such instances are doubtless too rare to form a frequent or an important exception to the general law, because it is only under very uncommon circumstances that rain water runs off over the surface of forest ground instead of sinking into it, and very rarely the case that such a soil as has just been supposed is covered by a layer of vegetable earth thick enough to retain, until it is evaporated, all the rain that falls upon it, without imparting any water to the strata below it.

If we look at the point under discussion as purely a question of fact, to be determined by positive evidence and not by argument, the observations of Boussingault are, both in the circumstances they detail, and in the weight of authority to be attached to the testimony, among the most important yet recorded. They are embodied in the fourth section of the twentieth chapter of that writer's *Économie Rurale*, and I have already referred to them on page 191 for another purpose. The interest of the question will justify me in giving, in Boussingault's own words, the facts and some of the remarks with which he accompanies the details of them: "In many localities," he observes,* "it has been thought that, within a certain number of years, a sensible diminution has been perceived in the volume of water of streams utilized as a motive power; at other points, there are grounds for believing that rivers have become shallower, and the increasing breadth of the belt of pebbles along their banks seems to prove the loss of a part of their water; and, finally, abundant springs have almost dried up. These observations have been principally made in valleys bounded by high mountains, and it is thought to have been noticed that this diminution of the waters has immediately followed the epoch when the inhabitants have begun to destroy, unsparingly, the woods which were spread over the face of the land.

"These facts would indicate that, where clearings have been made, it rains less than formerly, and this is the gener-

* *Économie Rurale*, p. 730.

ally received opinion. * * * But while the facts I have stated have been established, it has been observed, at the same time, that, since the clearing of the mountains, the rivers and the torrents, which seemed to have lost a part of their water, sometimes suddenly swell, and that, occasionally, to a degree which causes great disasters. Besides, after violent storms, springs which had become almost exhausted have been observed to burst out with impetuosity, and soon after to dry up again. These latter observations, it will be easily conceived, warn us not to admit hastily the common opinion that the felling of the woods lessens the quantity of rain; for not only is it very possible that the quantity of rain has not changed, but the mean volume of running water may have remained the same, in spite of the appearance of drought presented by the rivers and springs, at certain periods of the year. Perhaps the only difference would be that the flow of the same quantity of water becomes more irregular in consequence of clearing. For instance: if the low water of the Rhone during one part of the year were exactly compensated by a sufficient number of floods, it would follow that this river would convey to the Mediterranean the same volume of water which it carried to that sea in ancient times, before the period when the countries near its source were stripped of their woods, and when, probably, its mean depth was not subject to so great variations as in our days. If this were so, the forests would have this value—that of regulating, of economizing in a certain sort, the drainage of the rain water.

"If running streams really become rarer in proportion as clearing is extended, it follows either that the rain is less abundant, or that evaporation is greatly favored by a surface which is no longer protected by trees against the rays of the sun and the wind. These two causes, acting in the same direction, must often be cumulative in their effects, and before we attempt to fix the value of each, it is proper to inquire whether it is an established fact that running waters diminish on the surface of a country in which extensive clearing is going on; in a word, to examine whether an apparent fact has not been mistaken for a real one. And here lies the practical point of the question; for if it is once established that clearing diminishes the volume of streams, it is less important to know to what special cause this effect is due. * * * I shall attach no value except to facts which have taken place under the eye of man, as it is the influence of his labors on the meteorological condition of the atmosphere which I propose to estimate. What I am about to detail has been observed particularly in America, but I shall endeavor to establish, that what I believe to be true of America would be equally so for any other continent.

"One of the most interesting parts of Venezuela is, no doubt, the valley of Aragua. Situated at a short distance from the coast, and endowed, from its elevation, with various climates and a soil of unexampled fertility, its agriculture embraces at once the crops suited to tropical regions and to Europe. Wheat succeeds well on the heights of Victoria. Bounded on the north by the coast chain, on the south by a system of mountains connected with the Llanos, the valley is shut in on the east and the west by lines of hills which completely close it. In consequence of this singular configuration, the rivers which rise within it, having no outlet to the ocean, form, by their union, the beautiful Lake of Tacarigua or Valencia. This lake, according to Humboldt, is larger than that of Neufchâtel; it is at an elevation of 439 mètres [= 1,460 English feet] above the sea, and its greatest length does not exceed two leagues and a half [= seven English miles].

"At the time of Humboldt's visit to the valley of Aragua, the inhabitants were struck by the gradual diminution which the lake had been undergoing for thirty years. In fact, by comparing the descriptions given by historians with its actual condition, even making large allowance for exaggeration, it was easy to see that the level was considerably depressed. The facts spoke for themselves. Oviedo, who, toward the close of the sixteenth century, had often traversed the valley of Aragua, says positively that New Valencia was founded, in 1555, at half a league from the Lake of Tacarigua; in 1800,

Humboldt found this city 5,260 mètres [= $3\frac{1}{3}$ English miles] from the shore.

"The aspect of the soil furnished new proofs. Many hillocks on the plain retain the name of islands, which they more justly bore when they were surrounded by water. The ground laid bare by the retreat of the lake was converted into admirable plantations of cotton, bananas, and sugar cane; and buildings erected near the lake showed the sinking of the water from year to year. In 1796, new islands made their appearance. An important military point, a fortress built in 1740 on the island of Cabrera, was now on a peninsula; and, finally, on two granitic islands, those of Cura and Cabo Blanco, Humboldt observed among the shrubs, some mètres above the water, fine sand filled with helicites.

"These clear and positive facts suggested numerous explanations, all assuming a subterranean outlet, which permitted the discharge of the water to the ocean. Humboldt disposed of these hypotheses, and, after a careful examination of the locality, the distinguished traveller did not hesitate to ascribe the diminution of the waters of the lake to the numerous clearings which had been made in the valley of Aragua within half a century. * * *

"In 1800, the valley of Aragua possessed a population as dense as that of any of the best-peopled parts of France. * * * Such was the prosperous condition of this fine country when Humboldt occupied the Hacienda de Cura.

"Twenty-two years later, I explored the valley of Aragua, fixing my residence in the little town of Maracay. For some years previous, the inhabitants had observed that the waters of the lake were no longer retiring, but, on the contrary, were sensibly rising. Grounds, not long before occupied by plantations, were submerged. The islands of Nuevas Aparecidas, which appeared above the surface in 1796, had again become shoals dangerous to navigation. Cabrera, a tongue of land on the north side of the valley, was so narrow that the least rise of the water completely inundated it. A protracted north wind sufficed to flood the road between Maracay and New

Valencia. The fears which the inhabitants of the shores had so long entertained were reversed. * * * Those who had explained the diminution of the lake by the supposition of subterranean channels were suspected of blocking them up, to prove themselves in the right.

"During the twenty-two years which had elapsed, important political events had occurred. Venezuela no longer belonged to Spain. The peaceful valley of Aragua had been the theatre of bloody struggles, and a war of extermination had desolated these smiling lands and decimated their population. At the first cry of independence a great number of slaves found their liberty by enlisting under the banners of the new republic; the great plantations were abandoned, and the forest, which in the tropics so rapidly encroaches, had soon recovered a large proportion of the soil which man had wrested from it by more than a century of constant and painful labor.

"At the time of the growing prosperity of the valley of Aragua, the principal affluents of the lake were diverted, to serve for irrigation, and the rivers were dry for more than six months of the year. At the period of my visit, their waters, no longer employed, flowed freely."

Boussingault proceeds to state that two lakes near Ubate in New Granada, at an elevation of 2,562 mètres (= 8,500 English feet), where there is a constant temperature of 14° to 16° centigrade [= 57°, 61° Fahrenheit], had formed but one, a century before his visit; that the waters were gradually retiring, and the plantations extending over the abandoned bed; that, by inquiry of old hunters and by examination of parish records, he found that extensive clearings had been made and were still going on.

He found, also, that the length of the Lake of Fuquené, in the same valley, had, within two centuries, been reduced from ten leagues to one and a half, its breadth from three leagues to one. At the former period, timber was abundant, and the neighboring mountains were covered, to a certain height, with American oaks, laurels, and other trees of indigenous species; but at the time of his visit the mountains had been almost entirely stripped of their wood, chiefly to furnish fuel for salt-works. Our author adds that other cases, similar to those already detailed, might be cited, and he proceeds to show, by several examples, that the waters of other lakes in the same regions, where the valleys had always been bare of wood, or where the forests had not been disturbed, had undergone no change of level.

Boussingault further maintains that the lakes of Switzerland have sustained a depression of level since the too prevalent destruction of the woods, and arrives at the general conclusion, that, "in countries where great clearings have been made, there has most probably been a diminution in the living waters which flow upon the surface of the ground." This conclusion he further supports by two examples: one, where a fine spring, at the foot of a wooded mountain in the Island of Ascension, dried up when the mountain was cleared, but reappeared when the wood was replanted; the other at Marmato, in the province of Popayan, where the streams employed to drive machinery were much diminished in volume, within two years after the clearing of the heights from which they derived their supplies. This latter is an interesting case, because, although the rain gauges, established as soon as the decrease of water began to excite alarm, showed a greater fall of rain for the second year of observation than the first, yet there was no appreciable increase in the flow of the mill streams. From these cases, the distinguished physicist infers that very restricted local clearings may diminish and even suppress springs and brooks, without any reduction in the total quantity of rain.

It will have been noticed that these observations, with the exception of the last two cases, do not bear directly upon the question of the diminution of springs by clearings, but they logically infer it from the subsidence of the natural reservoirs which springs once filled. There is, however, no want of positive evidence on this subject.

Marschand cites the following instances: "Before the felling of the woods, within the last few years, in the valley of the Soulce, the Combe-ès-Mounin and the Little Valley, the Sorne furnished a regular and sufficient supply of water for the iron works of Unterwyl, which was almost unaffected by drought or by heavy rains. The Sorne has now become a torrent, every shower occasions a flood, and after a few days of fine weather, the current falls so low that it has been necessary to change the water wheels, because those of the old construction are no longer able to drive the machinery, and at last to introduce a steam engine to prevent the stoppage of the works for want of water.

"When the factory of St. Ursanne was established, the river that furnished its power was abundant, long known and tried, and had, from time immemorial, sufficed for the machinery of a previous factory. Afterward, the woods near its sources were cut. The supply of water fell off in consequence, the factory wanted water for half the year, and was at last obliged to stop altogether.

"The spring of Combefoulat, in the commune of Seleate, was well known as one of the best in the country; it was remarkably abundant and sufficient, in spite of the severest droughts, to supply all the fountains of the town; but, as soon as considerable forests were felled in Combe-de-pré Martin and in the valley of Combefoulat, the famous spring which lies below these woods has become a mere thread of water, and disappears altogether in times of drought.

"The spring of Varieux, which formerly supplied the castle of Pruntrut, lost more than half its water after the clearing of Varieux and Rougeoles. These woods have been replanted, the young trees are growing well, and with the woods, the waters of the spring are increasing.

"The Dog Spring between Pruntrut and Bressancourt has entirely vanished since the surrounding forests grounds were brought under cultivation.

"The Wolf Spring, in the commune of Soubey, furnishes a remarkable example of the influence of the woods upon fountains. A few years ago this spring did not exist. At the place where it now rises, a small thread of water was observed after very long rains, but the stream disappeared with the rain.

The spot is in the middle of a very steep pasture inclining to the south. Eighty years ago, the owner of the land, perceiving that young firs were shooting up in the upper part of it, determined to let them grow, and they soon formed a flourishing grove. As soon as they were well grown, a fine spring appeared in place of the occasional rill, and furnished abundant water in the longest droughts. For forty or fifty years, this spring was considered the best in the Clos du Doubs. A few years since, the grove was felled, and the ground turned again to a pasture. The spring disappeared with the wood, and is now as dry as it was ninety years ago." *

"The influence of the forest on springs," says Hummel, "is strikingly shown by an instance at Heilbronn. The woods on the hills surrounding the town are cut in regular succession every twentieth year. As the annual cuttings approach a certain point, the springs yield less water, some of them none at all; but as the young growth shoots up, they flow more and more freely, and at length bubble up again in all their original abundance." †

Piper states the following case: "Within about half a mile of my residence there is a pond upon which mills have been standing for a long time, dating back, I believe, to the first settlement of the town. These have been kept in constant operation until within some twenty or thirty years, when the supply of water began to fail. The pond owes its existence to a stream which has its source in the hills which stretch some miles to the south. Within the time mentioned, these hills, which were clothed with a dense forest, have been almost entirely stripped of trees; and to the wonder and loss of the mill owners, the water in the pond has failed, except in the season of freshets; and, what was never heard of before, the stream itself has been entirely dry. Within the last ten years a new growth of wood has sprung up on most of the land formerly occupied by the old forest; and now the water runs

* *Ueber die Entwaldung der Gebirge*, pp. 20 *et seqq.*

† *Physische Geographie*, p. 32.

through the year, notwithstanding the great droughts of the last few years, going back from 1856."

Dr. Piper quotes from a letter of William C. Bryant the following remarks: "It is a common observation that our summers are become drier, and our streams smaller. Take the Cuyahoga as an illustration. Fifty years ago large barges loaded with goods went up and down that river, and one of the vessels engaged in the battle of Lake Erie, in which the gallant Perry was victorious, was built at Old Portage, six miles north of Albion, and floated down to the lake. Now, in an ordinary stage of the water, a canoe or skiff can hardly pass down the stream. Many a boat of fifty tons burden has been built and loaded in the Tuscarawas, at New Portage, and sailed to New Orleans without breaking bulk. Now, the river hardly affords a supply of water at New Portage for the canal. The same may be said of other streams—they are drying up. And from the same cause—the destruction of our forests—our summers are growing drier, and our winters colder." *

No observer has more carefully studied the influence of the forest upon the flow of the waters, or reasoned more ably on the ascertained phenomena than Cantegril. The facts presented in the following case, communicated by him to the *Ami des Sciences* for December, 1859, are as nearly conclusive as any single instance well can be:

"In the territory of the commune of Labruguière, there is a forest of 1,834 hectares [4,530 acres], known by the name of the Forest of Montaut, and belonging to that commune. It extends along the northern slope of the Black Mountains. The soil is granitic, the maximum altitude 1,243 mètres [4,140 feet], and the inclination ranges between 15 and 60 to 100.

"A small current of water, the brook of Caunan, takes its rise in this forest, and receives the waters of two thirds of its surface. At the lower extremity of the wood and on the stream are several fulleries, each requiring a force of eight horse-power to drive the water wheels which work the stamp-

* *The Trees of America*, pp. 50, 51.

ers. The commune of Labruguière had been for a long time famous for its opposition to forest laws. Trespasses and abuses of the right of pasturage had converted the wood into an immense waste, so that this vast property now scarcely sufficed to pay the expense of protecting it, and to furnish the inhabitants with a meagre supply of fuel. While the forest was thus ruined, and the soil thus bared, the water, after every abundant rain, made an eruption into the valley, brought down a great quantity of pebbles which still clog the current of the Caunan. The violence of the floods was sometimes such that they were obliged to stop the machinery for some time. During the summer another inconvenience was felt. If the dry weather continued a little longer than usual, the delivery of water became insignificant. Each fullery could for the most part only employ a single set of stampers, and it was not unusual to see the work entirely suspended.

"After 1840, the municipal authority succeeded in enlightening the population as to their true interests. Protected by a more watchful supervision, aided by well-managed replantation, the forest has continued to improve to the present day. In proportion to the restoration of the forest, the condition of the manufactories has become less and less precarious, and the action of the water is completely modified. For example, there are, no longer, sudden and violent floods which make it necessary to stop the machinery. There is no increase in the delivery until six or eight hours after the beginning of the rain; the floods follow a regular progression till they reach their maximum, and decrease in the same manner. Finally, the fulleries are no longer forced to suspend work in summer; the water is always sufficiently abundant to allow the employment of two sets of stampers at least, and often even of three.

"This example is remarkable in this respect, that, all other circumstances having remained the same, the changes in the action of the stream can be attributed only to the restoration of the forest—changes which may be thus summed up: diminution of flood water during rains—increase of delivery at other seasons."

The Forest in Winter.

To estimate rightly the importance of the forest as a natural apparatus for accumulating the water that falls upon the surface and transmitting it to the subjacent strata, we must compare the condition and properties of its soil with those of cleared and cultivated earth, and examine the consequently different action of these soils at different seasons of the year. The disparity between them is greatest in climates where, as in the Northern American States and in the North of Europe, the open ground freezes and remains impervious to water during a considerable part of the winter; though, even in climates where the earth does not freeze at all, the woods have still an important influence of the same character. The difference is yet greater in countries which have regular wet and dry seasons, rain being very frequent in the former period, while, in the latter, it scarcely occurs at all. These countries lie chiefly in or near the tropics, but they are not wanting in higher latitudes; for a large part of Asiatic and even of European Turkey is almost wholly deprived of summer rains. In the principal regions occupied by European cultivation, and where alone the questions discussed in this volume are recognized as having, at present, any practical importance, rain falls at all seasons, and it is to these regions that, on this point as well as others, I chiefly confine my attention.

The influence of the forest upon the waters of the earth has been more studied in France than in any other part of the civilized world, because that country has, in recent times, suffered most severely from the destruction of the woods. But in the southern provinces of that empire, where the evils resulting from this cause are most sensibly felt, the winters are not attended with much frost, while, in Northern Europe, where the winters are rigorous enough to freeze the ground to the depth of some inches, or even feet, a humid atmosphere and frequent summer rains prevent the drying up of the springs observed in southern latitudes when the woods are gone. For these reasons, the specific character of the forest, as a winter reservoir of moisture in countries with a cold and dry atmosphere, has not attracted so much attention in France and Northern Europe as it deserves in the United States, where an excessive climate renders that function of the woods more important.

In New England, irregular as the climate is, the first autumnal snows usually fall before the ground is frozen at all, or when the frost extends at most to the depth of only a few inches. In the woods, especially those situated upon the elevated ridges which supply the natural irrigation of the soil and feed the perennial fountains and streams, the ground remains covered with snow during the winter; for the trees protect the snow from blowing from the general surface into the depressions, and new accessions are received before the covering deposited by the first fall is melted. Snow is of a color unfavorable for radiation, but, even when it is of considerable thickness, it is not wholly impervious to the rays of the sun, and for this reason, as well as from the warmth of lower strata, the frozen crust, if one has been formed, is soon thawed, and does not again fall below the freezing point during the winter.

The snow in contact with the earth now begins to melt, with greater or less rapidity, according to the relative temperature of the earth and the air, while the water resulting from its dissolution is imbibed by the vegetable mould, and carried off by infiltration so fast that both the snow and the layers of leaves in contact with it often seem comparatively dry, when, in fact, the under surface of the former is in a state of perpetual thaw. No doubt a certain proportion of the snow is returned to the atmosphere by direct evaporation, but in the woods it is partially protected from the action of the sun, and as very little water runs off in the winter by superficial watercourses, except in rare cases of sudden thaw, there can be no question that much the greater part of the snow deposited in the forest is slowly melted and absorbed by the earth.

The quantity of snow that falls in extensive forests, far from the open country, has seldom been ascertained by direct observation, because there are few meteorological stations in such situations. In the Northeastern border States of the American Union, the ground in the deep woods is covered with snow four or five months, and the proportion of water which falls in snow does not exceed one fifth of the total precipitation for the year.* Although, in the open grounds, snow and ice are evaporated with great rapidity in clear weather, even when the thermometer stands far below the freezing point, the surface of the snow in the woods does not indicate much loss in this way. Very small deposits of snowflakes remain unevaporated in the forest, for many days after snow let fall at the same time in the cleared field has disappeared without either a thaw to melt it or a wind powerful enough to drift it away. Even when bared of their leaves, the trees of a wood obstruct, in an important degree, both the direct action of the sun's rays on the snow, and the movement of drying and thawing winds.

Dr. Piper records the following observations: "A body of snow, one foot in depth, and sixteen feet square, was protected from the wind by a tight board fence about five feet high, while another body of snow, much more sheltered from the sun than the first, six feet in depth, and about sixteen feet square, was fully exposed to the wind. When the thaw came on, which lasted about a fortnight, the larger body of snow was entirely dissolved in less than a week, while the smaller body was not wholly gone at the end of the second week.

"Equal quantities of snow were placed in vessels of the same kind and capacity, the temperature of the air being seventy degrees. In the one case, a constant current of air was kept passing over the open vessel, while the other was protected by a cover. The snow in the first was dissolved in sixteen minutes, while the latter had a small unthawed proportion remaining at the end of eighty-five minutes." †

The snow in the woods is protected in the same way, though not literally to the same extent as by the fence in one

* THOMPSON'S *Vermont*, appendix, p. 8. † *Trees of America*, p. 48.

of these cases and the cover in the other. Little of the winter precipitation, therefore, is lost by evaporation, and as it slowly melts at bottom it is absorbed by the earth, and but a very small quantity of water runs off from the surface. The immense importance of the forest, as a reservoir of this stock of moisture, becomes apparent, when we consider that a large proportion of the summer rain either flows into the valleys and the rivers, because it falls faster than the ground can imbibe it; or, if absorbed by the warm superficial strata, is evaporated from them without sinking deep enough to reach wells and springs, which, of course, depend very much on winter rains and snows for their entire supply. This observation, though specially true of cleared and cultivated grounds, is not wholly inapplicable to the forest, particularly when, as is too often the case in Europe, the underwood and the decaying leaves are removed.

The general effect of the forest in cold climates is to assimilate the winter state of the ground to that of wooded regions under softer skies; and it is a circumstance well worth noting, that in Southern Europe, where nature has denied to the earth a warm winter-garment of flocculent snow, she has, by one of those compensations in which her empire is so rich, clothed the hillsides with umbrella pines, ilexes, cork oaks, and other trees of persistent foliage, whose evergreen leaves afford to the soil a protection analogous to that which it derives from snow in more northern climates.

The water imbibed by the soil in winter sinks until it meets a more or less impermeable, or a saturated stratum, and then, by unseen conduits, slowly finds its way to the channels of springs, or oozes out of the ground in drops which unite in rills, and so all is conveyed to the larger streams, and by them finally to the sea. The water, in percolating through the vegetable and mineral layers, acquires their temperature, and is chemically affected by their action, but it carries very little matter in mechanical suspension.

The process I have described is a slow one, and the supply of moisture derived from the snow, augmented by the rains of the following seasons, keeps the forest ground, where the surface is level or but moderately inclined, in a state of saturation through almost the whole year. The rivers fed by springs and shaded by woods are comparatively uniform in volume, in temperature, and in chemical composition. Their banks are little abraded, nor are their courses much obstructed by fallen timber, or by earth and gravel washed down from the highlands. Their channels are subject only to slow and gradual changes, and they carry down to the lakes and the sea no accumulation of sand or silt to fill up their outlets, and, by raising their beds, to force them to spread over the low grounds near their mouth.*

In this state of things, destructive tendencies of all sorts are arrested or compensated, and tree, bird, beast, and fish, alike, find a constant uniformity of condition most favorable to the regular and harmonious coexistence of them all.

General Consequences of the Destruction of the Forest.

With the disappearance of the forest, all is changed. At one season, the earth parts with its warmth by radiation to an open sky—receives, at another, an immoderate heat from the unobstructed rays of the sun. Hence the climate becomes excessive, and the soil is alternately parched by the fervors of

* Dumont, following Dansse, gives an interesting extract from the Misopogon of the Emperor Julian, showing that, in the fourth century, the Seine—the level of which now varies to the extent of thirty feet between extreme high and extreme low water mark—was almost wholly exempt from inundations, and flowed with a uniform current through the whole year. "Ego olim eram in hibernis apud caram Lutetiam, [sic] enim Galli Parisiorum oppidum appellant, quæ insula est non magna, in fluvio sita, qui eam omni ex parte cingit. Pontes sublicii utrinque ad eam ferunt, raròque fluvius minuitur ac crescit; sed qualis æstate, talis esse solet hyeme."—*Des Travaux Publics dans leur Rapports avec l'Agriculture*, p. 361, note.

As Julian was six years in Gaul, and his principal residence was at Paris, his testimony as to the habitual condition of the Seine, at a period when the provinces where its sources originate were well wooded, is very valuable.

summer, and seared by the rigors of winter. Bleak winds sweep unresisted over its surface, drift away the snow that sheltered it from the frost, and dry up its scanty moisture. The precipitation becomes as regular as the temperature; the melting snows and vernal rains, no longer absorbed by a loose and bibulous vegetable mould, rush over the frozen surface, and pour down the valleys seaward, instead of filling a retentive bed of absorbent earth, and storing up a supply of moisture to feed perennial springs. The soil is bared of its covering of leaves, broken and loosened by the plough, deprived of the fibrous rootlets which held it together, dried and pulverized by sun and wind, and at last exhausted by new combinations. The face of the earth is no longer a sponge, but a dust heap, and the floods which the waters of the sky pour over it hurry swiftly along its slopes, carrying in suspension vast quantities of earthy particles which increase the abrading power and mechanical force of the current, and, augmented by the sand and gravel of falling banks, fill the beds of the streams, divert them into new channels and obstruct their outlets. The rivulets, wanting their former regularity of supply and deprived of the protecting shade of the woods, are heated, evaporated, and thus reduced in their summer currents, but swollen to raging torrents in autumn and in spring. From these causes, there is a constant degradation of the uplands, and a consequent elevation of the beds of watercourses and of lakes by the deposition of the mineral and vegetable matter carried down by the waters. The channels of great rivers become unnavigable, their estuaries are choked up, and harbors which once sheltered large navies are shoaled by dangerous sandbars. The earth, stripped of its vegetable glebe, grows less and less productive, and, consequently, less able to protect itself by weaving a new network of roots to bind its particles together, a new carpeting of turf to shield it from wind and sun and scouring rain. Gradually it becomes altogether barren. The washing of the soil from the mountains leaves bare ridges of sterile rock, and the rich organic mould which covered them, now swept down into the dank low grounds, promotes a luxuriance of aquatic vegetation that breeds fever, and more insidious forms of mortal disease, by its decay, and thus the earth is rendered no longer fit for the habitation of man.*

To the general truth of this sad picture there are many exceptions, even in countries of excessive climates. Some of these are due to favorable conditions of surface, of geological structure, and of the distribution of rain; in many others, the evil consequences of man's improvidence have not yet been experienced, only because a sufficient time has not elapsed, since the felling of the forest, to allow them to develop themselves. But the vengeance of nature for the violation of her harmonies, though slow, is sure, and the gradual deterioration of soil and climate in such exceptional regions is as certain to result from the destruction of the woods as is any natural effect to follow its cause.

In the vast farrago of crudities which the elder Pliny's ambition of encyclopædic attainment and his ready credulity have gathered together, we meet some judicious observations. Among these we must reckon the remark with which he accompanies his extraordinary statement respecting the prevention of springs by the growth of forest trees, though, as is usual with him, his philosophy is wrong. "Destructive torrents are generally formed when hills are stripped of the trees which formerly confined and absorbed the rains." The absorption here referred to is not that of the soil, but of the roots, which, Pliny supposed, drank up the water to feed the growth of the trees.

Although this particular evil effect of too extensive clearing was so early noticed, the lesson seems to have been soon

* Almost every narrative of travel in those countries which were the earliest seats of civilization, contains evidence of the truth of these general statements, and this evidence is presented with more or less detail in most of the special works on the forest which I have occasion to cite. I may refer particularly to HOHENSTEIN, *Der Wald*, 1860, as full of important facts on this subject. See also CAIMI, *Cenni sulla Importanza dei Boschi*, for some statistics not readily found elsewhere, on this and other topics connected with the forest.

forgotten. The legislation of the Middle Ages in Europe is full of absurd provisions concerning the forests, which sovereigns sometimes destroyed because they furnished a retreat for rebels and robbers, sometimes protected because they were necessary to breed stags and boars for the chase, and sometimes spared with the more enlightened view of securing a supply of timber and of fuel to future generations.* It was reserved to later ages to appreciate their geographical importance, and it is only in very recent times, only in a few European countries, that the too general felling of the woods has been recognized as the most destructive among the many causes of the physical deterioration of the earth.

Condition of the Forest, and its Literature in different Countries.

The literature of the forest, which in England and America has not yet become sufficiently extensive to be known as a special branch of authorship, counts its thousands of volumes in Germany, Italy, and France. It is in the latter country, perhaps, that the relations of the woods to the regular drainage of the soil, and especially to the permanence of the natural configuration of terrestrial surface, have been most thoroughly investigated. On the other hand, the purely economical aspects of sylviculture have been most satisfactorily expounded,

* Stanley, citing SELDEN, *De Jure Naturali*, book vi, and FABRICIUS, *Cod. Pseudap. V. T.*, i, 874, mentions a remarkable Jewish tradition of uncertain but unquestionably ancient date, which is among the oldest evidences of public respect for the woods, and of enlightened views of their importance and proper treatment:

"To Joshua a fixed Jewish tradition ascribed ten decrees, laying down precise rules, which were instituted to protect the property of each tribe and of each householder from lawless depredation. Cattle, of a smaller kind, were to be allowed to graze in thick woods, not in thin woods; in woods, no kind of cattle without the owner's consent. Sticks and branches might be gathered by any Hebrew, but not cut. * * * Woods might be pruned, provided they were not olives or fruit trees, and that there was sufficient shade in the place."—*Lectures on the History of the Jewish Church*, part i, p. 271.

and that art has been most philosophically discussed, and most skilfully and successfully practised, in Germany.

The eminence of Italian theoretical hydrographers and the great ability of Italian hydraulic engineers are well known, but the specific geographical importance of the woods has not been so clearly recognized in Italy as in the states bordering it on the north and west. It is true that the face of nature has been as completely revolutionized by man, and that the action of torrents has created as wide and as hopeless devastation in that country as in France; but in the French Empire the desolation produced by clearing the forests is more recent,* has been more suddenly effected, and, therefore, excites a livelier and more general interest than in Italy, where public opinion does not so readily connect the effect with its true cause. Italy, too, from ancient habit, employs little wood in architectural construction; for generations she has maintained no military or commercial marine large enough to require exhaustive quantities of timber,† and the mildness of her climate makes

* There seems to have been a tendency to excessive clearing in Central and Western, earlier than in Southeastern France. Wise and good Bernard Palissy—one of those persecuted Protestants of the sixteenth century, whose heroism, virtue, refinement, and taste shine out in such splendid contrast to the brutality, corruption, grossness, and barbarism of their oppressors—in the *Recepte Véritable*, first printed in 1563, thus complains: "When I consider the value of the least clump of trees, or even of thorns, I much marvel at the great ignorance of men, who, as it seemeth, do nowadays study only to break down, fell, and waste the fair forests which their forefathers did guard so choicely. I would think no evil of them for cutting down the woods, did they but replant again some part of them; but they care nought for the time to come, neither reck they of the great damage they do to their children which shall come after them."—*Œuvres Complètes de Bernard Palissy*, 1844, p. 88.

† The great naval and commercial marines of Venice and of Genoa must have occasioned an immense consumption of lumber in the Middle Ages, and the centuries immediately succeeding those commonly embraced in that designation. The marine construction of that period employed larger timbers than the modern naval architecture of most commercial countries, but apparently without a proportional increase of strength. The old modes of ship building have been, to a considerable extent, handed down

small demands on the woods for fuel. Besides these circumstances, it must be remembered that the sciences of observation did not become knowledges of practical application till after the mischief was already mainly done and even forgotten in Alpine Italy, while its evils were just beginning to be sensibly felt in France when the claims of natural philosophy as a liberal study were first acknowledged in modern Europe. The former political condition of the Italian Peninsula would have effectually prevented the adoption of a general system of forest economy, however clearly the importance of a wise administration of this great public interest might have been understood. The woods which controlled and regulated the flow of the river sources were very often in one jurisdiction, the plains to be irrigated, or to be inundated by floods and desolated by torrents, in another. Concert of action on such a subject between a multitude of jealous petty sovereignties was obviously impossible, and nothing but the union of all the Italian states under a single government can render practicable the establishment of such arrangements for the conservation and restoration of the forests and the regulation of the flow of the waters as are necessary for the full development of the yet unexhausted resources of that fairest of lands, and even for the permanent maintenance of the present condition of its physical geography.

The denudation of the Central and Southern Apennines and of the Italian declivity of the Western Alps began at a period of unknown antiquity, but it does not seem to have been carried to a very dangerous length until the foreign con-

to the present day in the Mediterranean, and an American or an Englishman looks with astonishment at the huge beams and thick planks so often employed in the construction of very small vessels navigating that sea. According to Hummel, the desolation of the Karst, the high plateau lying north of Trieste, now one of the most parched and barren districts in Europe, is owing to the felling of its woods to build the navies of Venice. "Where the miserable peasant of the Karst now sees nothing but bare rock swept and scoured by the raging Bora, the fury of this wind was once subdued by mighty firs, which Venice recklessly cut down to build her fleets."—*Physische Geographie*, p. 32. See *Appendix*, No. 26.

quests and extended commerce of Rome created a greatly increased demand for wood for the construction of ships and for military material. The Eastern Alps, the Western Apennines, and the Maritime Alps retained their forests much later; but even here the want of wood, and the injury to the plains and the navigation of the rivers by sediment brought down by the torrents, led to some legislation for the protection of the forests, by the Republic of Venice in the fifteenth century, by that of Genoa as early at least as the seventeenth; and Marschand states that the latter Government passed laws requiring the proprietors of mountain lands to replant the woods. These, however, do not seem to have been effectually enforced. It is very common in Italy to ascribe to the French occupation under the first Empire all the improvements, and all the abuses of recent times, according to the political sympathies of the individual; and the French are often said to have prostrated every forest which has disappeared within a century.* But, however this may be, no energetic system of repression or restoration was adopted by any of the Italian states after the downfall of the Empire, and the taxes on forest property in some of them were so burdensome that rural municipalities sometimes proposed to cede their common woods to the Government, without any other compensation than the remission of the taxes imposed on forest lands.† Under such circumstances, woodlands would soon become disafforested, and where facilities of transportation and a good demand for timber have increased the inducements to fell it, as upon the borders of the Mediterranean, the destruction of the forest and all the evils which attend it have gone on at a seriously alarming rate. It has even been calculated that four tenths of the area of the Ligurian provinces have been washed away or rendered incapable of cultivation by the felling of the woods.‡

* *Le Alpi che cingono l'Italia*, i, p. 367.

† See the periodical *Politecnico*, published at Milan, for the month of May, 1862, p. 234.

‡ *Annali di Agricoltura, Industria e Commercio*, vol. i, p. 77.

The damp and cold climate of England requires the maintenance of household fires through a large part of the year. Contrivances for economizing fuel were of later introduction in that country than on the Continent. The soil, like the sky, was, in general, charged with humidity; its natural condition was unfavorable for common roads, and the transportation of so heavy a material as coal, by land, from the remote counties where alone it was mined in the Middle Ages, was costly and difficult. For all these reasons, the consumption of wood was large, and apprehensions of the exhaustion of the forests were excited at an early period. Legislation there, as elsewhere, proved ineffectual to protect them, and many authors of the sixteenth century express fears of serious evils from the wasteful economy of the people in this respect. Harrison, in his curious chapter "Of Woods and Marishes" in Holinshed's compilation, complains of the rapid decrease of the forests, and adds: "Howbeit thus much I dare affirme, that if woods go so fast to decaie in the next hundred yeere of Grace, as they haue doone and are like to doo in this, * * * it is to be feared that the fennie bote, broome, turfe, gall, heath, firze, brakes, whinnes, ling, dies, hassacks, flags, straw, sedge, réed, rush, and also *seacole*, will be good merchandize euen in the citie of London, whereunto some of them euen now haue gotten readie passage, and taken vp their innes in the greatest merchants' parlours. * * * I would wish that I might liue no longer than to sée foure things in this land reformed, that is: the want of discipline in the church: the couetous dealing of most of our merchants in the preferment of the commodities of other countries, and hinderance of their owne: the holding of faires and markets vpon the sundaie to be abolished and referred to the wednesdaies: and that euerie man, in whatsoeuer part of the champaine soile enioieth fortie acres of land, and vpwards, after that rate, either by frée deed, copie hold, or fee farme, might plant one acre of wood, or sowe the same with oke mast, hasell, béech, and sufficient prouision be made that it may be cherished and kept. But I feare me that I should then liue too long, and so long, that I should either be wearie of the world, or the world of me." * Evelyn's "Silva," the first edition of which appeared in 1664, rendered an extremely important service to the cause of the woods, and there is no doubt that the ornamental plantations in which England far surpasses all other countries, are, in some measure, the

* HOLINSHED, reprint of 1807, i, pp. 357, 358. It is evident from thi passage, and from another on page 397 of the same volume, that, though sea coal was largely exported to the Continent, it had not yet come into general use in England. It is a question of much interest, when coal was first employed in England for fuel. I can find no evidence that it was used as a combustible until more than a century after the Norman conquest. It has been said that it was known to the Anglo-Saxon population, but I am acquainted with no passage in the literature of that people which proves this. The dictionaries explain the Anglo-Saxon word *græfa* by sea coal. I have met with this word in no Anglo-Saxon work, except in the *Chronicle*, A. D. 852, from a manuscript certainly not older than the twelfth century, and in that passage it may as probably mean peat as coal, and quite as probably something else as either. Coal is not mentioned in King Alfred's Bede, in Glanville, or in Robert of Gloucester, though all these writers speak of jet as found in England, and are full in their enumeration of the mineral products of the island.

England was anciently remarkable for its forests, but Cæsar says it wanted the *fagus* and the *abies*. There can be no doubt that *fagus* means the beech, which, as the remains in the Danish peat mosses show, is a tree of late introduction into Denmark, where it succeeded the fir, a tree not now native to that country. The succession of forest crops seems to have been the same in England; for Harrison, p. 359, speaks of the "great store of firre" found lying "at their whole lengths" in the "fens and marises" of Lancashire and other counties, where not even bushes grew in his time. We cannot be sure what species of evergreen Cæsar intended by *abies*. The popular designations of spike-leaved trees are always more vague and uncertain in their application than those of broad-leaved trees. *Pinus*, *pine*, has been very loosely employed even in botanical nomenclature, and *Kiefer*, *Fichte*, and *Tanne* are often confounded in German.—ROSSMÄSSLER, *Der Wald*, pp. 256, 289, 324. If it were certain that the *abies* of Cæsar was the fir formerly and still found in peat mosses, and that he was right in denying the existence of the beech in England in his time, the observation would be very important, because it would fix a date at which the fir had become extinct, and the beech had not yet appeared in the island.

The English oak, though strong and durable, was not considered generally suitable for finer work in the sixteenth century. There were, how-

fruit of Evelyn's enthusiasm. In England, however, arboriculture, the planting and nursing of single trees, has, until recently, been better understood than sylviculture, the sowing and training of the forest. But this latter branch of rural improvement is now pursued on a very considerable scale, though, so far as I know, not by the National Government

The Influence of the Forest on Inundations.

Besides the climatic question, which I have already sufficiently discussed, and the obvious inconveniences of a scanty supply of charcoal, of fuel, and of timber for architectural and naval construction and for the thousand other uses to which wood is applied in rural and domestic economy, and in the

ever, exceptions. "Of all in Essex," observes HARRISON, *Holinshed*, i, p. 357, "that growing in Bardfield parke is the finest for ioiners craft: for oftentimes haue I seene of their workes made of that oke so fine and faire, as most of the wainescot that is brought hither out of Danske; for our wainescot is not made in England. Yet diuerse haue assaied to deale without [with our] okes to that end, but not with so good successe as they haue hoped, bicause the ab or iuice will not so soone be remoued and cleane drawne out, which some attribute to want of time in the salt water."

This passage is also of interest as showing that soaking in salt water, as a mode of seasoning, was practised in Harrison's time.

But the importation of wainscot, or boards for ceiling, panelling, and otherwise finishing rooms, which was generally of oak, commenced three centuries before the time of Harrison. On page 204 of the *Liber Albus*—a book which could have been far more valuable if the editor had given us the texts, with his learned notes, instead of a translation—mention is made of "squared oak timber," brought in from the country by carts, and of course of domestic growth, as free of city duty or octroi, and of "planks of oak" coming in in the same way as paying one plank a cartload. But in the chapter on the "Customs of Billyngesgate," pp. 208, 209, relating to goods imported from foreign countries, a duty of one halfpenny is imposed on every hundred of boards called "weynscotte," and of one penny on every hundred of boards called "Rygholt." The editor explains "Rygholt" as "wood of Riga." This was doubtless pine or fir. The year in which these provisions were made does not appear, but they belong to the reign of Henry III.

various industrial processes of civilized life, the attention of French foresters and public economists has been specially drawn to three points, namely: the influence of the forests on the permanence and regular flow of springs or natural fountains; on inundations by the overflow of rivers; and on the abrasion of soil and the transportation of earth, gravel, pebbles, and even of considerable masses of rock, from higher to lower levels, by torrents. There are, however, connected with this general subject, several other topics of minor or strictly local interest, or of more uncertain character, which I shall have occasion more fully to speak of hereafter.

The first of these three principal subjects—the influence of the woods on springs and other living waters—has been already considered; and if the facts stated in that discussion are well established, and the conclusions I have drawn from them are logically sound, it would seem to follow, as a necessary corollary, that the action of the forest is as important in diminishing the frequency and violence of river floods, as in securing the permanence and equability of natural fountains; for any cause which promotes the absorption and accumulation of the water of precipitation by the superficial strata of the soil, to be slowly given out by infiltration and percolation, must, by preventing the rapid flow of surface water into the natural channels of drainage, tend to check the sudden rise of rivers, and, consequently, the overflow of their banks, which constitutes what is called inundation. The mechanical resistance, too, offered by the trunks of trees and of undergrowth to the flow of water over the surface, tends sensibly to retard the rapidity of its descent down declivities, and to divert and divide streams which may have already accumulated from smaller threads of water.*

* In a letter addressed to the Minister of Public Works, after the terrible inundations of 1857, the Emperor thus happily expressed himself: "Before we seek the remedy for an evil, we inquire into its cause. Whence come the sudden floods of our rivers? From the water which falls on the mountains, not from that which falls on the plains. The waters which fall on our fields produce but few rivulets, but those which

Inundations are produced by the insufficiency of the natural channels of rivers to carry off the waters of their basins as fast as those waters flow into them. In accordance with the usual economy of nature, we should presume that she had everywhere provided the means of discharging, without disturbance of her general arrangements or abnormal destruction of her products, the precipitation which she sheds upon the face of the earth. Observation confirms this presumption, at least in the countries to which I confine my inquiries; for, so far as we know the primitive conditions of the regions brought under human occupation within the historical period, it appears that the overflow of river banks was much less frequent and destructive than at the present day, or, at least, that rivers rose and fell less suddenly before man had removed the natural checks to the too rapid drainage of the basins in which their tributaries originate. The banks of the rivers and smaller streams in the North American colonies were formerly little abraded by the currents. Even now the trees come down almost to the water's edge along the rivers, in the larger forests of the United States, and the surface of the streams seems liable to no great change in level or in rapidity of current. A circumstance almost conclusive as to the regularity of flow in forest rivers, is that they do not form large sedimentary deposits, at their points of discharge into lakes or larger streams, such accumulations beginning, or at least advancing far more rapidly, after the valleys are cleared.

In the Northern United States, although inundations are sometimes produced in the height of summer by heavy rains, it will be found generally true that the most rapid rise of the

fall on our roofs and are collected in the gutters, form small streams at once. Now, the roofs are mountains—the gutters are valleys."

"To continue the comparison," observes D'Héricourt, "roofs are smooth and impermeable, and the rain water pours rapidly off from their surfaces; but this rapidity of flow would be greatly diminished if the roofs were carpeted with mosses and grasses; more still, if they were covered with dry leaves, little shrubs, strewn branches, and other impediments—in short, if they were wooded."—*Annales Forestières, Déc.*, 1857, p. 311.

waters, and, of course, the most destructive "freshets," as they are called in America, are produced by the sudden dissolution of the snow before the open ground is thawed in the spring. It frequently happens that a powerful thaw sets in after a long period of frost, and the snow which had been months in accumulating is dissolved and carried off in a few hours. When the snow is deep, it, to use a popular expression, "takes the frost out of the ground" in the woods, and, if it lies long enough, in the fields also. But the heaviest snows usually fall after midwinter, and are succeeded by warm rains or sunshine, which dissolve the snow on the cleared land before it has had time to act upon the frost-bound soil beneath it. In this case, the snow in the woods is absorbed as fast as it melts, by the soil it has protected from freezing, and does not materially contribute to swell the current of the rivers. If the mild weather, in which great snowstorms usually occur, does not continue and become a regular thaw, it is almost sure to be followed by drifting winds, and the inequality with which they distribute the snow leaves the ridges comparatively bare, while the depressions are often filled with drifts to the height of many feet. The knolls become frozen to a great depth; succeeding partial thaws melt the surface snow, and the water runs down into the furrows of ploughed fields, and other artificial and natural hollows, and then often freezes to solid ice. In this state of things, almost the entire surface of the cleared land is impervious to water, and from the absence of trees and the general smoothness of the ground, it offers little mechanical resistance to superficial currents. If, under these circumstances, warm weather accompanied by rain occurs, the rain and melted snow are swiftly hurried to the bottom of the valleys and gathered to raging torrents.

It ought further to be considered that, though the lighter ploughed soils readily imbibe a great deal of water, yet the grass lands, and all the heavy and tenacious earths, absorb it in much smaller quantities, and less rapidly than the vegetable mould of the forest. Pasture, meadow, and clayey soils, taken together, greatly predominate over the sandy ploughed fields, in all large agricultural districts, and hence, even if, in the case we are supposing, the open ground chance to have been thawed before the melting of the snow which covers it, it is already saturated with moisture, or very soon becomes so, and, of course, cannot relieve the pressure by absorbing more water. The consequence is that the face of the country is suddenly flooded with a quantity of melted snow and rain equivalent to a fall of six or eight inches of the latter, or even more. This runs unobstructed to rivers often still bound with thick ice, and thus inundations of a fearfully devastating character are produced. The ice bursts, from the hydrostatic pressure from below, or is violently torn up by the current, and is swept by the impetuous stream, in large masses and with resistless fury, against banks, bridges, dams, and mills erected near them. The bark of the trees along the rivers is often abraded, at a height of many feet above the ordinary water level, by cakes of floating ice, which are at last stranded by the receding flood on meadow or ploughland, to delay, by their chilling influence, the advent of the tardy spring.

The surface of a forest, in its natural condition, can never pour forth such deluges of water as flow from cultivated soil. Humus, or vegetable mould, is capable of absorbing almost twice its own weight of water. The soil in a forest of deciduous foliage is composed of humus, more or less unmixed, to the depth of several inches, sometimes even of feet, and this stratum is usually able to imbibe all the water possibly resulting from the snow which at any one time covers it. But the vegetable mould does not cease to absorb water when it becomes saturated, for it then gives off a portion of its moisture to the mineral earth below, and thus is ready to receive a new supply; and, besides, the bed of leaves not yet converted to mould takes up and retains a very considerable proportion of snow water, as well as of rain.

In the warm climates of Southern Europe, as I have already said, the functions of the forest, so far as the disposal of the water of precipitation is concerned, are essentially the same at all seasons, and are analogous to those which it performs in the Northern United States in summer. Hence, in the former countries, the winter floods have not the characteristics which mark them in the latter, nor is the conservative influence of the woods in winter relatively so important, though it is equally unquestionable.

If the summer floods in the United States are attended with less pecuniary damage than those of the Loire and other rivers of France, the Po and its tributaries in Italy, the Emme and her sister torrents which devastate the valleys of Switzerland, it is partly because the banks of American rivers are not yet lined with towns, their shores and the bottoms which skirt them not yet covered with improvements whose cost is counted by millions, and, consequently, a smaller amount of property is exposed to injury by inundation. But the comparative exemption of the American people from the terrible calamities which the overflow of rivers has brought on some of the fairest portions of the Old World, is, in a still greater degree, to be ascribed to the fact that, with all our thoughtless improvidence, we have not yet bared all the sources of our streams, not yet overthrown all the barriers which nature has erected to restrain her own destructive energies. Let us be wise in time, and profit by the errors of our older brethren!

The influence of the forest in preventing inundations has been very generally recognized, both as a theoretical inference and as a fact of observation; but Belgrand and his commentator Vallès have deduced an opposite result from various facts of experience and from scientific considerations. They contend that the superficial drainage is more regular from cleared than from wooded ground, and that clearing diminishes rather than augments the intensity of inundations. Neither of these conclusions is warranted by their data or their reasoning, and they rest partly upon facts, which, truly interpreted, are not inconsistent with the received opinions on these subjects, partly upon assumptions which are contradicted by experience. Two of these latter are, first, that the fallen leaves in the forest constitute an impermeable covering of the soil over, not through, which the water of rains and of melting snows flows off, and secondly, that the roots of trees penetrate and choke up the fissures in the rocks, so as to impede the passage of water through channels which nature has provided for its descent to lower strata.

As to the first of these, we may appeal to familiar facts within the personal knowledge of every man acquainted with the operations of sylvan nature. I have before me a letter from an acute and experienced observer, containing this paragraph: "I think that rain water does not ever, except in very trifling quantities, flow over the leaves in the woods in summer or autumn. Water runs over them only in the spring, when they are pressed down smoothly and compactly, a state in which they remain only until they are dry, when shrinkage and the action of the wind soon roughen the surface so as effectually to stop, by absorption, all flow of water." I have observed that when a sudden frost succeeds a thaw at the close of the winter after the snow has principally disappeared, the water in and between the layers of leaves sometimes freezes into a solid crust, which allows the flow of water over it. But this occurs only in depressions and on a very small scale; and the ice thus formed is so soon dissolved that no sensible effect is produced on the escape of water from the general surface.

As to the influence of roots upon drainage, I believe there is no doubt that they, independently of their action as absorbents, mechanically promote it. Not only does the water of the soil follow them downward,* but their swelling growth powerfully tends to enlarge the crevices of rock into which they enter; and as the fissures in rocks are longitudinal, not mere circular orifices, every line of additional width gained by the growth of roots within them increases the area of the crev-

* "The roots of vegetables," says D'Héricourt, "perform the office of a perpendicular drainage analogous to that which has been practised with success in Holland and in some parts of the British Islands. This system consists in driving down three or four thousand stakes upon a hectare; the rain water filters down along the stakes, and, in certain cases, as favorable results are obtained by this method as by horizontal drains."—*Annales Forestières*, 1857, p. 312.

ice in proportion to its length. Consequently, the widening of a fissure to the extent of one inch might give an additional drainage equal to a square foot of open tubing.

The observations and reasonings of Belgrand and Vallès, though their conclusions have not been accepted by many, are very important in one point of view. These writers insist much on the necessity of taking into account, in estimating the relations between precipitation and evaporation, the abstraction of water from the surface and surface currents, by absorption and infiltration—an element unquestionably of great value, but hitherto much neglected by meteorological inquirers, who have very often reasoned as if the surface earth were either impermeable to water, or already saturated with it; whereas, in fact, it is a sponge, always imbibing humidity and always giving it off, not by evaporation only, but by infiltration and percolation.

The destructive effects of inundations considered simply as a mechanical power by which life is endangered, crops destroyed, and the artificial constructions of man overthrown, are very terrible. Thus far, however, the flood is a temporary and by no means an irreparable evil, for if its ravages end here, the prolific powers of nature and the industry of man soon restore what had been lost, and the face of the earth no longer shows traces of the deluge that had overwhelmed it. Inundations have even their compensations. The structures they destroy are replaced by better and more secure erections, and if they sweep off a crop of corn, they not unfrequently leave behind them, as they subside, a fertilizing deposit which enriches the exhausted field for a succession of seasons.* If,

* The productiveness of Egypt has been attributed too exclusively to the fertilizing effects of the slime deposited by the inundations of the Nile; for in that climate a liberal supply of water would produce good crops on almost any ordinary sand, while, without water, the richest soil would yield nothing. The sediment deposited annually is but a very small fraction of an inch in thickness. It is alleged that in quantity it would be hardly sufficient for a good top dressing, and that in quality it is not chemically distinguishable from the soil inches or feet below the surface. But

then, the too rapid flow of the surface waters occasioned no other evil than to produce, once in ten years upon the average, an inundation which should destroy the harvest of the low grounds along the rivers, the damage would be too inconsiderable, and of too transitory a character, to warrant the inconveniences and the expense involved in the measures which the most competent judges in many parts of Europe believe the respective governments ought to take to obviate it.

Destructive Action of Torrents.

But the great, the irreparable, the appalling mischiefs which have already resulted, and threaten to ensue on a still more extensive scale hereafter, from too rapid superficial drainage, are of a properly geographical character, and consist primarily in erosion, displacement, and transportation of the superficial strata, vegetable and mineral—of the integuments, so to speak, with which nature has clothed the skeleton framework of the globe. It is difficult to convey by description an idea of the desolation of the regions most exposed to the ravages of torrent and of flood; and the thousands, who, in these days of travel, are whirled by steam near or even through the theatres of these calamities, have but rare and imperfect opportunities of observing the destructive causes in action. Still more rarely can they compare the past with the actual condition of the provinces in question, and trace the progress of their conversion from forest-crowned hills, luxuriant pasture

to deny, as some writers have done, that the slime has any fertilizing properties at all, is as great an error as the opposite one of ascribing all the agricultural wealth of Egypt to that single cause of productiveness. Fine soils deposited by water are almost uniformly rich in all climates; those brought down by rivers, carried out into salt water, and then returned again by the tide, seem to be more permanently fertile than any others. The polders of the Netherland coast are of this character, and the meadows in Lincolnshire, which have been covered with slime by *warping*, as it is called, or admitting water over them at high tide, are remarkably productive. See *Appendix*, No. 27.

grounds, and abundant cornfields and vineyards well watered by springs and fertilizing rivulets, to bald mountain ridges, rocky declivities, and steep earth banks furrowed by deep ravines with beds now dry, now filled by torrents of fluid mud and gravel hurrying down to spread themselves over the plain, and dooming to everlasting barrenness the once productive fields. In traversing such scenes, it is difficult to resist the impression that nature pronounced the curse of perpetual sterility and desolation upon these sublime but fearful wastes, difficult to believe that they were once, and but for the folly of man might still be, blessed with all the natural advantages which Providence has bestowed upon the most favored climes. But the historical evidence is conclusive as to the destructive changes occasioned by the agency of man upon the flanks of the Alps, the Apennines, the Pyrenees, and other mountain ranges in Central and Southern Europe, and the progress of physical deterioration has been so rapid that, in some localities, a single generation has witnessed the beginning and the end of the melancholy revolution.

It is certain that a desolation, like that which has overwhelmed many once beautiful and fertile regions of Europe, awaits an important part of the territory of the United States, and of other comparatively new countries over which European civilization is now extending its sway, unless prompt measures are taken to check the action of destructive causes already in operation. It is vain to expect that legislation can do anything effectual to arrest the progress of the evil in those countries, except so far as the state is still the proprietor of extensive forests. Woodlands which have passed into private hands will everywhere be managed, in spite of legal restrictions, upon the same economical principles as other possessions, and every proprietor will, as a general rule, fell his woods, unless he believes that it will be for his pecuniary interest to preserve them. Few of the new provinces which the last three centuries have brought under the control of the European race, would tolerate any interference by the law-making power with what they regard as the most sacred of civil rights—the right, namely, of every man to do what he will with his own. In the Old World, even in France, whose people, of all European nations, love best to be governed and are least annoyed by bureaucratic supervision, law has been found impotent to prevent the destruction, or wasteful economy, of private forests; and in many of the mountainous departments of that country, man is at this moment so fast laying waste the face of the earth, that the most serious fears are entertained, not only of the depopulation of those districts, but of enormous mischiefs to the provinces contiguous to them.* The only legal provisions from which anything is to be hoped, are such as shall make it a matter of private advantage to the landholder to spare the trees upon his grounds, and promote the growth of the young wood. Something may be done by exempting standing forests from taxation, and by imposing taxes on wood felled for fuel or for timber, something by premiums or honorary distinctions for judicious management of the woods. It

* "The laws against clearing have never been able to prevent these operations when the proprietor found his advantage in them, and the long series of royal ordinances and decrees of parliaments, proclaimed from the days of Charlemagne to our own, with a view of securing forest property, have served only to show the impotence of legislative action on this subject."—CLAVÉ, *Études sur l'Économie Forestière*, p. 32.

"A proprietor can always contrive to clear his woods, whatever may be done to prevent him; it is a mere question of time, and a few imprudent cuttings, a few abuses of the right of pasturage, suffice to destroy a forest in spite of all regulations to the contrary."—DUNOYER, *De la Liberté du Travail*, ii, p. 452, as quoted by Clavé, p. 353.

Both authors agree that the preservation of the forests in France is practicable only by their transfer to the state, which alone can protect them and secure their proper treatment. It is much to be feared that even this measure would be inadequate to save the forests of the American Union. There is little respect for public property in America, and the Federal Government, certainly, would not be the proper agent of the nation for this purpose. It proved itself unable to protect the live-oak woods of Florida, which were intended to be preserved for the use of the navy, and it more than once paid contractors a high price for timber stolen from its own forests. The authorities of the individual States might be more efficient.

would be difficult to induce governments, general or local, to make the necessary appropriations for such purposes, but there can be no doubt that it would be sound economy in the end.

In countries where there exist municipalities endowed with an intelligent public spirit, the purchase and control of forests by such corporations would often prove advantageous; and in some of the provinces of Northern Lombardy, experience has shown that such operations may be conducted with great benefit to all the interests connected with the proper management of the woods. In Switzerland, on the other hand, except in some few cases where woods have been preserved as a defence against avalanches, the forests of the communes have been productive of little advantage to the public interests, and have very generally gone to decay. The rights of pasturage, everywhere destructive to trees, combined with toleration of trespasses, have so reduced their value, that there is, too often, nothing left that is worth protecting. In the canton of Ticino, the peasants have very frequently voted to sell the town woods and divide the proceeds among the corporators. The sometimes considerable sums thus received are squandered in wild revelry, and the sacrifice of the forests brings not even a momentary benefit to the proprietors.*

It is evidently a matter of the utmost importance that the public, and especially land owners, be roused to a sense of the dangers to which the indiscriminate clearing of the woods may expose not only future generations, but the very soil itself. Fortunately, some of the American States, as well as the governments of many European colonies, still retain the ownership of great tracts of primitive woodland. The State of New York, for example, has, in its northeastern counties, a vast extent of territory in which the lumberman has only here and there established his camp, and where the forest, though interspersed with permanent settlements, robbed of some of its finest pine groves, and often ravaged by devastating fires, still

* See the lively account of the sale of a communal wood in BERLEPSCH, *Die Alpen*, *Holzschläger und Flösser*.

covers far the largest proportion of the surface. Through this territory, the soil is generally poor, and even the new clearings have little of the luxuriance of harvest which distinguishes them elsewhere. The value of the land for agricultural uses is therefore very small, and few purchases are made for any other purpose than to strip the soil of its timber. It has been often proposed that the State should declare the remaining forest the inalienable property of the commonwealth, but I believe the motive of the suggestion has originated rather in poetical than in economical views of the subject. Both these classes of considerations have a real worth. It is desirable that some large and easily accessible region of American soil should remain, as far as possible, in its primitive condition, at once a museum for the instruction of the student, a garden for the recreation of the lover of nature, and an asylum where indigenous tree, and humble plant that loves the shade, and fish and fowl and four-footed beast, may dwell and perpetuate their kind, in the enjoyment of such imperfect protection as the laws of a people jealous of restraint can afford them. The immediate loss to the public treasury from the adoption of this policy would be inconsiderable, for these lands are sold at low rates. The forest alone, economically managed, would, without injury, and even with benefit to its permanence and growth, soon yield a regular income larger than the present value of the fee.

The collateral advantages of the preservation of these forests would be far greater. Nature threw up those mountains and clothed them with lofty woods, that they might serve as a reservoir to supply with perennial waters the thousand rivers and rills that are fed by the rains and snows of the Adirondacks, and as a screen for the fertile plains of the central counties against the chilling blasts of the north wind, which meet no other barrier in their sweep from the Arctic pole. The climate of Northern New York even now presents greater extremes of temperature than that of Southern France. The long continued cold of winter is far more intense, the short heats of summer not less fierce than in Provence, and hence the preservation of every influence that tends to maintain an equilibrium of temperature and humidity is of cardinal importance. The felling of the Adirondack woods would ultimately involve for Northern and Central New York consequences similar to those which have resulted from the laying bare of the southern and western declivities of the French Alps and the spurs, ridges, and detached peaks in front of them.

It is true that the evils to be apprehended from the clearing of the mountains of New York may be less in degree than those which a similar cause has produced in Southern France, where the intensity of its action has been increased by the inclination of the mountain declivities, and by the peculiar geological constitution of the earth. The degradation of the soil is, perhaps, not equally promoted by a combination of the same circumstances, in any of the American Atlantic States, but still they have rapid slopes and loose and friable soils enough to render widespread desolation certain, if the further destruction of the woods is not soon arrested. The effects of clearing are already perceptible in the comparatively unviolated region of which I am speaking. The rivers which rise in it flow with diminished currents in dry seasons, and with augmented volumes of water after heavy rains. They bring down much larger quantities of sediment, and the increasing obstructions to the navigation of the Hudson, which are extending themselves down the channel in proportion as the fields are encroaching upon the forest, give good grounds for the fear of serious injury to the commerce of the important towns on the upper waters of that river, unless measures are taken to prevent the expansion of "improvements" which have already been carried beyond the demands of a wise economy.

I have stated, in a general way, the nature of the evils in question, and of the processes by which they are produced; but I shall make their precise character and magnitude better understood by presenting some descriptive and statistical details of facts of actual occurrence. I select for this purpose the southeastern portion of France, not because that territory has suffered more severely than some others, but because its deterioration is comparatively recent, and has been watched and described by very competent and trustworthy observers, whose reports are more easily accessible than those published in other countries.*

The provinces of Dauphiny, Avignon, and Provence comprise a territory of fourteen or fifteen thousand square miles, bounded northwest by the Isere, northeast and east by the Alps, south by the Mediterranean, west by the Rhone, and extending from 42° to about 45° of north latitude. The surface is generally hilly and even mountainous, and several of the peaks in Dauphiny rise above the limit of perpetual snow. The climate, as compared with that of the United States in the same latitude, is extremely mild. Little snow falls, except upon the higher mountain ranges, the frosts are light, and the summers long, as might, indeed, be inferred from the vegetation; for in the cultivated districts, the vine and the fig everywhere flourish, the olive thrives as far north as 43½°, and upon the coast, grow the orange, the lemon, and the date palm. The forest trees, too, are of southern type, umbrella pines, various species of evergreen oaks, and many other trees and shrubs of persistent broad-leaved foliage, characterizing the landscape.

The rapid slope of the mountains naturally exposed these provinces to damage by torrents, and the Romans diminished their injurious effects by erecting, in the beds of ravines, barriers of rocks loosely piled up, which permitted a slow escape of the water, but compelled it to deposit above the dikes the

* Streffleur (*Ueber die Natur und die Wirkungen der Wildbäche*, p. 3) maintains that all the observations and speculations of French authors on the nature of torrents had been anticipated by Austrian writers. In proof of this assertion he refers to the works of Franz von Zallinger, 1778, Von Arretin, 1808, Franz Duile, 1826, all published at Innsbruck, and HAGEN'S *Beschreibung neuerer Wasserbauwerke*, Königsberg, 1826, none of which works are known to me. It is evident, however, that the conclusions of Surell and other French writers whom I cite, are original results of personal investigation, and not borrowed opinions.

earth and gravel with which it was charged.* At a later period the Crusaders brought home from Palestine, with much other knowledge gathered from the wiser Moslems, the art of securing the hillsides and making them productive by terracing and irrigation. The forests which covered the mountains secured an abundant flow of springs, and the process of clearing the soil went on so slowly that, for centuries, neither the want of timber and fuel, nor the other evils about to be depicted, were seriously felt. Indeed, throughout the Middle Ages, these provinces were well wooded, and famous for the fertility and abundance, not only of the low grounds, but of the hills.

Such was the state of things at the close of the fifteenth century. The statistics of the seventeenth show that while there had been an increase of prosperity and population in Lower Provence, as well as in the correspondingly situated parts of the other two provinces I have mentioned, there was an alarming decrease both in the wealth and in the population of Upper Provence and Dauphiny, although, by the clearing of the forests, a great extent of plough land and pasturage had been added to the soil before reduced to cultivation. It was found, in fact, that the augmented violence of the torrents had swept away, or buried in sand and gravel, more land than had been reclaimed by clearing; and the taxes computed by fires or habitations underwent several successive reductions in con-

* Whether Palissy was acquainted with this ancient practice, or whether it was one of those original suggestions of which his works are so full, I know not; but in his treatise, *Des Eaux et Fontaines*, he thus recommends it, by way of reply to the objections of "Théorique," who had expressed the fear that "the waters which rush violently down from the heights of the mountain would bring with them much earth, sand, and other things," and thus spoil the artificial fountain that "Practique" was teaching him to make: "And for hindrance of the mischiefs of great waters which may be gathered in few hours by great storms, when thou shalt have made ready thy parterre to receive the water, thou must lay great stones athwart the deep channels which lead to thy parterre. And so the force of the rushing currents shall be deadened, and thy water shall flow peacefully into his cisterns."—*Œuvres Complètes*, p. 173.

sequence of the gradual abandonment of the wasted soil by its starving occupants. The growth of the large towns on and near the Rhone and the coast, their advance in commerce and industry, and the consequently enlarged demand for agricultural products, ought naturally to have increased the rural population and the value of their lands; but the physical decay of the uplands was such that considerable tracts were deserted altogether, and in Upper Provence, the fires which in 1471 counted 897, were reduced to 747 in 1699, to 728 in 1733, and to 635 in 1776.

These facts I take from the *La Provence au point de vue des Bois, des Torrents et des Inondations*, of Charles de Ribbe, one of the highest authorities, and I add further details from the same source.

"Commune of Barles, 1707: Two hills have become connected by land slides, and have formed a lake which covers the best part of the soil. 1746: New slides buried twenty houses composing a village, no trace of which is left; more than one third of the land had disappeared.

"Monans, 1724: Deserted by its inhabitants and no longer cultivated.

"Gueydan, 1760: It appears by records that the best grounds have been swept off since 1756, and that ravines occupy their place.

"Digne, 1762: The river Bléone has destroyed the most valuable part of the territory.

"Malmaison, 1768: The inhabitants have emigrated, all their fields having been lost."

In the case of the commune of St. Laurent du Var, it appears that, after clearings in the Alps, succeeded by others in the common woods of the town, the floods of the torrent Var became more formidable, and had already carried off much land as early as 1708. "The clearing continued, and more soil was swept away in 1761. In 1762, after another destructive inundation, many of the inhabitants emigrated, and in 1765, one half of the territory had been laid waste.

"In 1766, the assessor Serraire said to the Assembly: 'As

to the damage caused by brooks and torrents, it is impossible to deny its extent Upper Provence is in danger of total destruction, and the waters which lay it waste threaten also the ruin of the most valuable grounds on the plain below. Villages have been almost submerged by torrents which formerly had not even names, and large towns are on the point of destruction from the same cause.' "

In 1776, Viscount Puget thus reported: "The mere aspect of Upper Provence is calculated to appal the patriotic magistrate. One sees only lofty mountains, deep valleys with precipitous sides, rivers with broad beds and little water, impetuous torrents, which in floods lay waste the cultivated land upon their banks and roll huge rocks along their channels; steep and parched hillsides, the melancholy consequences of indiscriminate clearing; villages whose inhabitants, finding no longer the means of subsistence, are emigrating day by day; houses dilapidated to huts, and but a miserable remnant of population."

"In a document of the year 1771, the ravages of the torrents were compared to the effects of an earthquake, half the soil in many communes seeming to have been swallowed up.

"Our mountains," said the administrators of the province of the Lower Alps in 1792, "present nothing but a surface of stony tufa; clearing is still going on, and the little rivulets are becoming torrents. Many communes have lost their harvests, their flocks, and their houses by floods. The washing down of the mountains is to be ascribed to the clearings and the practice of burning them over."

These complaints, it will be seen, all date before the Revolution, but the desolation they describe has since advanced with still swifter steps.

Surell—whose valuable work, *Étude sur les Torrents des Hautes Alpes*, published in 1841, presents the most appalling picture of the desolations of the torrent, and, at the same time, the most careful studies of the history and essential character of this great evil—in speaking of the valley of Dévoluy, on page 152, says · "Everything concurs to show that it was anciently wooded. In its peat bogs are found buried trunks of trees, monuments of its former vegetation. In the framework of old houses, one sees enormous timber, which is no longer to be found in the district. Many localities, now completely bare, still retain the name of 'wood,' and one of them is called, in old deeds, *Comba nigra* [Black forest or dell], on account of its dense woods. These and many other proofs confirm the local traditions which are unanimous on this point.

"There, as everywhere in the Upper Alps, the clearings began on the flanks of the mountains, and were gradually extended into the valleys and then to the highest accessible peaks. Then followed the Revolution, and caused the destruction of the remainder of the trees which had thus far escaped the woodman's axe."

In a note to this passage, the writer says: "Several persons have told me that they had lost flocks of sheep, by straying, in the forests of Mont Auroux, which covered the flanks of the mountain from La Cluse to Agnères. These declivities are now as bare as the palm of the hand."

The ground upon the steep mountains being once bared of trees, and the underwood killed by the grazing of horned cattle, sheep, and goats, every depression becomes a watercourse. "Every storm," says Surell, page 153, "gives rise to a new torrent. Examples of such are shown, which, though not yet three years old, have laid waste the finest fields of their valleys, and whole villages have narrowly escaped being swept into ravines formed in the course of a few hours. Sometimes the flood pours in a sheet over the surface, without ravine or even bed, and ruins extensive grounds, which are abandoned forever."

I cannot follow Surell in his description and classification of torrents, and I must refer the reader to his instructive work for a full exposition of the theory of the subject. In order, however, to show what a concentration of destructive energies may be effected by felling the woods that clothe and support the sides of mountain abysses, I cite his description of a valley descending from the Col Isoard, which he calls "a complete type of a basin of reception," that is, a gorge which serves as a common point of accumulation and discharge for the waters of several lateral torrents. "The aspect of the monstrous channel," says he, "is frightful. Within a distance of less than three kilomètres [= one mile and seven eighths English], more than sixty torrents hurl into the depths of the gorge the debris torn from its two flanks. The smallest of these secondary torrents, if transferred to a fertile valley, would be enough to ruin it."

The eminent political economist Blanqui, in a memoir read before the Academy of Moral and Political Science on the 25th of November, 1843, thus expresses himself: "Important as are the causes of impoverishment already described, they are not to be compared to the consequences which have followed from the two inveterate evils of the Alpine provinces of France, the extension of clearing and the ravages of torrents. * * The most important result of this destruction is this: that the agricultural capital, or rather the ground itself—which, in a rapidly increasing degree, is daily swept away by the waters—is totally lost. Signs of unparalleled destitution are visible in all the mountain zone, and the solitudes of those districts are assuming an indescribable character of sterility and desolation. The gradual destruction of the woods has, in a thousand localities, annihilated at once the springs and the fuel. Between Grenoble and Briançon in the valley of the Romanche, many villages are so destitute of wood that they are reduced to the necessity of baking their bread with sun-dried cowdung, and even this they can afford to do but once a year. This bread becomes so hard that it can be cut only with an axe, and I have myself seen a loaf of bread in September, at the kneading of which I was present the January previous.

"Whoever has visited the valley of Barcelonette, those of Embrun, and of Verdun, and that Arabia Petræa of the department of the Upper Alps, called Dévoluy, knows that there is no time to lose, that in fifty years from this date France will be separated from Savoy, as Egypt from Syria, by a desert." *

It deserves to be specially noticed that the district here referred to, though now among the most hopelessly waste in France, was very productive even down to so late a period as the commencement of the French Revolution. Arthur Young, writing in 1789, says: "About Barcelonette and in the highest parts of the mountains, the hill pastures feed a million of sheep, besides large herds of other cattle;" and he adds: "With such a soil, and in such a climate we are not to suppose a country barren because it is mountainous. The valleys I have visited are, in general, beautiful." † He ascribes the same character to the provinces of Dauphiny, Provence, and Auvergne, and, though he visited, with the eye of an attentive and practised observer, many of the scenes since blasted with the wild desolation described by Blanqui, the Durance and a part of the course of the Loire are the only streams he mentions as inflicting serious injury by their floods. The ravages of the torrents had, indeed, as we have seen, commenced earlier in some other localities, but we are authorized to infer that they were, in Young's time, too limited in range, and relatively too insignificant, to require notice in a general view of the provinces where they have now ruined so large a proportion of the soil.

But I resume my citations.

"I do not exaggerate," says Blanqui. "When I shall have finished my excursion and designated localities by their names,

* Ladoucette says the peasant of Dévoluy "often goes a distance of five hours over rocks and precipices for a single [man's] load of wood;" and he remarks on another page, that "the justice of peace of that canton had, in the course of forty-three years, but once heard the voice of the nightingale."—*Histoire, etc., des Hautes Alpes*, pp. 220, 434.

† The valley of Embrun, now almost completely devastated, was once remarkable for its fertility. In 1806, Héricart de Thury said of it: "In this magnificent valley nature had been prodigal of her gifts. Its inhabitants have blindly revelled in her favors, and fallen asleep in the midst of her profusion."—BECQUEREL, *Des Climats, etc.*, p. 314.

there will rise, I am sure, more than one voice from the spots themselves, to attest the rigorous exactness of this picture of their wretchedness. I have never seen its equal even in the Kabyle villages of the province of Constantine; for there you can travel on horseback, and you find grass in the spring, whereas in more than fifty communes in the Alps there is absolutely nothing.

"The clear, brilliant, Alpine sky of Embrun, of Gap, of Barcelonette, and of Digne, which for months is without a cloud, produces droughts interrupted only by diluvial rains like those of the tropics. The abuse of the right of pasturage and the felling of the woods have stripped the soil of all its grass and all its trees, and the scorching sun bakes it to the consistence of porphyry. When moistened by the rain, as it has neither support nor cohesion, it rolls down to the valleys, sometimes in floods resembling black, yellow, or reddish lava, sometimes in streams of pebbles, and even huge blocks of stone, which pour down with a frightful roar, and in their swift course exhibit the most convulsive movements. If you overlook from an eminence one of these landscapes furrowed with so many ravines, it presents only images of desolation and of death. Vast deposits of flinty pebbles, many feet in thickness, which have rolled down and spread far over the plain, surround large trees, bury even their tops, and rise above them, leaving to the husbandman no longer a ray of hope. One can imagine no sadder spectacle than the deep fissures in the flanks of the mountains, which seem to have burst forth in eruption to cover the plains with their ruins. These gorges, under the influence of the sun which cracks and shivers to fragments the very rocks, and of the rain which sweeps them down, penetrate deeper and deeper into the heart of the mountain, while the beds of the torrents issuing from them are sometimes raised several feet, in a single year, by the debris, so that they reach the level of the bridges, which, of course, are then carried off. The torrent beds are recognized at a great distance, as they issue from the mountains, and they spread themselves over the low grounds, in fan-shaped expansions, like a mantle of stone, sometimes ten thousand feet wide, rising high at the centre, and curving toward the circumference till their lower edges meet the plain.

"Such is their aspect in dry weather. But no tongue can give an adequate description of their devastations in one of those sudden floods which resemble, in almost none of their phenomena, the action of ordinary river water. They are now no longer overflowing brooks, but real seas, tumbling down in cataracts, and rolling before them blocks of stone, which are hurled forward by the shock of the waves like balls shot out by the explosion of gunpowder. Sometimes ridges of pebbles are driven down when the transporting torrent does not rise high enough to show itself, and then the movement is accompanied with a roar louder than the crash of thunder. A furious wind precedes the rushing water and announces its approach. Then comes a violent eruption, followed by a flow of muddy waves, and after a few hours all returns to the dreary silence which at periods of rest marks these abodes of desolation.

"This is but an imperfect sketch of this scourge of the Alps. Its devastations are increasing with the progress of clearing, and are every day turning a portion of our frontier departments into barren wastes.

"The unfortunate passion for clearing manifested itself at the beginning of the French Revolution, and has much increased under the pressure of immediate want. It has now reached an extreme point, and must be speedily checked, or the last inhabitant will be compelled to retreat when the last tree falls.

"The elements of destruction are increasing in violence. Rivers might be mentioned whose beds have been raised ten feet in a single year. The devastation advances in geometrical progression as the higher slopes are bared of their wood, and 'the ruin from above,' to use the words of a peasant, 'helps to hasten the desolation below.'

"The Alps of Provence present a terrible aspect. In the more equable climate of Northern France, one can form no conception of those parched mountain gorges where not even a bush can be found to shelter a bird, where, at most, the wanderer sees in summer here and there a withered lavender, where all the springs are dried up, and where a dead silence, hardly broken by even the hum of an insect, prevails. But if a storm bursts forth, masses of water suddenly shoot from the mountain heights into the shattered gulfs, waste without irrigating, deluge without refreshing the soil they overflow in their swift descent, and leave it even more seared than it was from want of moisture. Man at last retires from the fearful desert, and I have, the present season, found not a living soul in districts where I remember to have enjoyed hospitality thirty years ago."

In 1853, ten years after the date of Blanqui's memoir, M. de Bonville, prefect of the Lower Alps, addressed to the Government a report in which the following passages occur:

"It is certain that the productive mould of the Alps, swept off by the increasing violence of that curse of the mountains, the torrents, is daily diminishing with fearful rapidity. All our Alps are wholly, or in large proportion, bared of wood. Their soil, scorched by the sun of Provence, cut up by the hoofs of the sheep, which, not finding on the surface the grass they require for their sustenance, scratch the ground in search of roots to satisfy their hunger, is periodically washed and carried off by melting snows and summer storms.

"I will not dwell on the effects of the torrents. For sixty years they have been too often depicted to require to be further discussed, but it is important to show that their ravages are daily extending the range of devastation. The bed of the Durance, which now in some places exceeds 2,000 mètres [about 6,600 feet, or a mile and a quarter] in width, and, at ordinary times, has a current of water less than 10 mètres [about 33 feet] wide, shows something of the extent of the damage.* Where, ten years ago, there were still woods

* In the days of the Roman empire the Durance was a navigable river, with a commerce so important that the boatmen upon it formed a distinct corporation.—LADOUCETTE, *Histoire, etc., des Hautes Alpes*, p. 354.

Even as early as 1789, the Durance was computed to have already

and cultivated grounds to be seen, there is now but a vast torrent: there is not one of our mountains which has not at least one torrent, and new ones are daily forming.

"An indirect proof of the diminution of the soil is to be found in the depopulation of the country. In 1852, I reported to the General Council that, according to the census of that year, the population of the department of the Lower Alps had fallen off no less than 5,000 souls in the five years between 1846 and 1851.

"Unless prompt and energetic measures are taken, it is easy to fix the epoch when the French Alps will be but a desert. The interval between 1851 and 1856 will show a further decrease of population. In 1862, the ministry will announce a continued and progressive reduction in the number of acres devoted to agriculture; every year will aggravate the evil, and, in a half century, France will count more ruins, and a department the less."

Time has verified the predictions of De Bonville. The later census returns show a progressive diminution in the population of the departments of the Lower Alps, the Isère, the Drome, Ariège, the Upper and the Lower Pyrenees, the Lozère, the Ardennes, the Doubs, the Vosges, and, in short, in all the provinces formerly remarkable for their forests. This diminution is not to be ascribed to a passion for foreign emigration, as in Ireland, and in parts of Germany and of Italy; it is simply a transfer of population from one part of the empire to another, from soils which human folly has rendered uninhabitable, by ruthlessly depriving them of their natural advantages and securities, to provinces where the face of the earth was so formed by nature as to need no such safeguards, and where, consequently, she preserves her outlines in spite of the wasteful improvidence of man.*

covered with gravel and pebbles not less than 130,000 acres, "which, but for its inundations, would have been the finest land in the province."—ARTHUR YOUNG, *Travels in France*, vol. i, ch. i.

* Between 1851 and 1856 the population of Languedoc and Provence had increased by 101,000 souls. The augmentation, however, was wholly

Highly colored as these pictures seem, they are not exaggerated, although the hasty tourist through Southern France and Northern Italy, finding little in his high road experiences to justify them, might suppose them so. The lines of communication by locomotive train and diligence lead generally over safer ground, and it is only when they ascend the Alpine passes and traverse the mountain chains, that scenes somewhat resembling those just described fall under the eye of the ordinary traveller. But the extension of the sphere of devastation, by the degradation of the mountains and the transportation of their debris, is producing analogous effects upon the lower ridges of the Alps and the plains which skirt them; and even now one needs but an hour's departure from some great thoroughfares to reach sites where the genius of destruction revels as wildly as in the most frightful of the abysses which Blanqui has painted.*

in the provinces of the plains, where all the principal cities are found. In these provinces the increase was 204,000, while in the mountain provinces there was a diminution of 103,000. The reduction of the area of arable land is perhaps even more striking. In 1842, the department of the Lower Alps possessed 99,000 hectares, or nearly 245,000 acres, of cultivated soil. In 1852, it had but 74,000 hectares. In other words, in ten years 25,000 hectares, or 61,000 acres, had been washed away or rendered worthless for cultivation, by torrents and the abuses of pasturage.—CLAVÉ, *Études*, pp. 66, 67.

* The Skalära-Tobel, for instance, near Coire. See the description in BERLEPSCH, *Die Alpen*, pp. 169 *et seqq*, or in Stephen's English translation.

The recent change in the character of the Mella—a river anciently so remarkable for the gentleness of its current that it was specially noticed by Catullus as flowing *molli flumine*—deserves more than a passing remark. This river rises in the mountain chain east of Lake Iseo, and traversing the district of Brescia, empties into the Oglio after a course of about seventy miles. The iron works in the upper valley of the Mella had long created a considerable demand for wood, but their operations were not so extensive as to occasion any very sudden or general destruction of the forests, and the only evil experienced from the clearings was the gradual diminution of the volume of the river. Within the last twenty years, the superior quality of the arms manufactured at Brescia has greatly enlarged the sale of them, and very naturally stimulated the activity of both the forges and of the colliers who supply them, and the hillsides have been

There is one effect of the action of torrents which few travellers on the Continent are heedless enough to pass without notice. I refer to the elevation of the beds of mountain streams in consequence of the deposit of the debris with which they are charged. To prevent the spread of sand and gravel over the fields and the deluging overflow of the raging waters, the streams are confined by walls and embankments, which are gradually built higher and higher as the bed of the torrent is raised, so that, to reach a river, you ascend from the fields beside it; and sometimes the ordinary level of the stream is above the streets and even the roofs of the towns through which it passes.*

rapidly stripped of their timber. Up to 1850, no destructive inundation of the Mella had been recorded. Buildings in great numbers had been erected upon its margin, and its valley was conspicuous for its rural beauty and its fertility. But when the denudation of the mountains had reached a certain point, avenging nature began the work of retribution. In the spring and summer of 1850 several new torrents were suddenly formed in the upper tributary valleys, and on the 14th and 15th of August in that year, a fall of rain, not heavier than had been often experienced, produced a flood which not only inundated much ground never before overflowed, but destroyed a great number of bridges, dams, factories, and other valuable structures, and, what was a far more serious evil, swept off from the rocks an incredible extent of soil, and converted one of the most beautiful valleys of the Italian Alps into a ravine almost as bare and as barren as the savagest gorge of Southern France. The pecuniary damage was estimated at many millions of francs, and the violence of the catastrophe was deemed so extraordinary, even in a country subject to similar visitations, that the sympathy excited for the sufferers produced, in five months, voluntary contributions for their relief to the amount of nearly $200,000.—*Delle Inondazioni del Mella, etc., nella notte del* 14 *al* 15 *Agosto*, 1850.

The author of this remarkable pamphlet has chosen as a motto a passage from the Vulgate translation of Job, which is interesting as showing accurate observation of the action of the torrent: "Mons cadens definit, et saxum transfertur de loco suo; lapides excavant aquæ et alluvione paullatim terra consumitur."—*Job* xiv, 18, 19.

The English version is much less striking, and gives a different sense.

* Streffleur quotes from Duile the following observations: "The channel of the Tyrolese brooks is often raised much above the valleys through

The traveller who visits the depths of an Alpine ravine, observes the length and width of the gorge and the great height and apparent solidity of the precipitous walls which bound it, and calculates the mass of rock required to fill the vacancy, can hardly believe that the humble brooklet which purls at his feet has been the principal agent in accomplishing this tremendous erosion. Closer observation will often teach him, that the seemingly unbroken rock which overhangs the valley is full of cracks and fissures, and really in such a state of disintegration that every frost must bring down tons of it. If he compute the area of the basin which finds here its only discharge, he will perceive that a sudden thaw of the winter's deposit of snow, or one of those terrible discharges of rain so common in the Alps, must send forth a deluge mighty enough to sweep down the largest masses of gravel and of rock.*

which they flow. The bed of the Fersina is elevated high above the city of Trient, which lies near it. The Villerbach flows at a much more elevated level than that of the market place of Neumarkt and Vill, and threatens to overwhelm both of them with its waters. The Talfer at Botzen is at least even with the roofs of the adjacent town, if not above them. The tower steeples of the villages of Schlanders, Kortsch, and Laas, are lower than the surface of the Gadribach. The Saldurbach at Schluderns menaces the far lower village with destruction, and the chief town, Schwaz, is in similar danger from the Lahnbach."—STREFFLEUR, *Ueber die Wildbäche, etc.*, p. 7.

* The snow drifts into the ravines and accumulates to incredible depths, and the water resulting from its dissolution and from the deluging rains which fall in spring, and sometimes in the summer, being confined by rocky walls on both sides, rises to a very great height, and of course acquires an immense velocity and transporting power in its rapid descent to its outlet from the mountain. In the winter of 1842–'3, the valley of the Doveria, along which the Simplon road passes, was filled with solid snowdrifts to the depth of a hundred feet above the carriage road, and the sledge track by which passengers and the mails were carried ran at that height.

Other things being equal, the transporting power of the water is greatest where its flow is most rapid. This is usually in the direction of the axis of the ravine. As the current pours out of the gorge and escapes from the lateral confinement of its walls, it spreads and divides itself into numerous smaller streams, which shoot out from the mouth of the valley,

The simple measurement of the cubical contents of the semicircular hillock which he climbed before he entered the gorge, the structure and composition of which conclusively show that it must have been washed out of this latter by torrential action, will often account satisfactorily for the disposal of most of the matter which once filled the ravine.

It must further be remembered, that every inch of the violent movement of the rocks is accompanied with crushing concussion, or, at least, with great abrasion, and, as you follow the deposit along the course of the waters which transport it, you find the stones gradually rounding off in form, and diminishing in size until they pass successively into gravel, sand, impalpable slime.

as from a centre, in different directions, like the ribs of a fan from the pivot, each carrying with it its quota of stones and gravel. The plain below the point of issue from the mountain is rapidly raised by newly formed torrents, the elevation depending on the inclination of the bed and the form and weight of the matter transported. Every flood both increases the height of this central point and extends the entire circumference of the deposit. The stream retaining most nearly the original direction moves with the greatest momentum, and consequently transports the solid matter with which it is charged to the greatest distance.

The untravelled reader will comprehend this the better when he is informed that the southern slope of the Alps generally rises suddenly out of the plain, with no intervening hill to break the abruptness of the transition, except those consisting of comparatively small heaps of its own debris brought down by ancient glaciers or recent torrents. The torrents do not wind down valleys gradually widening to the rivers or the sea, but leap at once from the flanks of the mountains upon the plains below. This arrangement of surfaces naturally facilitates the formation of vast deposits at their points of emergence, and the centre of the accumulation in the case of very small torrents is not unfrequently a hundred feet high, and sometimes very much more.

Torrents and the rivers that receive them transport mountain debris to almost incredible distances. Lorentz, in an official report on this subject, as quoted by Marschand from the Memoirs of the Agricultural Society of Lyons, says: "The felling of the woods produces torrents which cover the cultivated soil with pebbles and fragments of rock, and they do not confine their ravages to the vicinity of the mountains, but extend them into the fertile fields of Provence and other departments, to the distance of forty or fifty leagues."—*Entwaldung der Gebirge*, p. 17.

I do not mean to assert that all the rocky valleys of the Alps have been produced by the action of torrents resulting from the destruction of the forests. All the greater, and many of the smaller channels, by which that chain is drained, owe their origin to higher causes. They are primitive fissures, ascribable to disruption in upheaval or other geological convulsion, widened and scarped, and often even polished, so to speak, by the action of glaciers during the ice period, and but little changed in form by running water in later eras.*

In these valleys of ancient formation, which extend into the very heart of the mountains, the streams, though rapid, have lost the true torrential character, if, indeed, they ever possessed it. Their beds have become approximately constant, and their walls no longer crumble and fall into the waters that wash their bases. The torrent-worn ravines, of which I have spoken, are of later date, and belong more properly to what may be called the crust of the Alps, consisting of loose rocks, of gravel, and of earth, strewed along the surface of the great declivities of the central ridge, and accumulated thickly between their solid buttresses. But it is on this crust that the mountaineer dwells. Here are his forests, here his pastures, and the ravages of the torrent both destroy his world, and convert it into a source of overwhelming desolation to the plains below.

Transporting Power of Rivers.

An instance that fell under my own observation in 1857, will serve to show something of the eroding and transporting

* The precipitous walls of the Val de Lys, and more especially of the Val Doveria, though here and there shattered, show in many places a smoothness of face over a large vertical plane, at the height of hundreds of feet above the bottom of the valley, which no known agency but glacier ice is capable of producing, and of course they can have undergone no sensible change at those points for a vast length of time. The beds of the rivers which flow through those valleys suffer lateral displacement occasionally, where there is room for the shifting of the channel; but if any elevation or depression takes place in them, it is too slow to be perceptible except in case of some merely temporary obstruction.

power of streams which, in these respects, fall incalculably below the torrents of the Alps. In a flood of the Ottauquechee, a small river which flows through Woodstock, Vermont, a milldam on that stream burst, and the sediment with which the pond was filled, estimated after careful measurement at 13,000 cubic yards, was carried down by the current. Between this dam and the slack water of another, four miles below, the bed of the stream, which is composed of pebbles interspersed in a few places with larger stones, is about sixty-five feet wide, though, at low water, the breadth of the current is considerably less. The sand and fine gravel were smoothly and evenly distributed over the bed to a width of fifty-five or sixty feet, and for a distance of about two miles, except at two or three intervening rapids, filled up all the interstices between the stones, covering them to the depth of nine or ten inches, so as to present a regularly formed concave channel, lined with sand, and reducing the depth of water, in some places, from five or six feet to fifteen or eighteen inches. Observing this deposit after the river had subsided and become so clear that the bottom could be seen, I supposed that the next flood would produce an extraordinary erosion of the banks and some permanent changes in the channel of the stream, in consequence of the elevation of the bed and the filling up of the spaces between the stones through which formerly much water had flowed; but no such result followed. The spring freshet of the next year entirely washed out the sand its predecessor had deposited, carried it to ponds and still-water reaches below, and left the bed of the river almost precisely in its former condition, though, of course, with the slight displacement of the pebbles which every flood produces in the channels of such streams. The pond, though often previously discharged by the breakage of the dam, had then been undisturbed for about twenty-five years, and its contents consisted almost entirely of sand, the rapidity of the current in floods being such that it would let fall little lighter sediment, even above an obstruction like a dam. The quantity I have mentioned evidently bears a very inconsiderable proportion to the total erosion of the stream during that period, because the wash of the banks consists chiefly of fine earth rather than of sand, and after the pond was once filled, or nearly so, even this material could no longer be deposited in it. The fact of the complete removal of the deposit I have described between the two dams in a single freshet, shows that, in spite of considerable obstruction from roughness of bed, large quantities of sand may be taken up and carried off by streams of no great rapidity of inclination; for the whole descent of the bed of the river between the two dams—a distance of four miles—is but sixty feet, or fifteen feet to the mile.

The Po and its Deposits.

The current of the river Po, for a considerable distance after its volume of water is otherwise sufficient for continuous navigation, is too rapid for that purpose until near Piacenza, where its velocity becomes too much reduced to transport great quantities of mineral matter, except in a state of minute division. Its southern affluents bring down from the Apennines a large quantity of fine earth from various geological formations, while its Alpine tributaries west of the Ticino are charged chiefly with rock ground down to sand or gravel.*

* Lombardini found, twenty years ago, that the mineral matter brought down to the Po by its tributaries was, in general, comminuted to about the same degree of fineness as the sands of its bed at their points of discharge. In the case of the Trebbia, which rises high in the Apennines and empties into the Po at Piacenza, it was otherwise, that river rolling pebbles and coarse gravel into the channel of the principal stream. The banks of the other affluents—excepting some of those which discharge their waters into the great lakes—then either retained their woods, or had been so long clear of them, that the torrents had removed most of the disintegrated and loose rock in their upper basins. The valley of the Trebbia had been recently cleared, and all the forces which tend to the degradation and transportation of rock were in full activity.—*Notice sur les Rivières de la Lombardie*, *Annales des Ponts et Chaussées*, 1847, 1er sémestre, p. 131.

Since the date of Lombardini's observations, many Alpine valleys have been stripped of their woods. It would be interesting to know whether any sensible change has been produced in the character or quantity of the matter transported by them to the Po.

The bed of the river has been somewhat elevated by the deposits in its channel, though not by any means above the level of the adjacent plains as has been so often represented. The dikes, which confine the current at high water, at the same time augment its velocity and compel it to carry most of its sediment to the Adriatic. It has, therefore, raised neither its own channel nor its alluvial shores, as it would have done if it had remained unconfined. But, as the surface of the water in floods is from six to fifteen feet above the general level of its banks, the Po can, at that period, receive no contributions of earth from the washing of the fields of Lombardy, and there is no doubt that a large proportion of the sediment it now deposits at its mouth descended from the Alps in the form of rock, though reduced by the grinding action of the waters, in its passage seaward, to the condition of fine sand, and often of silt.*

We know little of the history of the Po, or of the geography of the coast near the point where it enters the Adriatic, at any period more than twenty centuries before our own. Still less can we say how much of the plains of Lombardy had been formed by its action, combined with other causes, before man accelerated its levelling operations by felling the first woods on the mountains whence its waters are derived. But we know that since the Roman conquest of Northern Italy, its deposits have amounted to a quantity which, if recemented into rock, recombined into gravel, common earth, and vegetable mould, and restored to the situations where eruption or upheaval originally placed, or vegetation deposited it, would fill up hundreds of deep ravines in the Alps and Apennines, change the plan and profile of their chains, and give their

* In proportion as the dikes are improved, and breaches and the escape of the water through them are less frequent, the height of the annual inundations is increased. Many towns on the banks of the river, and of course within the system of parallel embankments, were formerly secure from flood by the height of the artificial mounds on which they were built; but they have recently been obliged to construct ring dikes for their protection.—BAUMGARTEN, after LOMBARDINI, in the paper last quoted, pp. 141, 147.

southern and northern faces respectively a geographical aspect very different from that they now present. Ravenna, forty miles south of the principal mouth of the Po, was built like Venice, in a lagoon, and the Adriatic still washed its walls at the commencement of the Christian era. The mud of the Po has filled up the lagoon, and Ravenna is now four miles from the sea. The town of Adria, which lies between the Po and the Adige, at the distance of some four or five miles from each, was once a harbor famous enough to have given its name to the Adriatic sea, and it was still a seaport in the time of Augustus. The combined action of the two rivers has so advanced the coast line that Adria is now about fourteen miles inland, and, in other places, the deposits made within the same period by these and other neighboring streams have a width of twenty miles.

What proportion of the earth with which they are charged these rivers have borne out into deep water, during the last two thousand years, we do not know, but as they still transport enormous quantities, as the North Adriatic appears to have shoaled rapidly, and as long islands, composed in great part of fluviatile deposits, have formed opposite their mouths, it must evidently have been very great. The floods of the Po occur but once, or sometimes twice in a year.* At other times, its waters are comparatively limpid and seem to hold no great amount of mud or fine sand in mechanical suspension; but at high water it contains a large proportion of solid matter, and according to Lombardini, it annually transports to the

* Three centuries ago, when the declivities of the mountains still retained a much larger proportion of their woods, the moderate annual floods of the Po were occasioned by the melting of the snows, and, as appears by a passage of Tasso quoted by Castellani (*Dell' Influenza delle Selve*, i, p. 58, note), they took place in May. The much more violent inundations of the present century are due to rains, the waters of which are no longer retained by a forest soil, but conveyed at once to the rivers—and they occur almost uniformly in the autumn or late summer. Castellani, on the page just quoted, says that even so late as about 1780, the Po required a heavy rain of a week to overflow its banks, but that forty years later, it was sometimes raised to full flood in a single day.

shores of the Adriatic not less than 42,760,000 cubic mètres, or very nearly 55,000,000 cubic yards, which carries the coast line out into the sea at the rate of more than 200 feet in a year.* The depth of the annual deposit is stated at eighteen centimètres, or rather more than seven inches, and it would cover an area of not much less than ninety square miles with a layer of that thickness. The Adige, also, brings every year to the Adriatic many million cubic yards of Alpine detritus, and the contributions of the Brenta from the same source are far from inconsiderable. The Adriatic, however, receives but a small proportion of the soil and rock washed away from the Italian slope of the Alps and the northern declivity of the Apennines by torrents. Nearly the whole of the debris thus removed from the southern face of the Alps between Monte Rosa and the sources of the Adda—a length of watershed not less than one hundred and fifty miles—is arrested by the still waters of the Lakes Maggiore and Como, and some smaller lacustrine reservoirs, and never reaches the sea. The Po is not continuously embanked except for the lower half of its course. Above Piacenza, therefore, it spreads and deposits sediment over a wide surface, and the water withdrawn from it for irrigation at lower points, as well as its inundations in the occasional ruptures of its banks, carry over the adjacent soil a large amount of slime.

If we add to the estimated annual deposits of the Po at its mouth, the earth and sand transported to the sea by the Adige, the Brenta, and other less important streams, the prodigious mass of detritus swept into Lago Maggiore by the Tosa, the Maggia, and the Ticino, into the lake of Como by the Maira

* This change of coast line cannot be ascribed to upheaval, for a comparison of the level of old buildings—as, for instance, the church of San Vitale and the tomb of Theodoric at Ravenna—with that of the sea, tends to prove a depression rather than an elevation of their foundations.

A computation by a different method makes the deposits at the mouth of the Po 2,123,000 mètres less; but as both of them omit the gravel and silt rolled, if not floated, down at ordinary and low water, we are safe in assuming the larger quantity.—*Article last quoted*, p. 174. (See note, p. 329)

and the Adda, into the lake of Garda by its affluents, and the yet vaster heaps of pebbles, gravel, and earth permanently deposited by the torrents near their points of eruption from mountain gorges, or spread over the wide plains at lower levels, we may safely assume that we have an aggregate of not less than four times the quantity carried to the Adriatic by the Po, or 220,000,000 cubic yards of solid matter, abstracted every year from the Italian Alps and the Apennines, and removed out of their domain by the force of running water.*

The present rate of deposit at the mouth of the Po has continued since the year 1600, the previous advance of the coast, after the year 1200, having been only one third as rapid. The great increase of erosion and transport is ascribed by Lombardini chiefly to the destruction of the forests in the basin of that river and the valleys of its tributaries, since the beginning of the seventeenth century.† We have no data to show the rate of deposit in any given century before the year 1200, and it doubtless varied according to the progress of population and the consequent extension of clearing and cultivation. The transporting power of torrents is greatest soon after their formation, because at that time their points of delivery are lower, and, of course, their general slope and velocity more rapid, than after years of erosion above, and deposit below, have depressed the beds of their mountain valleys, and elevated the channels of their lower course. Their eroding action also is most powerful at the same period, both because their mechanical force is then greatest, and because the loose earth and

* Mengotti estimated the mass of solid matter annually "united to the waters of the Po" at 822,000,000 cubic mètres, or nearly twenty times as much as, according to Lombardini, that river delivers into the Adriatic. Castellani supposes the computation of Mengotti to fall much below the truth, and there can be no doubt that a vastly larger quantity of earth and gravel is washed down from the Alps and the Apennines than is carried to the sea.—CASTELLANI, *Dell' Immediata Influenza delle Selve sul corso delle Acque*, i, pp. 42, 43.

I have contented myself with assuming less than one fifth of Mengotti's estimate.

† BAUMGARTEN, *An. des Ponts et Chaussées*, 1847, 1er sémestre, p. 175.

stones of freshly cleared forest ground are most easily removed. Many of the Alpine valleys west of the Ticino—that of the Dora Baltea for instance—were nearly stripped of their forests in the days of the Roman empire, others in the Middle Ages, and, of course, there must have been, at different periods before the year 1200, epochs when the erosion and transportation of solid matter from the Alps and the Apennines were as great as since the year 1600.

Upon the whole, we shall not greatly err if we assume that, for a period of not less than two thousand years, the walls of the basin of the Po—the Italian slope of the Alps, and the northern and northeastern declivities of the Apennines—have annually sent down into the Adriatic, the lakes, and the plains, not less than 150,000,000 cubic yards of earth and disintegrated rock. We have, then, an aggregate of 300,000,000,000 cubic yards of such material, which, allowing to the mountain surface in question an area of 50,000,000,000 square yards, would cover the whole to the depth of six yards.* There are very large portions of this area, where, as we know from ancient remains—roads, bridges, and the like—from other direct testimony, and from geological considerations, very little degradation has taken place within twenty centuries, and hence the quantity to be assigned to localities where the destructive causes have been most active is increased in proportion.

If this vast mass of pulverized rock and earth were restored to the localities from which it was derived, it certainly would not obliterate valleys and gorges hollowed out by great geological causes, but it would reduce the length and diminish the depth of ravines of later formation, modify the inclination of their walls, reclothe with earth many bare mountain ridges,

* The total superficies of the basin of the Po, down to Ponte Lagoscuro [Ferrara]—a point where it has received all its affluents—is 6,938,200 hectares, that is, 4,105,600 in mountain lands, 2,832,600 in plain lands.—Dumont, *Travaux Publics, etc.*, p. 272.

These latter two quantities are equal respectively to 10,145,348, and 6,999,638 acres, or 15,852 and 10,937 square miles.

essentially change the line of junction between plain and mountain, and carry back a long reach of the Adriatic coast many miles to the west.*

It is, indeed, not to be supposed that all the degradation

* I do not use the numbers I have borrowed or assumed as factors the value of which is precisely ascertained; nor, for the purposes of the present argument, is quantitative exactness important. I employ numerical statements simply as a means of aiding the imagination to form a general and certainly not extravagant idea of the extent of geographical revolutions which man has done much to accelerate, if not, strictly speaking, to produce.

There is an old proverb, *Dolus latet in generalibus*, and Arthur Young is not the only public economist who has warned his readers against the deceitfulness of round numbers. I think, on the contrary, that vastly more error has been produced by the affectation of precision in cases where precision is impossible. In all the great operations of terrestrial nature, the elements are so numerous and so difficult of exact appreciation, that, until the means of scientific observation and measurement are much more perfected than they now are, we must content ourselves with general approximations. I say *terrestrial* nature, because in cosmical movements we have fewer elements to deal with, and may therefore arrive at much more rigorous accuracy in determination of time and place than we can in fixing and predicting the quantities and the epochs of variable natural phenomena on the earth's surface.

The value of a high standard of accuracy in scientific observation can hardly be overrated; but habits of rigorous exactness will never be formed by an investigator who allows himself to trust implicitly to the numerical precision of the results of a few experiments. The wonderful accuracy of geodetic measurements in modern times is, in general, attained by taking the mean of a great number of observations at every station, and this final precision is but the mutual balance and compensation of numerous errors.

Travellers are often misled by local habits in the use of what may be called representative numbers, where a definite is put for an indefinite quantity. A Greek, who wished to express the notion of a great, but undetermined number, used "myriad, or ten thousand;" a Roman, "six hundred;" an Oriental, "forty," or, at present, very commonly, "fifteen thousand." Many a tourist has gravely repeated, as an ascertained fact, the vague statement of the Arabs and the monks of Mount Sinai, that the ascent from the convent of St. Catherine to the summit of Gebel Moosa counts "fifteen thousand" steps, though the difference of level is barely two thousand feet, and the "Forty" Thieves, the "forty" martyr monks

of the mountains is due to the destruction of the forests—that the flanks of every Alpine valley in Central Europe below the snow line were once covered with earth and green with woods, but there are not many particular cases, in which we can, with certainty, or even with strong probability, affirm the contrary.

of the convent of El Arbain—not to speak of a similar use of this numeral in more important cases—have often been understood as expressions of a known number, when in fact they mean simply *many*. The number "fifteen thousand" has found its way to Rome, and De Quincey seriously informs us, on the authority of a lady who had been at much pains to ascertain the *exact* truth, that, including closets large enough for a bed, the Vatican contains fifteen thousand rooms. Any one who has observed the vast dimensions of most of the apartments of that structure will admit that we make a very small allowance of space when we assign a square rod, sixteen and a half feet square, to each room upon the average. On an acre, there might be one hundred and sixty such rooms, including partition walls; and, to contain fifteen thousand of them, a building must cover more than nine acres, and be ten stories high, or possess other equivalent dimensions, which, as every traveller knows, many times exceeds the truth.

That most entertaining writer, About, reduces the number of rooms in the Vatican, but he compensates this reduction by increased dimensions, for he uses the word *salle*, which cannot be applied to closets barely large enough to contain a bed. According to him, there are in that "presbytère," as he irreverently calls it, twelve thousand large rooms [*salles*], thirty courts, and three hundred staircases.—*Rome Contemporaire*, p. 68.

The pretended exactness of statistical tables is generally little better than an imposture; and those founded not on direct estimation by competent observers, but on the report of persons who have no particular interest in knowing, but often have a motive for distorting, the truth—such as census returns—are commonly to be regarded as but vague guesses at the actual fact.

Fuller, who, for the combination of wit, wisdom, fancy, and personal goodness, stands first in English literature, thus remarks on the pretentious exactness of historical and statistical writers: "I approve the plain, country By-word, as containing much Innocent Simplicity therein,

'Almost and very nigh
Have saved many a Lie.'

So have the Latines their *prope*, *fere*, *juxta*, *circiter*, *plus minus*, used in matters of fact by the most authentic Historians. Yea, we may observe that the Spirit of Truth itself, where *Numbers* and *Measures* are concerned,

We cannot measure the share which human action has had in augmenting the intensity of causes of mountain degradation, but we know that the clearing of the woods has, in some cases, produced within two or three generations, effects as blasting as those generally ascribed to geological convulsions, and has laid waste the face of the earth more hopelessly than if it had been buried by a current of lava or a shower of volcanic sand. New torrents are forming every year in the Alps. Tradition, written records, and analogy concur to establish the belief that the ruin of most of the now desolate valleys in those mountains is to be ascribed to the same cause, and authentic descriptions of the irresistible force of the torrent show that, aided by frost and heat, it is adequate to level Mont Blanc and Monte Rosa themselves, unless new upheavals shall maintain their elevation.

It has been contended that all rivers which take their rise in mountains originated in torrents. These, it is said, have lowered the summits by gradual erosion, and, with the material thus derived, have formed shoals in the sea which once beat against the cliffs; then, by successive deposits, gradually raised them above the surface, and finally expanded them into broad plains traversed by gently flowing streams. If we could go back to earlier geological periods, we should find this theory often verified, and we cannot fail to see that the torrents go on

in Times, Places, and Persons, useth the aforesaid Modifications, save in such cases where some mystery contained in the number requireth a particular specification thereof:

In Times.	In Places.	In Persons.
Daniel, 5:33.	Luke, 24:13.	Exodus, 12:37.
Luke, 3:23.	John, 6:19.	Acts, 2:41.

None therefore can justly find fault with me, if, on the like occasion, I have secured myself with the same Qualifications. Indeed, such Historians who grind their Intelligence to the *powder of fraction*, pretending to *cleave the pin*, do sometimes *misse the But*. Thus, one reporteth, how in the Persecution under *Diocletian*, there were neither under nor over, but just *nine hundred ninety-nine* martyrs. Yea, generally those that trade in such *Retail-ware*, and deal in such small parcells, may by the ignorant be commended for their *care*, but condemned by the judicious for their ridiculous *curiosity*."—*The History of the Worthies of England*, i, p. 59.

at the present hour, depressing still lower the ridges of the Alps and the Apennines, raising still higher the plains of Lombardy and Provence, extending the coast still farther into the Adriatic and the Mediterranean, reducing the inclination of their own beds and the rapidity of their flow, and thus tending to become river-like in character.

There are cases where torrents cease their ravages of themselves, in consequence of some change in the condition of the basin where they originate, or of the face of the mountain at a higher level, while the plain or the sea below remains in substantially the same state as before. If a torrent rises in a small valley containing no great amount of earth and of disintegrated or loose rock, it may, in the course of a certain period, wash out all the transportable material, and if the valley is then left with solid walls, it will cease to furnish debris to be carried down by floods. If, in this state of things, a new channel be formed at an elevation above the head of the valley, it may divert a part, or even the whole of the rain water and melted snow which would otherwise have flowed into it, and the once furious torrent now sinks to the rank of a humble and harmless brooklet. "In traversing this department," says Surell, "one often sees, at the outlet of a gorge, a flattened hillock, with a fan-shaped outline and regular slopes; it is the bed of dejection of an ancient torrent. It sometimes requires long and careful study to detect the primitive form, masked as it is by groves of trees, by cultivated fields, and often by houses, but, when examined closely, and from different points of view, its characteristic figure manifestly appears, and its true history cannot be mistaken. Along the hillock flows a streamlet, issuing from the ravine, and quietly watering the fields. This was originally a torrent, and in the background may be discovered its mountain basin. Such *extinguished* torrents, if I may use the expression, are numerous." *

* SURELL, *Les Torrents des Hautes Alpes*, chap. xxiv. In such cases, the clearing of the ground, which, in consequence of a temporary diversion of the waters, or from some other cause, has become rewooded, sometimes renews the ravages of the torrent. Thus, on the left bank of the

But for the intervention of man and domestic animals, these latter beneficent revolutions would occur more frequently, proceed more rapidly. The new scarped mountains, the hillocks of debris, the plains elevated by sand and gravel spread over them, the shores freshly formed by fluviatile deposits, would clothe themselves with shrubs and trees, the intensity of the causes of degradation would be diminished, and nature would thus regain her ancient equilibrium. But these processes, under ordinary circumstances, demand, not years, generations, but centuries; * and man, who even now finds scarce breathing room on this vast globe, cannot retire from the Old World to some yet undiscovered continent, and wait for the slow action of such causes to replace, by a new creation, the Eden he has wasted.

Mountain Slides.

I have said that the mountainous regions of the Atlantic States of the American Union are exposed to similar ravages, and I may add that there is, in some cases, reason to apprehend from the same cause even more appalling calamities than those which I have yet described. The slide in the Notch of the White Mountains, by which the Willey family lost their lives, is an instance of the sort I refer to, though I am not able to say that in this particular case, the slip of the earth and

Durance, a wooded declivity had been formed by the debris brought down by torrents, which had extinguished themselves after having swept off much of the superficial strata of the mountain of Morgon. "All this district was covered with woods, which have now been thinned out and are perishing from day to day; consequently, the torrents have recommenced their devastations, and if the clearings continue, this declivity, now fertile, will be ruined, like so many others."—Id., p. 155.

* Where a torrent has not been long in operation, and earth still remains mixed with the rocks and gravel it heaps up at its point of eruption, vegetation soon starts up and prospers, if protected from encroachment. In Provence, "several communes determined, about ten years ago, to reserve the soils thus wasted, that is, to abandon them for a certain time, to spontaneous vegetation, which was not slow in making its appearance."—Becquerel, *Des Climats*, p. 315.

rock was produced by the denudation of the surface. It may have been occasioned by this cause, or by the construction of the road through the Notch, the excavations for which, perhaps, cut through the buttresses that supported the sloping strata above.

Not to speak of the fall of earth when the roots which held it together, and the bed of leaves and mould which sheltered it both from disintegrating frost and from sudden drenching and dissolution by heavy showers, are gone, it is easy to see that, in a climate with severe winters, the removal of the forest, and, consequently, of the soil it had contributed to form, might cause the displacement and descent of great masses of rock. The woods, the vegetable mould, and the soil beneath, protect the rocks they cover from the direct action of heat and cold, and from the expansion and contraction which accompany them. Most rocks, while covered with earth, contain a considerable quantity of water.* A fragment of rock pervaded with moisture cracks and splits, if thrown into a furnace, and sometimes with a loud detonation; and it is a familiar observation that the fire, in burning over newly cleared lands, breaks up and sometimes almost pulverizes the stones. This effect is due partly to the unequal expansion of the stone, partly to the action of heat on the water it contains in its pores. The sun, suddenly let in upon rock which had been covered with moist earth for centuries, produces more or less disintegration in the same way, and the stone is also exposed to chemical influences from which it was sheltered before. But in the climate of the United States as well as of the Alps, frost is a still more powerful agent in breaking up mountain masses.

* Rock is permeable by water to a greater extent than is generally supposed. Freshly quarried marble, and even granite, as well as most other stones, are sensibly heavier, as well as softer and more easily wrought, than after they are dried and hardened by air-seasoning. Many sandstones are porous enough to serve as filters for liquids, and much of that of Upper Egypt and Nubia hisses audibly when thrown into water, from the escape of the air forced out of it by hydrostatic pressure and the capillary attraction of the pores for water. See *Appendix*, No. 28.

The soil that protects the lime and sand stone, the slate and the granite from the influence of the sun, also prevents the water which filters into their crevices and between their strata from freezing in the hardest winters, and the moisture descends, in a liquid form, until it escapes in springs, or passes off by deep subterranean channels. But when the ridges are laid bare, the water of the autumnal rains fills the minutest pores and veins and fissures and lines of separation of the rocks, then suddenly freezes, and bursts asunder huge, and apparently solid blocks of adamantine stone.* Where the strata are inclined at a considerable angle, the freezing of a thin film of water over a large interstratal area might occasion a slide that should cover miles with its ruins; and similar results might be produced by the simple hydrostatic pressure of a column of water, admitted by the removal of the covering of earth to flow into a crevice faster than it could escape through orifices below.

Earth or rather mountain slides, compared to which the catastrophe that buried the Willey family in New Hampshire was but a pinch of dust, have often occurred in the Swiss, Italian, and French Alps. The land slip, which overwhelmed,

* Palissy had observed the action of frost in disintegrating rock, and he thus describes it, in his essay on the formation of ice: "I know that the stones of the mountains of Ardennes be harder than marble. Nevertheless, the people of that country do not quarry the said stones in winter, for that they be subject to frost; and many times the rocks have been seen to fall without being cut, by means whereof many people have been killed, when the said rocks were thawing." Palissy was ignorant of the expansion of water in freezing—in fact he supposed that the mechanical force exerted by freezing water was due to compression, not dilatation—and therefore he ascribes to thawing alone effects resulting not less from congelation.

Various forces combine to produce the stone avalanches of the higher Alps, the fall of which is one of the greatest dangers incurred by the adventurous explorers of those regions—the direct action of the sun upon the stone, the expansion of freezing water, and the loosening of masses of rock by the thawing of the ice which supported them or held them together.

and covered to the depth of seventy feet, the town of Plurs in the valley of the Maira, on the night of the 4th of September, 1618, sparing not a soul of a population of 2,430 inhabitants, is one of the most memorable of these catastrophes, and the fall of the Rossberg or Rufiberg, which destroyed the little town of Goldau in Switzerland, and 450 of its people, on the 2d of September, 1806, is almost equally celebrated. In 1771, according to Wessely, the mountain peak Piz, near Alleghe in the province of Belluno, slipped into the bed of the Cordevole, a tributary of the Piave, destroying in its fall three hamlets and sixty lives. The rubbish filled the valley for a distance of nearly two miles, and, by damming up the waters of the Cordevole, formed a lake about three miles long, and a hundred and fifty feet deep, which still subsists, though reduced to half its original length by the wearing down of its outlet.*

On the 14th of February, 1855, the hill of Belmonte, a little below the parish of San Stefano, in Tuscany, slid into the valley of the Tiber, which consequently flooded the village to the depth of fifty feet, and was finally drained off by a tunnel. The mass of debris is stated to have been about 3,500 feet long, 1,000 wide, and not less than 600 high.†

Such displacements of earth and rocky strata rise to the magnitude of geological convulsions, but they are of so rare occurrence in countries still covered by the primitive forest, so common where the mountains have been stripped of their native covering, and, in many cases, so easily explicable by the drenching of incohesive earth from rain, or the free admission of water between the strata of rocks—both of which a coating of vegetation would have prevented—that we are justified in ascribing them for the most part to the same cause as

* WESSELY, *Die Oesterreichischen Alpenländer und ihre Forste*, pp. 125, 126. Wessely records several other more or less similar occurrences in the Austrian Alps. Some of them, certainly, are not to be ascribed to the removal of the woods, but in most cases they are clearly traceable to that cause.

† BIANCHI, Appendix to the Italian translation of Mrs. SOMERVILLE's *Physical Geography*, p. xxxvi.

that to which the destructive effects of mountain torrents are chiefly due—the felling of the woods.

In nearly every case of this sort the circumstances of which are known, the immediate cause of the slip has been, either an earthquake, the imbibition of water in large quantities by bare earth, or its introduction between or beneath solid strata. If water insinuates itself between the strata, it creates a sliding surface, or it may, by its expansion in freezing, separate beds of rock, which had been nearly continuous before, widely enough to allow the gravitation of the superincumbent mass to overcome the resistance afforded by inequalities of face and by friction; if it finds its way beneath hard earth or rock reposing on clay or other bedding of similar properties, it converts the supporting layer into a semi-fluid mud, which opposes no obstacle to the sliding of the strata above.

The upper part of the mountain which buried Goldau was composed of a hard but brittle conglomerate, called *nagelflue*, resting on an unctuous clay, and inclining rapidly toward the village. Much earth remained upon the rock, in irregular masses, but the woods had been felled, and the water had free access to the surface, and to the crevices which sun and frost had already produced in the rock, and of course, to the slimy stratum beneath. The whole summer of 1806 had been very wet, and an almost incessant deluge of rain had fallen the day preceding the catastrophe, as well as on that of its occurrence. All conditions then, were favorable to the sliding of the rock, and, in obedience to the laws of gravitation, it precipitated itself into the valley as soon as its adhesion to the earth beneath it was destroyed by the conversion of the latter into a viscous paste. The mass that fell measured between two and a half and three miles in length by one thousand feet in width, and its average thickness is thought to have been about a hundred feet. The highest portion of the mountain was more than three thousand feet above the village, and the momentum acquired by the rocks and earth in their descent carried huge blocks of stone far up the opposite slope of the Rigi.

The Piz, which fell into the Cordevole, rested on a steeply

inclined stratum of limestone, with a thin layer of calcareous marl intervening, which, by long exposure to frost and the infiltration of water, had lost its original consistence, and become a loose and slippery mass instead of a cohesive and tenacious bed.

Protection against fall of Rocks and Avalanches by Trees.

Forests often subserve a valuable purpose in preventing the fall of rocks, by mere mechanical resistance. Trees, as well as herbaceous vegetation, grow in the Alps upon declivities of surprising steepness of inclination, and the traveller sees both luxuriant grass and flourishing woods on slopes at which the soil, in the dry air of lower regions, would crumble and fall by the weight of its own particles. When loose rocks lie scattered on the face of these declivities, they are held in place by the trunks of the trees, and it is very common to observe a stone that weighs hundreds of pounds, perhaps even tons, resting against a tree which has stopped its progress just as it was beginning to slide down to a lower level. When a forest in such a position is cut, these blocks lose their support, and a single wet season is enough not only to bare the face of a considerable extent of rock, but to cover with earth and stone many acres of fertile soil below.*

In Switzerland and other snowy and mountainous countries, forests render a most important service by preventing the formation and fall of destructive avalanches, and in many parts of the Alps exposed to this catastrophe, the woods are protected, though too often ineffectually, by law. No forest, indeed, could arrest a large avalanche once in motion, but the mechanical resistance afforded by the trees prevents their

* See in KOHL, *Alpenreisen*, i, 120, an account of the ruin of fields and pastures, and even of the destruction of a broad belt of forest, by the fall of rocks in consequence of cutting a few large trees. Cattle are very often killed in Switzerland by rock avalanches, and their owners secure themselves from loss by insurance against this risk as against damage by fire or hail.

formation, both by obstructing the wind, which gives to the dry snow of the *Staub-Lawine*, or dust avalanche, its first impulse, and by checking the disposition of moist snow to gather itself into what is called the *Rutsch-Lawine*, or sliding avalanche. Marschand states that, the very first winter after the felling of the trees on the higher part of a declivity between Saanen and Gsteig where the snow had never been known to slide, an avalanche formed itself in the clearing, thundered down the mountain, and overthrew and carried with it a hitherto unviolated forest to the amount of nearly a million cubic feet of timber.* The path once opened down the flanks of the mountain, the evil is almost beyond remedy. The snow sometimes carries off the earth from the face of the rock, or, if the soil is left, fresh slides every winter destroy the young plantations, and the restoration of the wood becomes impossible. The track widens with every new avalanche. Dwellings and their occupants are buried in the snow, or swept away by the rushing mass, or by the furious blasts it occasions through the displacement of the air; roads and bridges are destroyed; rivers blocked up, which swell till they overflow the valley above, and then, bursting their snowy barrier, flood the fields below with all the horrors of a winter inundation.†

Principal Causes of the Destruction of the Forest.

The needs of agriculture are the most familiar cause of the destruction of the forest in new countries; for not only does an increasing population demand additional acres to grow the

* *Entwaldung der Gebirge*, p. 41.

† The importance of the wood in preventing avalanches is well illustrated by the fact that, where the forest is wanting, the inhabitants of localities exposed to snow slides often supply the place of the trees by driving stakes through the snow into the ground, and thus checking its propensity to slip. The woods themselves are sometimes thus protected against avalanches originating on slopes above them, and as a further security, small trees are cut down along the upper line of the forest, and laid against the trunks of larger trees, transversely to the path of the slide, to serve as a fence or dam to the motion of an incipient avalanche,

vegetables which feed it and its domestic animals, but the slovenly husbandry of the border settler soon exhausts the luxuriance of his first fields, and compels him to remove his household gods to a fresher soil. With growing numbers, too, come the many arts for which wood is the material. The demands of the near and the distant market for this product excite the cupidity of the hardy forester, and a few years of that wild industry of which Springer's "Forest Life and Forest Trees" so vividly depicts the dangers and the triumphs, suffice to rob the most inaccessible glens of their fairest ornaments. The value of timber increases with its dimensions in almost geometrical proportion, and the tallest, most vigorous, and most symmetrical trees fall the first sacrifice. This is a fortunate circumstance for the remainder of the wood; for the impatient lumberman contents himself with felling a few of the best trees, and then hurries on to take his tithe of still virgin groves.

The unparalleled facilities for internal navigation, afforded by the numerous rivers of the present and former British colonial possessions in North America, have proved very fatal to the forests of that continent. Quebec has become a centre for a lumber trade, which, in the bulk of its material, and, consequently, in the tonnage required for its transportation, rivals the commerce of the greatest European cities. Immense rafts are collected at Quebec from the great Lakes, from the Ottawa, and from all the other tributaries which unite to swell the current of the St. Lawrence and help it to struggle against its mighty tides.* Ships, of burden formerly undreamed of, have been built to convey the timber to the markets of Europe, and during the summer months the St. Lawrence is almost as

which may by this means be arrested before it acquires a destructive velocity and force.

* The tide rises at Quebec to the height of twenty-five feet, and when it is aided by a northeast wind, it flows with almost irresistible violence. Rafts containing several hundred thousand cubic feet of timber are often caught by the flood tide, torn to pieces, and dispersed for miles along the shores.

crowded with vessels as the Thames.* Of late, Chicago, in Illinois, has been one of the greatest lumber as well as grain depots of the United States, and it receives and distributes contributions from all the forests in the States washed by Lake Michigan, as well as from some more distant points.

The operations of the lumberman involve other dangers to the woods besides the loss of the trees felled by him. The narrow clearings around his *shanties* † form openings which let in the wind, and thus sometimes occasion the overthrow of thousands of trees, the fall of which dams up small streams, and creates bogs by the spreading of the waters, while the decaying trunks facilitate the multiplication of the insects which breed in dead wood, and are, some of them, injurious to living trees. The escape and spread of camp fires, however, is the most devastating of all the causes of destruction that find their origin in the operations of the lumberman. The proportion of trees fit for industrial uses is small in all primitive woods. Only these fall before the forester's axe, but the fire destroys, indiscriminately, every age and every species of tree.‡

* One of these, the Baron of Renfrew—so named from one of the titles of the kings of England—built thirty or forty years ago, measured 5,000 tons. They were little else than rafts, being almost solid masses of timber designed to be taken to pieces and sold as lumber on arriving at their port of destination.

The lumber trade at Quebec is still very large. According to a recent article in the *Revue des Deux Mondes*, that city exported, in 1860, 30,000,000 cubic feet of squared timber, and 400,000,000 square feet of "planches.' The thickness of the boards is not stated, but I believe they are generally cut an inch and a quarter thick for the Quebec trade, and as they shrink somewhat in drying, we may estimate ten square for one cubic foot of boards. This gives a total of 70,000,000 cubic feet. The specific gravity of white pine is .554, and the weight of this quantity of lumber, very little of which is thoroughly seasoned, would exceed a million of tons, even supposing it to consist wholly of wood as light as pine. New Brunswick, too, exports a large amount of lumber.

† This name, from the French *chantier*, which has a wider meaning, is applied in America to temporary huts or habitations erected for the convenience of forest life, or in connection with works of material improvement.

‡ Trees differ much in their power of resisting the action of forest fires.

While, then, without much injury to the younger growths, the native forest will bear several "cuttings over" in a generation —for the increasing value of lumber brings into use, every four or five years, a quality of timber which had been before rejected as unmarketable—a fire may render the declivity of a mountain unproductive for a century.*

Different woods vary greatly in combustibility, and even when their bark is scarcely scorched, they are, partly in consequence of physiological character, and partly from the greater or less depth at which their roots habitually lie below the surface, very differently affected by running fires. The white pine, *Pinus strobus*, as it is the most valuable, is also perhaps the most delicate tree of the American forest, while its congener, the Northern pitch pine, *Pinus rigida*, is less injured by fire than any other tree of that country. I have heard experienced lumbermen maintain that the growth of this pine was even accelerated by a fire brisk enough to destroy all other trees, and I have myself seen it still flourishing after a conflagration which had left not a green leaf but its own in the wood, and actually throwing out fresh foliage, when the old had been quite burnt off and the bark almost converted into charcoal. The wood of the pitch pine is of comparatively little value for the joiner, but it is useful for very many purposes. Its rapidity of growth in even poor soils, its hardihood, and its abundant yield of resinous products, entitle it to much more consideration, as a plantation tree, than it has hitherto received in Europe or America.

* Between fifty and sixty years ago, a steep mountain with which I am very familiar, composed of metamorphic rock, and at that time covered with a thick coating of soil and a dense primeval forest, was accidentally burnt over. The fire took place in a very dry season, the slope of the mountain was too rapid to retain much water, and the conflagration was of an extraordinarily fierce character, consuming the wood almost entirely, burning the leaves and combustible portion of the mould, and in many places cracking and disintegrating the rock beneath. The rains of the following autumn carried off much of the remaining soil, and the mountain side was nearly bare of wood for two or three years afterward. At length, a new crop of trees sprang up and grew vigorously, and the mountain is now thickly covered again. But the depth of mould and earth is too small to allow the trees to reach maturity. When they attain to the diameter of about six inches, they uniformly die, and this they will no doubt continue to do until the decay of leaves and wood on the surface, and the decomposition of the subjacent rock, shall have formed, perhaps hundreds of years hence, a stratum of soil thick enough to support a full-grown forest.

American Forest Trees

The remaining forests of the Northern States and of Canada no longer boast the mighty pines which almost rivalled the gigantic Sequoia of California; and the growth of the larger forest trees is so slow, after they have attained to a certain size, that if every pine and oak were spared for two centuries, the largest now standing would not reach the stature of hundreds recorded to have been cut within two or three generations.* Dr. Williams, who wrote about sixty years ago, states the following as the dimensions of "such trees as are esteemed large ones of their kind in that part of America" [Vermont], qualifying his account with the remark that his measurements "do not denote the greatest which nature has produced of

* The growth of the white pine, on a good soil and in open ground, is rather rapid until it reaches the diameter of a couple of feet, after which it is much slower. The favorite habitat of this tree is light sandy earth. On this soil, and in a dense wood, it requires a century to attain the diameter of a yard. Emerson (*Trees of Massachusetts*, p. 65), says that a pine of this species, near Paris, "thirty years planted, is eighty feet high, with a diameter of three feet." He also states that ten white pines planted at Cambridge, Massachusetts, in 1809 or 1810, exhibited, in the winter of 1841 and 1842, an average of twenty inches diameter at the ground, the two largest measuring, at the height of three feet, four feet eight inches in circumference; and he mentions another pine growing in a rocky swamp, which, at the age of thirty-two years, "gave seven feet in circumference at the but, with a height of sixty-two feet six inches." This latter I suppose to be a seedling, the others *transplanted* trees, which might have been some years old when placed where they finally grew.

The following case came under my own observation: In 1824, a pine tree, so small that a young lady, with the help of a lad, took it up from the ground and carried it a quarter of a mile, was planted near a house in a town in Vermont. It was occasionally watered, but received no other special treatment. I measured this tree in 1860, and found it, at four feet from the ground, and entirely above the spread of the roots, two feet and four inches in diameter. It could not have been more than three inches through when transplanted, and must have increased its diameter twenty-five inches in thirty-six years.

their particular species, but the greatest which are to be found in most of our towns."

	Diameter.	Height.
Pine, ...	6 feet,	247 feet.
Maple, ...	5 " 9 inches,	From 100 to 200 feet.
Buttonwood, . . .	5 " 6 "	
Elm, ...	5 "	
Hemlock, ...	4 " 9 "	
Oak, ...	4 "	
Basswood, ...	4 "	
Ash, ...	4 "	
Birch, ...	4 "	

He adds a note saying that a white pine was cut in Dunstable, New Hampshire, in the year 1736, the diameter of which was seven feet and eight inches. Dr. Dwight says that a fallen pine in Connecticut was found to measure two hundred and forty-seven feet in height, and adds: "A few years since, such trees were in great numbers along the northern parts of Connecticut River." In another letter, he speaks of the white pine as "frequently six feet in diameter, and two hundred and fifty feet in height," and states that a pine had been cut in Lancaster, New Hampshire, which measured two hundred and sixty-four feet. Emerson wrote in 1846: "Fifty years ago, several trees growing on rather dry land in Blandford, Massachusetts, measured, after they were felled, two hundred and twenty-three feet. All these trees are surpassed by a pine felled at Hanover, New Hampshire, about a hundred years ago, and described as measuring two hundred and seventy-four feet.*

These descriptions, it will be noticed, apply to trees cut from sixty to one hundred years since. Persons, whom observation has rendered familiar with the present character of the American forest, will be struck with the smallness of the diameter which Dr. Williams and Dr. Dwight ascribe to trees

* WILLIAMS, *History of Vermont*, ii, p. 53. DWIGHT'S *Travels*, iv, p. 21, and iii, p. 36. EMERSON, *Trees of Massachusetts*, p. 61. PARISH, *Life of President Wheelock*, p. 56.

of such extraordinary height. Individuals of the several species mentioned in Dr. Williams's table, are now hardly to be found in the same climate, exceeding one half or at most two thirds of the height which he assigns to them; but, except in the case of the oak and the pine, the diameter stated by him would not be thought very extraordinary in trees of far less height, now standing. Even in the species I have excepted, those diameters, with half the heights of Dr. Williams, might perhaps be paralleled at the present time; and many elms, transplanted, at a diameter of six inches, within the memory of persons still living, measure six, and sometimes even seven feet through. For this change in the growth of forest trees there are two reasons: the one is, that the great commercial value of the pine and the oak have caused the destruction of all the best—that is, the tallest and straightest—specimens of both; the other, that the thinning of the woods by the axe of the lumberman has allowed the access of light and heat and air to trees of humbler worth and lower stature, which have survived their more towering brethren. These, consequently, have been able to expand their crowns and swell their stems to a degree not possible so long as they were overshadowed and stifled by the lordly oak and pine. While, therefore, the New England forester must search long before he finds a pine

> fit to be the mast
> Of some great ammiral,

beeches and elms and birches, as sturdy as the mightiest of their progenitors, are still no rarity.*

* The forest trees of the Northern States do not attain to extreme longevity in the dense woods. Dr. Williams found that none of the huge pines, the age of which he ascertained, exceeded three hundred and fifty or four hundred years, though he quotes a friend who thought he had noticed trees considerably older. The oak lives longer than the pine, and the hemlock spruce is perhaps equally long lived. A tree of this latter species, cut within my knowledge in a thick wood, counted four hundred and eighty-six, or, according to another observer, five hundred annual circles.

Great luxuriance of animal and vegetable production is not commonly accompanied by long duration of the individual. The oldest men are not

Another evil, sometimes of serious magnitude, which attends the operations of the lumberman, is the injury to the banks of rivers from the practice of floating. I do not here allude to rafts, which, being under the control of those who navigate them, may be so guided as to avoid damage to the shore, but to masts, logs, and other pieces of timber singly intrusted to the streams, to be conveyed by their currents to sawmill ponds, or to convenient places for collecting them into rafts. The lumbermen usually haul the timber to the banks of the rivers in the winter, and when the spring floods swell the streams and break up the ice, they roll the logs into the water, leaving them to float down to their destination. If the transporting stream is too small to furnish a sufficient channel for this rude navigation, it is sometimes dammed up, and the timber collected in the pond thus formed above the dam. When the pond is full, a sluice is opened, or the dam is blown up or otherwise suddenly broken, and the whole mass of lumber above it is hurried down with the rolling flood. Both of these modes of proceeding expose the banks of the rivers employed as channels of flotation to abrasion,* and in some of

found in the crowded city; and in the tropics, where life is prolific and precocious, it is also short. The most ancient forest trees of which we have accounts have not been those growing in thick woods, but isolated specimens, with no taller neighbor to intercept the light and heat and air, and no rival to share the nutriment afforded by the soil.

The more rapid growth and greater dimensions of trees standing near the boundary of the forest, are matters of familiar observation. "Long experience has shown that trees growing on the confines of the wood may be cut at sixty years of age as advantageously as others of the same species, reared in the depth of the forest, at a hundred and twenty. We have often remarked, in our Alps, that the trunk of trees upon the border of a grove is most developed or enlarged upon the outer or open side, where the branches extend themselves farthest, while the concentric circles of growth are most uniform in those entirely surrounded by other trees, or standing entirely alone."—A. and G. Villa, *Necessità dei Boschi*, pp. 17, 18.

* Caimi states that "a single flotation in the Valtelline in 1839, caused damages alleged to amount to more than $800,000, and actually appraised at $250,000."—*Cenni sulla Importanza e Coltura dei Boschi*, p. 65.

the American States it has been found necessary to protect, by special legislation, the lands through which they flow from the serious injury sometimes received through the practices I have described.*

Special Causes of the Destruction of European Woods.

The causes of forest waste thus far enumerated are more or less common to both continents; but in Europe extensive woods have, at different periods, been deliberately destroyed

* Most physicists who have investigated the laws of natural hydraulics maintain that, in consequence of direct obstruction and frictional resistance to the flow of the water of rivers along their banks, there is both an increased rapidity of current and an elevation of the water in the middle of the channel, so that a river presents always a convex surface. The lumbermen deny this. They affirm that, while rivers are rising, the water is highest in the middle of the channel, and tends to throw floating objects shoreward; while they are falling, it is lowest in the middle, and floating objects incline toward the centre. Logs, they say, rolled into the water during the rise, are very apt to lodge on the banks, while those set afloat during the falling of the waters keep in the current, and are carried without hindrance to their destination.

Foresters and lumbermen, like sailors and other persons whose daily occupations bring them into contact, and often into conflict, with great natural forces, have many peculiar opinions, not to say superstitions. In one of these categories we must rank the universal belief of lumbermen, that with a given head of water, and in a given number of hours, a sawmill cuts more lumber by night than by day. Having been personally interested in several sawmills, I have frequently conversed with sawyers on this subject, and have always been assured by them that their uniform experience established the fact that, other things being equal, the action of the machinery of sawmills is more rapid by night than by day. I am sorry—perhaps I ought to be ashamed—to say that my scepticism has been too strong to allow me to avail myself of my opportunities of testing this question by passing a night, watch in hand, counting the strokes of a millsaw. More unprejudiced, and I must add, very intelligent and credible persons have informed me that they have done so, and found the report of the sawyers abundantly confirmed. A land surveyor, who was also an experienced lumberman, sawyer, and machinist, a good mathematician and an exact observer, has repeatedly told me, that he had very often "timed" sawmills, and found the difference in favor of night work above thirty per cent. *Sed quære.*

by fire or the axe, because they afforded a retreat to enemies, robbers, and outlaws, and this practice is said to have been resorted to in the Mediterranean provinces of France as recently as the time of Napoleon I.* The severe and even sanguinary legislation, by which some of the governments of mediæval Europe, as well as of earlier ages, protected the woods, was dictated by a love of the chase, or the fear of a scarcity of fuel and timber. The laws of almost every European state more or less adequately secure the permanence of the forest; and I believe Spain is the only European land which has not made some public provision for the protection and restoration of the woods—the only country whose people systematically war upon the garden of God.†

* For many instances of this sort, see BECQUEREL, *Des Climats, etc.*, pp. 301–303. In 1664, the Swedes made an incursion into Jutland and felled a considerable extent of forest. After they retired, a survey of the damage was had, and the report is still extant. The number of trees cut was found to be 120,000, and as an account was kept of the numbers of each species of tree, the document is of interest in the history of the forest, as showing the relative proportions between the different trees which composed the wood. See VAUPELL, *Bögens Indvandring*, p. 35, and *Notes*, p. 55.

† Since writing this paragraph, I have fallen upon—and that in a Spanish author—one of those odd coincidences of thought which every man of miscellaneous reading so often meets with. Antonio Ponz (*Viage de España*, i, prólogo, p. lxiii), says: "Nor would this be so great an evil, were not some of them declaimers against *trees*, thereby proclaiming themselves, in some sort, enemies of the works of God, who gave us the leafy abode of Paradise to dwell in, where we should be even now sojourning, but for the first sin, which expelled us from it."

I do not know at what period the two Castiles were bared of their woods, but the Spaniard's proverbial "hatred of a tree" is of long standing. Herrera vigorously combats this foolish prejudice; and Ponz, in the prologue to the ninth volume of his journey, says that many carried it so far as wantonly to destroy the shade and ornamental trees planted by the municipal authorities. "Trees," they contended, and still believe, "breed birds, and birds eat up the grain." Our author argues against the supposition of the "breeding of birds by trees," which, he says, is as absurd as to believe that an elm tree can yield pears; and he charitably suggests that the expression is, perhaps, a *manière de dire*, a popular phrase, signifying simply that trees harbor birds.

Royal Forests and Game Laws.

The French authors I have quoted, as well as many other writers of the same nation, refer to the French Revolution as having given a new impulse to destructive causes which were already threatening the total extermination of the woods.* The general crusade against the forests, which accompanied that important event, is to be ascribed, in a considerable degree, to political resentments. The forest codes of the mediæval kings, and the local "coutumes" of feudalism contained many severe and even inhuman provisions, adopted rather for the preservation of game than from any enlightened views of the more important functions of the woods. Ordericus Vitalis informs us that William the Conqueror destroyed sixty parishes, and drove out their inhabitants, in order that he might turn their lands into a forest,† to be reserved as a hunting ground for himself and his posterity, and he punished with death the killing of a deer, wild boar, or even a hare. His successor, William Rufus, according to the *Histoire des Ducs de Normandie et des Rois d'Angleterre*, p. 67, "was hunting one day in a new forest, which he had caused to be made out of eighteen parishes that he had destroyed, when, by mis-

* Religious intolerance had produced similar effects in France at an earlier period. "The revocation of the edict of Nantes and the dragonnades occasioned the sale of the forests of the unhappy Protestants, who fled to seek in foreign lands the liberty of conscience which was refused to them in France. The forests were soon felled by the purchasers, and the soil in part brought under cultivation."—BECQUEREL, *Des Climats, etc.*, p. 303.

† The American reader must be reminded that, in the language of the chase and of the English law, a "forest" is not necessarily a wood. Any large extent of ground, withdrawn from cultivation, reserved for the pleasures of the chase, and allowed to clothe itself with a spontaneous growth, serving as what is technically called "cover" for wild animals, is, in the dialects I have mentioned, a forest. When, therefore, the Norman kings afforested the grounds referred to in the text, it is not to be supposed that they planted them with trees, though the protection afforded to them by the game laws would, if cattle had been kept out, soon have converted them into real woods.

chance, he was killed by an arrow wherewith Tyreus de Rois [Sir Walter Tyrell] thought to slay a beast, but missed the beast, and slew the king, who was beyond it. And in this very same forest, his brother Richard ran so hard against a tree that he died of it. And men commonly said that these things were because they had so laid waste and taken the said parishes."

These barbarous acts, as Bonnemère observes,* were simply the transfer of the customs of the French kings, of their vassals, and even of inferior gentlemen, to conquered England. "The death of a hare," says our author, "was a hanging matter, the murder of a plover a capital crime. Death was inflicted on those who spread nets for pigeons; wretches who had drawn a bow upon a stag were to be tied to the animal alive; and among the seigniors it was a standing excuse for having killed game on forbidden ground, that they aimed at a serf." The feudal lords enforced these codes with unrelenting rigor, and not unfrequently took the law into their own hands. In the time of Louis IX, according to William of Nangis, "three noble children, born in Flanders, who were sojourning at the abbey of St. Nicholas in the Wood, to learn the speech of France, went out into the forest of the abbey, with their bows and iron-headed arrows, to disport them in shooting hares, chased the game, which they had started in the wood of the abbey, into the forest of Enguerrand, lord of Coucy, and were taken by the sergeants which kept the wood. When the fell and pitiless Sir Enguerrand knew this, he had the children straightway hanged without any manner of trial." † The

* *Histoire des Paysans*, ii, p. 190. The work of Bonnemère is of great value to those who study the history of mediæval Europe from a desire to know its real character, and not in the hope of finding apparent facts to sustain a false and dangerous theory. Bonnemère is one of the few writers who, like Michelet, have been honest enough and bold enough to speak the truth with regard to the relations between the church and the people in the Middle Ages.

† It is painful to add that a similar outrage was perpetrated a very few years ago, in one of the European states, by a prince of a family now de-

matter being brought to the notice of good King Louis, Sir Enguerrand was summoned to appear, and, finally, after many feudal shifts and dilatory pleas, brought to trial before Louis himself and a special council. Notwithstanding the opposition of the other seigniors, who, it is needless to say, spared no efforts to save a peer, probably not a greater criminal than themselves, the king was much inclined to inflict the punishment of death on the proud baron. "If he believed," said he, "that our Lord would be as well content with hanging as with pardoning, he would hang Sir Enguerrand in spite of all his barons;" but noble and clerical interests unfortunately prevailed. The king was persuaded to inflict a milder retribution, and the murderer was condemned to pay ten thousand livres in coin, and to "build for the souls of the three children two chapels wherein mass should be said every day." * The hope of shortening the purgatorial term of the young persons, by the religious rites to be celebrated in the chapels, was doubtless the consideration which operated most powerfully on the mind of the king; and Europe lost a great example for the sake of a mass.

The desolation and depopulation, resulting from the exten-

throned. In this case, however, the prince killed the trespasser with his own hand, his sergeants refusing to execute his mandate.

* Guillaume de Nangis, as quoted in the notes to Joinville, *Nouvelle Collection des Mémoires, etc.*, par Michaud et Poujoulat, première série, i, p. 335.

Persons acquainted with the character and influence of the mediæval clergy will hardly need to be informed that the ten thousand livres never found their way to the royal exchequer. It was easy to prove to the simple-minded king that, as the profits of sin were a monopoly of the church, he ought not to derive advantage from the commission of a crime by one of his subjects; and the priests were cunning enough both to secure to themselves the amount of the fine, and to extort from Louis large additional grants to carry out the purposes to which they devoted the money. "And though the king did take the moneys," says the chronicler, "he put them not into his treasury, but turned them into good works; for he builded therewith the maison-Dieu of Pontoise, and endowed the same with rents and lands; also the schools and the dormitory of the friars preachers of Paris, and the monastery of the Minorite friars."

sion of the forest and the enforcement of the game laws, induced several of the French kings to consent to some relaxation of the severity of these latter. Francis I, however, revived their barbarous provisions, and, according to Bonnemère, even so good a monarch as Henry IV reënacted them, and "signed the sentence of death upon peasants guilty of having defended their fields against devastation by wild beasts." "A fine of twenty livres," he continues, "was imposed on every one shooting at pigeons, which, at that time, swooped down by thousands upon the new-sown fields and devoured the seed. But let us count even this a progress, for we have seen that the murder of a pigeon had been a capital crime." *

Not only were the slightest trespasses on the forest domain —the cutting of an oxgoad, for instance—severely punished, but game animals were still sacred when they had wandered from their native precincts and were ravaging the fields of the peasantry. A herd of deer or of wild boars often consumed or trod down a harvest of grain, the sole hope of the year for a whole family; and the simple driving out of such animals from this costly pasturage brought dire vengeance on the head of the rustic, who had endeavored to save his children's bread from their voracity. "At all times," says Paul Louis Courier, speaking in the name of the peasants of Chambord, in the "Simple Discours," "the game has made war upon us. Paris was blockaded eight hundred years by the deer, and its environs, now so rich, so fertile, did not yield bread enough to support the gamekeepers." †

In the popular mind, the forest was associated with all the

* *Histoire des Paysans*, ii, p. 200.

† The following details from Bonnemère will serve to give a more complete idea of the vexatious and irritating nature of the game laws of France. The officers of the chase went so far as to forbid the pulling up of thistles and weeds, or the mowing of any unenclosed ground before St. John's day [24th June], in order that the nests of game birds might not be disturbed. It was unlawful to fence-in any grounds in the plains where royal residences were situated; thorns were ordered to be planted in all fields of

abuses of feudalism, and the evils the peasantry had suffered from the legislation which protected both it and the game it sheltered, blinded them to the still greater physical mischiefs which its destruction was to entail upon them. No longer protected by law, the crown forests and those of the great lords were attacked with relentless fury, unscrupulously plundered and wantonly laid waste, and even the rights of property in small private woods were no longer respected.* Various absurd theories, some of which are not even yet exploded, were propagated with regard to the economical advantages of converting the forest into pasture and plough-

wheat, barley, or oats, to prevent the use of ground nets for catching the birds which consumed, or were believed to consume, the grain, and it was forbidden to cut or pull stubble before the first of October, lest the partridge and the quail might be deprived of their cover. For destroying the eggs of the quail, a fine of one hundred livres was imposed for the first offence, double that amount for the second, and for the third the culprit was flogged and banished for five years to a distance of six leagues from the forest.—*Histoire des Paysans*, ii, p. 202, text and notes.

Neither these severe penalties, nor any provisions devised by the ingenuity of modern legislation, have been able effectually to repress poaching. "The game laws," says Clavé, "have not delivered us from the poachers, who kill twenty times as much game as the sportsmen. In the forest of Fontainebleau, as in all those belonging to the state, poaching is a very common and a very profitable offence. It is in vain that the gamekeepers are on the alert night and day, they cannot prevent it. Those who follow the trade begin by carefully studying the habits of the game. They will lie motionless on the ground, by the roadside or in thickets, for whole days, watching the paths most frequented by the animals," &c.—*Revue des Deux Mondes*, Mai, 1863, p. 160.

The writer adds many details on this subject, and it appears that, as there are "beggars on horseback" in South America, there are poachers in carriages in France.

* "Whole trees were sacrificed for the most insignificant purposes; the peasants would cut down two firs to make a single pair of wooden shoes." —Michelet, as quoted by Clavé, *Études*, p. 24.

A similar wastefulness formerly prevailed in Russia, though not from the same cause. In St. Pierre's time, the planks brought to St. Petersburg were not sawn, but hewn with the axe, and a tree furnished but a single plank.

land, its injurious effects upon climate, health, facility of internal communication, and the like. Thus resentful memory of the wrongs associated with the forest, popular ignorance, and the cupidity of speculators cunning enough to turn these circumstances to profitable account, combined to hasten the sacrifice of the remaining woods, and a waste was produced which hundreds of years and millions of treasure will hardly repair.

Small Forest Plants, and Vitality of Seed.

Another function of the woods to which I have barely alluded deserves a fuller notice than can be bestowed upon it in a treatise the scope of which is purely economical. The forest is the native habitat of a large number of humbler plants, to the growth and perpetuation of which its shade, its humidity, and its vegetable mould appear to be indispensable necessities.* We cannot positively say that the felling of the

* "A hundred and fifty paces from my house is a hill of drift sand, on which stood a few scattered pines. *Pinus sylvestris*, and *Sempervivum tectorum* in abundance, *Statice armeria*, *Ammone vernalis*, *Dianthus carthusianorum*, with other sand plants, were growing there. I planted the hill with a few birches, and all the plants I have mentioned completely disappeared, though there were many naked spots of sand between the trees. It should be added, however, that the hillock is more thickly wooded than before. * * * It seems then that *Sempervivum tectorum*, &c., will not bear the neighborhood of the birch, though growing well near the *Pinus sylvestris*. I have found the large red variety of *Agaricus deliciosus* only among the roots of the pine; the greenish-blue *Agaricus deliciosus* among alder roots, but not near any other tree. Birds have their partialities among trees and shrubs. The *Silviæ* prefer the *Pinus Larix* to other trees. In my garden this *Pinus* is never without them, but I never saw a bird perch on *Thuja occidentalis* or *Juniperus sabina*, although the thick foliage of these latter trees affords birds a better shelter than the loose leafage of other trees. Not even a wren ever finds its way to one of them. Perhaps the scent of the *Thuja* and the *Juniperus* is offensive to them. I have spoiled one of my meadows by cutting away the bushes. It formerly bore grass four feet high, because many umbelliferous plants, such as *Heracleum spondylium*, *Spiræa ulmaria*, *Laserpitium latifolia*, &c., grew in it. Under the shelter of the bushes these plants ripened and bore seed, but they gradually disappeared as the shrubs were extirpated, and the grass now does

woods in a given vegetable province would involve the final extinction of the smaller plants which are found only within their precincts. Some of these, though not naturally prop gating themselves in the open ground, may perhaps germina and grow under artificial stimulation and protection, and finally become hardy enough to maintain an independent existence in very different circumstances from those which at present seem essential to their life.

not grow to the height of more than two feet, because it is no longer obliged to keep pace with the umbellifera which flourished among it." See a paper by J. G. BÜTTNER, of Kurland, in BERGHAUS' *Geographisches Jahrbuch*, 1852, No. 4, pp. 14, 15.

These facts are interesting as illustrating the multitude of often obscure conditions upon which the life or vigorous growth of smaller organisms depends. Particular species of truffles and of mushrooms are found associated with particular trees, without being, as is popularly supposed, parasites deriving their nutriment from the dying or dead roots of those trees. The success of Rousseau's experiments seem decisive on this point, for he obtains larger crops of truffles from ground covered with young seedling oaks than from that filled with roots of old trees. See an article on Mont Ventoux, by Charles Martins, in the *Revue des Deux Mondes*, Avril, 1863, p. 626.

It ought to be much more generally known than it is that most, if not all mushrooms, even of the species reputed poisonous, may be rendered harmless and healthful as food by soaking them for two hours in acidulated or salt water. The water requires two or three spoonfuls of vinegar or two spoonfuls of gray salt to the quart, and a quart of water is enough for a pound of sliced mushrooms. After thus soaking, they are well washed in fresh water, thrown into cold water, which is raised to the boiling point, and, after remaining half an hour, taken out and again washed. Gérard, to prove that "crumpets is wholesome," ate one hundred and seventy-five pounds of the most poisonous mushrooms thus prepared, in a single month, fed his family *ad libitum* with the same, and finally administered them, in heroic doses, to the members of a committee appointed by the Council of Health of the city of Paris. See FIGUIER, *L'Année Scientifique*, 1862, pp. 353, 384. See *Appendix*, No. 29.

It has long been known that the Russian peasantry eat, with impunity, mushrooms of species everywhere else regarded as very poisonous. Is it not probable that the secret of rendering them harmless—which was known to Pliny, though since forgotten in Italy—is possessed by the rustic Muscovites?

Besides this, although the accounts of the growth of seeds, which have lain for ages in the ashy dryness of Egyptian catacombs, are to be received with great caution, or, more probably, to be rejected altogether, yet their vitality seems almost imperishable while they remain in the situations in which nature deposits them. When a forest old enough to have witnessed the mysteries of the Druids is felled, trees of other species spring up in its place; and when they, in their turn, fall before the axe, sometimes even as soon as they have spread their protecting shade over the surface, the germs which their predecessors had shed years, perhaps centuries before, sprout up, and in due time, if not choked by other trees belonging to a later stage in the order of natural succession, restore again the original wood. In these cases, the seeds of the new crop may often have been brought by the wind, by birds, by quadrupeds, or by other causes; but, in many instances, this explanation is not probable.

When newly cleared ground is burnt over in the United States, the ashes are hardly cold before they are covered with a crop of fire weed, a tall herbaceous plant, very seldom seen growing under other circumstances, and often not to be found for a distance of many miles from the clearing. Its seeds, whether the fruit of an ancient vegetation or newly sown by winds or birds, require either a quickening by a heat which raises to a certain high point the temperature of the stratum where they lie buried, or a special pabulum furnished only by the combustion of the vegetable remains that cover the ground in the woods. Earth brought up from wells or other excavations soon produces a harvest of plants often very unlike those of the local flora.

Moritz Wagner, as quoted by Wittwer,* remarks in his description of Mount Ararat: "A singular phenomenon to which my guide drew my attention is the appearance of several plants on the earth-heaps left by the last catastrophe [an earthquake], which grow nowhere else on the mountain, and

* *Physikalische Geographie*, p. 486.

had never been observed in this region before. The seeds of these plants were probably brought by birds, and found in the loose, clayey soil remaining from the streams of mud, the conditions of growth which the other soil of the mountain refused them." This is probable enough, but it is hardly less so that the flowing mud brought them up to the influence of air and sun, from depths where a previous convulsion had buried them ages before. Seeds of small sylvan plants, too deeply buried by successive layers of forest foliage and the mould resulting from its decomposition to be reached by the plough when the trees are gone and the ground brought under cultivation, may, if a wiser posterity replants the wood which sheltered their parent stems, germinate and grow, after lying for generations in a state of suspended animation.

Darwin says: "In Staffordshire, on the estate of a relation, where I had ample means of investigation, there was a large and extremely barren heath, which had never been touched by the hand of man, but several hundred acres of exactly the same nature had been enclosed twenty-five years previously and planted with Scotch fir. The change in the native vegetation of the planted part of the heath was most remarkable—more than is generally seen in passing from one quite different soil to another; not only the proportional numbers of the heath plants were wholly changed, but *twelve species* of plants (not counting grasses and sedges) flourished in the plantation which could not be found on the heath." * Had the author informed us that these twelve plants belonged to a species whose seeds enter into the nutriment of the birds which appeared with the young wood, we could easily account for their presence in the soil; but he says distinctly that the birds were of insectivorous species, and it therefore seems more probable that the seeds had been deposited when an ancient forest protected the growth of the plants which bore them, and that they sprang up to new life when a return of favorable conditions awaked them from a sleep of centuries. Darwin

* *Origin of Species*, American edition, p. 69.

indeed says that the heath "had never been touched by the hand of man." Perhaps not, after it became a heath; but what evidence is there to control the general presumption that this heath was preceded by a forest, in whose shade the vegetables which dropped the seeds in question might have grown? *

Although, therefore, the destruction of a wood and the reclaiming of the soil to agricultural uses suppose the death of its smaller dependent flora, these revolutions do not exclude the possibility of its resurrection. In a practical view of the subject, however, we must admit that when the woodman fells a tree he sacrifices the colony of humbler growths which had

* Writers on vegetable physiology record numerous instances where seeds have grown after lying dormant for ages. The following cases, mentioned by Dr. Dwight (*Travels*, ii, pp. 438, 439), may be new to many readers:

"The lands [in Panton, Vermont], which have here been once cultivated, and again permitted to lie waste for several years, yield a rich and fine growth of hickory [*Carya porcina*]. Of this wood there is not, I believe, a single tree in any original forest within fifty miles from this spot. The native growth was here white pine, of which I did not see a single stem in a whole grove of hickory."

The hickory is a walnut, bearing a fruit too heavy to be likely to be carried fifty miles by birds, and besides, I believe it is not eaten by any bird indigenous to Vermont.

"A field, about five miles from Northampton, on an eminence called Rail Hill, was cultivated about a century ago. The native growth here, and in all the surrounding region, was wholly oak, chestnut, &c. As the field belonged to my grandfather, I had the best opportunity of learning its history. It contained about five acres, in the form of an irregular parallelogram. As the savages rendered the cultivation dangerous, it was given up. On this ground there sprang up a grove of white pines covering the field and retaining its figure exactly. So far as I remember, there was not in it a single oak or chestnut tree. * * * There was not a single pine whose seeds were, or, probably, had for ages been, sufficiently near to have been planted on this spot. The fact that these white pines covered this field exactly, so as to preserve both its extent and its figure, and that there were none in the neighborhood, are decisive proofs that cultivation brought up the seeds of a former forest within the limits of vegetation, and gave them an opportunity to germinate."

vegetated under its protection. Some wood plants are known to possess valuable medicinal properties, and experiment may show that the number of these is greater than we now suppose. Few of them, however, have any other economical value than that of furnishing a slender pasturage to cattle allowed to roam in the woods; and even this small advantage is far more than compensated by the mischief done to the young trees by browsing animals. Upon the whole, the importance of this class of vegetables, as physic or as food, is not such as to furnish a very telling popular argument for the conservation of the forest as a necessary means of their perpetuation. More potent remedial agents may supply their place in the *materia medica*, and an acre of grass land yields more nutriment for cattle than a range of a hundred acres of forest. But he whose sympathies with nature have taught him to feel that there is a fellowship between all God's creatures; to love the brilliant ore better than the dull ingot, iodic silver and crystallized red copper better than the shillings and the pennies forged from them by the coiner's cunning; a venerable oak tree than the brandy cask whose staves are split out from its heart wood; a bed of anemones, hepaticas, or wood violets than the leeks and onions which he may grow on the soil they have enriched and in the air they made fragrant—he who has enjoyed that special training of the heart and intellect which can be acquired only in the unviolated sanctuaries of nature, "where man is distant, but God is near"—will not rashly assert his right to extirpate a tribe of harmless vegetables, barely because their products neither tickle his palate nor fill his pocket; and his regret at the dwindling area of the forest solitude will be augmented by the reflection that the nurselings of the woodland perish with the pines, the oaks, and the beeches that sheltered them.*

* Quaint old Valvasor had observed the subduing influence of nature's solitudes. In describing the lonely Canker-Thal, which, though rocky, was in his time well wooded with "fir, larches, beeches, and other trees," he says: "Gladsomeness and beauty, which dwell in many valleys, may not be looked for there. The journey through it is cheerless, melancholy.

Although, as I have said, birds do not frequent the deeper recesses of the wood,* yet a very large proportion of them build their nests in trees, and find in their foliage and branches a secure retreat from the inclemencies of the seasons and the pursuit of the reptiles and quadrupeds which prey upon them. The borders of the forests are vocal with song; and when the gray morning calls the creeping things of the earth out of their night cells, it summons from the neighboring wood legions of their winged enemies, which swoop down upon the fields to save man's harvests by devouring the destroying worm, and surprising the lagging beetle in his tardy retreat to the dark cover where he lurks through the hours of daylight.

The insects most injurious to rural industry do not multiply in or near the woods. The locust, which ravages the East with its voracious armies, is bred in vast open plains which admit the full heat of the sun to hasten the hatching of the eggs, gather no moisture to destroy them, and harbor no bird to feed upon the larvæ.† It is only since the felling of the forests of Asia Minor and Cyrene that the locust has become so fearfully destructive in those countries; and the grasshopper, which now threatens to be almost as great a pest to the agriculture of some North American soils, breeds in seriously

wearisome, and serveth to temper and mortify over-joyousness of thought. * * * In sum it is a very wild, wherein the wildness of human pride doth grow tame."—*Ehre der Crain*, i, p. 136, b.

* Valvasor says, in the same paragraph from which I have just quoted, "In my many journeys through this valley, I did never have sight of so much as a single bird."

† Smela, in the government of Kiew, has, for some years, not suffered at all from the locusts, which formerly came every year in vast swarms, and the curculio, so injurious to the turnip crops, is less destructive there than in other parts of the province. This improvement is owing partly to the more thorough cultivation of the soil, partly to the groves which are interspersed among the plough lands. * * * When in the midst of the plains woods shall be planted and filled with insectivorous birds, the locusts will cease to be a plague and a terror to the farmer.—Rentzsch, *Der Wald*, pp. 45, 46.

injurious numbers only where a wide extent of surface is bare of woods.

Utility of the Forest.

In most parts of Europe, the woods are already so nearly extirpated that the mere protection of those which now exist is by no means an adequate remedy for the evils resulting from the want of them; and besides, as I have already said, abundant experience has shown that no legislation can secure the permanence of the forest in private hands. Enlightened individuals in most European states, governments in others, have made very extensive plantations,* and France has now set herself energetically at work to restore the woods in the southern provinces, and thereby to prevent the utter depopulation and waste with which that once fertile soil and delicious climate are threatened.

The objects of the restoration of the forest are as multifarious as the motives that have led to its destruction, and as the evils which that destruction has occasioned. It is hoped that the planting of the mountains will diminish the frequency and violence of river inundations, prevent the formation of torrents, mitigate the extremes of atmospheric temperature, humidity, and precipitation, restore dried-up springs, rivulets, and sources of irrigation, shelter the fields from chilling and from parching winds, prevent the spread of miasmatic effluvia,

* England is, I believe, the only country where private enterprise has pursued sylviculture on a really great scale, though admirable examples have been set in many others on both sides of the Atlantic. In England the law of primogeniture, and other institutions and national customs which tend to keep large estates long undivided and in the same line of inheritance, the wealth of the landholders, and the difficulty of finding safe and profitable investments of capital, combine to afford encouragements for the plantation of forests, which nowhere else exist in the same degree. The climate of England, too, is very favorable to the growth of forest trees, though the character of surface secures a large part of the island from the evils which have resulted from the destruction of the woods elsewhere, and therefore their restoration is a matter of less geographical importance in England than on the Continent.

and, finally, furnish an inexhaustible and self-renewing supply of a material indispensable to so many purposes of domestic comfort, to the successful exercise of every art of peace, every destructive energy of war.*

But our enumeration of the uses of trees is not yet complete. Besides the influence of the forest, in mountain ranges, as a means of preventing the scooping out of ravines and the accumulations of water which fill them, trees subserve a valuable purpose, in lower positions, as barriers against the spread of floods and of the material they transport with them; but this will be more appropriately considered in the chapter on the waters; and another very important use of trees, that of fixing movable sand-dunes, and reclaiming them to profitable cultivation, will be pointed out in the chapter on the sands.

The vast extension of railroads, of manufactures and the mechanical arts, of military armaments, and especially of the commercial fleets and navies of Christendom within the present century, has greatly augmented the demand for wood,†

* The preservation of the woods on the eastern frontier of France, as a kind of natural abattis, is also recognized by the Government of that country as an important measure of military defence, though there have been conflicting opinions on the subject.

† Let us take the supply of timber for railroad ties. According to Clavé (p. 248), France has 9,000 kilomètres of railway in operation, 7,000 in construction, half of which is built with a double track. Adding turnouts and extra tracks at stations, the number of ties required for a single track is stated at 1,200 to the kilomètre, or, as Clavé computes, for the entire network of France, 58,000,000. As the schoolboys say, "this sum does not prove;" for 16,000 + 8,000 for the double track halfway = 24,000, and 24,000 × 1,200 = 28,800,000. According to Bigelow (*Les États Unis en* 1863, p. 439), the United States had in operation or construction on the first of January, 1862, 51,000 miles, or about 81,000 kilomètres of railroad, and the military operations of the present civil war are rapidly extending the system. Allowing the same proportion as in France, the American railroads required 97,200,000 ties in 1862. The consumption of timber in Europe and America during the present generation, occasioned by this demand, has required the sacrifice of many hundred thousand acres of forest, and if we add the quantity employed for telegraph posts, we have an amount of destruction, for entirely new purposes, which is really appalling.

The consumption of wood for lucifer matches is enormous, and I have

and, but for improvements in metallurgy which have facilitated the substitution of iron for that material, the last twenty-five years would almost have stripped Europe of her only

heard of several instances where tracts of pine forest, hundreds and even thousands of acres in extent, have been purchased and felled, solely to supply timber for this purpose.

The demand for wood for small carvings and for children's toys is incredibly large. Rentzsch states the export of such objects from the town of Sonneberg alone to have amounted, in 1853, to 60,000 centner, or three thousand tons' weight.—*Der Wald*, p. 68. See *Appendix*, No. 32.

The importance of so managing the forest that it may continue indefinitely to furnish an adequate supply of material for naval architecture is well illustrated by some remarks of the same author in the valuable little work just cited. He suggests that the prosperity of modern England is due, in no small degree, to the supplies of wood and other material for building and equipping ships, received from the forests of her colonies and of other countries with which she has maintained close commercial relations, and he adds: "Spain, which by her position seemed destined for universal power, and once, in fact, possessed it, has lost her political rank, because during the unwise administration of the successors of Philip II, the empty exchequer could not furnish the means of building new fleets; for the destruction of the forests had raised the price of timber above the resources of the state."—*Der Wald*, p. 63.

The market price of timber, like that of all other commodities, may be said, in a general way, to be regulated by the laws of demand and supply, but it is also controlled by those seemingly unrelated accidents which so often disappoint the calculations of political economists in other branches of commerce. A curious case of this sort is noticed by Cerini, *Dell' Impianto e Conservazione dei Boschi*, p. 17: "In the mountains on the Lago Maggiore, in years when maize is cheap, the woodcutters can provide themselves with corn meal enough for a week by three days' labor, and they refuse to work the remaining four. Hence the dealers in wood, not being able to supply the demand, for want of laborers, are obliged to raise the price for the following season, both for timber and for firewood; so that a low price of grain occasions a high price of building lumber and of fuel. The consequence is, that though the poor have supplied themselves cheaply with food, they must pay dear for firewood, and they cannot get work, because the high price of lumber has discouraged repairs and building, the expense of which landed proprietors cannot undertake when their incomes have been reduced by sales of grain at low rates, and hence there is not demand enough for lumber to induce the timber merchants to furnish employment to the woodmen."

remaining trees fit for such uses.* The walnut trees alone felled in Europe within two years to furnish the armies of

* Besides the substitution of iron for wood, a great saving of consumption of this latter material has been effected by the revival of ancient methods of increasing its durability, and the invention of new processes for the same purpose. The most effectual preservative yet discovered for wood employed on land, is sulphate of copper, a solution of which is introduced into the pores of the wood while green, by soaking, by forcing-pumps, or, most economically, by the simple pressure of a column of the fluid in a small pipe connected with the end of the piece of timber subjected to the treatment. Clavé (*Études Forestières*, pp. 240–249) gives an interesting account of the various processes employed for rendering wood imperishable, and states that railroad ties injected with sulphate of copper in 1846, were found absolutely unaltered in 1855; and telegraphic posts prepared two years earlier, are now in a state of perfect preservation.

For many purposes, the method of injection is too expensive, and some simpler process is much to be desired. The question of the proper time of felling timber is not settled, and the best modes of air, water, and steam seasoning are not yet fully ascertained. Experiments on these subjects would be well worth the patronage of governments in new countries, where they can be very easily made, without the necessity of much waste of valuable material, and without expensive arrangements for observation.

The practice of stripping living trees of their bark some years before they are felled, is as old as the time of Vitruvius, but is much less followed than it deserves, partly because the timber of trees so treated inclines to crack and split, and partly because it becomes so hard as to be wrought with considerable difficulty.

In America, economy in the consumption of fuel has been much promoted by the substitution of coal for wood, the general use of stoves both for wood and coal, and recently by the employment of anthracite in the furnaces of stationary and locomotive steam-engines. All the objections to the use of anthracite for this latter purpose appear to have been overcome, and the improvements in its combustion have been attended with a great pecuniary saving, and with much advantage to the preservation of the woods.

The employment of coal has produced a great reduction in the consumption of fire wood in Paris. In 1815, the supply of fire wood for the city required 1,200,000 stères, or cubic mètres; in 1859, it had fallen to 501,805, while, in the mean time, the consumption of coal had risen from 600,000 to 432,000,000 metrical quintals. See Clavé, *Études*, p. 212.

I think there must be some error in this last sum, as 432 millions of metrical quintals would amount to 43 millions of tons, a quantity which

America with gunstocks, would form a forest of no inconsiderable extent.*

The Forests of Europe.

Mirabeau estimated the forests of France in 1750 at seventeen millions of hectares [42,000,000 acres]; in 1860 they were reduced to eight millions [19,769,000 acres]. This would be at the rate of 82,000 hectares [202,600 acres] per year. Troy, from whose valuable pamphlet, *Étude sur le Reboisement des Montagnes*, I take these statistical details, supposes that Mirabeau's statement may have been an extravagant one, but it still remains certain that the waste has been enormous; for it is known that, in some departments, that of Ariège, for instance, clearing has gone on during the last half century at the rate of three thousand acres a year,† and in all parts of the empire trees have been felled faster than they have grown. The total area of France, excluding Savoy, is about one hundred and thirty-one millions of acres. The extent of forest supposed by Mirabeau would be about thirty-two per cent. of the whole territory.‡ In a country and a climate where the conservative influences of the forest are so necessary as in France, trees must cover a large surface and be

it is difficult to suppose could be consumed in the city of Paris. The price of fire wood has scarcely advanced at all in Paris for half a century, though that of timber generally has risen enormously.

* In the first two years of the present civil war in the United States, twenty-eight thousand walnut trees were felled to supply a single European manufactory of gunstocks for the American market.

† Among the indirect proofs of the comparatively recent existence of extensive forests in France, may be mentioned the fact, that wolves were abundant, not very long since, in parts of the empire where there are now neither wolves nor woods to shelter them. Arthur Young more than once speaks of the "innumerable multitudes" of these animals which infested France in 1789, and George Sand states, in the *Histoire de ma Vie*, that some years after the restoration of the Bourbons, they chased travellers on horseback in the Southern provinces, and literally knocked at the doors of her father-in-law's country seat.

‡ In the *Recepte Véritable*, Palissy having expressed his indignation at the folly of men in destroying the woods, his interlocutor defends the

grouped in large masses, in order to discharge to the best advantage the various functions assigned to them by nature. The consumption of wood is rapidly increasing in that empire, and a large part of its territory is mountainous, sterile, and otherwise such in character or situation that it can be more profitably devoted to the growth of wood than to any agricultural use. Hence it is evident that the proportion of forest in 1750, taking even Mirabeau's large estimate, was not very much too great for permanent maintenance, though doubtless the distribution was so unequal that it would have been sound policy to fell the woods and clear land in some provinces, while large forests should have been planted in others.* Du-

policy of felling them, by citing the example of "divers bishops, cardinals, priors, abbots, monkeries, and chapters, which, by cutting their woods, have made three profits," the sale of the timber, the rent of the ground, and the "good portion" they received of the grain grown by the peasants upon it. To this argument, Palissy replies: "I cannot enough detest this thing, and I call it not an error, but a curse and a calamity to all France; for when forests shall be cut, all arts shall cease, and they which practise them shall be driven out to eat grass with Nebuchadnezzar and the beasts of the field. I have divers times thought to set down in writing the arts which shall perish when there shall be no more wood; but when I had written down a great number, I did perceive that there could be no end of my writing, and having diligently considered, I found there was not any which could be followed without wood." * * "And truly I could well allege to thee a thousand reasons, but 'tis so cheap a philosophy, that the very chamber wenches, if they do but think, may see that without wood, it is not possible to exercise any manner of human art or cunning." —*Œuvres de* BERNARD PALISSY, p. 89.

* Since writing the above paragraph, I have found the view I have taken of this point confirmed by the careful investigations of Rentzsch, who estimates the proper proportion of woodland to entire surface at twenty-three per cent. for the interior of Germany, and supposes that near the coast, where the air is supplied with humidity by evaporation from the sea, it might safely be reduced to twenty per cent. See Rentzsch's very valuable prize essay, *Der Wald im Haushalt der Natur und der Volkswirthschaft*, cap. viii.

The due proportion in France would considerably exceed that for the German States, because France has relatively more surface unfit for any growth but that of wood, because the form and geological character of her

ring the period in question, France neither exported manufactured wood or rough timber, nor derived important collateral advantages of any sort from the destruction of her forests. She is consequently impoverished and crippled to the extent of the difference between what she actually possesses of wooded surface and what she ought to have retained.

Italy and Spain are bared of trees in a greater degree than France, and even Russia, which we habitually consider as substantially a forest country, is beginning to suffer seriously for want of wood. Jourdier, as quoted by Clavé, observes: "Instead of a vast territory with immense forests, which we expect to meet, one sees only scattered groves thinned by the wind or by the axe of the *moujik*, grounds cut over and more or less recently cleared for cultivation. There is probably not a single district in Russia which has not to deplore the ravages of man or of fire, those two great enemies of Muscovite sylviculture. This is so true, that clear-sighted men already foresee a crisis which will become terrible, unless the discovery of great deposits of some new combustible, as pit coal or anthracite, shall diminish its evils." *

mountains expose her territory to much greater injury from torrents, and because at least her southern provinces are more frequently visited both by extreme drought and by deluging rains.

* *Études sur l'Économie Forestière*, p. 261. Clavé adds (p. 262): "The Russian forests are very unequally distributed through the territory of this vast empire. In the north they form immense masses, and cover whole provinces, while in the south they are so completely wanting that the inhabitants have no other fuel than straw, dung, rushes, and heath." * * * "At Moscow, firewood costs thirty per cent. more than at Paris, while, at the distance of a few leagues, it sells for a tenth of that price."

This state of things is partly due to the want of facilities of transportation, and some parts of the United States are in a similar condition. During a severe winter, six or seven years ago, the sudden freezing of the canals and rivers, before a large American town had received its usual supply of fuel, occasioned an enormous rise in the price of wood and coal, and the poor suffered severely for want of it. Within a few hours of the city were large forests and an abundant stock of firewood felled and prepared for burning. This might easily have been carried to town by the

Germany, from character of surface and climate, and from the attention which has long been paid in all the German States to sylviculture, is, taken as a whole, in a far better condition in this respect than its more southern neighbors; but in the Alpine provinces of Bavaria and Austria, the same improvidence which marks the rural economy of the corresponding districts of Switzerland, Italy, and France, is producing effects hardly less disastrous. As an instance of the scarcity of fuel in some parts of the territory of Bavaria, where, not long since, wood abounded, I may mention the fact that the water of salt springs is, in some instances, conveyed to the distance of sixty

railroads which passed through the woods; but the managers of the roads refused to receive it as freight, because the opening of a new market for wood might raise the price of the fuel they employed for their locomotives.

Hohenstein, who was long professionally employed as a forester in Russia, describes the consequences of the general war upon the woods in that country as already most disastrous, and as threatening still more ruinous evils. The river Volga, the life artery of Russian internal commerce, is drying up from this cause, and the great Muscovite plains are fast advancing to a desolation like that of Persia.—*Der Wald*, p. 223.

The level of the Caspian Sea is eighty-three feet lower than that of the Sea of Azoff, and the surface of Lake Aral is fast sinking. Von Baer maintains that the depression of the Caspian was produced by a sudden subsidence, from geological causes, and not gradually by excess of evaporation over supply. See *Kaspische Studien*, p. 25. But this subsidence diminished the area and consequently the evaporation of that sea, and the rivers which once maintained its ancient equilibrium ought to raise it to its former level, if their own flow had not been diminished. It is, indeed, not proved that the laying bare of a wooded country diminishes the total annual precipitation upon it; but it is certain that the summer evaporation from the surface of a champaign region, like that through which the Volga, its tributaries, and the feeders of Lake Aral flow, is increased by the removal of its woods. Hence, though as much rain may still fall in the valleys of those rivers as when their whole surface was covered with forests, a less quantity of water may be delivered by them since their basins were cleared, and therefore the present condition of the inland waters in question may be due to the removal of the forests in their basins.

miles, in iron pipes, to reach a supply of fuel for boiling it down.*

Forests of the United States and Canada.

The vast forests of the United States and Canada cannot long resist the improvident habits of the backwoodsman and the increased demand for lumber. According to the census of the former country for 1860, which gives returns of the

* Rentzsch (*Der Wald, etc.*, pp. 123, 124) states the proportions of woodland in different European countries as follows:

	Per cent.	Acres per head of population.		Per cent.	Acres per head of population.
Germany.............	26.58	0.6638	Switzerland...........	15.	0.396
Great Britain.........	5.	0.1	Holland...............	7.10	0.12
France...............	16.79	0.3766	Belgium...............	18.52	0.186
Russia...............	30.90	4.28	Spain.................	5.52	0.291
Sweden...............	60.	8.55	Portugal..............	4.40	0.182
Norway...............	66.	24.61	Sardinia..............	12.29	0.223
Denmark.............	5.50	0.22	Naples................	9.43	0.138

Probably no European countries can so well dispense with the forests, in their capacity of conservative influences, as England and Ireland. Their insular position and latitude secure an abundance of atmospheric moisture, and the general inclination of surface is not such as to expose it to special injury from torrents. The due proportion of woodland in England and Ireland is, therefore, almost purely an economical question, to be decided by the comparative direct pecuniary return from forest growth, pasturage, and plough land.

In Scotland, where the country is for the most part more broken and mountainous, the general destruction of the forests has been attended with very serious evils, and it is in Scotland that many of the most extensive British forest plantations have now been formed. But although the inclination of surface in Scotland is rapid, the geological constitution of the soil is not of a character to promote such destructive degradation by running water as in Southern France, and it has not to contend with the parching droughts by which the devastations of the torrents are rendered more injurious in that part of the French empire.

In giving the proportion of woodland to population, I compute Rentzsch's Morgen at .3882 of an English acre, because I find, by Alexander's most accurate and valuable Dictionary of Weights and Measures, that this is the value of the Dresden Morgen, and Rentzsch is a Saxon writer. In the different German States, there are more than twenty dif

"sawed and planed lumber" alone, timber for framing and for a vast variety of mechanical purposes being omitted altogether, the value of the former material prepared for market in the United States was, in 1850, $58,521,976; in 1860, $95,912,286. The quantity of unsawed lumber is not likely to have increased in the same proportion, because comparatively little is exported in that condition, and because masonry is fast taking the place of carpentry in building, and stone, brick, and iron are used instead of timber more largely than they were ten years ago. Still a much greater quantity of unsawed lumber must have been marketed in 1860 than in 1850. It must further be admitted that the price of lumber rose considerably between those dates, and consequently that the increase in quantity is not to be measured by the increase in pecuniary value. Perhaps this rise of prices may even be sufficient to make the entire difference between the value of "sawed and planed lumber" produced in the ten years in question by the six New England States (21 per cent.), and the six Middle States (15 per cent.); but the amount produced by the Western and by the Southern States had doubled, and that returned from the Pacific States and Territories had trebled in value in the same interval, so that there was certainly, in those States, a large increase in the actual quantity prepared for sale.

I greatly doubt whether any one of the American States, except, perhaps, Oregon, has, at this moment, more woodland than it ought permanently to preserve, though, no doubt, a different distribution of the forests in all of them might be highly advantageous. It is a great misfortune to the American Union that the State Governments have so generally disposed of their original domain to private citizens. It is

ferent land measures known by the name of Morgen, varying from about one third of an acre to more than three acres in value. When will the world be wise enough to unite in adopting the French metrical and monetary systems? As to the latter, never while Christendom continues to be ruled by money changers, who can compel you to part with your sovereigns in France at twenty-five francs, and in England to accept fifteen shillings for your napoleons. I speak as a sufferer. *Experto crede Roberto.*

true that public property is not sufficiently respected in the United States; and it is also true that, within the memory of almost every man of mature age, timber was of so little value in that country, that the owners of private woodlands submitted, almost without complaint, to what would be regarded elsewhere as very aggravated trespasses upon them.* Under

* According to the maxims of English jurisprudence, the common law consists of general customs so long established that "the memory of man runneth not to the contrary." In other words, long custom makes law. In new countries, the change of circumstances creates new customs, and, in time, new law, without the aid of legislation. Had the American colonists observed a more sparing economy in the treatment of their woods, a new code of customary forest law would have sprung up and acquired the force of a statute. Popular habit was fast elaborating the fundamental principles of such a code, when the rapid increase in the value of timber, in consequence of the reckless devastation of the woodlands, made it the interest of the proprietors to interfere with this incipient system of forest jurisprudence, and appeal to the rules of English law for the protection of their woods. The courts have sustained these appeals, and forest property is now legally as inviolable as any other, though common opinion still combats the course of judicial decision on such questions.

In the United States, swarms of honey bees, on leaving the parent hive, often take up their quarters in hollow trees in the neighboring woods. By the early customs of New England, the finder of a "bee tree" on the land of another owner was regarded as entitled to the honey by right of discovery; and as a necessary incident of that right, he might cut the tree, at the proper season, without asking permission of the proprietor of the soil. The quantity of "wild honey" in a tree was often large, and "bee hunting" was so profitable that it became almost a regular profession. The "bee hunter" sallied forth with a small box containing honey and a little vermilion. The bees which were attracted by the honey marked themselves with the vermilion, and hence were more readily followed in their homeward flight, and recognized when they returned a second time for booty. When loaded with spoil, this insect returns to his hive by the shortest route, and hence a straight line is popularly called in America a "bee line." By such a line, the hunter followed the bees to their sylvan hive, marked the tree with his initials, and returned to secure his prize in the autumn. When the right of the "bee hunter" was at last disputed by the land proprietors, it was with difficulty that judgments could be obtained, in inferior courts, in favor of the latter, and it was only after repeated decisions of the higher legal tribunals that the superior right of the owner of the soil was at last acquiesced in.

such circumstances, it is difficult to protect the forest, whether it belong to the state or to individuals. Property of this kind would be subject to much plunder, as well as to frequent damage by fire. The destruction from these causes would, indeed, considerably lessen, but would not wholly annihilate the climatic and geographical influences of the forest, or ruinously diminish its value as a regular source of supply of fuel and timber. For prevention of the evils upon which I have so long dwelt, the American people must look to the diffusion of general intelligence on this subject, and to the enlightened self interest, for which they are remarkable, not to the action of their local or general legislatures. Even in France, government has moved with too slow and hesitating a pace, and preventive measures do not yet compensate destructive causes. The judicious remarks of Troy on this point may well be applied to other countries than France, other measures of public policy than the preservation of the woods. "To move softly," says he, "is to commit the most dangerous, the most unpardonable of imprudences; it diminishes the prestige of authority; it furnishes a triumph to the sneerer and the incredulous; it strengthens opposition and encourages resistance; it ruins the administration in the opinion of the people, weakens its power and depresses its courage." *

The Economy of the Forest.

The legislation of European states upon sylviculture, and the practice of that art, divide themselves into two great branches—the preservation of existing forests, and the creation of new. From the long operation of causes already set forth, what is understood in America and other new countries by the "primitive forest," no longer exists in the territories which were the seats of ancient civilization and empire, except upon a small scale, and in remote and almost inaccessible glens quite out of the reach of ordinary observation. The oldest European woods, indeed, are native, that is, sprung from self-sown seed,

* *Étude sur le Reboisement des Montagnes*, p. 5.

or from the roots of trees which have been felled for human purposes; but their growth has been controlled, in a variety of ways, by man and by domestic animals, and they always present more or less of an artificial character and arrangement. Both they and planted forests, which, though certainly not few, are of recent date in Europe, demand, as well for protection as for promotion of growth, a treatment different in some respects from that which would be suited to the character and wants of the virgin wood.

On this latter branch of the subject, experience and observation have not yet collected a sufficient stock of facts to serve for the construction of a complete system of sylviculture; but the management of the forest as it exists in France—the different zones and climates of which country present many points of analogy with those of the United States and some of the British colonies—has been carefully studied, and several manuals of practice have been prepared for the foresters of that empire. I believe the best of these is the *Cours Élémentaire de Culture des Bois créé à l'École Forestière de Nancy, par M. Lorentz, complété et publié par A. Parade*, with a supplement under the title of *Cours d'Aménagement des Forêts, par Henri Nanquette.* The *Études sur l'Économie Forestière, par Jules Clavé*, which I have often quoted, presents a great number of interesting views on this subject, and well deserves to be translated for the use of the English and American reader; but it is not designed as a practical guide, and it does not profess to be sufficiently specific in its details to serve that purpose. Notwithstanding the difference of conditions between the aboriginal and the trained forest, the judicious observer who aims at the preservation of the former will reap much instruction from the treatises I have cited, and I believe he will be convinced that the sooner a natural wood is brought into the state of an artificially regulated one, the better it is for all the multiplied interests which depend on the wise administration of this branch of public economy.*

* "In America," says Clavé (p. 124, 125), "where there is a vast extent of land almost without pecuniary value, but where labor is dear and

One consideration bearing on this subject has received less attention than it merits, because most persons interested in such questions have not opportunities for the comparison I refer to. I mean the great general superiority of cultivated timber to that of strictly spontaneous growth. I say *general* superiority, because there are exceptions to the rule. The white pine, *Pinus strobus*, for instance, and other trees of similar character and uses, require, for their perfect growth, a density of forest vegetation around them, which protects them from too much agitation by wind, and from the persistence of the lateral branches which fill the wood with knots. A pine which has grown under those conditions possesses a tall, straight stem, admirably fitted for masts and spars, and, at the same time, its wood is almost wholly free from knots, is regular in annular structure, soft and uniform in texture, and, consequently, superior to almost all other timber for joinery. If, while a large pine is spared, the broad-leaved or other

the rate of interest high, it is profitable to till a large surface at the least possible cost; *extensive* cultivation is there the most advantageous. In England, France, and Germany, where every corner of soil is occupied, and the least bit of ground is sold at a high price, but where labor and capital are comparatively cheap, it is wisest to employ *intensive* cultivation. * * * All the efforts of the cultivator ought to be directed to the obtaining of a given result with the least sacrifice, and there is equally a loss to the commonwealth if the application of improved agricultural processes be neglected where they are advantageous, or if they be employed where they are not required. * * * In this point of view, sylviculture must follow the same laws as agriculture, and, like it, be modified according to the economical conditions of different states. In countries abounding in good forests, and thinly peopled, elementary and cheap methods must be pursued; in civilized regions, where a dense population requires that the soil shall be made to produce all it can yield, the regular artificial forest, with all the processes that science teaches, should be cultivated. It would be absurd to apply to the endless woods of Brazil and of Canada the method of the Spessart by " double stages," and not less so in our country, where every yard of ground has a high value, to leave to nature the task of propagating trees, and to content ourselves with cutting, every twenty or twenty-five years, the meagre growths that chance may have produced."

smaller trees around it are felled, the swaying of the tree from the action of the wind mechanically produces separations between the layers of annual growth, and greatly diminishes the value of the timber.

The same defect is often observed in pines which, from some accident of growth, have much overtopped their fellows in the virgin forest. The white pine, growing in the fields, or in open glades in the woods, is totally different from the true forest tree, both in general aspect and in quality of wood. Its stem is much shorter, its top less tapering, its foliage denser and more inclined to gather into tufts, its branches more numerous and of larger diameter, its wood shows much more distinctly the divisions of annual growth, is of coarser grain, harder and more difficult to work into mitre joints. Intermixed with the most valuable pines in the American forests, are met many trees of the character I have just described. The lumbermen call them "saplings," and generally regard them as different in species from the true white pine, but botanists are unable to establish a distinction between them, and as they agree in almost all respects with trees grown in the open grounds from known white-pine seedlings, I believe their peculiar character is due to unfavorable circumstances in their early growth. The pine, then, is an exception to the general rule as to the inferiority of the forest to the open-ground tree. The pasture oak and pasture beech, on the contrary, are well known to produce far better timber than those grown in the woods, and there are few trees to which the remark is not equally applicable.*

* It is often laid down as a universal law, that the wood of trees of slow vegetation is superior to that of quick growth. This is one of those commonplaces by which men love to shield themselves from the labor of painstaking observation. It has, in fact, so many exceptions, that it may be doubted whether it is in any sense true. Most of the cedars are slow of growth; but while the timber of some of them is firm and durable, that of others is light, brittle, and perishable. The hemlock spruce is slower of growth than the pines, but its wood is of very little value. The pasture oak and beech show a breadth of grain—and, of course, an annual incre-

Another advantage of the artificially regulated forest is, that it admits of such grading of the ground as to favor the retention or discharge of water at will, while the facilities it affords for selecting and duly proportioning, as well as properly spacing, the trees which compose it, are too obvious to require to be more than hinted at. In conducting these operations, we must have a diligent eye to the requirements of nature, and must remember that a wood is not an arbitrary assemblage of trees to be selected and disposed according to the caprice of its owner. "A forest," says Clavé, "is not, as is often supposed, a simple collection of trees succeeding each other in long perspective, without bond of union, and capable of isolation from each other; it is, on the contrary, a whole, the different parts of which are interdependent upon each other, and it constitutes, so to speak, a true individuality. Every forest has a special character, determined by the form of the surface it grows upon, the kinds of trees that compose it, and the manner in which they are grouped." *

ment—twice as great as trees of the same species grown in the woods; and the American locust, *Robinia pseudacacia*, the wood of which is of extreme toughness and durability, is, of all trees indigenous to Northeastern America, by far the most rapid in growth.

As an illustration of the mutual interdependence of the mechanic arts, I may mention that in Italy, where stone, brick, and plaster are almost the only materials used in architecture, and where the "hollow ware" kitchen implements are of copper or of clay, the ordinary tools for working wood are of a very inferior description, and the locust timber is found too hard for their temper. Southey informs us, in "Espriella's Letters," that when a small quantity of mahogany was brought to England, early in the last century, the cabinetmakers were unable to use it, from the defective temper of their tools, until the demand for furniture from the new wood compelled them to improve the quality of their implements. In America, the cheapness of wood long made it the preferable material for almost all purposes to which it could by any possibility be applied. The mechanical cutlery and artisans' tools of the United States are of admirable temper, finish, and convenience, and no wood is too hard, or otherwise too refractory, to be wrought with great facility, both by hand tools and by the multitude of ingenious machines which the Americans have invented for this purpose.

* *Études Forestières*, p. 7.

European and American Trees compared.

The woods of North America are strikingly distinguished from those of Europe by the vastly greater variety of species they contain. According to Clavé, there are in "France and in most parts of Europe" only about twenty forest trees, five or six of which are spike-leaved and resinous, the remainder broad-leaved." * Our author, however, doubtless means genera, though he uses the word *espèces.* Rossmässler enumerates fifty-seven species of forest trees as found in Germany, but some of these are mere shrubs, some are fruit and properly garden trees, and some others are only varieties of familiar species. The valuable manual of Parade describes about the same number, including, however, two of American origin—the locust, *Robinia pseudacacia*, and the Weymouth or white pine, *Pinus strobus*—and the cedar of Lebanon from Asia, though it is indigenous in Algeria also. We may then safely say that Europe does not possess above forty or fifty trees of such economical value as to be worth the special care of the forester, while the oak alone numbers not less than thirty species in the United States,† and some other North American genera are almost equally diversified.‡

* *Études Forestières*, p. 7.

† For very full catalogues of American forest trees, and remarks on their geographical distribution, consult papers on the subject by Dr. J. G. Cooper, in the Report of the Smithsonian Institution for 1858, and the Report of the United States Patent Office, Agricultural Division, for 1860.

‡ Although Spenser's catalogue of trees occurs in the first canto of the first book of the "Faëry Queene"—the only canto of that exquisite poem actually read by most students of English literature—it is not so generally familiar as to make the quotation of it altogether superfluous:

VII.

Enforst to seeke some covert nigh at hand,
A shadie grove not farr away they spide,
That promist ayde the tempest to withstand;
Whose loftie trees, yclad with sommers pride,
Did spred so broad, that heavens light did hide,
Not perceable with power of any starr:
And all within were pathes and alleies wide,
With footing worne, and leading inward farr;
Faire harbour that them seems; so in they entred ar.

Few European trees, except those bearing edible fruit, have been naturalized in the United States, while the American forest flora has made large contributions to that of Europe. It is a very poor taste which has led to the substitution of the less picturesque European for the graceful and majestic American elm, in some public grounds in the United States. On the other hand, the European mountain ash—which in beauty and healthfulness of growth is superior to our own—the horse chestnut, and the abele, or silver poplar, are valuable additions to the ornamental trees of North America. The Swiss arve or zirbelkiefer, *Pinus cembra*, which yields a well-flavored edible seed and furnishes excellent wood for carving, the umbrella pine which also bears a seed agreeable to the taste, and which, from the color of its foliage and the beautiful form of its dome-like crown, is among the most elegant of trees, the white birch of Central Europe, with its pendulous branches almost rivalling those of the weeping willow in length, flexibility, and gracefulness of fall, and, especially, the "cypresse funerall," might be introduced into the United States with great advantage to the landscape. The European beech and chestnut furnish timber of far better quality than that of their American congeners. The fruit of the European chestnut,

VIII.

And foorth they passe, with pleasure forward led,
Joying to heare the birdes sweete harmony,
Which therein shrouded from the tempest dred,
Seemd in their song to scorne the cruell sky.
Much can they praise the trees so straight and hy,
The sayling pine; the cedar stout and tall;
The vine-propp elm; the poplar never dry;
The builder oake, sole king of forrests all;
The aspine good for staves; the cypresse funerall;

IX.

The laurell, meed of mightie conquerours
And poets sage; the firre that weepeth still;
The willow, worne of forlorn paramours;
The eugh, obedient to the benders will;
The birch for shaftes; the sallow for the mill;
The mirrhe sweete-bleeding in the bitter wound;
The warlike beech; the ash for nothing ill;
The fruitfull olive; and the platane round;
The carver holme; the maple seeldom inward sound.

though inferior to the American in flavor, is larger, and is an important article of diet among the French and Italian peasantry. The walnut of Europe, though not equal to some of the American species in beauty of growth or of wood, or to others in strength and elasticity of fibre, is valuable for its timber and its oil.* The maritime pine, which has proved of such immense use in fixing drifting sands in France, may perhaps be better adapted to this purpose than any of the pines of the New World, and it is of great importance for its turpentine, resin, and tar. The épicéa, or common fir, *Abies picea*, *Abies*

* The walnut is a more valuable tree than is generally supposed. It yields one third of the oil produced in France, and in this respect occupies an intermediate position between the olive of the south, and the oleaginous seeds of the north. A hectare (about two and a half acres), will produce nuts to the value of five hundred francs a year, which cost nothing but the gathering. Unfortunately, its maturity must be long waited for, and more nut-trees are felled than planted. The demand for its wood in cabinet work is the principal cause of its destruction. See LAVERGNE, *Économie Rurale de la France*, p. 253.

According to Cosimo Ridolfi (Lezioni Orali, ii. p. 424), France obtains three times as much oil from the walnut as from the olive, and nearly as much as from all oleaginous seeds together. He states that the walnut bears nuts at the age of twenty years, and yields its maximum product at seventy, and that a hectare of ground, with thirty trees, or twelve to the acre, is equal to a capital of twenty-five hundred francs.

The nut of this tree is known in the United States as the "English walnut." The fruit and the wood much resemble those of the American black walnut, *Juglans nigra*, but for cabinet work the American is the more beautiful material, especially when the large knots are employed. The timber of the European species, when straight grained, and *clear*, or free from knots, is, for ordinary purposes, better than that of the American black walnut, but bears no comparison with the wood of the hickory, when strength combined with elasticity is required, and its nut is very inferior in taste to that of the shagbark, as well as to the butternut, which it somewhat resembles.

"The chestnut is more valuable still, for it produces on a sterile soil, which, without it, would yield only ferns and heaths, an abundant nutriment for man."—LAVERGNE, *Économie Rurale de la France*, p. 253.

I believe the varieties developed by cultivation are less numerous in the walnut than in the chestnut, which latter tree is often grafted in Southern Europe.

excelsa, *Picea excelsa*, abundant in the mountains of France and the contiguous country, is known for its product, Burgundy pitch, and, as it flourishes in a greater variety of soil and climate than almost any other spike-leaved tree, it might be well worth transplantation.* The cork oak has been introduced into the United States, I believe, and would undoubtedly thrive in the Southern section of the Union.†

In the walnut, the chestnut, the cork oak, the mulberry, the olive, the orange, the lemon, the fig, and the multitude of other trees which, by their fruit, or by other products, yield an annual revenue, nature has provided Southern Europe with

* This fir is remarkable for its tendency to cicatrize or heal over its stumps, a property which it possesses in common with some other firs, the maritime pine, and the European larch. When these trees grow in thick clumps, their roots are apt to unite by a species of natural grafting, and if one of them be felled, although its own proper rootlets die, the stump may continue, sometimes for a century, to receive nourishment from the radicles of the surrounding trees, and a dome of wood and bark of considerable thickness be formed over it. The cicatrization is, however, only apparent, for the entire stump, except the outside ring of annual growth, soon dies, and even decays within its covering, without sending out new shoots.

† At the age of twelve or fifteen years, the cork tree is stripped of its outer bark for the first time. This first yield is of inferior quality, and is employed for floats for nets and buoys, or burnt for lampblack. After this, a new layer of cork, an inch or an inch and a quarter in thickness, is formed about once in ten years, and is removed in large sheets without injury to the tree, which lives a hundred and fifty years or more. According to Clavé (p. 252), the annual product of a forest of cork oaks is calculated at about 660 kilogrammes, worth 150 francs, to the hectare, which, deducting expenses, leaves a profit of 100 francs. This is about equal to 250 pound weight, and eight dollars profit to the acre. The cork oaks of the national domain in Algeria cover about 500,000 acres, and are let to individuals at rates which are expected, when the whole is rented, to yield to the state a revenue of about $2,000,000.

George Sand, in the *Histoire de ma Vie*, speaks of the cork forests in Southern France as among the most profitable of rural possessions, and states, what I do not remember to have seen noticed elsewhere, that Russia is the best customer for cork. The large sheets taken from the trees are slit into thin plates, and used to line the walls of apartments in that cold climate.

a partial compensation for the loss of the native forest. It is true that these trees, planted as most of them are at such distances as to admit of cultivation, or of the growth of grass among them, are but an inadequate substitute for the thick and shady wood; but they perform to a certain extent the same offices of absorption and transpiration, they shade the surface of the ground, they serve to break the force of the wind, and on many a steep declivity, many a bleak and barren hillside, the chestnut binds the soil together with its roots, and prevents tons of earth and gravel from washing down upon the fields and the gardens. Fruit trees are not wanting, certainly, north of the Alps. The apple, the pear, and the prune are important in the economy both of man and of nature, but they are far less numerous in Switzerland and Northern France than are the trees I have mentioned in Southern Europe, both because they are in general less remunerative, and because the climate, in higher latitudes, does not permit the free introduction of shade trees into grounds occupied for agricultural purposes.*

The multitude of species, intermixed as they are in their spontaneous growth, gives the American forest landscape a variety of aspect not often seen in the woods of Europe, and

* The walnut, the chestnut, the apple, and the pear are common to the border between the countries I have mentioned, but the range of the other trees is bounded by the Alps, and by a well-defined and sharply drawn line to the west of those mountains. I cannot give statistical details as to the number of any of the trees in question, or as to the area they would cover if brought together in a given country. From some peculiarity in the sky of Europe, cultivated plants will thrive, in Northern Italy, in Southern France, and even in Switzerland, under a depth of shade where no crop, not even grass, worth harvesting, would grow in the United States with an equally high summer temperature. Hence the cultivation of all these trees is practicable in Europe to a greater extent than would be supposed reconcilable with the interests of agriculture. Some idea of the importance of the olive orchards may be formed from the fact that Sicily alone, an island scarcely exceeding 10,000 square miles in area, of which one third at least is absolutely barren, has exported to the single port of Marseilles more than 2,000,000 pounds weight of olive oil per year, for the last twenty years

the gorgeous tints, which nature repeats from the dying dolphin to paint the falling leaf of the American maples, oaks, and ash trees, clothe the hillsides and fringe the watercourses with a rainbow splendor of foliage, unsurpassed by the brightest groupings of the tropical flora. It must be admitted, however, that both the northern and the southern declivities of the Alps exhibit a nearer approximation to this rich and multifarious coloring of autumnal vegetation than most American travellers in Europe are willing to allow; and, besides, the small deciduous shrubs which often carpet the forest glades of these mountains are dyed with a ruddy and orange glow, which, in the distant landscape, is no mean substitute for the scarlet and crimson and gold and amber of the transatlantic woodland.

No American evergreen known to me resembles the umbrella pine sufficiently to be a fair object of comparison with it.* A cedar, very common above the Highlands on the Hudson, is extremely like the cypress, straight, slender, with erect, compressed ramification, and feathered to the ground, but its foliage is neither so dark nor so dense, the tree does not attain the majestic height of the cypress, nor has it the lithe flexibility of that tree. In mere shape, the Lombardy poplar nearly resembles this latter, but it is almost a profanation to compare the two, especially when they are agitated by the wind; for under such circumstances, the one is the most majestic, the other the most ungraceful, or—if I may apply such an expression to anything but human affectation of movement —the most awkward of trees. The poplar trembles before the blast, flutters, struggles wildly, dishevels its foliage, gropes around with its feeble branches, and hisses as in impotent passion. The cypress gathers its limbs still more closely to its stem, bows a gracious salute rather than an humble obeisance

* It is hard to say how far the peculiar form of the graceful crown of this pine is due to pruning. It is true that the extremities of the topmost branches are rarely lopped, but the lateral boughs are almost uniformly removed to a very considerable height, and it is not improbable that the shape of the top is thereby affected.

to the tempest, bends to the wind with an elasticity that assures you of its prompt return to its regal attitude, and sends from its thick leaflets a murmur like the roar of the far-off ocean.

The cypress and the umbrella pine are not merely conventional types of the Italian landscape. They are essential elements in a field of rural beauty which can be seen in perfection only in the basin of the Mediterranean, and they are as characteristic of this class of scenery as the date palm is of the oases of the desert. There is, however, this difference: a single cypress or pine is often enough to shed beauty over a wide area; the palm is a social tree, and its beauty is not so much that of the individual as of the group. The frequency of the cypress and the pine—combined with the fact that the other trees of Southern Europe which most interest a stranger from the north, the orange and the lemon, the cork oak, the ilex, the myrtle, and the laurel, are evergreens—goes far to explain the beauty of the winter scenery of Italy. Indeed it is only in the winter that a tourist who confines himself to wheel carriages and high roads can acquire any notion of the face of the earth, and form any proper geographical image of that country. At other seasons, not high walls only, but equally impervious hedges, and now, unhappily, acacias thickly planted along the railway routes, confine the view so completely, that the arch of a tunnel, or a night cap over the traveller's eyes, is scarcely a more effectual obstacle to the gratification of his curiosity.*

* Besides this, in a country so diversified in surface—I wish we could with the French say *accidented*—as Italy with the exception of the champaign region drained by the Po, every new field of view requires either an extraordinary *coup d'œil* in the spectator, or a long study, in order to master its relief, its plans, its salient and retreating angles. In summer, the universal greenery confounds light and shade, distance and foreground; and though the impression upon a traveller, who journeys for the sake of "sensations," may be strengthened by the mysterious annihilation of all standards for the measurement of space, yet the superior intelligibility of the winter scenery of Italy is more profitable to those who see with a view to analyze.

Sylviculture.

The art, or, as the Continental foresters call it, the science of sylviculture has been so little pursued in England and America, that its nomenclature has not been introduced into the English vocabulary, and I shall not be able to describe its processes with technical propriety of language, without occasionally borrowing a word from the forest literature of France and Germany. A full discussion of the methods of sylviculture would, indeed, be out of place in a work like the present, but the almost total want of conveniently accessible means of information on the subject, in English-speaking countries, will justify me in presenting it with somewhat more of detail than would otherwise be pertinent.

The two best known methods are those distinguished as the *taillis*, copse or coppice treatment,* and the *futaie*, for which I find no English equivalent, but which may not inappropriately be called the *full-growth* system. A *taillis*, copse, or coppice, is a wood composed of shoots from the roots of trees previously cut for fuel and timber. The shoots are thinned out from time to time, and finally cut, either after a fixed number of years, or after the young trees have attained to certain dimensions, their roots being then left to send out a new progeny as before. This is the cheapest method of management, and therefore the best wherever the price of labor and of capital bears a high proportion to that of land and of timber; but it is essentially a wasteful economy. If the woodland is, in the first place, completely cut over, as is found most convenient in practice, the young shoots have neither the shade nor the protection from wind so important to forest growth, and their progress is comparatively slow, while, at the same time, the thick clumps they form choke the seedlings that may

* Copse, or coppice, from the French *couper*, to cut, signifies properly a wood the trees of which are cut at certain periods of immature growth, and allowed to shoot up again from the roots; but it has come to signify, very commonly, a young wood, grove, or thicket, without reference to its origin, or to its character of a forest crop.

have sprouted near them. If domestic animals of any species are allowed to roam in the wood, they browse upon the terminal buds and the tender branches, thereby stunting, if they do not kill, the young trees, and depriving them of all beauty and vigor of growth. The evergreens, once cut, do not shoot up again,* and the mixed character of the forest—in many respects an important advantage, if not an indispensable condition of growth—is lost; † and besides this, large wood of

* It has been recently stated, upon the evidence of the Government foresters of Greece, and of the queen's gardener, that a large wood has been discovered in Arcadia, consisting of a fir which has the property of sending up both vertical and lateral shoots from the stump of felled trees and forming a new crown. It was at first supposed that this forest grew only on the "mountains," of which the hero of About's most amusing story, *Le Roi des Montagnes*, was "king;" but it is now said that small stumps, with the shoots attached, have been sent to Germany, and recognized by able botanists as true natural products.

† Natural forests are rarely, if ever, composed of trees of a single species, and experience has shown that oaks and other broad-leaved trees, planted as artificial woods, require to be mixed, or associated with others of different habits.

In the forest of Fontainebleau, "oaks, mingled with beeches in due proportion," says Clavé, "may arrive at the age of five or six hundred years in full vigor, and attain dimensions which I have never seen surpassed; when, however, they are wholly unmixed with other trees, they begin to decay and die at the top, at the age of forty or fifty years, like men, old before their time, weary of the world, and longing only to quit it. This has been observed in most of the oak plantations of which I have spoken, and they have not been able to attain to full growth. When the vegetation was perceived to languish, they were cut, in the hope that this operation would restore their vigor, and that the new shoots would succeed better than the original trees; and, in fact, they seemed to be recovering for the first few years. But the shoots were soon attacked by the same decay, and the operation had to be renewed at shorter and shorter intervals, until at last it was found necessary to treat as coppices plantations originally designed for the full-growth system. Nor was this all: the soil, periodically bared by these cuttings, became impoverished, and less and less suited to the growth of the oak. * * * It was then proposed to introduce the pine and plant with it the vacancies and glades. * * * By this means, the forest was saved from the ruin which threatened it, and now more than 10,000 acres of pines, from fifteen to thirty

any species cannot be grown in this method, because trees which shoot from decaying stumps and their dying roots, become hollow or otherwise unsound before they acquire their full dimensions. A more fatal objection still, is, that the roots of trees will not bear more than two or three, or at most four cuttings of their shoots before their vitality is exhausted, and the wood can then be restored only by replanting entirely. The period of cutting coppices varies in Europe from fifteen to forty years, according to soil, species, and rapidity of growth.

In the *futaie*, or full-growth system, the trees are allowed to stand as long as they continue in healthy and vigorous growth. This is a shorter period than would be at first supposed, when we consider the advanced age and great dimensions to which, under favorable circumstances, many forest trees attain in temperate climates. But, as every observing person familiar with the natural forest is aware, these are exceptional cases, just as are instances of great longevity or of gigantic stature among men. Able vegetable physiologists have maintained that the tree, like most reptiles, has no natural limit of life or of growth, and that the only reason why our oaks and our pines do not reach the age of twenty centuries and the height of a hundred fathoms, is, that in the multitude of accidents to which they are exposed, the chances of their attaining to such a length of years and to such dimensions of growth are a million to one against them. But another explanation of this fact is possible. In trees affected by no discoverable external cause of death, decay begins at the topmost branches, which seem to wither and die for want of nutriment. The mysterious force by which the sap is carried

years old, are disseminated at various points, sometimes intermixed with broad-leaved trees, sometimes forming groves by themselves."—*Revue des Deux Mondes*, Mai, 1863, pp. 153, 154.

The forests of Denmark, which, in modern times, have been succeeded by the beech—a species more inclined to be exclusive than any other broad-leaved tree—were composed of birches, oaks, firs, aspens, willows, hazel, and maple, the first three being the leading species. At present, the beech greatly predominates.—VAUPELL, *Bögens Indvandring*, pp. 19, 20.

from the roots to the utmost twigs, cannot be conceived to be unlimited in power, and it is probable that it differs in different species, so that while it may suffice to raise the fluid to the height of five hundred feet in the sequoia, it may not be able to carry it beyond one hundred and fifty in the oak. The limit may be different, too, in different trees of the same species, not from defective organization in those of inferior growth, but from more or less favorable conditions of soil, nourishment, and exposure. Whenever a tree attains to the limit beyond which its circulating fluids cannot rise, we may suppose that decay begins, and death follows, from the same causes which bring about the same results in animals of limited size—such, for example, as the interruption of functions essential to life, in consequence of the clogging up of ducts by matter assimilable in the stage of growth, but no longer so when increment has ceased.

In the natural woods, we observe that, though, among the myriads of trees which grow upon a square mile, there are several vegetable giants, yet the great majority of them begin to decay long before they have attained their maximum of stature, and this seems to be still more emphatically true of the artificial forest. In France, according to Clavé, "oaks, in a suitable soil, may stand, without exhibiting any sign of decay, for two or three hundred years; the pines hardly exceed one hundred and twenty, and the soft or white woods [*bois blancs*], in wet soils, languish and die before reaching the fiftieth year." * These ages are certainly below the average of those of American forest trees, and are greatly exceeded in very numerous well-attested instances of isolated trees in Europe.

The former mode of treating the futaie, called the garden system, was to cut the trees individually as they arrived at maturity, but, in the best regulated forests, this practice has been abandoned for the German method, which embraces not only the securing of the largest immediate profit, but the re-

* *Études Forestières*, p. 89.

planting of the forest, and the care of the young growth. This is effected in the case of a forest, whether natural or artificial, which is to be subjected to regular management, by three operations. The first of these consists in felling about one third of the wood, in such way as to leave convenient spaces for the growth of young trees. The remaining two-thirds are relied upon to replant the vacancies, by natural sowing, which they seldom or never fail to do. The seedlings are watched, are thinned out when too dense, the ill formed and sickly, as well as those of inferior value, and the shrubs and thorns which might otherwise choke or too closely shade them, are pulled up. When they have attained sufficient strength and development of foliage to bear or to require more light and air, the second step is taken, by removing a suitable proportion of the old trees which had been spared at the first cutting; and when, finally, they are hardened enough to bear frost and sun without other protection than that which they mutually give to each other, the remainder of the original forest is felled, and the wood now consists wholly of young and vigorous trees. This result is obtained after about twenty years. At convenient periods afterward, the unhealthy stocks and those injured by wind or other accidents are removed, and in some instances the growth of the remainder is promoted by irrigation or by fertilizing applications.* When the forest is ap-

* The grounds which it is most important to clothe with wood as a conservative influence, and which, also, can best be spared from agricultural use, are steep hillsides. But the performance of all the offices of the forester to the tree—seeding, planting, thinning, and finally felling and removing for consumption—is more laborious upon a rapid declivity than on a level soil, and at the same time it is difficult to apply irrigation or manures to trees so situated. Experience has shown that there is great advantage in terracing the face of a hill before planting it, both as preventing the wash of the earth by checking the flow of water down its slope, and as presenting a surface favorable for irrigation, as well as for manuring and cultivating the tree. But even without so expensive a process, very important results have been obtained by simply ditching declivities. "In order to hasten the growth of wood on the flanks of a mountain, Mr. Eugène Chevandier divided the slope into zones forty or fifty

proaching to maturity, the original processes already described are repeated; and as, in different parts of an extensive forest,

feet wide, by horizontal ditches closed at both ends, and thereby obtained, from firs of different ages, shoots double the dimensions of those which grew on a dry soil of the same character, where the water was allowed to run off without obstruction."—DUMONT, *Des Travaux Publics, etc.*, pp. 94–96.

The ditches were about two feet and a half deep, and three feet and a half wide, and they cost about forty francs the hectare, or three dollars the acre. This extraordinary growth was produced wholly by the retention of the rain water in the ditches, whence it filtered through the whole soil and supplied moisture to the roots of the trees. It may be doubted whether in a climate cold enough to freeze the entire contents of the ditches in winter, it would not be expedient to draw off the water in the autumn, as the presence of so large a quantity of ice in the soil might prove injurious to trees too young and small to shelter the ground effectually against frost.

Chevandier computes that, if the annual growth of the pine in the marshy soil of the Vosges be represented by one, it will equal two in dry ground, four or five on slopes so ditched or graded as to retain the water flowing upon them from roads or steep declivities, and six where the earth is kept constantly moist by infiltration from running brooks.—*Comptes Rendus à l'Académie des Sciences*—t. xix, Juillet, Dec., 1844, p. 167.

The effect of accidental irrigation is well shown in the growth of the trees planted along the canals of irrigation which traverse the fields in many parts of Italy. They flourish most luxuriantly, in spite of continual lopping, and yield a very important contribution to the stock of fuel for domestic use; while trees, situated so far from canals as to be out of the reach of infiltration from them, are of much slower growth, under circumstances otherwise equally favorable.

In other experiments of Chevandier, under better conditions, the yield of wood was increased, by judicious irrigation, in the ratio of seven to one, the profits in that of twelve to one. At the Exposition of 1855, Chambrelent exhibited young trees, which, in four years from the seed, had grown to the height of sixteen and twenty feet, and the diameter of ten and twelve inches. Chevandier experimented with various manures, and found that some of them might be profitably applied to young, but not to old trees, the quantity required in the latter case being too great. Wood ashes and the refuse of soda factories are particularly recommended. I have seen an extraordinary growth produced in fir trees by the application of soapsuds.

they would take place in different zones, it would afford indefinitely an annual crop of firewood and timber.

The duties of the forester do not end here. It sometimes happens that the glades left by felling the older trees are not sufficiently seeded, or that the species, or *essences*, as the French oddly call them, are not duly proportioned in the new crop. In this case, seed must be artificially sown, or young trees planted in the vacancies.

One of the most important rules in the administration of the forest is the absolute exclusion of domestic quadrupeds from every wood which is not destined to be cleared. No growth of young trees is possible where cattle are admitted to pasture at any season of the year, though they are undoubtedly most destructive while trees are in leaf.*

* Although the economy of the forest has received little attention in the United States, no lover of American nature can have failed to observe a marked difference between a native wood from which cattle are excluded and one where they are permitted to browse. A few seasons suffice for the total extirpation of the "underbrush," including the young trees on which alone the reproduction of the forest depends, and all the branches of those of larger growth which hang within reach of the cattle are stripped of their buds and leaves, and soon wither and fall off. These effects are observable at a great distance, and a wood pasture is recognized, almost as far as it can be seen, by the regularity with which its lower foliage terminates at what Ruskin somewhere calls the "cattle line." This always runs parallel to the surface of the ground, and is determined by the height to which domestic quadrupeds can reach to feed upon the leaves. In describing a visit to the grand-ducal farm of San Rossore near Pisa, where a large herd of camels is kept, Chateauvieux says: "In passing through a wood of evergreen oaks, I observed that all the twigs and foliage of the trees were clipped up to the height of about twelve feet above the ground, without leaving a single spray below that level. I was informed that the browsing of the camels had trimmed the trees as high as they could reach."—LULLIN DE CHATEAUVIEUX, *Lettres sur l'Italie*, p. 113.

The removal of the shelter afforded by the brushwood and the pendulous branches of trees permits drying and chilling winds to parch and cool the ground, and of course injuriously affects the growth of the wood. But this is not all. The tread of quadrupeds exposes and bruises the roots of the trees, which often die from this cause, as any one may observe by following the paths made by cattle through woodlands.

It is often necessary to take measures for the protection of young trees against the rabbit, the mole, and other rodent quadrupeds, and of older ones against the damage done by the larvæ of insects hatched upon the surface or in the tissues of the bark, or even in the wood itself. The much greater liability of the artificial than of the natural forest to injury from this cause is perhaps the only point in which the superiority of the former to the latter is not as marked as that of any domesticated vegetable to its wild representative. But the better quality of the wood and the much more rapid growth of the trained and regulated forest are abundant compensations for the loss thus occasioned, and the progress of entomological science will, perhaps, suggest new methods of preventing the ravages of insects. Thus far, however, the collection and destruction of the eggs, by simple but expensive means, has proved the only effectual remedy.*

It is common in Europe to permit the removal of the fallen leaves and fragments of bark and branches with which the forest soil is covered, and sometimes the cutting of the lower

* I have remarked elsewhere that most insects which deposit and hatch their eggs in the wood of the natural forest confine themselves to dead trees. Not only is this the fact, but it is also true that many of the borers attack only freshly cut timber. Their season of labor is a short one, and unless the tree is cut during this period, it is safe from them. In summer you may hear them plying their augers in the wood of a young pine with soft green bark, as you sit upon its trunk, within a week after it has been felled, but the windfalls of the winter lie uninjured by the worm and even undecayed for centuries. In the pine woods of New England, after the regular lumberman has removed the standing trees, these old trunks are hauled out from the mosses and leaves which half cover them, and often furnish excellent timber. The slow decay of such timber in the woods, it may be remarked, furnishes another proof of the uniformity of temperature and humidity in the forest, for the trunk of a tree lying on grass or plough land, and of course exposed to all the alternations of climate, hardly resists complete decomposition for a generation. The forests of Europe exhibit similar facts. Wessely, in a description of the primitive wood of Neuwald in Lower Austria, says that the windfalls required from 150 to 200 years for entire decay.—*Die Oesterreichischen Alpenländer und ihre Forste*, p. 312.

twigs of evergreens. The leaves and twigs are principally used as litter for cattle, and finally as manure, the bark and wind-fallen branches as fuel. By long usage, sometimes by express grant, this privilege has become a vested right of the population in the neighborhood of many public, and even large private forests; but it is generally regarded as a serious evil. To remove the leaves and fallen twigs is to withdraw much of the pabulum upon which the tree was destined to feed. The small branches and leaves are the parts of the tree which yield the largest proportion of ashes on combustion, and of course they supply a great amount of nutriment for the young shoots. "A cubic foot of twigs," says Vaupell, "yields four times as much ashes as a cubic foot of stem wood. * * For every hundred weight of dried leaves carried off from a beech forest, we sacrifice a hundred and sixty cubic feet of wood. The leaves and the mosses are a substitute, not only for manure, but for ploughing. The carbonic acid given out by decaying leaves, when taken up by water, serves to dissolve the mineral constituents of the soil, and is particularly active in disintegrating feldspar and the clay derived from its decomposition. * * * The leaves belong to the soil. Without them it cannot preserve its fertility, and cannot furnish nutriment to the beech. The trees languish, produce seed incapable of germination, and the spontaneous self-sowing, which is an indispensable element in the best systems of sylviculture, fails altogether in the bared and impoverished soil." *

* VAUPELL, *Bögens Indvandring i de Danske Skove*, pp. 29, 46. Vaupell further observes, on the page last quoted: "The removal of leaves is injurious to the forest, not only because it retards the growth of trees, but still more because it disqualifies the soil for the production of particular species. When the beech languishes, and the development of its branches is less vigorous and its crown less spreading, it becomes unable to resist the encroachments of the fir. This latter tree thrives in an inferior soil, and being no longer stifled by the thick foliage of the beech, it spreads gradually through the wood, while the beech retreats before it and finally perishes."

The study of the natural order of succession in forest trees is of the utmost importance in sylviculture, because it guides us in the selection of

Besides these evils, the removal of the leaves deprives the soil of that spongy character which gives it such immense value as a reservoir of moisture and a regulator of the flow of springs; and, finally, it exposes the surface roots to the drying influence of sun and wind, to accidental mechanical injury from the tread of animals or men, and, in cold climates, to the destructive effects of frost.

The annual lopping and trimming of trees for fuel, so common in Europe, is fatal to the higher uses of the forest, but where small groves are made, or rows of trees planted, for no other purpose than to secure a supply of firewood, or to serve

the species to be employed in planting a new or restoring a decayed forest. When ground is laid bare both of trees and of vegetable mould, and left to the action of unaided and unobstructed nature, she first propagates trees which germinate and grow only under the influence of a full supply of light and air, and then, in succession, other species, according to their ability to bear the shade and their demand for more abundant nutriment. In Northern Europe, the larch, the white birch, the aspen, first appear; then follow the maple, the alder, the ash, the fir; then the oak and the linden; and then the beech. The trees called by these respective names in the United States are not specifically the same as their European namesakes, nor are they always even the equivalents of these latter, and therefore the order of succession in America would not be precisely as indicated by the foregoing list, but it nevertheless very nearly corresponds to it.

It is thought important to encourage the growth of the beech in Denmark and Northern Germany, because it upon the whole yields better returns than other trees, and particularly because it appears not to exhaust, but on the contrary to enrich the soil; for by shedding its leaves it returns to it most of the nutriment it has drawn from it, and at the same time furnishes a solvent which aids materially in the decomposition of its mineral constituents.

When the forest is left to itself, the order of succession is constant, and its occasional inversion is always explicable by some human interference. It is curious that the trees which require most light are content with the poorest soils, and *vice versa*. The trees which first appear are also those which propagate themselves farthest to the north. The birch, the larch, and the fir bear a severer climate than the oak, the oak than the beech. "These parallelisms," says Vaupell, "are very interesting, because they are entirely independent of each other," and each prescribes the same order of succession.—*Bögens Indvandring*, p. 42.

as supports for the vine, it is often very advantageous. The willows, and many other trees, bear polling for a long series of years without apparent diminution of growth of branches, and though certainly a polled, or, to use an old English word, a doddered tree, is in general a melancholy object, yet it must be admitted that the aspect of some species—the American locust, *Robinia pseudacacia*, for instance—when young, is improved by this process.*

I have spoken of the needs of agriculture as a principal cause of the destruction of the forest, and of domestic cattle as particularly injurious to the growth of young trees. But these animals affect the forest, indirectly, in a still more important way, because the extent of cleared ground required for agricultural use depends very much on the number and kinds of the cattle bred. We have seen, in a former chapter, that, in the United States, the domestic quadrupeds amount to more than a hundred millions, or three times the number of the human population of the Union. In many of the Western States, the swine subsist more or less on acorns, nuts, and other products of the woods, and the prairies, or natural meadows of the Mississippi valley, yield a large amount of food for beast, as well as for man. With these exceptions, all this vast army of quadrupeds is fed wholly on grass, grain, pulse, and roots grown on soil reclaimed from the forest by European settlers. It is true that the flesh of domestic quadrupeds enters very largely into the aliment of the American people, and greatly reduces the quantity of vegetable nutriment which they would otherwise consume, so that a smaller amount of agricultural product is required for immediate human food, and, of course, a smaller extent of cleared land is needed for the growth of that product, than if no domestic animals existed. But the flesh of the horse, the ass, and the mule is not

* When vigorous young locusts, of two or three inches in diameter, are polled, they throw out a great number of very thick-leaved shoots, which arrange themselves in a globular head, so unlike the natural crown of the acacia, that persons familiar only with the untrained tree often take them for a different species.

consumed by man, and the sheep is reared rather for its fleece than for food. Besides this, the ground required to produce the grass and grain consumed in rearing and fattening a grazing quadruped, would yield a far larger amount of nutriment, if devoted to the growing of breadstuffs, than is furnished by his flesh; and, upon the whole, whatever advantages may be reaped from the breeding of domestic cattle, it is plain that the cleared land devoted to their sustenance in the originally wooded part of the United States, after deducting a quantity sufficient to produce an amount of aliment equal to their flesh, still greatly exceeds that cultivated for vegetables, directly consumed by the people of the same regions; or, to express a nearly equivalent idea in other words, the meadow and the pasture, taken together, much exceed the plough land.*

In fertile countries, like the United States, the foreign demand for animal and vegetable aliment, for cotton, and for tobacco, much enlarges the sphere of agricultural operations, and, of course, prompts further encroachments upon the forest. The commerce in these articles, therefore, constitutes in America a special cause of the destruction of the woods, which does not exist in the numerous states of the Old World that derive the raw material of their mechanical industry from distant lands, and import many articles of vegetable food or luxury which their own climates cannot advantageously produce.

* The two ideas expressed in the text are not exactly equivalent, because, though the consumption of animal food diminishes the amount of vegetable aliment required for human use, yet the animals themselves consume a great quantity of grain and roots grown on ground ploughed and cultivated as regularly and as laboriously as any other.

The 170,000,000 bushels of oats raised in the United States in 1860, and fed to the 6,000,000 horses, the potatoes, the turnips, and the maize employed in fattening the oxen, the sheep, and the swine slaughtered the same year, occupied an extent of ground which, cultivated by hand labor and with Chinese industry and skill, would probably have produced a quantity of vegetable food equal in alimentary power to the flesh of the quadrupeds killed for domestic use. Hence, so far as the naked question of *amount* of aliment is concerned, the meadows and the pastures might as well have remained in the forest condition.

The growth of arboreal vegetation is so slow that, though he who buries an acorn may hope to see it shoot up to a miniature resemblance of the majestic tree which shall shade his remote descendants, yet the longest life hardly embraces the seedtime and the harvest of a forest. The planter of a wood must be actuated by higher motives than those of an investment the profits of which consist in direct pecuniary gain to himself or even to his posterity; for if, in rare cases, an artificial forest may, in two or three generations, more than repay its original cost, still, in general, the value of its timber will not return the capital expended and the interest accrued.* But when we consider the immense collateral advantages derived from the presence, the terrible evils necessarily resulting from the destruction of the forest, both the preservation of existing woods, and the far more costly extension of them where they have been unduly reduced, are among the most obvious of the duties which this age owes to those that are to come after it. Especially is this obligation incumbent upon Americans. No civilized people profits so largely from the toils and sacrifices of its immediate predecessors as they; no generations have

* According to Clavé (*Études*, p. 159), the net revenue from the forests of the state in France, making no allowance for interest on the capital represented by the forest, is two dollars per acre. In Saxony it is about the same, though the cost of administration is twice as much as in France; in Würtemberg it is about a dollar an acre; and in Prussia, where half the income is consumed in the expenses of administration, it sinks to less than half a dollar. This low rate in Prussia is partly explained by the fact that a considerable proportion of the annual product of wood is either conceded to persons claiming prescriptive rights, or sold, at a very small price, to the poor. Taking into account the capital invested in forest land, and adding interest upon it, Pressler calculates that a pine wood, managed with a view to felling it when eighty years old, would yield only one eighth of one per cent. annual profit; a fir wood, at one hundred years, one sixth of one per cent.; a beech wood, at one hundred and twenty years, one fourth of one per cent. The same author (p. 335) gives the net income of the New forest in England, over and above expenses, interest not computed, at twenty-five cents per acre only. In America, where no expense is bestowed upon the woods, the annual growth would generally be estimated much higher.

ever sown so liberally, and, in their own persons, reaped so scanty a return, as the pioneers of Anglo-American social life. We can repay our debt to our noble forefathers only by a like magnanimity, by a like self-forgetting care for the moral and material interests of our own posterity.

Instability of American Life.

All human institutions, associate arrangements, modes of life, have their characteristic imperfections. The natural, perhaps the necessary defect of ours, is their instability, their want of fixedness, not in form only, but even in spirit. The face of physical nature in the United States shares this incessant fluctuation, and the landscape is as variable as the habits of the population. It is time for some abatement in the restless love of change which characterizes us, and makes us almost a nomade rather than a sedentary people.* We have now felled forest enough everywhere, in many districts far too much. Let us restore this one element of material life to its normal

* It is rare that a middle-aged American dies in the house where he was born, or an old man even in that which he has built; and this is scarcely less true of the rural districts, where every man owns his habitation, than of the city, where the majority live in hired houses. This life of incessant flitting is unfavorable for the execution of permanent improvements of every sort, and especially of those which, like the forest, are slow in repaying any part of the capital expended in them. It requires a very generous spirit in a landholder to plant a wood on a farm he expects to sell, or which he knows will pass out of the hands of his descendants at his death. But the very fact of having begun a plantation would attach the proprietor more strongly to the soil for which he had made such a sacrifice; and the paternal acres would have a greater value in the eyes of a succeeding generation, if thus improved and beautified by the labors of those from whom they were inherited. Landed property, therefore, the transfer of which is happily free from every legal impediment or restriction in the United States, would find, in the feelings thus prompted, a moral check against a too frequent change of owners, and would tend to remain long enough in one proprietor or one family to admit of gradual improvements which would increase its value both to the possessor and to the state.

proportions, and devise means for maintaining the permanence of its relations to the fields, the meadows, and the pastures, to the rain and the dews of heaven, to the springs and rivulets with which it waters the earth. The establishment of an approximately fixed ratio between the two most broadly characterized distinctions of rural surface—woodland and plough land—would involve a certain persistence of character in all the branches of industry, all the occupations and habits of life, which depend upon or are immediately connected with either, without implying a rigidity that should exclude flexibility of accommodation to the many changes of external circumstance which human wisdom can neither prevent nor foresee, and would thus help us to become, more emphatically, a well-ordered and stable commonwealth, and, not less conspicuously, a people of progress.

NOTE on word *watershed*, omitted on p. 257.—Sir John F. W. Herschel (*Physical Geography*, 137, and elsewhere) spells this word *water-sched*, because he considers it a translation, or rather an adoption of the German "Wasser-scheide, separation of the waters, not water-*shed*, the slope *down which* the waters run." As a point of historical etymology, it is probable that the word in question was suggested to those who first used it by the German *Wasserscheide;* but the spelling *water-sched*, proposed by Herschel, is objectionable, both because *sch* is a combination of letters wholly unknown to modern English orthography and properly representing no sound recognized in English orthoepy, and for the still better reason that *water-shed*, in the sense of *division-of-the-waters*, has a legitimate English etymology.

The Anglo-Saxon *sceadan* meant both to separate or divide, and to shade or shelter. It is the root of the English verbs *to shed* and *to shade*, and in the former meaning is the A. S. equivalent of the German verb *scheiden.*

Shed in Old English had the meaning *to separate* or *distinguish.* It is so used in the *Owl and the Nightingale*, v. 197. Palsgrave (*Lesclarcissement, etc.*, p. 717) defines *I shede*, I departe thinges asonder; and the word still means *to divide* in several English local dialects. Hence, *watershed*, the division or separation of the waters, is good English both in sense and spelling.

CHAPTER IV.

THE WATERS.

LAND ARTIFICIALLY WON FROM THE WATERS: *a*, EXCLUSION OF THE SEA BY DIKING; *b*, DRAINING OF LAKES AND MARSHES; *c*, GEOGRAPHICAL INFLUENCE OF SUCH OPERATIONS—LOWERING OF LAKES—MOUNTAIN LAKES—CLIMATIC EFFECTS OF DRAINING LAKES AND MARSHES—GEOGRAPHICAL AND CLIMATIC EFFECTS OF AQUEDUCTS, RESERVOIRS, AND CANALS—SURFACE AND UNDERDRAINING, AND THEIR CLIMATIC AND GEOGRAPHICAL EFFECTS—IRRIGATION AND ITS CLIMATIC AND GEOGRAPHICAL EFFECTS.

INUNDATIONS AND TORRENTS: *a*, RIVER EMBANKMENTS; *b*, FLOODS OF THE ARDÈCHE; *c*, CRUSHING FORCE OF TORRENTS; *d*, INUNDATIONS OF 1856 IN FRANCE; *e*, REMEDIES AGAINST INUNDATIONS—CONSEQUENCES IF THE NILE HAD BEEN CONFINED BY LATERAL DIKES.

IMPROVEMENTS IN THE VAL DI CHIANA—IMPROVEMENTS IN THE TUSCAN MAREMME—OBSTRUCTION OF RIVER MOUTHS—SUBTERRANEAN WATERS—ARTESIAN WELLS—ARTIFICIAL SPRINGS—ECONOMIZING PRECIPITATION.

Land artificially won from the Waters.

MAN, as we have seen, has done much to revolutionize the solid surface of the globe, and to change the distribution and proportions, if not the essential character, of the organisms which inhabit the land and even the waters. Besides the influence thus exerted upon the life which peoples the sea, his action upon the land has involved a certain amount of indirect encroachment upon the territorial jurisdiction of the ocean. So far as he has increased the erosion of running waters by the destruction of the forest, he has promoted the deposit of solid matter in the sea, thus reducing its depth, advancing the coast line, and diminishing the area covered by the waters. He has gone beyond this, and invaded the realm of the ocean by constructing within its borders wharves, piers, lighthouses, breakwaters, fortresses, and other facilities for his commercial and military operations; and in some countries he has permanently rescued from tidal overflow, and even from the very bed of the deep, tracts of ground extensive enough to constitute valuable additions to his agricultural domain. The quantity of soil gained from the sea by these different modes of acquisition is, indeed, too inconsiderable to form an appreciable element in the comparison of the general proportion between the two great forms of terrestrial surface, land and water; but the results of such operations, considered in their physical and their moral bearings, are sufficiently important to entitle them to special notice in every comprehensive view of the relations between man and nature.

There are cases, as on the western shores of the Baltic, where, in consequence of the secular elevation of the coast, the sea appears to be retiring; others, where, from the slow sinking of the land, it seems to be advancing. These movements depend upon geological causes wholly out of our reach, and man can neither advance nor retard them. There are also cases where similar apparent effects are produced by local oceanic currents, by river deposit or erosion, by tidal action, or by the influence of the wind upon the waves and the sands of the sea beach. A regular current may drift suspended earth and seaweed along a coast until they are caught by an eddy and finally deposited out of the reach of further disturbance, or it may scoop out the bed of the sea and undermine promontories and headlands; a powerful river, as the wind changes the direction of its flow at its outlet, may wash away shores and sandbanks at one point to deposit their material at another; the tide or waves, stirred to unusual depths by the wind, may gradually wear down the line of coast, or they may form shoals and coast dunes by depositing the sand they have rolled up from the bottom of the ocean. These latter modes of action are slow in producing effects sufficiently important to be noticed in general geography, or even to be visible in the representations of coast line laid down in ordinary maps; but they nevertheless form conspicuous features in local topography, and they are attended with consequences of great moment to the material and the moral interests of men.

The forces which produce these results are all in a considerable degree subject to control, or rather to direction and resistance, by human power, and it is in guiding and combating them that man has achieved some of his most remarkable and honorable conquests over nature. The triumphs in question, or what we generally call harbor and coast improvements, whether we estimate their value by the money and labor expended upon them, or by their bearing upon the interests of commerce and the arts of civilization, must take a very high rank among the great works of man, and they are fast assuming a magnitude greatly exceeding their former relative importance. The extension of commerce and of the military marine, and especially the introduction of vessels of increased burden and deeper draught of water, have imposed upon engineers tasks of a character which a century ago would have been pronounced, and, in fact, would have been impracticable; but necessity has stimulated an ingenuity which has contrived means of executing them, and which gives promise of yet greater performance in time to come.

Men have ceased to admire the power which heaped up the great pyramid to gratify the pride of a despot with a giant sepulchre; for many great harbors, many important lines of internal communication, in the civilized world, now exhibit works which surpass the vastest remains of ancient architectural art in mass and weight of matter, demand the exercise of far greater constructive skill, and involve a much heavier pecuniary expenditure than would now be required for the building of the tomb of Cheops. It is computed that the great pyramid, the solid contents of which when complete were about 3,000,000 cubic yards, could be erected for a million of pounds sterling. The breakwater at Cherbourg, founded in rough water sixty feet deep, at an average distance of more than two miles from the shore, contains double the mass of the pyramid, and many a comparatively unimportant railroad has been constructed at twice the cost which would now build that stupendous monument. Indeed, although man, detached from the solid earth, is almost powerless to struggle against the sea, he is fast becoming invincible by it so long as his foot is planted on the shore, or even on the bottom of the rolling ocean; and though on some battle fields between the waters and the land, he is obliged slowly to yield his ground, yet he retreats still facing the foe, and will finally be able to say to the sea: "Thus far shalt thou come and no farther, and here shall thy proud waves be stayed!"

The description of works of harbor and coast improvement which have only an economical value, not a true geographical importance, does not come within the plan of the present volume, and in treating this branch of my subject, I shall confine myself to such as are designed either to gain new soil by excluding the waters from grounds which they had permanently or occasionally covered, or to resist new encroachments of the sea upon the land.

a. *Exclusion of the Sea by Diking.*

The draining of the Lincolnshire fens in England, which converted about 400,000 acres of marsh, pool, and tide-washed flat into plough land and pasturage, is a work, or rather series of works, of great magnitude, and it possesses much economical, and, indeed, no trifling geographical importance. Its plans and methods were, at least in part, borrowed from the example of like improvements in Holland, and it is, in difficulty and extent, inferior to works executed for the same purpose on the opposite coast of the North Sea, by Dutch, Frisic, and Low German engineers. The space I can devote to such operations will be better employed in describing the latter, and I content myself with the simple statement I have already made of the quantity of worthless and even pestilential land which has been rendered both productive and salubrious in Lincolnshire, by diking out the sea, and the rivers which traverse the fens of that country.

The almost continued prevalence of west winds upon both coasts of the German Ocean occasions a constant set of the currents of that sea to the east, and both for this reason and on account of the greater violence of storms from the former quarter, the English shores are much less exposed to invasion by the waves than those of the Netherlands and the provinces contiguous to them on the north. The old Netherlandish chronicles are filled with the most startling accounts of the damage done by the irruptions of the ocean, from west winds or extraordinarily high tides, at times long before any considerable extent of seacoast was diked. Several hundreds of these terrible inundations are recorded, and in very many of them the loss of human lives is estimated as high as one hundred thousand. It is impossible to doubt that there must be enormous exaggeration in these numbers; for, with all the reckless hardihood shown by men in braving the dangers and privations attached by nature to their birthplace, it is inconceivable that so dense a population as such wholesale destruction of life supposes could find the means of subsistence, or content itself to dwell, on a territory liable, a dozen times in a century, to such fearful devastation. There can be no doubt, however, that the low continental shores of the German Ocean very frequently suffered immense injury from inundation by the sea, and it is natural, therefore, that the various arts of resistance to the encroachments of the ocean, and, finally, of aggressive warfare upon its domain, and of permanent conquest of its territory, should have been earlier studied and carried to higher perfection in the latter countries, than in England, which had much less to lose or to gain by the incursions or the retreat of the waters.

Indeed, although the confinement of swelling rivers by artificial embankments is of great antiquity, I do not know that the defence or acquisition of land from the sea by diking was ever practised on a large scale until systematically undertaken by the Netherlanders, a few centuries after the commencement of the Christian era. The silence of the Roman historians affords a strong presumption that this art was unknown to the inhabitants of the Netherlands at the time of the Roman invasion, and the elder Pliny's description of the mode of life along the coast which has now been long diked in, applies precisely to the habits of the people who live on the low islands and mainland flats lying outside of the chain of dikes, and wholly unprotected by embankments of any sort.

It has been conjectured, and not without probability, that the causeways built by the Romans across the marshes of the Low Countries, in their campaigns against the Germanic tribes, gave the natives the first hint of the utility which might be derived from similar constructions applied to a different purpose.* If this is so, it is one of the most interesting among the many instances in which the arts and enginery of war have been so modified as to be eminently promotive of the blessings of peace, thereby in some measure compensating the wrongs and sufferings they have inflicted on humanity.† The Low-

* It has been often asserted by eminent writers that a part of the fens in Lincolnshire was reclaimed by sea dikes under the government of the Romans. I have found no ancient authority in support of this allegation, nor can I refer to any passage in Roman literature in which sea dikes are expressly mentioned otherwise than as walls or piers, except that in Pliny (*Hist. Nat.* xxxvi, 24), where it is said that the Tyrrhenian sea was excluded from the Lucrine lake by dikes.

† A friend has recently suggested to me an interesting illustration of the applicability of military instrumentalities to pacific art. The sale of gunpowder in the United States, he informs me, is smaller since the commencement of the present rebellion than before, because the war has caused the suspension of many public and private improvements, in the execution of which great quantities of powder were used for blasting.

It is alleged that the same observation was made in France during the Crimean war, and that, in general, not ten per cent. of the powder manufactured on either side of the Atlantic is employed for military purposes.

It is a fact not creditable to the moral sense of modern civilization, that very many of the most important improvements in machinery and the working of metals have originated in the necessities of war, and that man's highest ingenuity has been shown, and many of his most remarkable triumphs over natural forces achieved, in the contrivance of engines for the destruction of his fellow man. The military material employed by the first Napoleon has become, in less than two generations, nearly as obsolete as the sling and stone of the shepherd, and attack and defence now begin

landers are believed to have secured some coast and bay islands by ring dikes, and to have embanked some fresh water channels, as early as the eighth or ninth century; but it does not appear that sea dikes, important enough to be noticed in historical records, were constructed on the mainland before the thirteenth century. The practice of draining inland accumulation of water, whether fresh or salt, for the purpose of bringing under cultivation the ground they cover, is of later origin, and is said not to have been adopted until after the middle of the fifteenth century.*

The total amount of surface gained to the agriculture of the Netherlands by diking out the sea and by draining shallow bays and lakes, is estimated by Staring at three hundred and fifty-five thousand *bunder* or hectares, equal to eight hundred and seventy-seven thousand two hundred and forty acres, which is one tenth of the area of the kingdom.† In very many instances, the dikes have been partially, in some particularly exposed localities totally destroyed by the violence of the sea, and the drained lands again flooded. In some cases, the soil thus painfully won from the ocean has been entirely lost; in others it has been recovered by repairing or rebuilding the dikes and pumping out the water. Besides this, the weight

at distances to which, half a century ago, military reconnoissances hardly extended. Upon a partial view of the subject, the human race seems destined to become its own executioner—on the one hand, exhausting the capacity of the earth to furnish sustenance to her taskmaster; on the other, compensating diminished production by inventing more efficient methods of exterminating the consumer.

But war develops great civil virtues, and brings into action a degree and kind of physical energy which seldom fails to awaken a new intellectual life in a people that achieves great moral and political results through great heroism and endurance and perseverance. Domestic corruption has destroyed more nations than foreign invasion, and a people is rarely conquered till it has deserved subjugation.

* Staring, *Voormaals en Thans*, p. 150.

† Idem, p. 163. Much the largest proportion of the lands so reclaimed, though for the most part lying above low-water tidemark, are at a lower level than the Lincolnshire fens, and more subject to inundation from the irruptions of the sea.

of the dikes gradually sinks them into the soft soil beneath, and this loss of elevation must be compensated by raising the surface, while the increased burden thus added tends to sink them still lower. "Tetens declares," says Kohl, "that in some places the dikes have gradually sunk to the depth of sixty or even a hundred feet." * For these reasons, the processes of dike building have been almost everywhere again and again repeated, and thus the total expenditure of money and of labor upon the works in question is much greater than would appear from an estimate of the actual cost of diking-in a given extent of coast land and draining a given area of water surface.†

On the other hand, by erosion of the coast line, the drifting of sand dunes into the interior, and the drowning of fens and morasses by incursions of the sea—all caused, or at least greatly aggravated, by human improvidence—the Netherlands have lost a far larger area of land since the commencement of the Christian era than they have gained by diking and draining. Staring despairs of the possibility of calculating the loss from the first-mentioned two causes of destruction, but he esti-

* *Die Inseln und Marschen der Herzogthümer Schleswig und Holstein* iii, p. 151.

† The purely agricultural island of Pelworm, off the coast of Schleswig, containing about 10,000 acres, annually expends for the maintenance of its dikes not less than £6,000 sterling, or nearly $30,000.—J. G. KOHL, *Inseln und Marschen Schleswig's und Holstein's*, ii, p. 394.

The original cost of the dikes of Pelworm is not stated.

"The greatest part of the province of Zeeland is protected by dikes measuring 250 miles in length, the maintenance of which costs, in ordinary years, more than a million guilders [above $400,000]. * * * The annual expenditure for dikes and hydraulic works in Holland is from five to seven million guilders" [$2,000,000 to $2,800,000].—WILD, *Die Niederlande*, i, p. 62.

One is not sorry to learn that the Spanish tyranny in the Netherlands had some compensations. The great chain of ring dikes which surrounds a large part of Zeeland is due to the energy of Caspar de Robles, the Spanish governor of that province, who in 1570 ordered the construction of these works at the public expense, as a substitute for the private embankments which had previously partially served the same purpose.—WILD, *Die Niederlande*, i, p. 62.

mates that not less than six hundred and forty thousand bunder, or one million five hundred and eighty-one thousand acres, of fen and marsh have been washed away, or rather deprived of their vegetable surface and covered by water, and thirty-seven thousand bunder, or ninety-one thousand four hundred acres of recovered land, have been lost by the destruction of the dikes which protected them.* The average value of land gained from the sea is estimated at about nineteen pounds sterling, or ninety dollars, per acre; while the lost fen and morass was not worth more than one twenty-fifth part of the same price. The ground buried by the drifting of the dunes appears to have been almost entirely of this latter character, and, upon the whole, there is no doubt that the soil added by human industry to the territory of the Netherlands, within the historical period, greatly exceeds in pecuniary value that which has fallen a prey to the waves during the same era.

Upon most low and shelving coasts, like those of the Netherlands, the maritime currents are constantly changing, in consequence of the variability of the winds, and the shifting of the sandbanks, which the currents themselves now form and now displace. While, therefore, at one point the sea is advancing landward, and requiring great effort to prevent the undermining and washing away of the dikes, it is shoaling at another by its own deposits, and exposing, at low water, a gradually widening belt of sands and ooze. The coast lands selected for diking-in are always at points where the sea is depositing productive soil. The Eider, the Elbe, the Weser, the Ems, the Rhine, the Maas, and the Schelde bring down large quantities of fine earth. The prevalence of west winds prevents the waters from carrying this material far out from the coast, and it is at last deposited northward or southward from the mouth of the rivers which contribute it, according to the varying drift of the currents.

The process of natural deposit which prepares the coast for diking-in is thus described by Staring: "All sea-deposited soil

* Staring, *Voormaals en Thans*, p. 163.

is composed of the same constituents. First comes a stratum of sand, with marine shells, or the shells of mollusks living in brackish water. If there be tides, and, of course, flowing and ebbing currents, mud is let fall upon the sand only after the latter has been raised above low-water mark; for then only, at the change from flood to ebb, is the water still enough to form a deposit of so light a material. Where mud is found at greater depths, as, for example, in a large proportion of the Ij, it is a proof that at this point there was never any considerable tidal flow or other current. * * * The powerful tidal currents, flowing and ebbing twice a day, drift sand with them. They scoop out the bottom at one point, raise it at another, and the sandbanks in the current are continually shifting. As soon as a bank raises itself above low-water mark, flags and reeds establish themselves upon it. The mechanical resistance of these plants checks the retreat of the high water and favors the deposit of the earth suspended in it, and the formation of land goes on with surprising rapidity. When it has risen to high-water level, it is soon covered with grasses, and becomes what is called *schor* in Zeeland, *kwelder* in Friesland. Such grounds are the foundation or starting point of the process of diking. When they are once elevated to the flood-tide level, no more mud is deposited upon them except by extraordinary high tides. Their further rise is, accordingly, very slow, and it is seldom advantageous to delay longer the operation of diking." *

The formation of new banks by the sea is constantly going on at points favorable for the deposit of sand and earth, and hence opportunity is continually afforded for enclosure of new land outside of that already diked in, the coast is fast advancing seaward, and every new embankment increases the security of former enclosures. The province of Zeeland consists of islands washed by the sea on their western coasts, and separated by the many channels through which the Schelde and some other rivers find their way to the ocean. In the twelfth

* *Voormaals en Thans*, pp. 150, 151.

century these islands were much smaller and more numerous than at present. They have been gradually enlarged, and, in several instances, at last connected by the extension of their system of dikes. Walcheren is formed of ten islets united into one about the end of the fourteenth century. At the middle of the fifteenth century, Goeree and Overflakkee consisted of separate islands, containing altogether about ten thousand acres; by means of above sixty successive advances of the dikes, they have been brought to compose a single island, whose area is not less than sixty thousand acres.*

In the Netherlands—which the first Napoleon characterized as a deposit of the Rhine, and as, therefore, by natural law, rightfully the property of him who controlled the sources of that great river—and on the adjacent Frisic, Low German and Danish shores and islands, sea and river dikes have been constructed on a grander and more imposing scale than in any other country. The whole economy of the art has been there most thoroughly studied, and the literature of the subject is very extensive. For my present aim, which is concerned with results rather than with processes, it is not worth while to refer to professional treatises, and I shall content myself with presenting such information as can be gathered from works of a more popular character.†

The superior strata of the lowlands upon and near the coast are, as we have seen, principally composed of soil

* STARING, *Voormaals en Thans*, p. 152. Kohl states that the peninsula of Diksand on the coast of Holstein consisted, at the close of the last century, of several islands measuring together less than five thousand acres. In 1837 they had been connected with the mainland, and had nearly doubled in area.—*Inseln u. Marschen Schlesw. Holst.*, iii, p. 262.

† The most instructive and entertaining of tourists, J. G. Kohl—so aptly characterized by Davies as the "Herodotus of modern Europe"—furnishes a great amount of interesting information on the dikes of the Low German seacoast, in his *Inseln und Marschen der Herzogthümer Schleswig und Holstein.* I am acquainted with no popular work on this subject which the reader can consult with greater profit. See also STARING, *Voormaals en Thans*, and *De Bodem van Nederland*, on the dikes of the Netherlands.

brought down by the great rivers I have mentioned, and either directly deposited by them upon the sands of the bottom, or carried out to sea by their currents, and then, after a shorter or longer exposure to the chemical and mechanical action of salt water and marine currents, restored again to the land by tidal overflow and subsidence from the waters in which it was suspended. At a very remote period, the coast flats were, at many points, raised so high by successive alluvious or tidal deposits as to be above ordinary high water level, but they were still liable to occasional inundation from river floods, and from the sea water also, when heavy or long-continued west winds drove it landward. The extraordinary fertility of this soil and its security as a retreat from hostile violence attracted to it a considerable population, while its want of protection against inundation exposed it to the devastations of which the chroniclers of the Middle Ages have left such highly colored pictures. The first permanent dwellings on the coast flats were erected upon artificial mounds, and many similar precarious habitations still exist on the unwalled islands and shores beyond the chain of dikes. River embankments, which, as is familiarly known, have from the earliest antiquity been employed in many countries where sea dikes are unknown, were probably the first works of this character constructed in the Low Countries, and when two neighboring streams of fresh water had been embanked, the next step in the process would naturally be to connect the river walls together by a transverse dike or raised causeway, which would serve to secure the intermediate ground both against the backwater of river floods and against overflow by the sea. The oldest true sea dikes described in historical records, however, are those enclosing islands in the estuaries of the great rivers, and it is not impossible that the double character they possess as a security against maritime floods and as a military rampart, led to their adoption upon those islands before similar constructions had been attempted upon the mainland.

At some points of the coast, various contrivances, such as piers, piles, and, in fact, obstructions of all sorts to the ebb of the current, are employed to facilitate the deposit of slime, before a regular enclosure is commenced. Usually, however, the first step is to build low and cheap embankments, extending from an older dike, or from high ground, around the parcel of flat intended to be secured. These are called summer dikes (*sommer-deich*, pl. *sommer-deiche*, German; *zomerkaai*, *zomerkade*, pl. *zomerkaaie*, *zomerkaden*, Dutch). They are erected when a sufficient extent of ground to repay the cost has been elevated enough to be covered with coarse vegetation fit for pasturage. They serve both to secure the ground from overflow by the ordinary flood tides of mild weather, and to retain the slime deposited by very high water, which would otherwise be partly carried off by the retreating ebb. The elevation of the soil goes on slowly after this; but when it has at last been sufficiently enriched, and raised high enough to justify the necessary outlay, permanent dikes are constructed by which the water is excluded at all seasons. These embankments are constructed of sand from the coast dunes or from sandbanks, and of earth from the mainland or from flats outside the dikes, bound and strengthened by fascines, and provided with sluices, which are generally founded on piles and of very expensive construction, for drainage at low water. The outward slope of the sea dikes is gentle, experience having shown that this form is least exposed to injury both from the waves and from floating ice, and the most modern dikes are even more moderate in the inclination of the seaward scarp than the older ones.* The crown of the dike, however, for the last three or four feet of its height, is much steeper, being intended rather as a protection against the spray than against the waves, and the inner slope is always comparatively abrupt.

The height and thickness of dikes varies according to the elevation of the ground they enclose, the rise of the tides, the direction of the prevailing winds, and other special causes of exposure, but it may be said that they are, in general, raised from fifteen to twenty feet above ordinary high-water mark.

* The inclination varies from one foot rise in four of base to one foot in fourteen.—KOHL, iii, p. 210.

The water slopes of river dikes are protected by plantations of willows or strong semi-aquatic shrubs or grasses, but as these will not grow upon banks exposed to salt water, sea dikes must be faced with stone, fascines, or some other *revêtement.** Upon the coast of Schleswig and Holstein, where the people have less capital at their command, they defend their embankments against ice and the waves by a coating of twisted straw or reeds, which must be renewed as often as once, sometimes twice a year. The inhabitants of these coasts call the chain of dikes "the golden border," a name it well deserves, whether we suppose it to refer to its enormous cost, or, as is more probable, to its immense value as a protection to their fields and their firesides.

When outlying flats are enclosed by building new embankments, the old interior dikes are suffered to remain, both as an additional security against the waves, and because the removal of them would be expensive. They serve, also, as roads or causeways, a purpose for which the embankments nearest the sea are seldom employed, because the whole structure might be endangered from the breaking of the turf by wheels and the hoofs of horses. Where successive rows of dikes have been

* The dikes are sometimes founded upon piles, and sometimes protected by one or more rows of piles driven deeply down into the bed of the sea in front of them. "Triple rows of piles of Scandinavian pine," says Wild, "have been driven down along the coast of Friesland, where there are no dunes, for a distance of one hundred and fifty miles. The piles are bound together by strong cross timbers and iron clamps, and the interstices filled with stones. The ground adjacent to the piling is secured with fascines, and at exposed points heavy blocks of stone are heaped up as an additional protection. The earth dike is built behind the mighty bulwark of this breakwater, and its foot also is fortified with stones." * * * "The great Helder dike is about five miles long and forty feet wide at the top, along which runs a good road. It slopes down two hundred feet into the sea, at an angle of forty degrees. The highest waves do not reach the summit, the lowest always cover its base. At certain distances, immense buttresses, of a height and width proportioned to those of the dike, and even more strongly built, run several hundred feet out into the rolling sea. This gigantic artificial coast is entirely composed of Norwegian granite."—WILD, *Die Niederlande*, i, pp. 61, 62.

thus constructed, it is observed that the ground defended by the more ancient embankments is lower than that embraced within the newer enclosures, and this depression of level has been ascribed to a general subsidence of the coast from geological causes; but the better opinion seems to be that it is, in most cases, due merely to the consolidation and settling of the earth from being more effectually dried, from the weight of the dikes, from the tread of men and cattle, and from the movement of the heavy wagons which carry off the crops.*

* The shaking of the ground, even when loaded with large buildings, by the passage of heavy carriages or artillery, or by the march of a body of cavalry or even infantry, shows that such causes may produce important mechanical effects on the condition of the soil. The bogs in the Netherlands, as in most other countries, contain large numbers of fallen trees, buried to a certain depth by earth and vegetable mould. When the bogs are dry enough to serve as pastures, it is observed that trunks of these ancient trees rise of themselves to the surface. Staring ascribes this singular phenomenon to the agitation of the ground by the tread of cattle. "When roadbeds," observes he, "are constructed of gravel and pebbles of different sizes, and these latter are placed at the bottom without being broken and rolled hard together, they are soon brought to the top by the effect of travel on the road. Lying loosely, they undergo some motion from the passage of every wagon wheel and the tread of every horse that passes over them. This motion is an oscillation or partial rolling, and as one side of a pebble is raised, a little fine sand or earth is forced under it, and the frequent repetition of this process by cattle or carriages moving in opposite directions brings it at last to the surface. We may suppose that a similar effect is produced on the stems of trees in the bogs by the tread of animals."—*De Bodem van Nederland*, i, pp. 75, 76.

It is observed in the Northern United States, that when soils containing pebbles are cleared and cultivated, and the stones removed from the surface, new pebbles, and even bowlders of many pounds weight, continue to show themselves above the ground, every spring, for a long series of years. In clayey soils the fence posts are thrown up in a similar way, and it is not uncommon to see the lower rail of a fence thus gradually raised a foot or even two feet above the ground. This rising of stones and fences is popularly ascribed to the action of the severe frosts of that climate. The expansion of the ground, in freezing, it is said, raises its surface, and, with the surface, objects lying near or connected with it. When the soil thaws in the spring, it settles back again to its former level, while the pebbles and posts are prevented from sinking as low as before by loose earth which

Notwithstanding this slow sinking, most of the land enclosed by dikes is still above low-water mark, and can, therefore, be wholly or partially freed from rain water, and from that received by infiltration from higher ground, by sluices opened at the ebb of the tide. For this purpose, the land is carefully ditched, and advantage is taken of every favorable occasion for discharging the water through the sluices. But the ground cannot be effectually drained by this means, unless it is elevated four or five feet, at least, above the level of the ebb tide, because the ditches would not otherwise have a sufficient descent to carry the water off in the short interval between ebb and flow, and because the moisture of the saturated sub soil is always rising by capillary attraction. Whenever, therefore, the soil has sunk below the level I have mentioned, and in cases where its surface has never been raised above it, pumps, worked by wind or some other mechanical power, must be very frequently employed to keep the land dry enough for pasturage and cultivation.*

has fallen under them. The fact that the elevation spoken of is observed only in the spring, gives countenance to this theory, which is perhaps applicable also to the cases stated by Staring, and it is probable that the two causes above assigned concur in producing the effect.

The question of the subsidence of the Netherlandish coast has been much discussed. Not to mention earlier geologists, Venema, in several essays, and particularly in *Het Dalen van de Noordelijke Kuststreken van ons Land*, 1854, adduces many facts and arguments to prove a slow sinking of the northern provinces of Holland. Laveleye (*Affaissement du sol et envasement des fleuves survenus dans les temps historiques*, 1859), upon a still fuller investigation, arrives at the same conclusion. The eminent geologist Staring, however, who briefly refers to the subject in *De Bodem van Nederland*, i, p. 356 *et seqq.*, does not consider the evidence sufficient to prove anything more than the sinking of the surface of the polders from drying and consolidation.

* The elevation of the lands enclosed by dikes—or *polders*, as they are called in Holland—above low water mark, depends upon the height of the tides, or, in other words, upon the difference between ebb and flood. The tide cannot deposit earth higher than it flows, and after the ground is once enclosed, the decay of the vegetables grown upon it and the addition of manures do not compensate the depression occasioned by drying and con-

b. *Draining of Lakes and Marshes.*

The substitution of steam engines for the feeble and uncertain action of windmills, in driving pumps, has much facilitated the removal of water from the polders and the draining of lakes, marshes, and shallow bays, and thus given such an impulse to these enterprises, that not less than one hundred and ten thousand acres were reclaimed from the waters, and added to the agricultural domain of the Netherlands, between 1815 and 1858. The most important of these undertakings was the draining of the Lake of Haarlem, and for this purpose some of the most powerful hydraulic engines ever constructed were designed and executed.* The origin of this lake is unknown. It is supposed by some geographers to be a part of an ancient bed of the Rhine, the channel of which, as there is good reason to believe, has undergone great changes since the Roman invasion of the Netherlands; by others it is thought to have once formed an inland marine channel, separated from the sea by a chain of low islands, which the sand washed up by the tides has since connected with the mainland and converted into a continuous line of coast. The best authorities, however, find geological evidence that the surface occupied by the lake was originally a marshy tract containing within its limits little solid ground, but many ponds and inlets, and much floating as well as fixed fen.

In consequence of the cutting of turf for fuel, and the de-

solidation. On the coast of Zeeland and the islands of South Holland, the tides, and of course the surface of the lands deposited by them, are so high that the polders can be drained by ditching and sluices, but at other points, as in the enclosed grounds of North Holland on the Zuiderzee, where the tide rises but three feet or even less, pumping is necessary from the beginning.—STARING, *Voormaals en Thans*, p. 152.

* The principal engine—called the Leeghwater, from the name of an engineer who had proposed the draining of the lake in 1641—was of 500 horse power, and drove eleven pumps making six strokes per minute. Each pump raised six cubic mètres, or nearly eight cubic yards of water to the stroke, amounting in all to 23,760 cubic mètres, or above 31,000 cubic yards, the hour.—WILD, *Die Niederlande*, i, p. 87.

struction of the few trees and shrubs which held the loose soil together with their roots, the ponds are supposed to have gradually extended themselves, until the action of the wind upon their enlarged surface gave their waves sufficient force to overcome the resistance of the feeble barriers which separated them, and to unite them all into a single lake. Popular tradition, it is true, ascribes the formation of the Lake of Haarlem to a single irruption of the sea, at a remote period, and connects it with one or another of the destructive inundations of which the Netherland chronicles describe so many; but on a map of the year 1531, a chain of four smaller waters occupies nearly the ground afterward covered by the Lake of Haarlem, and they have more probably been united by gradual encroachments resulting from the improvident practices above referred to, though no doubt the consummation may have been hastened by floods, and by the neglect to maintain dikes, or the intentional destruction of them, in the long wars of the sixteenth century.

The Lake of Haarlem was a body of water not far from fifteen miles in length, by seven in greatest width, lying between the cities of Amsterdam and Leyden, running parallel with the coast of Holland at the distance of about five miles from the sea, and covering an area of about 45,000 acres. By means of the Ij, it communicated with the Zuiderzee, the Mediterranean of the Netherlands, and its surface was little above the mean elevation of that of the sea. Whenever, therefore, the waters of the Zuiderzee were acted upon by strong northwest winds, those of the Lake of Haarlem were raised proportionally and driven southward, while winds from the south tended to create a flow in the opposite direction. The shores of the lake were everywhere low, and though in the course of the eighty years between 1767 and 1848 more than £350,000 or $1,700,000 had been expended in checking its encroachments, it often burst its barriers, and produced destructive inundations. On the 29th of November, 1836, a south wind brought its waters to the very gates of Amsterdam, and on the 26th of December of the same year, in a northwest gale, they overflowed twenty thousand acres of land at the southern extremity of the lake, and flooded a part of the city of Leyden. The depth of water did not, in general, exceed fourteen feet, but the bottom was a semi-fluid ooze or slime, which partook of the agitation of the waves, and added considerably to their mechanical force. Serious fears were entertained that the lake would form a junction with the inland waters of the Legmeer and Mijdrecht, swallow up a vast extent of valuable soil, and finally endanger the security of a large proportion of the land which the industry of Holland had gained in the course of centuries from the ocean.

For this reason, and for the sake of the large addition the bottom of the lake would make to the cultivable soil of the state, it was resolved to drain it, and the preliminary steps for that purpose were commenced in the year 1840. The first operation was to surround the entire lake with a ring canal and dike, in order to cut off the communication with the Ij, and to exclude the water of the streams and morasses which discharged themselves into it from the land side. The dike was composed of different materials, according to the means of supply at different points, such as sand from the coast dunes, earth and turf excavated from the line of the ring canal, and floating turf,* fascines being everywhere used to bind and compact the

* In England and New England, where the marshes have been already drained or are of comparatively small extent, the existence of large floating islands seems incredible, and has sometimes been treated as a fable, but no geographical fact is better established. Kohl (*Inseln und Marschen Schleswig-Holsteins*, iii, p. 309) reminds us that Pliny mentions among the wonders of Germany the floating islands, covered with trees, which met the Roman fleets at the mouths of the Elbe and the Weser. Our author speaks also of having visited, in the territory of Bremen, floating moors, bearing not only houses but whole villages. At low stages of the water these moors rest upon a bed of sand, but are raised from six to ten feet by the high water of spring, and remain afloat until, in the course of the summer, the water beneath is exhausted by evaporation and drainage, when they sink down upon the sand again. See *Appendix*, No. 38.

Staring explains, in an interesting way, the whole growth, formation, and functions of floating fens or bogs, in his very valuable work, *De Bodem van*

mass together. This operation was completed in 1848, and three steam pumps were then employed for five years in discharging the water. The whole enterprise was conducted at the expense of the state, and in 1853 the recovered lands were

Nederland, i, pp. 36–43. The substance of his account is as follows: The first condition for the growth of the plants which compose the substance of turf and the surface of the fens, is stillness of the water. Hence they are not found in running streams, nor in pools so large as to be subject to frequent agitation by the wind. For example, not a single plant grew in the open part of the Lake of Haarlem, and fens cease to form in all pools as soon as, by the cutting of the turf for fuel or other purposes, their area is sufficiently enlarged to be much acted on by wind. When still water above a yard deep is left undisturbed, aquatic plants of various genera, such as Nuphar, Nymphæa, Limnanthemum, Stratiotes, Polygonum, and Potamogeton, fill the bottom with roots and cover the surface with leaves. Many of the plants die every year, and prepare at the bottom a soil fit for the growth of a higher order of vegetation, Phragmites, Acorus, Sparganium, Rumex, Lythrum, Pedicularis, Spiræa, Polystichum, Comarum, Caltha, &c., &c. In the course of twenty or thirty years the muddy bottom is filled with roots of aquatic and marsh plants, which are lighter than water, and if the depth is great enough to give room for detaching this vegetable network, a couple of yards for example, it rises to the surface, bearing with it, of course, the soil formed above it by decay of stems and leaves. New genera now appear upon the mass, such as Carex, Menyanthes, and others, and soon thickly cover it. The turf has now acquired a thickness of from two to four feet, and is called in Groningen *lad;* in Friesland, *til*, *tilland*, or *drijftil;* in Overijssel, *krag;* and in Holland, *rietzod.* It floats about as driven by the wind, gradually increasing in thickness by the decay of its annual crops of vegetation, and in about half a century reaches the bottom and becomes fixed. If it has not been invaded in the mean time by men or cattle, trees and arborescent plants, Alnus, Salix, Myrica, &c. appear, and these contribute to hasten the attachment of the turf to the bottom, both by their weight and by sending their roots quite through into the ground.

This is the regular method employed by nature for the gradual filling up of shallow lakes and pools, and converting them first into morass and then into dry land. Whenever therefore man removes the peat or turf, he exerts an injurious geographical agency, and, as I have already said, there is no doubt that the immense extension of the inland seas of Holland in modern times is owing to this and other human imprudences. "Hundreds of hectares of floating pastures," says our author, "which have nothing in their appearance to distinguish them from grass lands resting on solid bog,

offered for sale for its benefit. Up to 1858, forty-two thousand acres had been sold at not far from sixteen pounds sterling or seventy-seven dollars an acre, amounting altogether to £661,000 sterling or $3,200,000. The unsold lands were valued at more than £6,000 or nearly $30,000, and as the total cost was £764,500 or about $3,700,000, the direct loss to the state, exclusive of interest on the capital expended, may be stated at £100,000 or something less than $500,000.

In a country like the United States, of almost boundless extent of sparsely inhabited territory, such an expenditure for such an object would be poor economy. But Holland has a narrow domain, great pecuniary resources, an excessively crowded population, and a consequent need of enlarged room and opportunity for the exercise of industry. Under such circumstances, and especially with an exposure to dangers so formidable, there is no question of the wisdom of the measure. It has already provided homes and occupation for more than

are found in Overijssel, in North Holland and near Utrecht. In short, they occur in all deep bogs, and wherever deep water is left long undisturbed."

In one case, a floating island, which had attached itself to the shore, continued to float about for a long time after it was torn off by a flood, and was solid enough to keep a pond of fresh water upon it sweet, though the water in which it was swimming had become brackish from the irruption of the sea. After the hay is cut, cattle are pastured upon these islands, and they sometimes have large trees growing upon them.

When the turf or peat has been cut, leaving water less than a yard deep, Equisetum limosum grows at once, and is followed by the second class of marsh plants mentioned above. Their roots do not become detached from the bottom in such shallow water, but form ordinary turf or peat. These processes are so rapid that a thickness of from three to six feet of turf is formed in half a century, and many men have lived to mow grass where they had fished in their boyhood, and to cut turf twice in the same spot.

Captain Gilliss says that before Lake Taguataga in Chili was drained, there were in it islands composed of dead plants matted together to a thickness of from four to six feet, and with trees of medium size growing upon them. These islands floated before the wind "with their trees and browsing cattle."—*United States Naval Astronomical Expedition to the Southern Hemisphere*, i, pp. 16, 17.

five thousand citizens, and furnished a profitable investment for a capital of not less than £400,000 sterling or $2,000,000, which has been expended in improvements over and above the purchase money of the soil; and the greater part of this sum, as well as of the cost of drainage, has been paid as a compensation for labor. The excess of governmental expenditure over the receipts, if employed in constructing ships of war or fortifications, would have added little to the military strength of the kingdom; but the increase of territory, the multiplication of homes and firesides which the people have an interest in defending, and the augmentation of agricultural resources, constitute a stronger bulwark against foreign invasion than a ship of the line or a fortress armed with a hundred cannon.

The bearing of the works I have noticed, and of others similar in character, upon the social and moral, as well as the purely economical interests of the people of the Netherlands, has induced me to describe them more in detail than the general purpose of this volume may be thought to justify; but if we consider them simply from a geographical point of view, we shall find that they are possessed of no small importance as modifications of the natural condition of terrestrial surface. There is good reason to believe that before the establishment of a partially civilized race upon the territory now occupied by Dutch, Frisic, and Low German communities, the grounds not exposed to inundation were overgrown with dense woods, that the lowlands between these forests and the sea coasts were marshes, covered and partially solidified by a thick matting of peat plants and shrubs interspersed with trees, and that even the sand dunes of the shore were protected by a vegetable growth which, in a great measure, prevented the drifting and translocation of them.

The present causes of river and coast erosion existed, indeed, at the period in question; but some of them must have acted with less intensity, there were strong natural safeguards against the influence of marine and fresh-water currents, and the conflicting tendencies had arrived at a condition of approximate equilibrium, which permitted but slow and gradual changes in the face of nature. The destruction of the forests around the sources and along the valleys of the rivers by man gave them a more torrential character. The felling of the trees, and the extirpation of the shrubbery upon the fens by domestic cattle, deprived the surface of cohesion and consistence, and the cutting of peat for fuel opened cavities in it, which, filling at once with water, rapidly extended themselves by abrasion of their borders, and finally enlarged to pools, lakes, and gulfs, like the Lake of Haarlem and the northern part of the Zuiderzee. The cutting of the wood and the depasturing of the grasses upon the sand dunes converted them from solid bulwarks against the ocean to loose accumulations of dust, which every sea breeze drove farther landward, burying, perhaps, fertile soil and choking up watercourses on one side, and exposing the coast to erosion by the sea upon the other.

c. *Geographical Influence of such Operations.*

The changes which human action has produced within twenty centuries in the Netherlands and the neighboring provinces, are certainly of no small geographical importance, considered simply as a direct question of loss and gain of territory. They have also undoubtedly been attended with some climatic consequences, they have exercised a great influence on the spontaneous animal and vegetable life of this region, and they cannot have failed to produce effects upon tidal and other oceanic currents, the range of which may be very extensive. The force of the tidal wave, the height to which it rises, the direction of its currents, and, in fact, all the phenomena which characterize it, as well as all the effects it produces, depend as much upon the configuration of the coast it washes, and the depth of water, and form of bottom near the shore, as upon the attraction which occasions it. Every one of the terrestrial conditions which affect the character of tidal and other marine currents has been very sensibly modified by the operations I have described, and on this coast, at least, man has acted almost as powerfully on the physical geography of the sea as on that of the land.

Lowering of Lakes.

The hydraulic works of the Netherlands and of the neighboring states are of such magnitude, that they quite throw into the shade all other known artificial arrangements for defending the land against the encroachments of the rivers and the sea, and for reclaiming to the domain of agriculture and civilization soil long covered by the waters. But although the recovery and protection of lands flooded by the sea seems to be an art wholly of Netherlandish origin, we have abundant evidence, that in ancient as well as in comparatively modern times, great enterprises more or less analogous in character have been successfully undertaken, both in inland Europe and in the less familiar countries of the East.

One of the best known of these is the tunnel which serves to discharge the surplus waters of the Lake of Albano, about fourteen miles from Rome. This lake, about six miles in circuit, occupies one of the craters of an extinct volcanic range, and the surface of its waters is about nine hundred feet above the sea. It is fed by rivulets and subterranean springs originating in the Alban Mount, or Monte Cavo, the most elevated peak of the volcanic group just mentioned, which rises to the height of about three thousand feet. At present the lake has no discoverable natural outlet, but it is not known that the water ever stood at such a height as to flow regularly over the lip of the crater. It seems that at the earliest period of which we have any authentic memorials, its level was usually kept by evaporation, or by discharge through subterranean channels, considerably below the rim of the basin which encompassed it, but in the year 397 B. C., the water, either from the obstruction of such channels, or in consequence of increased supplies from unknown sources, rose to such a height as to flow over the edge of the crater, and threaten inundation to the country below by bursting through its walls. To obviate this danger, a tunnel for carrying off the water was pierced at a level much below the height to which it had risen. This gallery, cut entirely with the chisel through the rock for a distance of six thousand feet, or nearly a mile and one seventh, is still in so good condition as to serve its original purpose. The fact that this work was contemporaneous with the siege of Veii, has given to ancient annalists occasion to connect the two events, but modern critics are inclined to reject Livy's account of the matter, as one of the many improbable fables which disfigure the pages of that historian. It is, however, repeated by Cicero and by Dionysius of Halicarnassus, and it is by no means impossible that, in an age when priests and soothsayers monopolized both the arts of natural magic and the little which yet existed of physical science, the Government of Rome, by their aid, availed itself at once of the superstition and of the military ardor of its citizens to obtain their sanction to an enterprise which sounder arguments might not have induced them to approve.

Still more remarkable is the tunnel cut by the Emperor Claudius to drain the Lake Fucinus, now Lago di Celano, in the Neapolitan territory, about fifty miles eastward of Rome. This lake, as far as its history is known, has varied very considerably in its dimensions at different periods, according to the character of the seasons. It has no visible outlet, but was originally either drained by natural subterranean conduits, or kept within certain extreme limits by evaporation. In years of uncommon moisture, it spread over the adjacent soil and destroyed the crops; in dry seasons, it retreated, and produced epidemic disease by poisonous exhalations from the decay of vegetable and animal matter upon its exposed bed. Julius Cæsar had proposed the construction of a tunnel to drain the lake, but the enterprise was not actually undertaken until the reign of Claudius, when—after a temporary failure, from errors in levelling by the engineers, as was pretended at the time, or, as now appears certain, in consequence of frauds by the contractors in the execution of the work—it was at least partially completed. From this imperfect construction, it soon got out of repair, but was restored by Hadrian, and seems to have answered its design for some centuries. In the barbarism which followed the downfall of the empire, it again fell into decay, and though numerous attempts were made to repair it during the Middle Ages, no tolerable success seems to have attended any of these efforts, until the present generation.

Works have now been some years in progress for restoring, or rather enlarging and rebuilding this ancient tunnel, upon a scale of grandeur which does infinite honor to the liberality and public spirit of the projectors, and with an ingenuity of design and a constructive skill which reflect the highest credit upon the professional ability of the engineers who have planned the works and directed their execution. The length of this tunnel is 18,634 feet, or rather more than three miles and a half. Of course, it is one of the longest subterranean galleries yet executed in Europe, and it offers many curious particulars in its original design which cannot here be described. The difference between the highest and the lowest known levels of the surface of the lake amounts to at least forty feet, and the difference of area covered at these respective stages is not much less than eight thousand acres. The tunnel will reduce the water to a much lower point, and it is computed that, including the lands occasionally overflowed, not less than forty thousand acres of as fertile soil as any in Italy will be recovered from the lake and permanently secured from inundation by its waters.

Many similar enterprises have been conceived and executed in modern times, both for the purpose of reclaiming land covered by water and for sanitary reasons.* They are sometimes attended with wholly unexpected evils, as, for example, in the case of Barton Pond, in Vermont, and in that of the Lake Storsjö, in Sweden, already mentioned on a former

* A considerable work of this character is mentioned by Captain Gilliss as having been executed in Chili, a country to which we should have hardly looked for an improvement of such a nature. The Lake Taguataga was partially drained by cutting through a narrow ridge of land, not at the natural outlet, but upon one side of the lake, and eight thousand acres of land covered by it were gained for cultivation.—*U. S. Naval Astronomical Expedition to the Southern Hemisphere*, i, pp. 16, 17.

page. Another still less obvious consequence of the withdrawal of the waters has occasionally been observed in these operations. The hydrostatic force with which the water, in virtue of its specific gravity, presses against the banks that confine it, has a tendency to sustain them whenever their composition and texture are not such as to expose them to softening and dissolution by the infiltration of the water. If then, the slope of the banks is considerable, or if the earth of which they are composed rests on a smooth and slippery stratum inclining toward the bed of the lake, they are liable to fall or slide forward when the mechanical support of the water is removed, and this sometimes happens on a considerable scale. A few years ago, the surface of the Lake of Lungern, in the Canton of Unterwalden, in Switzerland, was lowered by driving a tunnel about a quarter of a mile long through the narrow ridge, called the Kaiserstuhl, which forms a barrier at the north end of the basin. When the water was drawn off, the banks, which are steep, cracked and burst, several acres of ground slid down as low as the water receded, and even the whole village of Lungern was thought to be in no small danger.

Other inconveniences of a very serious character have often resulted from the natural wearing down, or, much more frequently, the imprudent destruction, of the barriers which confine mountain lakes. In their natural condition, such basins serve both to receive and retain the rocks and other detritus brought down by the torrents which empty into them, and to check the impetus of the rushing waters by bringing them to a temporary pause; but if the outlets are lowered so as to drain the reservoirs, the torrents continue their rapid flow through the ancient bed of the basins, and carry down with them the sand and gravel with which they are charged, instead of depositing their burden as before in the still waters of the lakes.

Mountain Lakes.

It is a common opinion in America that the river meadows, bottoms, or *intervales*, as they are popularly called, are generally the beds of ancient lakes which have burst their barriers and left running currents in their place. It was shown by Dr. Dwight, many years ago, that this is very far from being universally true; but there is no doubt that mountain lakes were of much more frequent occurrence in primitive than in modern geography, and there are many chains of such still existing in regions where man has yet little disturbed the original features of the earth. In the long valleys of the Adirondack range in Northern New York, and in the mountainous parts of Maine, eight, ten, and even more lakes and lakelets are sometimes found in succession, each emptying into the next lower pool, and so all at last into some considerable river. When the mountain slopes which supply these basins shall be stripped of their woods, the augmented swelling of the lakes will break down their barriers, their waters will run off, and the valleys will present successions of flats with rivers running through them, instead of chains of lakes connected by natural canals.

A similar state of things seems to have existed in the ancient geography of France. "Nature," says Lavergne, "has not excavated on the flanks of our Alps reservoirs as magnificent as those of Lombardy; she had, however, constructed smaller, but more numerous lakes, which the negligence of man has permitted to disappear. Auguste de Gasparin, brother of the illustrious agriculturist, demonstrated more than thirty years ago, in an original paper, that many natural dikes formerly existed in the mountain valleys, which have been swept away by the waters. He proposed to rebuild and to multiply them. This interesting suggestion has reappeared several times since, but has met with strong opposition from skilful engineers. It would, nevertheless, be well to try the experiment of creating artificial lakes which should fill themselves with the water of melting snows and deluging rains, to be drawn out in times of drought. If this plan has able opposers, it has also warm advocates. Experience alone can decide the question." *

* *Économie Rurale de la France*, p. 289.

Climatic Effects of Draining Lakes and Marshes.

The draining of lakes, marshes, and other superficial accumulations of moisture, reduces the water surface of a country, and, of course, the evaporation from it. Lakes, too, in elevated positions, lose a part of their water by infiltration, and thereby supply other lakes, springs, and rivulets at lower levels. Hence, it is evident that the draining of such waters, if carried on upon a large scale, must affect both the humidity and the temperature of the atmosphere, and the permanent supply of water for extensive districts.*

Geographical and Climatic Effects of Aqueducts, Reservoirs, and Canals.

Many processes of internal improvement, such as aqueducts for the supply of great cities, railroad cuts and embankments, and the like, divert water from its natural channels, and affect its distribution and ultimate discharge. The collecting of the waters of a considerable district into reservoirs, to be thence carried off by means of aqueducts, as, for example, in the forest of Belgrade, near Constantinople, deprives the grounds originally watered by the springs and rivulets of the necessary moisture, and reduces them to barrenness. Similar effects must have followed from the construction of the numerous aqueducts which supplied ancient Rome with such a profuse abundance of water. On the other hand, the filtration of water through the banks or walls of an aqueduct car-

* In a note on a former page of this volume I noticed an observation of Jacini, to the effect that the great Italian lakes discharge themselves partly by infiltration beneath the hills which bound them. The amount of such infiltration must depend much upon the hydrostatic pressure on the walls of the lake basins, and, of course, the lowering of the surface of these lakes, by diminishing that pressure, would diminish also the infiltration. It is now proposed to lower the level of the Lake of Como some feet by deepening its outlet. It is possible that the effect of this may manifest itself in a diminution of the water in springs and *fontanili* or artesian wells in Lombardy. See *Appendix*, No. 41.

ried upon a high level across low ground, often injures the adjacent soil, and is prejudicial to the health of the neighboring population; and it has been observed in Switzerland, that fevers have been produced by the stagnation of the water in excavations from which earth had been taken to form embankments for railways.

If we consider only the influence of physical improvements on civilized life, we shall perhaps ascribe to navigable canals a higher importance, or at least a more diversified influence, than to any other works of man designed to control the waters of the earth, and to affect their distribution. They bind distant regions together by social ties, through the agency of the commerce they promote; they facilitate the transportation of military stores and engines, and of other heavy material connected with the discharge of the functions of government; they encourage industry by giving marketable value to raw material and to objects of artificial elaboration which would otherwise be worthless on account of the cost of conveyance; they supply from their surplus waters means of irrigation and of mechanical power; and, in many other ways, they contribute much to advance the prosperity and civilization of nations. Nor are they wholly without geographical importance. They sometimes drain lands by conveying off water which would otherwise stagnate on the surface, and, on the other hand, like aqueducts, they render the neighboring soil cold and moist by the percolation of water through their embankments; * they dam

* Simonde, speaking of the Tuscan canals, observes: "But inundations are not the only damage caused by the waters to the plains of Tuscany. Raised, as the canals are, above the soil, the water percolates through their banks, penetrates every obstruction, and, in spite of all the efforts of industry, sterilizes and turns to morasses fields which nature and the richness of the soil seemed to have designed for the most abundant harvests. In ground thus pervaded with moisture, or rendered *cold*, as the Tuscans express it, by the filtration of the canal water, the vines and the mulberries, after having for a few years yielded fruit of a saltish taste, rot and perish. The wheat decays in the ground, or dies as soon as it sprouts. Winter crops are given up, and summer cultivation tried for a time; but the increasing humidity, and the saline matter communicated to the earth

up, check, and divert the course of natural currents, and deliver them at points opposite to, or distant from, their original outlets; they often require extensive reservoirs to feed them, thus retaining through the year accumulations of water—which would otherwise run off, or evaporate in the dry season—and thereby enlarging the evaporable surface of the country; and we have already seen that they interchange the flora and the fauna of provinces widely separated by nature. All these modes of action certainly influence climate and the character of terrestrial surface, though our means of observation are not yet perfected enough to enable us to appreciate and measure their effects.

Climatic and Geographical Effects of Surface and Underground Draining.

I have commenced this chapter with a description of the dikes and other hydraulic works of the Netherland engineers, because the geographical results of such operations are more obvious and more easily measured, though certainly not more important, than those of the older and more widely diffused modes of resisting or directing the flow of waters, which have been practised from remote antiquity in the interior of all civilized countries. Draining and irrigation are habitually regarded as purely agricultural processes, having little or no relation to technical geography; but we shall find that they exert a powerful influence on soil, climate, and animal and vegetable life, and may, therefore, justly claim to be regarded as geographical elements.

Surface and Under-draining and their Effects.

Superficial draining is a necessity in all lands newly reclaimed from the forest. The face of the ground in the woods

—which affects the taste of all its products, even to the grasses, which the cattle refuse to touch—at last compel the husbandman to abandon his fields, and leave uncultivated a soil that no longer repays his labor."—*Tableau de l'Agriculture Toscane*, pp. 11, 12.

is never so regularly inclined as to permit water to flow freely over it. There are, even on the hillsides, many small ridges and depressions, partly belonging to the original distribution of the soil, and partly occasioned by irregularities in the growth and deposit of vegetable matter. These, in the husbandry of nature, serve as dams and reservoirs to collect a larger supply of moisture than the spongy earth can at once imbibe. Besides this, the vegetable mould is, even under the most favorable circumstances, slow in parting with the humidity it has accumulated under the protection of the woods, and the infiltration from neighboring forests contributes to keep the soil of small clearings too wet for the advantageous cultivation of artificial crops. For these reasons, surface draining must have commenced with agriculture itself, and there is probably no cultivated district, one may almost say no single field, which is not provided with artificial arrangements for facilitating the escape of superficial water, and thus carrying off moisture which, in the natural condition of the earth, would have been imbibed by the soil.

The beneficial effects of surface drainage, the necessity of extending the fields as population increased, and the inconveniences resulting from the presence of marshes in otherwise improved regions, must have suggested at a very early period of human industry the expediency of converting bogs and swamps into dry land by drawing off their waters; and it would not be long after the introduction of this practice before further acquisition of agricultural territory would be made by lowering the outlet of small ponds and lakes, and adding the ground they covered to the domain of the husbandman.

All these processes belong to the incipient civilization of the ante-historical periods, but the construction of subterranean channels for the removal of infiltrated water marks ages and countries distinguished by a great advance in agricultural theory and practice, a great accumulation of pecuniary capital, and a density of population which creates a ready demand and a high price for all products of rural industry. Under-draining, too, would be most advantageous in damp and cool climates, where evaporation is slow, and upon soils where the natural inclination of surface does not promote a very rapid flow of the surface waters. All the conditions required to make this mode of rural improvement, if not absolutely necessary, at least apparently profitable, exist in Great Britain, and it is, therefore, very natural that the wealthy and intelligent farmers of England should have carried this practice farther, and reaped a more abundant pecuniary return from it, than those of any other country.

Besides superficial and subsoil drains, there is another method of disposing of superfluous surface water, which, however, can rarely be practised, because the necessary conditions for its employment are not of frequent occurrence. Whenever a tenacious water-holding stratum rests on a loose, gravelly bed, so situated as to admit of a free discharge of water from or through it by means of the outcropping of the bed at a lower level, or of deep-lying conduits leading to distant points of discharge, superficial waters may be carried off by opening a passage for them through the impervious into the permeable stratum. Thus, according to Bischof, as early as the time of King Réné, in the first half of the fifteenth century, the plain of Paluns, near Marseilles, was laid dry by boring, and Wittwer informs us that draïnage is effected at Munich by conducting the superfluous water into large excavations, from which it filters through into a lower stratum of pebble and gravel lying a little above the level of the river Isar.* So at Washington, in the western part of the city, which lies high above the rivers Potomac and Rock Creek, many houses are provided with dry wells for draining their cellars and foundations. These extend through hard tenacious earth to the depth of thirty or forty feet, when they strike a stratum of gravel, through which the water readily passes off.

This practice has been extensively employed at Paris, not merely for carrying off ordinary surface water, but for the dis-

* *Physikalische Geographie*, p. 288. Draining by driving down stakes, mentioned in a note in a chapter on the woods, *ante*, is a process of the same nature.

charge of offensive and deleterious fluids from chemical and manufacturing establishments. A well of this sort received, in the winter of 1832–'33, twenty thousand gallons per day of the foul water from a starch factory, and the same process was largely used in other factories. The apprehension of injury to common and artesian wells and springs led to an investigation on this subject, in behalf of the municipal authorities, by Girard and Parent Duchatelet, in the latter year. The report of these gentlemen, published in the *Annales des Ponts et Chaussées* for 1833, second half year, is full of curious and instructive facts respecting the position and distribution of the subterranean waters under and near Paris; but it must suffice to say that the report came to the conclusion that, in consequence of the absolute immobility of these waters, and the relatively small quantity of noxious fluid to be conveyed to them, there was no danger of the diffusion of this latter, if discharged into them. This result will not surprise those who know that, in another work, Duchatelet maintains analogous opinions as to the effect of the discharge of the city sewers into the Seine upon the waters of that river. The quantity of matter delivered by them he holds to be so nearly infinitesimal, as compared with the volume of water of the Seine, that it cannot possibly affect it to a sensible degree. I would, however, advise determined water drinkers living at Paris to adopt his conclusions, without studying his facts and his arguments; for it is quite possible that he may convert his readers to a faith opposite to his own, and that they will finally agree with the poet who held water an "ignoble beverage."

Climatic and Geographical Effects of Surface Draining.

When we remove water from the surface, we diminish the evaporation from it, and, of course, the refrigeration which accompanies all evaporation is diminished in proportion. Hence superficial draining ought to be attended with an elevation of atmospheric temperature, and, in cold countries, it might be expected to lessen the frequency of frosts. Accordingly, it is a fact of experience that, other things being equal, dry soils, and the air in contact with them, are perceptibly warmer during the season of vegetation, when evaporation is most rapid, than moist lands and the atmospheric stratum resting upon them. Instrumental observation on this special point has not yet been undertaken on a very large scale, but still we have thermometric data sufficient to warrant the general conclusion, and the influence of drainage in diminishing the frequency of frost appears to be even better established than a direct increase of atmospheric temperature. The steep and dry uplands of the Green Mountain range in New England often escape frosts when the Indian corn harvest on moister grounds, five hundred or even a thousand feet lower, is destroyed or greatly injured by them. The neighborhood of a marsh is sure to be exposed to late spring and early autumnal frosts, but they cease to be feared after it is drained, and this is particularly observable in very cold climates, as, for example, in Lapland.*

In England, under-drains are not generally laid below the reach of daily variations of temperature, or below a point from which moisture might be brought to the surface by capillary attraction and evaporated by the heat of the sun. They, therefore, like surface drains, withdraw from local solar action much moisture which would otherwise be vaporized by it, and, at the same time, by drying the soil above them, they increase its effective hygroscopicity, and it consequently absorbs from the atmosphere a greater quantity of water than it did when, for want of under-drainage, the subsoil was always humid, if not saturated. Under-drains, then, contribute to the dryness as well as to the warmth of the atmosphere, and, as dry ground is more readily heated by the rays of the sun than wet, they tend also to raise the mean, and especially the summer temperature of the soil.

* "The simplest backwoodsman knows by experience that all cultivation is impossible in the neighborhood of bogs and marshes. Why is a crop near the borders of a marsh cut off by frost, while a field upon a hillock, a few stone's throws from it, is spared?"—LARS LEVI LÆSTADIUS, *Om Uppodlingar i Lappmarken*, pp. 69, 74.

So far as respects the immediate improvement of soil and climate, and the increased abundance of the harvests, the English system of surface and subsoil drainage has fully justified the eulogiums of its advocates; but its extensive adoption appears to have been attended with some altogether unforeseen and undesirable consequences, very analogous to those which I have described as resulting from the clearing of the forests. The under-drains carry off very rapidly the water imbibed by the soil from precipitation, and through infiltration from neighboring springs or other sources of supply. Consequently, in wet seasons, or after heavy rains, a river bordered by artificially drained lands receives in a few hours, from superficial and from subterranean conduits, an accession of water which, in the natural state of the earth, would have reached it only by small instalments after percolating through hidden paths for weeks or even months, and would have furnished perennial and comparatively regular contributions, instead of swelling deluges, to its channel. Thus, when human impatience rashly substitutes swiftly acting artificial contrivances for the slow methods by which nature drains the surface and superficial strata of a river basin, the original equilibrium is disturbed, the waters of the heavens are no longer stored up in the earth to be gradually given out again, but are hurried out of man's domain with wasteful haste; and while the inundations of the river are sudden and disastrous, its current, when the drains have run dry, is reduced to a rivulet, it ceases to supply the power to drive the machinery for which it was once amply sufficient, and scarcely even waters the herds that pasture upon its margin.*

Irrigation and its Climatic and Geographical Effects.

We know little of the history of the extinct civilizations which preceded the culture of the classic ages, and no nation has, in modern times, spontaneously emerged from barbarism,

* Babinet condemns even the general draining of marshes. "Draining," says he, "has been much in fashion for some years. It has been a special object to dry and fertilize marshy grounds. My opinion has always

and created for itself the arts of social life.* The improvements of the savage races whose history we can distinctly trace are borrowed and imitative, and our theories as to the origin and natural development of industrial art are conjectural. Of course, the relative antiquity of particular branches of human industry depends much upon the natural character of soil, climate, and spontaneous vegetable and animal life in different countries; and while the geographical influence of man would, under given circumstances, be exerted in one direction, it would, under different conditions, act in an opposite or a diverging line. I have given some reasons for thinking that in the climates to which our attention has been chiefly directed, man's first interference with the natural arrangement and disposal of the waters was in the way of drainage of surface. But if we are to judge from existing remains alone, we should probably conclude that irrigation is older than drainage; for, in the regions regarded by general tradition as the cradle of the human race, we find traces of canals evidently constructed for the former purpose at a period long preceding the ages of which we have any written memorials. There are, in ancient Armenia, extensive districts which were already abandoned to desolation at the earliest historical epoch, but which, in a yet remoter antiquity, had been irrigated by a complicated and highly artificial system of canals, the lines of which can still be followed; and there are, in all the highlands where the sources of the Euphrates rise, in Persia, in Egypt, in India,

been that excessive dryness is thus produced, and that other soils in the neighborhood are sterilized in proportion."

* I ought perhaps to except the Mexicans and the Peruvians, whose arts and institutions are not yet shown to be historically connected with those of any more ancient people. The lamentable destruction of so many memorials of these tribes, by the ignorance and bigotry of the so-called Christian barbarians who conquered them, has left us much in the dark as to many points of their civilization; but they seem to have reached that stage where continued progress in knowledge and in power over nature is secure, and a few more centuries of independence might have brought them to originate for themselves most of the great inventions which the last four centuries have bestowed upon man.

and in China, works of this sort which must have been in existence before man had begun to record his own annals.

In warm countries, such as most of those just mentioned, the effects I have described as usually resulting from the clearing of the forests would very soon follow. In such climates, the rains are inclined to be periodical; they are also violent, and for these reasons the soil would be parched in summer and liable to wash in winter. In these countries, therefore, the necessity for irrigation must soon have been felt, and its introduction into mountainous regions like Armenia must have been immediately followed by a system of terracing, or at least scarping the hillsides. Pasture and meadow, indeed, may be irrigated even when the surface is both steep and irregular, as may be observed abundantly on the Swiss as well as on the Piedmontese slope of the Alps; but in dry climates, plough land and gardens on hilly grounds require terracing, both for supporting the soil and for administering water by irrigation, and it should be remembered that terracing, of itself, even without special arrangements for controlling the distribution of water, prevents or at least checks the flow of rain water, and gives it time to sink into the ground instead of running off over the surface.

There are few things in Continental husbandry which surprise English or American observers so much as the extent to which irrigation is employed in agriculture, and that, too, on soils, and with a temperature, where their own experience would have led them to suppose it would be injurious to vegetation rather than beneficial to it. The summers in Northern Italy, though longer, are very often not warmer than in New England; and in ordinary years, the summer rains are as frequent and as abundant in the former country as in the latter. Yet in Piedmont and Lombardy, irrigation is bestowed upon almost every crop, while in New England it is never employed at all in farming husbandry, or indeed for any purpose except in kitchen gardens, and possibly, in rare cases, in some other small branch of agricultural industry.*

* The necessity of irrigation in the great alluvial plain of Northern

The summers in Egypt, in Syria, and in Asia Minor and even Rumelia, are almost rainless. In such climates, the necessity of irrigation is obvious, and the loss of the ancient means of furnishing it readily explains the diminished fertility of most of the countries in question.* The surface of Pales-

Italy is partly explained by the fact that the superficial stratum of fine earth and vegetable mould is very extensively underlaid by beds of pebbles and gravel brought down by mountain torrents at a remote epoch. The water of the surface soil drains rapidly down into these loose beds, and passes off by subterranean channels to some unknown point of discharge; but this circumstance alone is not a sufficient solution. Is it not possible that the habits of vegetables, grown in countries where irrigation has been immemorially employed, have been so changed that they require water under conditions of soil and climate where their congeners, which have not been thus indulgently treated, do not?

There are some atmospheric phenomena in Northern Italy, which an American finds it hard to reconcile with what he has observed in the United States. To an American eye, for instance, the sky of Piedmont, Lombardy, and the northern coast of the Mediterranean, is always whitish and curdled, and it never has the intensity and fathomless depth of the blue of his native heavens. And yet the heat of the sun's rays, as measured by sensation, and, at the same time, the evaporation, are greater than they would be with the thermometer at the same point in America. I have frequently felt in Italy, with the mercury below 60° Fahrenheit, and with a mottled and almost opaque sky, a heat of solar irradiation which I can compare to nothing but the scorching sensation experienced in America at a temperature twenty degrees higher, during the intervals between showers, or before a rain, when the clear blue of the sky seems infinite in depth and transparency. Such circumstances may create a necessity for irrigation where it would otherwise be superfluous, if not absolutely injurious.

In speaking of the superior apparent clearness of the *sky* in America, I confine myself to the concave vault of the heavens, and do not mean to assert that terrestrial objects are generally visible at greater distances in the United States than in Italy. Indeed I am rather disposed to maintain the contrary; for though I know that the lower strata of the atmosphere in Europe never equal in transparency the air near the earth in New Mexico, Peru, and Chili, yet I think the accidents of the coast line of the Riviera, as, for example, between Nice and La Spezia, and those of the incomparable Alpine parorama seen from Turin, are distinguishable at greater distances than they would be in the United States.

* In Egypt, evaporation and absorption by the earth are so rapid, that

tine, for example, is composed, in a great measure, of rounded limestone hills, once, no doubt, covered with forests. These were partially removed before the Jewish conquest.* When the soil began to suffer from drought, reservoirs to retain the waters of winter were hewn in the rock near the tops of the hills, and the declivities were terraced. So long as the cisterns were in good order, and the terraces kept up, the fertility of Palestine was unsurpassed, but when misgovernment and for-

all annual crops require irrigation during the whole period of their growth. As fast as the water retires by the subsidence of the annual inundation, the seed is sown upon the still moist uncovered soil, and irrigation begins at once. Upon the Nile, you hear the creaking of the water wheels, and sometimes the movement of steam pumps, through the whole night, while the poorer cultivators unceasingly ply the simple *shadoof*, or bucket-and-sweep, laboriously raising the water from trough to trough by as many as six or seven stages when the river is low. The bucket is of flexible leather, with a stiff rim, and is emptied into the trough, not by inverting it like a wooden bucket, but by putting the hand beneath and pushing the bottom up till the water all runs out over the brim, or, in other words, by turning the vessel inside out.

The quantity of water thus withdrawn from the Nile is enormous. Most of this is evaporated directly from the surface or the superficial strata, but some moisture percolates down and oozes through the banks into the river again, while a larger quantity sinks till it joins the slow current of infiltration by which the Nile water pervades the earth of the valley to the distance, at some points, of not less than fifty miles.

* "Forests," "woods," and "groves," are very frequently mentioned in the Old Testament as existing at particular places, and they are often referred to by way of illustration, as familiar objects. "Wood" is twice spoken of as a material in the New Testament, but otherwise—at least according to Cruden—not one of the above words occurs in that volume.

This interesting fact, were other evidence wanting, would go far to prove that a great change had taken place in this respect between the periods when the Old Testament and the New were respectively composed; for the scriptural writers, and the speakers introduced into their narratives, are remarkable for their frequent allusions to the natural objects and the social and industrial habits which characterized their ages and their country. See *Appendix*, No. 42.

Solomon anticipated Chevandier in the irrigation of forest trees: "I made me pools of water, to water therewith the wood that bringeth forth trees."—*Ecclesiastes* ii, 6.

eign and intestine war occasioned the neglect or destruction of these works—traces of which still meet the traveller's eye at every step,—when the reservoirs were broken and the terrace walls had fallen down, there was no longer water for irrigation in summer, the rains of winter soon washed away most of the thin layer of earth upon the rocks, and Palestine was reduced almost to the condition of a desert.

The course of events has been the same in Idumæa. The observing traveller discovers everywhere about Petra, particularly if he enters the city by the route of Wadi Ksheibeh, very extensive traces of ancient cultivation, and upon the neighboring ridges are the ruins of numerous cisterns evidently constructed to furnish a supply of water for irrigation.* In

* One of these, upon Mount Hor, two stories in height, is still in such preservation that I found not less than ten feet of water in it in the month of June, 1851.

The brook Ain Musa, which runs through the city of Petra and finally disappears in the sands of Wadi el Araba, is a considerable river in winter, and the inhabitants of that town were obliged to excavate a tunnel through the rock near the right bank, just above the upper entrance of the Sik, to discharge a part of its swollen current. The sagacity of Dr. Robinson detected the necessity of this measure, though the tunnel, the mouth of which was hidden by brushwood, was not discovered till some time after his visit. I even noticed unequivocal remains of a sluice by which the water was diverted to the tunnel near the arch that crosses the Sik. Immense labor was also expended in widening the natural channel at several points below the town, to prevent the damming up and setting back of the water—a fact I believe not hitherto noticed by travellers.

The Fellahheen above Petra still employ the waters of Ain Musa for irrigation, and in summer the superficial current is wholly diverted from its natural channel for that purpose. At this season, the bed of the brook, which is composed of pebbles, gravel, and sand, is dry in the Sik and through the town; but the infiltration is such that water is generally found by digging to a small depth in the channel. Observing these facts in a visit to Petra in the summer, I was curious to know whether the subterranean waters escaped again to daylight, and I followed the ravine below the town for a long distance. Not very far from the upper entrance of the ravine, arborescent vegetation appeared upon its bottom, and as soon as the ground was well shaded, a thread of water burst out. This was joined by others a little lower down, and, at the distance of a mile from the town, a strong current was formed and ran down toward Wadi el Araba.

primitive ages, the precipitation of winter in these hilly countries was, in great part, retained for a time in the superficial soil, first by the vegetable mould of the forests, and then by the artificial arrangements I have described. The water imbibed by the earth was partly taken up by direct evaporation, partly absorbed by vegetation, and partly carried down by infiltration to subjacent strata which gave it out in springs at lower levels, and thus a fertility of soil and a condition of the atmosphere were maintained sufficient to admit of the dense population that once inhabited those now arid wastes. At present, the rain water runs immediately off from the surface and is carried down to the sea, or is drunk up by the sands of the wadis, and the hillsides which once teemed with plenty are bare of vegetation, and seared by the scorching winds of the desert.

In Southern Europe, in the Turkish Empire, and in many other countries, a very large proportion of the surface is, if not absolutely flooded, at least thoroughly moistened by irrigation, a great number of times in the course of every season, and this, especially, at periods when it would otherwise be quite dry, and when, too, the power of the sun and the capacity of the air for absorbing moisture are greatest. Hence it is obvious that the amount of evaporation from the earth in these countries, and, of course, the humidity and the temperature of both the soil and the atmosphere in contact with it, must be much affected by the practice of irrigation. The cultivable area of Egypt, or the space accessible to cultivation, between desert and desert, is more than seven thousand square statute miles. Much of the surface, though not out of the reach of irrigation, lies too high to be economically watered, and irrigation and cultivation are therefore confined to an area of five or six thousand square miles, nearly the whole of which is regularly and constantly watered when not covered by the inundation, except in the short interval between the harvest and the rise of the waters. For nearly half of the year, then, irrigation adds five or six thousand square miles, or more than a square equatorial degree, to the evaporable surface of the Nile valley, or, in other words, more than decuples the area from which an appreciable quantity of moisture would otherwise be evaporated; for after the Nile has retired within its banks, its waters by no means cover one tenth of the space just mentioned.* The fresh-water canals now constructing, in connec-

* The authorities differ as to the extent of the cultivable and the cultivated soil of Egypt. Lippincott's, or rather Thomas and Baldwin's, *Gazetteer*—a work of careful research—estimates "the whole area comprised in the valley [below the first cataract] and delta," at 11,000 square miles. Smith's *Dictionary of the Bible*, article "Egypt," says: "Egypt has a superficies of about 9,582 square geographical miles of soil, which the Nile either does or can water and fertilize. This computation includes the river and lakes as well as sundry tracts which can be inundated, and the whole space either cultivated or fit for cultivation is no more than about 5,626 square miles." By geographical mile is here meant, I suppose, the nautical mile of sixty to an equatorial degree, or about 2,025 yards. The whole area, then, by this estimate, is 12,682 square statute or English miles, that of the space "cultivated or fit for cultivation," 7,447. Smith's *Dictionary of Greek and Roman Geography*, article "Ægyptus," gives 2,255 square miles as the area of the valley between Syene and the bifurcation of the Nile, exclusive of the Fayoom, which is estimated at 340. The area of the Delta is stated at 1,976 square miles between the main branches of the river, and, including the irrigated lands east and west of those branches, at 4,500 square miles. This latter work does not inform us whether these are statute or nautical miles, but nautical miles must be intended.

Other writers give estimates differing considerably from those just cited. The latest computations I have seen are those in the first volume of Kremer's *Ægypten*, 1863. This author (pp. 6, 7) assigns to the Delta an area of 200 square German geographical miles (fifteen to the degree); to all Lower Egypt, including, of course, the Delta, 400 such miles. These numbers are equal, respectively, to 4,239 and 8,478 square statute miles, and the great lagoons are embraced in the areas computed. Upper Egypt (above Cairo) is said (p. 11) to contain 4,000,000 feddan of *culturfläche*, or cultivable land. The feddan is stated (p. 37) to contain 7,333 square piks, the pik being 75 centimètres, and it therefore corresponds almost exactly to the English acre. Hence, according to Kremer, the cultivable soil of Upper Egypt is 6,250 square statute miles, or twice as much as the whole area of the valley between Syene and the bifurcation of the Nile, according to Smith's *Dictionary of Greek and Roman Geography*. I suspect that 4,000,000 feddan is erroneously given as the cultivable area of Upper Egypt alone, when in fact it should be taken for the arable surface of both

tion with the works for the Suez canal, will not only restore the long abandoned fields east of the Nile, but add to the arable soil of Egypt hundreds of square miles of newly reclaimed desert, and thus still further increase the climatic effects of irrigation.*

The Nile receives not a single tributary in its course through Egypt; there is not so much as one living spring in the whole land,† and, with the exception of a narrow strip of coast, where the annual precipitation is said to amount to six inches, the fall of rain in the territory of the Pharaohs is not two inches in the year. The subsoil of the whole valley is pervaded with moisture by infiltration from the Nile, and water can everywhere be found at the depth of a few feet. Were irrigation suspended, and Egypt abandoned, as in that case it must be,

Lower and Upper Egypt; for from the statistical tables in the same volume, it appears that 3,317,125 feddan, or 5,253 square statute miles, were cultivated, in both geographical divisions, in the year referred to in the tables, the date of which is not stated.

The area which the Nile would now cover at high water, if left to itself, is greater than in ancient times, because the bed of the river has been elevated, and consequently the lateral spread of the inundation increased. See SMITH's *Dictionary of Geography*, article "Ægyptus." But the industry of the Egyptians in the days of the Pharaohs and the Ptolomies carried the Nile-water to large provinces which have now been long abandoned and have relapsed into the condition of a desert. "Anciently," observes the writer of the article "Egypt" in Smith's *Dictionary of the Bible*, "2,735 square miles more [about 3,700 square statute miles] may have been cultivated. In the best days of Egypt, probably all the land was cultivated that could be made available for agricultural purposes, and hence we may estimate the ancient arable area of that country at not less than 11,000 square statute miles, or fully double its present extent."

* A canal has been constructed, and new ones are in progress, to convey water from the Nile to the city of Suez, and to various points on the line of the ship canal, with the double purpose of supplying fresh water to the inhabitants and laborers, and of irrigating the adjacent soil. The area of land which may be thus reclaimed and fertilized is very large, but the actual quantity which it will be found economically expedient to bring under cultivation cannot now be determined.

† The so-called spring at Heliopolis is only a thread of water infiltrated from the Nile or the canals.

to the operations of nature, there is no doubt that trees, the roots of which penetrate deeply, would in time establish themselves on the deserted soil, fill the valley with verdure, and perhaps at last temper the climate, and even call down abundant rain from the heavens.* But the immediate effect of discontinuing irrigation would be, first, an immense reduction of the evaporation from the valley in the dry season, and then a greatly augmented dryness and heat of the atmosphere. Even the almost constant north wind—the strength of which would be increased in consequence of these changes—would little reduce the temperature of the narrow cleft between the burning mountains which hem in the channel of the Nile, so that a single year would transform the most fertile of soils to the most barren of deserts, and render uninhabitable a territory that irrigation makes capable of sustaining as dense a population as has ever existed in any part of the world.† Whether man found the valley of the Nile a forest, or such a waste as I have just described, we do not historically know. In either case, he has not simply converted a wilderness into

* The date and the doum palm, the *sont* and many other acacias, the caroub, the sycamore, and other trees, grow well in Egypt without irrigation, and would doubtless spread through the entire valley in a few years.

† Wilkinson has shown that the cultivable soil of Egypt has not been diminished by encroachment of the desert sands, or otherwise, but that, on the contrary, it must have been increased since the age of the Pharaohs. The Gotha *Almanac* for 1862 states the population of Egypt in 1859 at 5,125,000 souls; but this must be a great exaggeration, even supposing the estimate to include the inhabitants of Nubia, and of much other territory not geographically belonging to Egypt. In general, the population of that country has been estimated at something more than three millions, or about six hundred to the square mile; but with a better government and better social institutions, the soil would sustain a much greater number, and in fact it is believed that in ancient times its inhabitants were twice, perhaps even thrice, as numerous as at present.

Wilkinson (*Handbook for Travellers in Egypt*, p. 10) observes that the total population, which two hundred years ago was estimated at 4,000,000, amounted till lately only to about 1,800,000 souls, having been reduced since 1800 from 2,500,000 to that number.

a garden, but has unquestionably produced extensive climatic change.*

The fields of Egypt are more regularly watered than those of any other country bordering on the Mediterranean, except the rice grounds in Italy, and perhaps the *marcite* or winter meadows of Lombardy; but irrigation is more or less employed throughout almost the entire basin of that sea, and is everywhere attended with effects which, if less in degree, are analogous in character to those resulting from it in Egypt. In general, it may be said that the soil is nowhere artificially watered except when it is so dry that little moisture would be evaporated from it, and, consequently, every acre of irrigated ground is so much added to the evaporable surface of the country. When the supply of water is unlimited, it is allowed, after serving its purpose on one field, to run into drains, canals, or rivers. But in most regions where irrigation is regularly employed, it is necessary to economize the water; after passing over or through one parcel of ground, it is conducted to

* Ritter supposes Egypt to have been a sandy desert when it was first occupied by man. "The first inhabitant of the sandy valley of the Nile was a desert dweller, as his neighbors right and left, the Libyan, the nomade Arab, still are. But the civilized people of Egypt transformed, by canals, the waste into the richest granary of the world; they liberated themselves from the shackles of the rock and sand desert, in the midst of which, by a wise distribution of the fluid through the solid geographical form, by irrigation in short, they created a region of culture most rich in historical monuments."—*Einleitung zur allgemeinen vergleichenden Geographie*, pp. 165, 166.

This view seems to me highly improbable; for though, by canals and embankments, man has done much to modify the natural distribution of the waters of the Nile, and possibly has even transferred its channel from one side of the valley to the other, yet the annual inundation is not his work, and the river must have overflowed its banks and carried spontaneous vegetation with its waters, as well before as since Egypt was first occupied by the human family. There is, indeed, some reason to suppose that man lived upon the banks of the Nile when its channel was much lower, and the spread of its inundations much narrower than at present; but wherever its flood reached, there the forest would propagate itself, and its shores are much more likely to have been morasses than sands.

another; no more is withdrawn from the canals at any one point than is absorbed by the soil it irrigates, or evaporated from it, and, consequently, it is not restored to liquid circulation, except by infiltration or precipitation. We are safe, then, in saying that the humidity evaporated from any artificially watered soil is increased by a quantity bearing a large proportion to the whole amount distributed over it; for most even of that which is absorbed by the earth is immediately given out again either by vegetables or by evaporation.

It is not easy to ascertain precisely either the extent of surface thus watered, or the amount of water supplied, in any given country, because these quantities vary with the character of the season; but there are not many districts in Southern Europe where the management of the arrangements for irrigation is not one of the most important branches of agricultural labor. The eminent engineer Lombardini describes the system of irrigation in Lombardy as, "every day in summer, diffusing over 550,000 hectares of land 45,000,000 cubic mètres of water, which is equal to the entire volume of the Seine, at an ordinary flood, or a rise of three mètres above the hydrometer at the bridge of La Tournelle at Paris." * Niel states the quantity of land irrigated in the former kingdom of Sardinia, including Savoy, in 1856, at 240,000 hectares, or not much less than 600,000 acres. This is about four thirteenths of the cultivable soil of the kingdom. According to the same author, the irrigated lands in France did not exceed 100,000 hectares, or 247,000 acres, while those in Lombardy amounted to 450,000 hectares, more than 1,100,000 acres.† In these three states alone, then, there were more than three thousand square miles of artificially watered land, and if we add the irrigated soils of the rest of Italy, of the Mediterranean islands, of the Spanish peninsula, of Turkey in Europe and in Asia Minor, of Syria, of Egypt and the remainder of Northern Africa, we shall see that irrigation increases the evaporable surface of the

* *Memorie sui progetti per l'estensione dell' Irrigazione, etc., il Politecnico*, for January, 1863, p. 6.

† Niel, *L'Agriculture des États Sardes*, p. 232.

Mediterranean basin by a quantity bearing no inconsiderable proportion to the area naturally covered by water within it. As near as can be ascertained, the amount of water applied to irrigated lands is scarcely anywhere less than the total precipitation during the season of vegetable growth, and in general it much exceeds that quantity. In grass grounds and in field culture it ranges from 27 or 28 to 60 inches, while in smaller crops, tilled by hand labor, it is sometimes carried as high as 300 inches.* The rice grounds and the *marcite* of

* NIEL, *Agriculture des États Sardes*, p. 237. Lombardini's computation just given allows eighty-one cubic mètres per day to the hectare, which, supposing the season of irrigation to be one hundred days, is equal to a precipitation of thirty-two inches. But in Lombardy, water is applied to some crops during a longer period than one hundred days; and in the *marcite* it flows over the ground even in winter.

According to Boussingault (*Économie Rurale*, ii, p. 246) grass grounds ought to receive, in Germany, twenty-one centimètres of water per week, and with less than half that quantity it is not advisable to incur the expense of supplying it. The ground is irrigated twenty-five or thirty times, and if the full quantity of twenty-one centimètres is applied, it receives about two hundred inches of water, or six times the total amount of precipitation. Puvis, quoted by Boussingault, after much research comes to the conclusion that a proper quantity is twenty centimètres applied twenty-five or thirty times, which corresponds with the estimate just stated. Puvis adds—and, as our author thinks, with reason—that this amount might be doubled without disadvantage.

Boussingault observes that rain water is vastly more fertilizing than the water of irrigating canals, and therefore the supply of the latter must be greater. This is explained partly by the different character of the substances held in solution or suspension by the waters of the earth and of the sky, partly by the higher temperature of the latter, and, possibly, partly also by the mode of application—the rain being finely divided in its fall or by striking plants on the ground, river water flowing in a continuous sheet.

The temperature of the water is thought even more important than its composition. The sources which irrigate the *marcite* of Lombardy—meadows so fertile that less than an acre furnishes grass for a cow the whole year—are very warm. The ground watered by them never freezes, and a first crop, for soiling, is cut from it in January or February. The Canal Cavour, just now commenced—which is to take its supply from the Po at Chivasso, fourteen or fifteen miles below Turin—will furnish water of much higher fertilizing power than that derived from the Dora Baltea

Lombardy are not included in these estimates of the amount of water applied. Arrangements are concluded, and new plans proposed, for an immense increase of the lands fertilized by irrigation in France and Italy, and there is every reason to believe that the artificially watered soil of the latter country will be doubled, that of France quadrupled, before the end of this century. There can be no doubt that by these operations man is exercising a powerful influence on soil, on vegetable and animal life, and on climate, and hence that in this, as in many other fields of industry, he is truly a geographical agency.*

and the Sesia, both because it is warmer, and because it transports a more abundant and a richer sediment than the latter streams, which are fed by Alpine icefields and melting snows, and which flow, for long distances, in channels ground smooth and bare by ancient glaciers, and not now contributing much vegetable mould or fine slime to their waters.

* It belongs rather to agriculture than to geography to discuss the quality of the crops obtained by irrigation, or the permanent effects produced by it on the productiveness of the soil. There is no doubt, however, that all crops which can be raised without watering are superior in flavor and in nutritive power to those grown by the aid of irrigation. Garden vegetables, particularly, profusely watered, are so insipid as to be hardly eatable. Wherever irrigation is practised, there is an almost irresistible tendency, especially among ignorant cultivators, to carry it to excess; and in Piedmont and Lombardy, if the supply of water is abundant, it is so liberally applied as sometimes not only to injure the quality of the product, but to drown the plants and diminish the actual weight of the crop.

Professor Liebig, in his *Modern Agriculture*, says: "There is not to be found in chemistry a more wonderful phenomenon, one which more confounds all human wisdom, than is presented by the soil of a garden or field. By the simplest experiment, any one may satisfy himself that rain water filtered through field or garden soil does not dissolve out a trace of potash, silicic acid, ammonia, or phosphoric acid. The soil does not give up to the water one particle of the food of plants which it contains. The most continuous rains cannot remove from the field, except mechanically, any of the essential constituents of its fertility."

"The soil not only retains firmly all the food of plants which is actually in it, but its power to preserve all that may be useful to them extends much farther. If rain or other water holding in solution ammonia, potash, and phosphoric and silicic acids, be brought in contact with soil, these sub-

The quantity of water artificially withdrawn from running streams for the purpose of irrigation is such as very sensibly to affect their volume, and it is, therefore, an important element in the geography of rivers. Brooks of no trifling current

stances disappear almost immediately from the solution; the soil withdraws them from the water. Only such substances are completely withdrawn by the soil as are indispensable articles of food for plants; all others remain wholly or in part in solution."

The first of the paragraphs just quoted is not in accordance with the alleged experience of agriculturists in those parts of Italy where irrigation is most successfully applied. They believe that the constituents of vegetable growth are washed out of the soil by excessive and long-continued watering. They consider it also established as a fact of observation, that water which has flowed through or over rich ground is far more valuable for irrigation than water from the same source, which has not been impregnated with fertilizing substances by passing through soils containing them; and, on the other hand, that water, rich in the elements of vegetation, parts with them in serving to irrigate a poor soil, and is therefore less valuable as a fertilizer of lower grounds to which it may afterward be conducted.

The practice of irrigation—except in mountainous countries where springs and rivulets are numerous—is attended with very serious economical, social, and political evils. The construction of canals and their immensely ramified branches, and the grading and scarping of the ground to be watered, are always expensive operations, and they very often require an amount of capital which can be commanded only by the state, by moneyed corporations, or by very wealthy proprietors; the capacity of the canals must be calculated with reference to the area intended to be irrigated, and when they and their branches are once constructed, it is very difficult to extend them, or to accommodate any of their original arrangements to changes in the condition of the soil, or in the modes or objects of cultivation; the flow of the water being limited by the abundance of the source or the capacity of the canals, the individual proprietor cannot be allowed to withdraw water at will, according to his own private interest or convenience, but both the time and the quantity of supply must be regulated by a general system applicable, as far as may be, to the whole area irrigated by the same canal, and every cultivator must conform his industry to a plan which may be quite at variance with his special objects or with his views of good husbandry. The clashing interests and the jealousies of proprietors depending on the same means of supply are a source of incessant contention and litigation, and the caprices or partialities of the officers who control, or of contractors who farm the canals, are often wholly diverted from their natural channels to supply the canals, and their entire mass of water completely absorbed, so that it does not reach the river which it naturally feeds, except in such proportion as it is conveyed to it by infiltration. Irrigation, therefore, diminishes great rivers in warm countries by cutting off their sources of supply as well as by direct abstraction of water from their channels. We have just seen that the system of irrigation in Lombardy deprives the Po of a quantity of water equal to the total delivery of the Seine at ordinary flood, or, in other words, of the equivalent of a tributary navigable for hundreds of miles by vessels of considerable burden. The new canals commenced and projected will greatly increase the loss. The water required for irrigation in Egypt is less than would be supposed from the exceeding rapidity of evaporation in that arid climate; for the soil is thoroughly saturated during the inundation, and infiltration from the Nile continues to supply a considerable amount of humidity in the dryest season. Linant Bey computed that twenty-nine cubic mètres per day sufficed to irrigate a hectare in the Delta.* This is equivalent to a fall of rain of two millimètres and nine tenths per day, or, if we suppose water to be applied for one hundred and fifty days during the dry season, to a total precipitation of 435 millimètres, about seventeen inches and one third. Taking the area of actually cultivated soil in Egypt at the low estimate of 3,600,000 acres, and the average amount of water daily applied

lead not unfrequently to ruinous injustice toward individual landholders. These circumstances discourage the division of the soil into small properties, and there is a constant tendency to the accumulation of large estates of irrigated land in the hands of great capitalists, and consequently to the dispossession of the small cultivators, who pass from the condition of owners of the land to that of hireling tillers. The farmers are no longer yeomen, but peasants. Having no interest in the soil which composes their country, they are virtually expatriated, and the middle class, which ought to constitute the real physical and moral strength of the land, ceases to exist as a rural estate, and is found only among the professional, the mercantile, and the industrial population of the cities.

* BOUSSINGAULT, *Économie Rurale*, ii, pp. 248, 249.

in both Upper and Lower Egypt at twelve hundredths of an inch in depth, we have an abstraction of 61,000,000 cubic yards, which—the mean daily delivery of the Nile being in round numbers 320,000,000 cubic yards—is nearly one fifth of the average quantity of water contributed to the Mediterranean by that river.

Irrigation, as employed for certain special purposes in Europe and America, is productive of very prejudicial climatic effects. I refer particularly to the cultivation of rice in the Slave States of the American Union and in Italy. The climate of the Southern States is not necessarily unhealthy for the white man, but he can scarcely sleep a single night in the vicinity of the rice grounds without being attacked by a dangerous fever.* The neighborhood of the rice fields is less pestilential in Lombardy and Piedmont than in South Carolina and Georgia, but still very insalubrious to both man and beast. "Not only does the population decrease where rice is grown," says Escourrou Milliago, "but even the flocks are attacked by typhus. In the rice grounds, the soil is divided into compartments rising in gradual succession to the level of the irrigating canal, in order that the water, after having flowed one field, may be drawn off to another, and thus a single current serve for several compartments, the lowest field, of course, still being higher than the ditch which at last drains both it and the adjacent soil. This arrangement gives a cer-

* The cultivation of rice is so prejudicial to health everywhere that nothing but the necessities of a dense population can justify the sacrifice of life it costs in countries where it is pursued.

It has been demonstrated by actual experiment, that even in Mississippi, cotton can be advantageously raised by the white man without danger to health; and in fact, a great deal of the cotton brought to the Vicksburg market for some years past has been grown exclusively by white labor. There is no reason why the cultivation of cotton should be a more unhealthy occupation in America than it is in other countries where it was never dreamed of as dangerous, and no well-informed American, in the Slave States or out of them, believes that the abolition of slavery in the South would permanently diminish the cotton crop of those States.

tain force of hydrostatic pressure to the water with which the rice is irrigated, and the infiltration from these fields is said to extend through neighboring grounds, sometimes to the distance of not less than a myriamètre, or six English miles, and to be destructive to crops and even trees reached by it. Land thus affected can no longer be employed for any purpose but growing rice, and when prepared for that crop, it propagates still further the evils under which it had itself suffered, and, of course, the mischief is a growing one." *

The attentive traveller in Egypt and Nubia cannot fail to notice many localities, generally of small extent, where the soil is rendered infertile by an excess of saline matter in its composition. In many cases, perhaps in all, these barren spots lie rather above the level usually flooded by the inundations of the Nile, and yet they exhibit traces of former cultivation. Recent observations in India, a notice of which I find in an account of a meeting of the Asiatic Society in the Athenæum of December 20, 1862, No. 1834, suggest a possible explanation of this fact. At this meeting, Professor Medlicott read an essay on "the saline efflorescence called 'Reh' and 'Kuller,'" which is gradually invading many of the most fertile districts of Northern and Western India, and changing them into sterile deserts. It consists principally of sulphate of soda (Glauber's salts), with varying proportions of common salt. Mr. Medlicott pronounces "these salts (which, in small quantities are favorable to fertility of soil) to be the gradual result of concentration by evaporation of river and canal waters, which contain them in very minute quantities, and with which the lands are either irrigated or occasionally overflowed." The river inundations in hot countries usually take place but once in a year, and, though the banks remain submerged for days or even weeks, the water at that period, being derived principally from rains and snows, must be less highly charged with mineral matter than at lower stages, and besides, it is always in motion. The water of irrigation, on the other hand, is applied

* *L'Italie à propos de l'Exposition de Paris*, p. 92.

for many months in succession, it is drawn from rivers at the seasons when their proportion of salts is greatest, and it either sinks into the superficial soil, carrying with it the saline substances it holds in solution, or is evaporated from the surface, leaving them upon it. Hence irrigation must impart to the soil more salts than natural inundation. The sterilized grounds in Egypt and Nubia lying above the reach of the floods, as I have said, we may suppose them to have been first cultivated in that remote antiquity when the Nile valley received its earliest inhabitants. They must have been artificially irrigated from the beginning; they may have been under cultivation many centuries before the soil at a lower level was invaded by man, and hence it is natural that they should be more strongly impregnated with saline matter than fields which are exposed every year, for some weeks, to the action of running water so nearly pure that it would be more likely to dissolve salts than to deposit them.

INUNDATIONS AND TORRENTS.

In pointing out in a former chapter the evils which have resulted from the too extensive destruction of the forests, I dwelt at some length on the increased violence of river inundations, and especially on the devastations of torrents, in countries improvidently deprived of their woods, and I spoke of the replanting of the forests as the only effectual method of preventing the frequent recurrence of disastrous floods. There are many regions where, from the loss of the superficial soil, from financial considerations, and from other causes, the restoration of the woods is not, under present circumstances, to be hoped for. Even where that measure is feasible and in actual process of execution, a great number of years must elapse before the action of the destructive causes in question can be arrested or perhaps even sensibly mitigated by it. Besides this, leaving out of view the objections urged by Belgrand and his followers to the generally received opinions concerning the beneficial influence of the forest as respects river inundations—for no one disputes its importance in preventing the formation and limiting the ravages of mountain torrents—floods will always occur in years of excessive precipitation, whether the surface of the soil be generally cleared or generally wooded.

Physical improvement in this respect, then, cannot be confined to preventive measures, but, in countries subject to damage by inundation, means must be contrived to obviate dangers and diminish injuries to which human life and all the works of human industry will occasionally be exposed, in spite of every effort to lessen the frequency of their recurrence by acting directly on the causes that produce them. As every civilized country is, in some degree, subject to inundation by the overflow of rivers, the evil is a familiar one, and needs no general description. In discussing this branch of the subject, therefore, I may confine myself chiefly to the means that have been or may be employed to resist the force and limit the ravages of floods, which, left wholly unrestrained, would not only inflict immense injury upon the material interests of man, but produce geographical revolutions of no little magnitude.

a. *River Embankments.*

The most obvious and doubtless earliest method of preventing the escape of river waters from their natural channels, and the overflow of fields and towns by their spread, is that of raised embankments along their course. The necessity of such embankments usually arises from the gradual elevation of the bed of running streams in consequence of the deposit of the earth and gravel they are charged with in high water; and, as we have seen, this elevation is rapidly accelerated when the highlands around the headwaters of rivers are cleared of their forests. When a river is embanked at a given point, and, consequently, the water of its floods, which would otherwise spread over a wide surface, is confined within narrow limits, the velocity of the current and its transporting power are augmented, and its burden of sand and gravel is deposited at some lower point, where the rapidity of its flow is checked by a diminution in the inclination of the bed, by a wider channel, or finally by a lacustrine or marine basin which receives its waters. Wherever it lets fall solid material, its channel is raised in consequence, and the declivity of the whole bed between the head of the embankment and the slack of the stream is reduced. Hence the current, at first accelerated by confinement, is afterward checked by the mechanical resistance of the matter deposited, and by the diminished inclination of its channel, and then begins again to let fall the earth it holds in suspension, and to raise its bed at the point where its overflow had been before prevented by embankment. The bank must now be raised in proportion, and these processes would be repeated and repeated indefinitely, had not nature provided a remedy in floods, which sweep out recent deposits, burst the bonds of the river and overwhelm the adjacent country with final desolation, or divert the current into a new channel, destined to become, in its turn, the scene of a similar struggle between man and the waters.

Few rivers, like the Nile, more than compensate by the fertilizing properties of their water and their slime for the damage they may do in inundations, and, consequently, there are few whose floods are not an object of dread, few whose encroachments upon their banks are not a source of constant anxiety and expense to the proprietors of the lands through which they flow. River dikes, for confining the spread of currents at high water, are of great antiquity in the East, and those of the Po and its tributaries were begun before we have any trustworthy physical or political annals of the provinces upon their borders. From the earliest ages, the Italian hydraulic engineers have stood in the front rank of their profession, and the Italian literature of this branch of material improvement is exceedingly voluminous. But the countries for which I write have no rivers like the Po, no plains like those of Lombardy, and the dangers to which the inhabitants of English and American river banks are exposed are more nearly analogous to those that threaten the soil and population in the valleys and plains of France, than to the perils and losses of the Lombard. The writings of the Italian hydrographers, too, though rich in professional instruction, are less accessible to foreigners and less adapted to popular use than those of French engineers.* For these reasons I shall take my citations principally from French authorities, though I shall occasionally allude to Italian writers on the floods of the Tiber, of the Arno, and some other Italian streams which much resemble those of the rivers of England and the United States.

b. *Floods of the Ardèche.*

The floods of mountain streams are attended with greater immediate danger to life and property than those of rivers of less rapid flow, because their currents are more impetuous, and they rise more suddenly and with less previous warning. At the same time, their ravages are confined within narrower limits, the waters retire sooner to their accustomed channel, and the danger is more quickly over, than in the case of inundations of larger rivers. The Ardèche, which has given its name to a department in France, drains a basin of 600,238 acres, or a little less than nine hundred and thirty-eight square miles. Its remotest source is about seventy-five miles, in a straight line, from its junction with the Rhone, and springs at an elevation of four thousand feet above that point. At the lowest stage of the river, the bed of the Chassezac, its largest and longest tributary, is in many places completely dry on the

* The very valuable memoirs of Lombardini, *Cenni idrografi sulla Lombardia*, *Intorno al sistema idraulico del Po*, and other papers on similar subjects, were published in periodicals little known out of Italy; and the *Idraulica Pratica* of Mari has not, I believe, been translated into French or English. These works, and other sources of information equally inaccessible out of Italy, have been freely used by Baumgarten, in a memoir entitled *Notice sur les Rivières de la Lombardie*, in the *Annales des Ponts et Chaussées*, 1847, 1er sémestre, pp. 129 *et seqq.*, and by Dumont, *Des Travaux Publics dans leurs Rapports avec l'Agriculture*, note, viii, pp. 269 *et seqq.* For the convenience of my readers, I shall use these two articles instead of the original authorities on which they are founded.

surface—the water being sufficient only to supply the subterranean channels of infiltration—and the Ardèche itself is almost everywhere fordable, even below the mouth of the Chassezac. But in floods, the river has sometimes risen more than sixty feet at the Pont d'Arc, a natural arch of two hundred feet chord, which spans the stream below its junction with all its important affluents. At the height of the inundation of 1827, the quantity of water passing this point—after deducting thirty per cent. for material transported with the current and for irregularity of flow—was estimated at 8,845 cubic yards to the second, and between twelve at noon on the 10th of September of that year and ten o'clock the next morning, the water discharged through the passage in question amounted to more than 450,000,000 cubic yards. This quantity, distributed equally through the basin of the river, would cover its entire area to a depth of more than five inches.

The Ardèche rises so suddenly that, in the inundation of 1846, the women who were washing in the bed of the river had not time to save their linen, and barely escaped with their lives, though they instantly fled upon hearing the roar of the approaching flood. Its waters and those of its affluents fall almost as rapidly, for in less than twenty-four hours after the rain has ceased in the Cévennes, where it rises, the Ardèche returns within its ordinary channel, even at its junction with the Rhone. In the flood of 1772, the water at La Beaume de Ruoms, on the Beaume, a tributary of the Ardèche, rose thirty-five feet above low water, but the stream was again fordable on the evening of the same day. The inundation of 1827 was, in this respect, exceptional, for it continued three days, during which period the Ardèche poured into the Rhone 1,305,000,000 cubic yards of water.

The Nile delivers into the sea 101,000 cubic feet or 3,741 cubic yards per second, on an average of the whole year.*

* Sir John F. W. Herschel, citing Talabot as his authority, *Physical Geography* (24).

In an elaborate paper on "Irrigation," printed in the *United States Patent Report* for 1860, p. 169, it is stated that the volume of water poured

This is equal to 323,222,400 cubic yards per day. In a single day of flood, then, the Ardèche, a river too insignificant to be known except in the local topography of France, contributed to the Rhone once and a half, and for three consecutive days once and one third, as much as the average delivery of the Nile during the same periods, though the basin of the latter river contains 500,000 square miles of surface, or more than five hundred times as much as that of the former.

The average annual precipitation in the basin of the Ardèche is not greater than in many other parts of Europe, but excessive quantities of rain frequently fall in that valley in the autumn. On the 9th of October, 1827, there fell at Joyeuse, on the Beaume, no less than thirty-one inches between three o'clock in the morning and midnight. Such facts as this explain the extraordinary suddenness and violence of the floods of the Ardèche, and the basins of many other tributaries of the Rhone exhibit meteorological phenomena not less remarkable.* The inundation of the 10th September, 1857, was accompanied with a terrific hurricane, which passed along the

into the Mediterranean by the Nile in twenty-four hours, at low water, is 150,566,392,368 cubic mètres; at high water, 705,514,667,440 cubic mètres. Taking the mean of these two numbers, the average daily delivery of the Nile would be 428,081,059,808 cubic mètres, or more than 550,000,000,000 cubic yards. There is some enormous mistake, probably a typographical error, in this statement, which makes the delivery of the Nile seventeen hundred times as great as computed by Talabot, and many times more than any physical geographer has ever estimated the quantity supplied by all the rivers on the face of the globe.

* The Drac, a torrent emptying into the Isère a little below Grenoble, has discharged 5,200, the Isère, which receives it, 7,800 cubic yards, and the Durance an equal quantity, per second.—MONTLUISANT, *Note sur les Desséchements, etc.*, *Annales des Ponts et Chaussées*, 1833, 2me sémestre, p. 288.

The floods of some other French rivers scarcely fall behind those of the Rhone. The Loire, above Roanne, has a basin of 2,471 square miles, or about twice and a half the area of that of the Ardèche. In some of its inundations it has delivered above 9,500 cubic yards per second.—BELGRAND, *De l'Influence des Forêts, etc.*, *Annales des Ponts et Chaussées*, 1854, 1er sémestre, p. 15, note.

eastern slope of the high grounds where the Ardèche and several other western affluents of the Rhone take their rise. The wind tore up all the trees in its path, and the rushing torrents bore their trunks down to the larger streams, which again transported them to the Rhone in such rafts that one might almost have crossed that river by stepping from trunk to trunk.* The Rhone, therefore, is naturally subject to great and sudden inundations, and the same remark may be applied to most of the principal rivers of France, because the geographical character of all of them is approximately the same.

The height and violence of the inundations of most great rivers are determined by the degree in which the floods of the different tributaries are coincident in time. Were all the affluents of the Rhone to pour their highest annual floods into its channel at once, were a dozen Niles to empty themselves into its bed at the same moment, its water would rise to a height and rush with an impetus that would sweep into the Mediterranean the entire population of its banks, and all the works that man has erected upon the plains which border it. But such a coincidence can never happen. The tributaries of this river run in very different directions, and some of them are swollen principally by the melting of the snows about their sources, others almost exclusively by heavy rains. When a damp southeast wind blows up the valley of the Ardèche, its moisture is condensed, and precipitated in a deluge upon the mountains which embosom the headwaters of that stream, thus producing a flood, while a neighboring basin, the axis of

* The original forests in which the basin of the Ardèche was rich have been rapidly disappearing, for many years, and the terrific violence of the inundations which are now laying it waste is ascribed, by the ablest investigators, to that cause. In an article inserted in the *Annales Forestières* for 1843, quoted by Hohenstein, *Der Wald*, p. 177, it is said that about one third of the area of the department had already become absolutely barren, in consequence of clearing, and that the destruction of the woods was still going on with great rapidity. New torrents were constantly forming, and they were estimated to have covered more than 70,000 acres of good land, or one eighth of the surface of the department, with sand and gravel.

which lies transversely or obliquely to that of the Ardèche, is not at all affected.*

It is easy to see that the damage occasioned by such floods as I have described must be almost incalculable, and it is by no means confined to the effects produced by overflow and the mechanical force of the superficial currents. In treating of the devastations of torrents in a former chapter, I confined myself principally to the erosion of surface and the transportation of mineral matter to lower grounds by them. The general action of torrents, as there shown, tends to the ultimate elevation of their beds by the deposit of the earth, gravel, and stone conveyed by them; but until they have thus raised their outlets so as sensibly to diminish the inclination of their channels—and sometimes when extraordinary floods give the torrents momentum enough to sweep away the accumulations which they have themselves heaped up—the swift flow of their currents, aided by the abrasion of the rolling rocks and gravel, scoops their beds constantly deeper, and they consequently not only undermine their banks, but frequently sap the most solid foundations which the art of man can build for the support of bridges and hydraulic structures.†

* "There is no example of a coincidence between great floods of the Ardèche and of the Rhone, all the known inundations of the former having taken place when the latter was very low."—MARDIGNY, *Mémoire sur les Inondations des Rivières de l'Ardèche*, p. 26.

I take this occasion to acknowledge myself indebted to the interesting memoir just quoted for all the statements I make respecting the floods of the Ardèche, except the comparison of the volume of its waters with that of the Nile, and the computation with respect to the capacity required for reservoirs to be constructed in its basin.

† In some cases where the bed of rapid Alpine streams is composed of very hard rock—as is the case in many of the valleys once filled by ancient glaciers—and especially where they are fed by glaciers not overhung by crumbling cliffs, the channel may remain almost unchanged for centuries. This is observable in many of the tributaries of the Dora Baltea, which drains the valley of the Aosta. Several of these small rivers are spanned by more or less perfect Roman bridges—one of which, that over the Lys at Pont St. Martin, is still in good repair and in constant use, An examination

In the inundation of 1857, the Ardèche destroyed a stone bridge near La Beaume, which had been built about eighty years before. The resistance of the piers, which were erected on piles, the channel at that point being of gravel, produced an eddying current that washed away the bed of the river above them, and the foundation, thus deprived of lateral support, yielded to the weight of the bridge, and the piles and piers fell up stream.

By a curious law of compensation, the stream which, at flood, scoops out cavities in its bed, often fills them up again as soon as the diminished velocity of the current allows it to let fall the sand and gravel with which it is charged, so that when the waters return to their usual channel, the bottom shows no sign of having been disturbed. In a flood of the Escontay, a tributary of the Rhone, in 1846, piles driven sixteen feet into its gravelly bed for the foundation of a pier were torn up and carried off, and yet, when the river had fallen to low-water mark, the bottom at that point appeared to have been raised higher than it was before the flood, by new deposits of sand and gravel, while the cut stones of the half-built pier were found buried to a great depth in the excavation which the water had first washed out. The gravel with which rivers thus restore the level of their beds is principally derived from the crushing of the rocks brought down by the mountain torrents, and the destructive effects of inundations are immensely diminished by this reduction of large stones to minute fragments. If the blocks hurled down from the cliffs were transported unbroken to the channels of large rivers, the mechanical force of their movement would be irresistible. They would overthrow the strongest barriers, spread themselves

of the rocks on which the abutments of this and some other similar structures are founded, and of the channels of the rivers they cross, shows that the beds of the streams cannot have been much elevated or depressed since the bridges were built. In other cases, as at the outlet of the Val Tournanche at Chatillon, where a single rib of a Roman bridge still remains, there is nothing to forbid the supposition that the deep excavation of the channel may have been partly effected at a much later period. See *App.*, No. 45.

over a surface as wide as the flow of the waters, and convert the most smiling valleys into scenes of the wildest desolation.

c. *Crushing Force of Torrents.*

There are few operations of nature where the effect seems more disproportioned to the cause than in the comminution of rock in the channel of swift waters. Igneous rocks are generally so hard as to be wrought with great difficulty, and they bear the weight of enormous superstructures without yielding to the pressure; but to the torrent they are as wheat to the millstone. The streams which pour down the southern scarp of the Mediterranean Alps along the Riviera di Ponente, near Genoa, have short courses, and a brisk walk of a couple of hours or even less takes you from the sea beach to the headspring of many of them. In their heaviest floods, they bring rounded masses of serpentine quite down to the sea, but at ordinary high water their lower course is charged only with finely divided particles of that rock. Hence, while, near their sources, their channels are filled with pebbles and angular fragments, intermixed with a little gravel, the proportions are reversed near their mouths, and, just above the points where their outlets are partially choked by the rolling shingle of the beach, their beds are composed of sand and gravel to the almost total exclusion of pebbles. The greatest depth of the basin of the Ardèche is seventy-five miles, but most of its tributaries have a much shorter course. "These affluents," says Mardigny, "hurl into the bed of the Ardèche enormous blocks of rock, which this river, in its turn, bears onward, and grinds down, at high water, so that its current rolls only gravel at its confluence with the Rhone." *

* *Mémoire sur les Inondations des Rivières de l'Ardèche*, p. 16. "The terrific roar, the thunder of the raging torrents proceeds principally from the stones which are rolled along in the bed of the stream. This movement is attended with such powerful attrition that, in the Southern Alps, the atmosphere of valleys where the limestone contains bitumen, has, at the time of floods, the marked bituminous smell produced by rubbing

Guglielmini argued that the gravel and sand of the beds of running streams were derived from the trituration of rocks by the action of the currents, and inferred that this action was generally sufficient to reduce hard rock to sand in its passage from the source to the outlet of rivers. Frisi controverted this opinion, and maintained that river sand was of more ancient origin, and he inferred from experiments in artificially grinding stones that the concussion, friction, and attrition of rock in the channel of running waters were inadequate to its comminution, though he admitted that these same causes might reduce silicious sand to a fine powder capable of transportation to the sea by the currents.* Frisi's experiments were tried upon rounded and polished river pebbles, and prove nothing with regard to the action of torrents upon the irregular, more or less weathered, and often cracked and shattered rocks which lie loose in the ground at the head of mountain valleys. The fury of the waters and of the wind which accompanies them in the floods of the French Alpine torrents is such, that large blocks of stone are hurled out of the bed of the stream to the height of twelve or thirteen feet. The impulse of masses driven with such force overthrows the most solid masonry, and their concussion cannot fail to be attended with the crushing of the rocks themselves.†

d. *Inundations of* 1856 *in France.*

The month of May, 1856, was remarkable for violent and almost uninterrupted rains, and most of the river basins of France were inundated to an extraordinary height. In the valleys of the Loire and its affluents, about a million of acres, including many towns and villages, were laid under water, and the amount of pecuniary damage was almost incalculable.‡ The flood was not less destructive in the valley of

pieces of such limestone together."—WESSELY, *Die Oesterreichischen Alpenländer*, i, p. 113. See *Appendix*, No. 46.

* FRISI, *Del modo di regolare i Fiumi e i Torrenti*, pp. 4–19.

† SURELL, *Étude sur les Torrents*, pp. 31–36.

‡ CHAMPION, *Les Inondations en France*, iii, p. 156, note.

the Rhone, and in fact an invasion by a hostile army could hardly have been more disastrous to the inhabitants of the plains than was this terrible deluge. There had been a flood of this latter river in the year 1840, which, for height and quantity of water, was almost as remarkable as that of 1856, but it took place in the month of November, when the crops had all been harvested, and the injury inflicted by it upon agriculturists was, therefore, of a character to be less severely and less immediately felt than the consequences of the inundation of 1856.*

In the fifteen years between these two great floods, the population and the rural improvements of the river valleys had much increased, common roads, bridges, and railways had been multiplied and extended, telegraph lines had been constructed, all of which shared in the general ruin, and hence greater and more diversified interests were affected by the catastrophe of 1856 than by any former like calamity. The great flood of 1840 had excited the attention and roused the sympathies of the French people, and the subject was invested with new interest by the still more formidable character of the inundations of 1856. It was felt that these scourges had ceased to be a matter of merely local concern, for, although they bore most heavily on those whose homes and fields were situated within the immediate reach of the swelling waters, yet they frequently destroyed harvests valuable enough to be a matter of national interest, endangered the personal security of the population of important political centres, interrupted communication for days and even weeks together on great lines of

* Notwithstanding this favorable circumstance, the damage done by the inundation of 1840 in the valley of the Rhone was estimated at seventy-two millions of francs.—CHAMPION, *Les Inondations en France*, iv, p. 124.

Several smaller floods of the Rhone, experienced at a somewhat earlier season of the year in 1846, occasioned a loss of forty-five millions of francs. "What if," says Dumont, "instead of happening in October, that is between harvest and seedtime, they had occurred before the crops were secured? The damage would have been counted by hundreds of millions." —*Des Travaux Publics*, p. 99, note.

traffic and travel—thus severing as it were all Southwestern France from the rest of the empire—and finally threatened to produce great and permanent geographical changes. The well-being of the whole commonwealth was seen to be involved in preventing the recurrence, and in limiting the range of such devastations. The Government encouraged scientific investigation of the phenomena and their laws. Their causes, their history, their immediate and remote consequences, and the possible safeguards to be employed against them, have been carefully studied by the most eminent physicists, as well as by the ablest theoretical and practical engineers of France. Many hitherto unobserved facts have been collected, many new hypotheses suggested, and many plans, more or less original in character, have been devised for combating the evil; but thus far, the most competent judges are not well agreed as to the mode, or even the possibility, of applying a remedy.

e. *Remedies against Inundations.*

Perhaps no one point has been more prominent in the discussions than the influence of the forest in equalizing and regulating the flow of the water of precipitation. As we have already seen, opinion is still somewhat divided on this subject, but the conservative action of the woods in this respect has been generally recognized by the public of France, and the Government of the empire has made this principle the basis of important legislation for the protection of existing forests, and for the formation of new. The clearing of woodland, and the organization and functions of a police for its protection, are regulated by a law bearing date June 18th, 1859, and provision was made for promoting the restoration of private woods by a statute adopted on the 28th of July, 1860. The former of these laws passed the legislative body by a vote of 246 against 4, the latter with but a single negative voice. The influence of the government, in a country where the throne is so potent as in France, would account for a large majority, but when it is considered that both laws, the former especially, interfere very materially with the rights of private domain, the almost entire unanimity with which they were adopted is proof of a very general popular conviction, that the protection and extension of the forests is a measure more likely than any other to check the violence, if not to prevent the recurrence, of destructive inundations. The law of July 28th, 1860, appropriated 10,000,000 francs, to be expended, at the rate of 1,000,000 francs per year, in executing or aiding the replanting of woods. It is computed that this appropriation will secure the creation of new forest to the extent of about 250,000 acres, or one eleventh part of the soil where the restoration of the forest is thought feasible and, at the same time, specially important as a security against the evils ascribed in a great measure to its destruction.

The provisions of the laws in question are preventive rather than remedial; but some immediate effect may be expected to result from them, particularly if they are accompanied with certain other measures, the suggestion of which has been favorably received. The strong repugnance of the mountaineers to the application of a system which deprives them of a part of their pasturage—for the absolute exclusion of domestic animals is indispensable to the maintenance of an existing forest and to the formation of a new—is the most formidable obstacle to the execution of the laws of 1859–'60. It is proposed to compensate this loss by a cheap system of irrigation of lower pasture grounds, consisting in little more than in running horizontal furrows along the hillsides, thus converting the scarp of the hills into a succession of small terraces which, when once turfed over, are very permanent. Experience is said to have demonstrated that this simple process suffices to retain the water of rains, of snows, and of small springs and rivulets, long enough for the irrigation of the soil, thus increasing its product of herbage in a fivefold proportion, and that it partially checks the too rapid flow of surface water into the valleys, and, consequently, in some measure obviates one of the most prominent causes of inundations.* It is evi-

* Troy, *Étude sur le Reboisement des Montagnes*, §§ 6, 7, 21.

dent that if such results are produced by this method, its introduction upon an extensive scale must also have the same climatic effects as other systems of irrigation.

Whatever may be the ultimate advantages of reclothing a large extent of the territory of France with wood, or of so shaping its surface as to prevent the too rapid flow of water over it, the results to be obtained by such processes can be realized in an adequate measure only after a long succession of years. Other steps must be taken, both for the immediate security of the lives and property of the present generation, and for the prevention of yet greater and remoter evils which are inevitable unless means to obviate them are found before it is forever too late. The frequent recurrence of inundations like those of 1856, for a single score of years, in the basins of the Rhone and the Loire, with only the present securities against them, would almost depopulate the valleys of those rivers, and produce physical revolutions in them, which, like revolutions in the political world, could never be made to "go backward."

Destructive inundations are seldom, if ever, produced by precipitation within the limits of the principal valley, but almost uniformly by sudden thaws or excessive rains on the mountain ranges where the tributaries take their rise. It is therefore plain that any measures which shall check the flow of surface waters into the channels of the affluents, or which shall retard the delivery of such waters into the principal stream by its tributaries, will diminish in the same proportion the dangers and the evils of inundation by great rivers. The retention of the surface waters upon or in the soil can hardly be accomplished except by the methods already mentioned, replanting of forests, and furrowing or terracing. The current of mountain streams can be checked by various methods, among which the most familiar and obvious is the erection of barriers or dams across their channels, at points convenient for forming reservoirs large enough to retain the superfluous waters of great rains and thaws. Besides the utility of such basins in preventing floods, the construction of them is recommended by very strong considerations, such as the meteorological effects of increased evaporable surface, the furnishing of a constant supply of water for agricultural and mechanical purposes, and, finally, their value as ponds for breeding and rearing fish, and, perhaps, for cultivating aquatic vegetables.

The objections to the general adoption of the system of reservoirs are these: the expense of their construction and maintenance; the reduction of cultivable area by the amount of surface they must cover; the interruption they would occasion to free communication; the probability that they would soon be filled up with sediment, and the obvious fact that when full of earth or even water, they would no longer serve their principal purpose; the great danger to which they would expose the country below them in case of the bursting of their barriers; * the evil consequences they would occasion by prolonging the flow of inundations in proportion as they diminished their height; the injurious effects it is supposed they would produce upon the salubrity of the neighboring districts; and, lastly, the alleged impossibility of constructing artificial basins sufficient in capacity to prevent, or in any considerable measure to mitigate, the evils they are intended to guard against.

The last argument is more easily reduced to a numerical question than the others. The mean and extreme annual precipitation of all the basins where the construction of such works would be seriously proposed is already approximately known by meteorological tables, and the quantity of water, delivered by the greatest floods which have occurred within the memory of man, may be roughly estimated from their visible traces. From these elements, or from recorded observations, the capacity of the necessary reservoirs can be calculated. Let us take the case of the Ardèche. In the inundation of 1857, that river poured into the Rhone 1,305,000,000 cubic yards of water in three days. If we suppose that half

* For accounts of damage from the bursting of reservoirs, see VALLÉE, *Mémoire sur les Reservoirs d'Alimentation des Canaux*, *Annales des Ponts et Chaussées*, 1833, 1er sémestre, p. 261.

this quantity might have been suffered to flow down its channel without inconvenience, we shall have about 650,000,000 cubic yards to provide for by reservoirs. The Ardèche and its principal affluent, the Chassezac, have, together, about twelve considerable tributaries rising near the crest of the mountains which bound the basin. If reservoirs of equal capacity were constructed upon all of them, each reservoir must be able to contain 54,000,000 cubic yards, or, in other words, must be equal to a lake 3,000 yards long, 1,000 yards wide, and 18 yards deep, and besides, in order to render any effectual service, the reservoirs must all have been empty at the commencement of the rains which produced the inundation.

Thus far, I have supposed the swelling of the waters to be uniform throughout the whole basin; but such was by no means the fact in the inundation of 1857, for the rise of the Chassezac, which is as large as the Ardèche proper, did not exceed the limits of ordinary floods, and the dangerous excess came solely from the headwaters of the latter stream. Hence reservoirs of double the capacity I have supposed would have been necessary upon the tributaries of that river, to prevent the injurious effects of the inundation. It is evident that the construction of reservoirs of such magnitude for such a purpose is financially, if not physically, impracticable, and when we take into account a point I have just suggested, namely, that the reservoirs must be empty at all times of apprehended flood, and, of course, their utility limited almost solely to the single object of preventing inundations, the total inapplicability of such a measure in this particular case becomes still more glaringly manifest.

Another not less conclusive fact is that the valleys of all the upland tributaries of the Ardèche descend so rapidly, and have so little lateral expansion, as to render the construction of capacious reservoirs in them quite impracticable. Indeed, engineers have found but two points in the whole basin suitable for that purpose, and the reservoirs admissible at these would have only a joint capacity of about 70,000,000 cubic yards, or less than one ninth part of what I suppose to be required. The case of the Ardèche is no doubt an extreme one, both in the topographical character of its basin and in its exposure to excessive rains; but all destructive inundations are, in a certain sense, extreme cases also, and this of the Ardèche serves to show that the construction of reservoirs is not by any means to be regarded as a universal panacea against floods.

Nor, on the other hand, is this measure to be summarily rejected. Nature has adopted it on a great scale, on both flanks of the Alps, and on a smaller, on those of the Adirondacks and lower chains, and in this as in many other instances, her processes may often be imitated with advantage. The validity of the remaining objections to the system under discussion depends on the topography, geology, and special climate of the regions where it is proposed to establish such reservoirs. Many upland streams present numerous points where none of these objections, except those of expense and of danger from the breaking of dams, could have any application. Reservoirs may be so constructed as to retain the entire precipitation of the heaviest thaws and rains, leaving only the ordinary quantity to flow along the channel; they may be raised to such a height as only partially to obstruct the surface drainage; or they may be provided with sluices by means of which their whole contents can be discharged in the dry season and a summer crop be grown upon the ground they cover at high water. The expediency of employing them and the mode of construction depend on local conditions, and no rules of universal applicability can be laid down on the subject.

It is remarkable that nations which we, in the false pride of our modern civilization, so generally regard as little less than barbarian, should have long preceded Christian Europe in the systematic employment of great artificial basins for the various purposes they are calculated to subserve. The ancient Peruvians built strong walls, of excellent workmanship, across the channels of the mountain sources of important streams, and the Arabs executed immense works of similar description, both in the great Arabian peninsula and in all the provinces of Spain which had the good fortune to fall under their sway. The Spaniards of the fifteenth and sixteenth centuries, who, in many points of true civilization and culture, were far inferior to the races they subdued, wantonly destroyed these noble monuments of social and political wisdom, or suffered them to perish, because they were too ignorant to appreciate their value, or too unskilful as practical engineers to be able to maintain them, and some of their most important territories were soon reduced to sterility and poverty in consequence.

Another method of preventing or diminishing the evils of inundation by torrents and mountain rivers, analogous to that employed for the drainage of lakes, consists in the permanent or occasional diversion of their surplus waters, or of their entire currents, from their natural courses, by tunnels or open channels cut through their banks. Nature, in many cases, resorts to a similar process. Most great rivers divide themselves into several arms in their lower course, and enter the sea by different mouths. There are also cases where rivers send off lateral branches to convey a part of their waters into the channel of other streams.* The most remarkable of these is the junction between the Amazon and the Orinoco by the natural canal of the Cassiquiare and the Rio Negro. In India, the Cambodja and the Menam are connected by the Anam; the Saluen and the Irawaddi by the Panlaun. There are similar examples, though on a much smaller scale, in Europe. The Torneå and the Calix rivers in Lapland communicate by the Tarando, and in Westphalia, the Else, an arm of the Haase, falls into the Weser.

The change of bed in rivers by gradual erosion of their banks is familiar to all, but instances of the sudden abandonment of a primitive channel are by no means wanting. At a

* Some geographical writers apply the term *bifurcation* exclusively to this intercommunication of rivers; others, with more etymological propriety, use it to express the division of great rivers into branches at the head of their deltas. A technical term is wanting to designate the phenomenon mentioned in the text.

26

period of unknown antiquity, the Ardèche pierced a tunnel 200 feet wide and 100 high, through a rock, and sent its whole current through it, deserting its former bed, which gradually filled up, though its course remained traceable. In the great inundation of 1827, the tunnel proved insufficient for the discharge of the water, and the river burst through the obstructions which had now choked up its ancient channel, and resumed its original course.*

It was probably such facts as these that suggested to ancient engineers the possibility of like artificial operations, and there are numerous instances of the execution of works for this purpose in very remote ages. The Bahr Jusef, the great stream which supplies the Fayoum with water from the Nile, has been supposed, by some writers, to be a natural channel; but both it and the Bahr el Wady are almost certainly artificial canals constructed to water that basin, to regulate the level of Lake Moeris, and possibly, also, to diminish the dangers resulting from excessive inundations of the Nile, by serving as waste-weirs to discharge a part of its surplus waters. Several of the seven ancient mouths of the Nile are believed to be artificial channels, and Herodotus even asserts that King Menes diverted the entire course of that river from the Libyan to the Arabian side of the valley. There are traces of an ancient river bed along the western mountains, which give some countenance to this statement. But it is much more probable that the works of Menes were designed rather to prevent a natural, than to produce an artificial, change in the channel of the river.

Two of the most celebrated cascades in Europe, those of the Teverone at Tivoli and of the Velino at Terni, owe, if not their existence, at least their position and character, to the diversion of their waters from their natural beds into new channels, in order to obviate the evils produced by their frequent floods. Remarkable works of the same sort have been executed in Switzerland, in very recent times. Until the year

* Mardigny, *Mémoire sur les Inondations de l'Ardèche*, p. 13.

1714, the Kander, which drains several large Alpine valleys, ran, for a considerable distance, parallel with the Lake of Thun, and a few miles below the city of that name emptied into the river Aar. It frequently flooded the flats along the lower part of its course, and it was determined to divert it into the Lake of Thun. For this purpose, two parallel tunnels were cut through the intervening rock, and the river turned into them. The violence of the current burst up the roof of the tunnels, and, in a very short time, wore the new channel down not less than one hundred feet, and even deepened the former bed at least fifty feet, for a distance of two or three miles above the tunnel. The lake was two hundred feet deep at the point where the river was conducted into it, but the gravel and sand carried down by the Kander has formed at its mouth a delta containing more than a hundred acres, which is still advancing at the rate of several yards a year. The Linth, which formerly sent its waters directly to the Lake of Zurich, and often produced very destructive inundations, was turned into the Wallensee about forty years ago, and in both these cases a great quantity of valuable land was rescued both from flood and from insalubrity.

In Switzerland, the most terrible inundations often result from the damming up of deep valleys by ice slips or by the gradual advance of glaciers, and the accumulation of great masses of water above the obstructions. The ice is finally dissolved by the heat of summer or the flow of warm waters, and when it bursts, the lake formed above is discharged almost in an instant, and all below is swept down to certain destruction. In 1595, about a hundred and fifty lives and a great amount of property were lost by the eruption of a lake formed by the descent of a glacier into the valley of the Drance, and a similar calamity laid waste a considerable extent of soil in the year 1818. On this latter occasion, the barrier of ice and snow was 3,000 feet long, 600 thick, and 400 high, and the lake which had formed above it contained not less than 800,000,000 cubic feet. A tunnel was driven through the ice, and about 300,000,000 cubic feet of water safely drawn off by it, but the thawing of the walls of the tunnel rapidly enlarged it, and before the lake was half drained, the barrier gave way and the remaining 500,000,000 cubic feet of water were discharged in half an hour. The recurrence of these floods has since been prevented by directing streams of water, warmed by the sun, upon the ice in the bed of the valley, and thus thawing it before it accumulates in sufficient mass to threaten serious danger.

In the cases of diversion of streams above mentioned, important geographical changes have been directly produced by those operations. By the rarer process of draining glacier lakes, natural eruptions of water, which would have occasioned not less important changes in the face of the earth, have been prevented by human agency.

The principal means hitherto relied upon for defence against river inundations has been the construction of dikes along the banks of the streams, parallel to the channel and generally separated from each other by a distance not much greater than the natural width of the bed.* If such walls are high enough to confine the water and strong enough to resist its pressure, they secure the lands behind them from all the evils of inundation except those resulting from infiltration; but such ramparts are enormously costly in original construction and maintenance, and, as we have already seen, the filling up of the bed of the river in its lower course, by sand and gravel, involves the necessity of occasionally incurring new expenditures in increasing the height of the banks.† They

* In the case of rivers flowing through wide alluvial plains and much inclined to shift their beds, like the Po, the embankments often leave a very wide space between them. The dikes of the Po are sometimes three or four miles apart.—Baumgarten, after Lombardini, *Annales des Ponts et Chaussées*, 1847, 1er sémestre, p. 149.

† It appears from the investigations of Lombardini that the rate of elevation of the bed of the Po has been much exaggerated by earlier writers, and in some parts of its course the change is so slow that its level may be regarded as nearly constant.—Baumgarten, volume before cited, pp. 175, *et seqq.* See *Appendix*, No. 47.

If the western coast of the Adriatic is undergoing a secular depression,

are attended, too, with some collateral disadvantages. They deprive the earth of the fertilizing deposits of the waters, which are powerful natural restoratives of soils exhausted by cultivation; they accelerate the rapidity and transporting power of the current at high water by confining it to a narrower channel, and it consequently conveys to the sea the earthy matter it holds in suspension, and chokes up harbors with a deposit which it would otherwise have spread over a wide surface; they interfere with roads and the convenience of river navigation, and no amount of cost or care can secure them from occasional rupture, in case of which the rush of the waters through the breach is more destructive than the natural flow of the highest inundation.*

as many circumstances concur to prove, the sinking of the plain near the coast may both tend to prevent the deposit of sediment in the river bed by increasing the velocity of its current, and compensate the elevation really produced by deposits, so that no sensible elevation would result, though much gravel and slime might be let fall.

* To secure the city of Sacramento in California from the inundations to which it is subject, a dike or levée was built upon the bank of the river and raised to an elevation above that of the highest known floods, and it was connected, below the town, with grounds lying considerably above the river. On one occasion a breach in the dike occurred above the town at a very high stage of the flood. The water poured in behind it, and overflowed the lower part of the city, which remained submerged for some time after the river had retired to its ordinary level, because the dike, which had been built to keep the water *out*, now kept it *in*.

According to Arthur Young, on the lower Po, where the surface of the river has been elevated much above the level of the adjacent fields by diking, the peasants in his time frequently endeavored to secure their grounds against threatened devastation through the bursting of the dikes, by crossing the river when the danger became imminent and opening a cut in the opposite bank, thus saving their own property by flooding their neighbors'. He adds, that at high water the navigation of the river was absolutely interdicted, except to mail and passenger boats, and that the guards fired upon all others; the object of the prohibition being to prevent the peasants from resorting to this measure of self-defence.—*Travels in Italy and Spain*, Nov. 7, 1789.

In a flood of the Po in 1839, a breach of the embankment took place at Bonizzo. The water poured through and inundated 116,000 acres, or 181

For these reasons, many experienced engineers are of opinion that the system of longitudinal dikes ought to be abandoned, or, where that cannot be done without involving too great a sacrifice of existing constructions, their elevation should be much reduced, so as to present no obstruction to the lateral spread of extraordinary floods, and they should be provided with sluices to admit the water without violence whenever they are likely to be overflowed. Where dikes have not been erected, and where they have been reduced in height, it is proposed to construct, at convenient intervals, transverse embankments of moderate height running from the banks of the river across the plains to the hills which bound them. These measures, it is argued, will diminish the violence of inundations by permitting the waters to extend themselves over a greater surface and thus retarding the flow of the river currents, and will, at the same time, secure the deposit of fertilizing slime upon all the soil covered by the flood.

Rozet, an eminent French engineer, has proposed a method of diminishing the ravages of inundations, which aims to combine the advantages of all other systems, and at the same time to obviate the objections to which they are all more or less liable.* The plan of Rozet is recommended by its simplicity and cheapness as well as its facility and rapidity of execution, and is looked upon with favor by many persons very competent to judge in such matters. He proposes to commence with the amphitheatres in which mountain torrents so often rise, by covering their slopes and filling their beds with loose blocks of rock, and by constructing at their outlets, and at other narrow points in the channels of the torrents, permeable barriers of the same material promiscuously heaped up, much according to the method employed by the ancient Romans in their northern provinces for a similar purpose. By this means, he

square miles, of the plain, to the depth of from twenty to twenty-three feet in its lower parts.—Baumgarten, after Lombardini, volume before cited, p. 152.

* Moyens *de forcer les Torrents de rendre une partie du sol qu'ils ravagent, et d'empêcher les grandes Inondations.*

supposes, the rapidity of the current would be checked, and the quantity of transported pebbles and gravel much diminished.

When the stream has reached that part of its course where it is bordered by soil capable of cultivation, and worth the expense of protection, he proposes to place along one or both sides of the stream, according to circumstances, a line of cubical blocks of stone or pillars of masonry three or four feet high and wide, and at the distance of about eleven yards from each other. The space between the two lines, or between a line and the opposite high bank, would, of course, be determined by observation of the width of the swift-water current at high floods. As an auxiliary measure, small ditches and banks, or low walls of pebbles, should be constructed from the line of blocks across the grounds to be protected, nearly at right angles to the current, but slightly inclining downward, and at convenient distances from each other. Rozet thinks the proper interval would be 300 yards, and it is evident that, if he is right in his main principle, hedges, rows of trees, or even common fences, would in many cases answer as good a purpose as banks and trenches or low walls. The blocks or pillars of stone would, he contends, check the lateral currents so as to compel them to let fall all their pebbles and gravel in the main channel—where they would be rolled along until ground down to sand or silt—and the transverse obstructions would detain the water upon the soil long enough to secure the deposit of its fertilizing slime. Numerous facts are cited in support of the author's views, and I imagine there are few residents of rural districts whose own observation will not furnish testimony confirmatory of their soundness.*

* The effect of trees and other detached obstructions in checking the flow of water is particularly noticed by Palissy in his essay on *Waters and Fountains*, p. 173, edition of 1844. "There be," says he, "in divers parts of France, and specially at Nantes, wooden bridges, where, to break the force of the waters and of the floating ice, which might endamage the piers of the said bridges, they have driven upright timbers into the bed of the rivers above the said piers, without the which they should abide but little. And in like wise, the trees which be planted along the mountains do much deaden the violence of the waters that flow from them."

The deposit of slime by rivers upon the flats along their banks not only contributes greatly to the fertility of the soil thus flowed, but it subserves a still more important purpose in the general economy of nature. All running streams begin with excavating channels for themselves, or deepening the natural depressions in which they flow; * but in proportion as their outlets are raised by the solid material transported by their currents, their velocity is diminished, they deposit gravel and sand at constantly higher and higher points, and so at last elevate, in the middle and lower part of their course, the beds they had previously scooped out.† The raising of the chan-

* I do not mean to say that all rivers excavate their own valleys, for I have no doubt that in the majority of cases such depressions of the surface originate in higher geological causes, and hence the valley makes the river, not the river the valley. But even if we suppose a basin of the hardest rock to be elevated at once, completely formed, from the submarine abyss where it was fashioned, the first shower of rain that falls upon it after it rises to the air, while its waters will follow the lowest lines of the surface, will cut those lines deeper, and so on with every successive rain. The disintegrated rock from the upper part of the basin forms the lower by alluvial deposit, which is constantly transported farther and farther until the resistance of gravitation and cohesion balances the mechanical force of the running water. Thus plains, more or less steeply inclined, are formed, in which the river is constantly changing its bed, according to the perpetually varying force and direction of its currents, modified as they are by ever-fluctuating conditions. Thus the Po is said to have long inclined to move its channel southward in consequence of the superior mechanical force of its northern affluents. A diversion of these tributaries from their present beds, so that they should enter the main stream at other points and in different directions, might modify the whole course of that great river. But the mechanical force of the tributary is not the only element of its influence on the course of the principal stream. The deposits it lodges in the bed of the latter, acting as simple obstructions or causes of diversion, are not less important agents of change.

† The distance to which a new obstruction to the flow of a river, whether by a dam or by a deposit in its channel, will retard its current, or, in popular phrase, "set back the water," is a problem of more difficult practical solution than almost any other in hydraulics. The elements—such as straightness or crookedness of channel, character of bottom and banks, volume and previous velocity of current, mass of water far above the obstruction, extraordinary drought or humidity of seasons, relative

nels is compensated in part by the simultaneous elevation of their banks and the flats adjoining them, from the deposit of the finer particles of earth and vegetable mould brought down from the mountains, without which elevation the low grounds bordering all rivers would be, as in many cases they in fact are, mere morasses.

All arrangements which tend to obstruct this process of raising the flats adjacent to the channel, whether consisting in dikes which confine the waters, and, at the same time, augment the velocity of the current, or in other means of producing the last-mentioned effect, interfere with the restorative economy of nature, and at last occasion the formation of marshes where, if left to herself, she would have accumulated

extent to which the river may be affected by the precipitation in its own basin, and by supplies received through subterranean channels from sources so distant as to be exposed to very different meteorological influences, effects of clearing and other improvements always going on in new countries—are all extremely difficult, and some of them impossible, to be known and measured. In the American States, very numerous watermills have been erected within a few years, and there is scarcely a stream in the settled portion of the country which has not several milldams upon it. When a dam is raised—a process which the gradual diminution of the summer currents renders frequently necessary—or when a new dam is built, it often happens that the meadows above are flowed, or that the retardation of the stream extends back to the dam next above. This leads to frequent lawsuits. From the great uncertainty of the facts, the testimony is more conflicting in these than in any other class of cases, and the obstinacy with which "water causes" are disputed has become proverbial.

The subterranean courses of the waters form a subject very difficult of investigation, and it is only recently that its vast importance has been recognized. The interesting observations of Schmidt on the caves of the Karst and their rivers throw much light on the underground hydrography of limestone districts, and serve to explain how, in the low peninsula of Florida, rivers, which must have their sources in mountains a hundred or more miles distant, can pour out of the earth in currents large enough to admit of steamboat navigation to their very basins of eruption. Artesian wells are revealing to us the existence of subterranean lakes and rivers sometimes superposed one above another in successive sheets; but the still more important subject of the absorption of water by earth and its transmission by infiltration is yet wrapped in great obscurity.

inexhaustible stores of the richest soil, and spread them out in plains above the reach of ordinary floods.*

Consequences if the Nile had been Diked.

If a system of continuous lateral dikes, like those of the Po, had been adopted in Egypt in the early dynasties, when the power and the will to undertake the most stupendous material enterprises were so eminently characteristic of the government of that country, and the waters of the annual inundation consequently prevented from flooding the land, it is conceivable that the productiveness of the small area of cultivable soil in the Nile valley might have been long kept up by artificial irrigation and the application of manures. But nature would have rebelled at last, and centuries before our time the mighty river would have burst the fetters by which impotent man had vainly striven to bind his swelling floods, the fertile fields of Egypt would have been converted into dank morasses, and then, perhaps, in some distant future, when the expulsion of man should have allowed the gradual restoration of the primitive equilibrium, would be again transformed into luxuriant garden and plough land. Fortunately, the "wisdom of Egypt" taught her children better things. They invited and welcomed, not repulsed, the slimy embraces of Nilus, and his favors have been, from the hoariest antiquity, the greatest material blessing ever bestowed upon a people.†

* The sediment of the Po has filled up some lagoons and swamps in its delta, and converted them into comparatively dry land; but, on the other hand, the retardation of the current from the lengthening of its course, and the diminution of its velocity by the deposits at its mouth, have forced its waters at some higher points to spread in spite of embankments, and thus fertile fields have been turned into unhealthy and unproductive marshes.—See BOTTER, *Sulla condizione dei Terreni Maremmani nel Ferrarese. Annali di Agricoltura, etc.*, Fasc. v, 1863.

† Deep borings have not detected any essential difference in the quantity or quality of the deposits of the Nile for forty or fifty, or, as some compute, for a hundred centuries. From what vast store of rich earth does this river derive the three or four inches of fertilizing material which

The valley of the Po has probably not been cultivated or inhabited so long as that of the Nile, but embankments have been employed on its lower course for at least two thousand years, and for many centuries they have been connected in a continuous chain. I have pointed out in a former chapter the effects produced on the geography of the Adriatic by the deposit of river sediment in the sea at the mouths of the Po, the Adige, and the Brenta. If these rivers had been left unconfined, like the Nile, and allowed to spread their muddy waters at will, according to the laws of nature, the slime they have carried to the coast would have been chiefly distributed over the plains of Lombardy. Their banks would have risen as fast as their beds, the coast line would not have been extended so far into the Adriatic, and, the current of the streams being consequently shorter, the inclination of their channel and the rapidity of their flow would not have been so greatly diminished. Had man spared a reasonable proportion of the forests of the Alps, and not attempted to control the natural drainage of the surface, the Po would resemble the Nile in all its essential characteristics, and, in spite of the difference of climate, perhaps be regarded as the friend and ally, not the enemy and the invader, of the population which dwells upon its banks.*

it spreads over the soil of Egypt every hundred years? Not from the White Nile, for that river drops nearly all its suspended matter in the broad expansions and slow current of its channel south of the tenth degree of north latitude. Nor does it appear that much sediment is contributed by the Bahr-el-Azrek, which flows through forests for a great part of its course. I have been informed by an old European resident of Egypt who is very familiar with the Upper Nile, that almost the whole of the earth with which its waters are charged is brought down by the Takazzé.

* It is very probably true that, as Lombardini supposes, the plain of Lombardy was anciently covered with forests and morasses (Baumgarten, l. c. p. 156); but, had the Po remained unconfined, its deposits would have raised its banks as fast as its bed, and there is no obvious reason why this plain should be more marshy than other alluvial flats traversed by great rivers. Its lower course would possibly have become more marshy than at present, but the banks of its middle and upper course would have been in a better condition for agricultural use than they now are.

The Nile is larger than all the rivers of Lombardy together,* it drains a basin twenty times as extensive, its banks have been occupied by man probably twice as long. But its geographical character has not been much changed in the whole period of recorded history, and, though its outlets have somewhat fluctuated in number and position, its historically known encroachments upon the sea are trifling compared with those of the Po and the neighboring streams. The deposits of the Nile are naturally greater in Upper than in Lower Egypt. They are found to have raised the soil at Thebes about seven feet within the last seventeen hundred years, and in the Delta the rise has been certainly more than half as great.

We shall, therefore, not exceed the truth if we suppose the annually inundated surface of Egypt to have been elevated, upon an average, ten feet, within the last 5,000 years, or twice and a half the period during which the history of the Po is known to us.†

We may estimate the present actually cultivated area of Egypt at about 5,500 square statute miles. As I have computed in a note on page 372, that area is not more than half as extensive as under the dynasties of the Pharaohs and the Ptolemies; for—though, in consequence of the elevation of the river bed, the inundations now have a wider *natural* spread—the industry of the ancient Egyptians conducted the

* From daily measurements during a period of fourteen years—1827 to 1840—the mean delivery of the Po at Ponte Lagoscuro, below the entrance of its last tributary, is found to be 1,720 cubic mètres, or 60,745 cubic feet, per second. Its smallest delivery is 186 cubic mètres, or 6,569 cubic feet, its greatest 5,156 cubic mètres, or 182,094 cubic feet.—Baumgarten, following Lombardini, volume before cited, p. 159.

The average delivery of the Nile being 101,000 cubic feet per second, it follows that the Po contributes to the Adriatic six tenths as much water as the Nile to the Mediterranean—a result which will surprise most readers.

† We are quite safe in supposing that the valley of the Nile has been occupied by man at least 5,000 years. The dates of Egyptian chronology are uncertain, but I believe no inquirer estimates the age of the great pyramids at less than forty centuries, and the construction of such works implies an already ancient civilization.

Nile water over a great extent of soil it does not now reach. We may, then, adopt a mean between the two quantities, and we shall probably come near the truth if we assume the convenient number of 7,920 square statute miles as the average measure of the inundated land during the historical period. Taking the deposit on this surface at ten feet, the river sediment let fall on the soil of Egypt within the last fifty centuries would amount to fifteen cubic miles.

Had the Nile been banked in, like the Po, all this deposit, except that contained in the water diverted by canals or otherwise drawn from the river for irrigation and other purposes, would have been carried out to sea.* This would have been a considerable quantity; for the Nile holds earth in suspension even at low water, a much larger proportion during the flood, and irrigation must have been carried on during the whole year. The precise amount which would have been thus distributed over the soil is matter of conjecture, but three cubic miles is certainly a liberal estimate. This would leave twelve cubic miles as the quantity which embankments would have compelled the Nile to transport to the Mediterranean over and above what it has actually deposited in that sea. The Mediterranean is shoal for some miles out to sea along the whole coast of the Delta, and the large bays or lagoons within the coast line, which communicate both with the river and the sea, have little depth of water. These lagoons the river deposits would have filled up, and there would still have been surplus earth enough to extend the Delta far into the Mediterranean.†

* There are many dikes in Egypt, but they are employed in but a very few cases to exclude the waters of the inundation. Their office is to retain the water received at high Nile into the inclosures formed by them until it shall have deposited its sediment or been drawn out for irrigation; and they serve also as causeways for interior communication during the floods. The Egyptian dikes, therefore, instead of forcing the river, like those of the Po, to transport its sediment to the sea, help to retain the slime, which, if the flow of the current over the land were not obstructed, might be carried back into the channel, and at last to the Mediterranean.

† The Mediterranean front of the Delta may be estimated at one hundred and fifty miles in length. Two cubic miles of earth would more than

Deposits of the Tuscan Rivers.

The Arno, and all the rivers rising on the western slopes and spurs of the Apennines, carry down immense quantities of mud to the Mediterranean. There can be no doubt that the volume of earth so transported is very much greater than it would have been had the soil about the headwaters of those rivers continued to be protected from wash by forests; and there is as little question that the quantity borne out to sea by the rivers of Western Italy is much increased by artificial embankments, because they are thereby prevented from spreading over the surface the sedimentary matter with which they are charged. The western coast of Tuscany has advanced some miles seaward within a very few centuries. The bed of the sea, for a long distance, has been raised, and of course the relative elevation of the land above it lessened; harbors have been filled up and destroyed; long lines of coast dunes have been formed, and the diminished inclination of the beds of the rivers near their outlets has caused their waters to overflow their banks and convert them into pestilential marshes. The territorial extent of Western Italy has thus been considerably increased, but the amount of soil habitable and cultivable by man has been, in a still higher proportion, diminished. The coast of ancient Etruria was filled with great commercial towns, and their rural environs were occupied by a large and prosperous population. But maritime Tuscany has long been one of the most unhealthy districts in Christendom; the famous mart of Populonia has not an inhabitant; the coast is

fill up the lagoons on the coast, and the remaining ten, even allowing the mean depth of the water to be twenty fathoms, which is beyond the truth, would have been sufficient to extend the coast line about three miles farther seaward, and thus, including the land gained by the filling up of the lagoons, to add more than five hundred square miles to the area of Egypt. Nor is this all; for the retardation of the current, by lengthening the course and consequently diminishing the inclination of the channel, would have increased the deposit of suspended matter, and proportionally augmented the total effect of the embankment.

almost absolutely depopulated, and the malarious fevers have extended their ravages far into the interior.

These results are certainly not to be ascribed wholly to human action. They are, in a large proportion, due to geological causes over which man has no control. The soil of much of Tuscany becomes pasty, almost fluid even, as soon as it is moistened, and when thoroughly saturated with water, it flows like a river. Such a soil as this would not be completely protected by woods, and, indeed, it would now be difficult to confine it long enough to allow it to cover itself with forest vegetation. Nevertheless, it certainly was once chiefly wooded, and the rivers which flow through it must then have been much less charged with earthy matter than at present, and they must have carried into the sea a smaller proportion of their sediment when they were free to deposit it on their banks than since they have been confined by dikes.*

* For the convenience of navigation, and to lessen the danger of inundation by giving greater directness, and, of course, rapidity to the current, bends in rivers are sometimes cut off and winding channels made straight. This process has the same general effects as diking, and therefore cannot be employed without many of the same results.

This practice has often been resorted to on the Mississippi with advantage to navigation, but it is quite another question whether that advantage has not been too dearly purchased by the injury to the banks at lower points. If we suppose a river to have a navigable course of 1,600 miles as measured by its natural channel, with a descent of 800 feet, we shall have a fall of six inches to the mile. If the length of channel be reduced to 1,200 miles by cutting off bends, the fall is increased to eight inches per mile. The augmentation of velocity consequent upon this increase of inclination is not computable without taking into account other elements, such as depth and volume of water, diminution of direct resistance, and the like, but in almost any supposable case, it would be sufficient to produce great effects on the height of floods, the deposit of sediment in the channel, on the shores, and at the outlet, the erosion of banks and other points of much geographical importance.

The Po, in those parts of its course where the embankments leave a wide space between, often cuts off bends in its channel and straightens its course. These short cuts are called *salti*, or leaps, and sometimes reduce the distance between their termini by several miles. In 1777, the salto of Cottaro shortened a distance of 7,000 mètres by 5,000, or, in other

It is, in general, true, that the intervention of man has hitherto seemed to insure the final exhaustion, ruin, and desolation of every province of nature which he has reduced to his dominion. Attila was only giving an energetic and picturesque expression to the tendencies of human action, as personified in himself, when he said that "no grass grew where his horse's hoofs had trod." The instances are few, where a second civilization has flourished upon the ruins of an ancient culture, and lands once rendered uninhabitable by human acts or neglect have generally been forever abandoned as hopelessly irreclaimable. It is, as I have before remarked, a question of vast importance, how far it is practicable to restore the garden we have wasted, and it is a problem on which experience throws little light, because few deliberate attempts have yet been made at the work of physical regeneration, on a scale large enough to warrant general conclusions in any one class of cases.

The valleys and shores of Tuscany form, however, a striking exception to this remark. The success with which human guidance has made the operations of nature herself available for the restoration of her disturbed harmonies, in the Val di Chiana and the Tuscan Maremma, is among the noblest, if not the most brilliant achievements of modern engineering, and, regarded in all its bearings on the great question of which I have just spoken, it is, as an example, of more importance to the general interests of humanity than the proudest work of internal improvement that mechanical means have yet constructed. The operations in the Val di Chiana have consisted chiefly in so regulating the flow of the surface waters into and through it, as to compel them to deposit their sedimentary matter at the will of the engineers, and thereby to raise grounds rendered insalubrious and unfit for agricultural use by stagnating water; the improvements in the Maremma have embraced both this method of elevating the level of the soil,

words, reduced the length of the channel more than three miles; and in 1807 and 1810 the two salti of Mezzanone effected a reduction of distance to the amount of between seven and eight miles.—BAUMGARTEN, l. c. p. 138.

and the prevention of the mixture of salt water with fresh in the coast marshes and shallow bays, which is a very active cause of the development of malarious influences.*

Improvements in the Val di Chiana.

For twenty miles or more after the remotest headwaters of the Arno have united to form a considerable stream, this river flows southeastward to the vicinity of Arezzo. It here sweeps round to the northwest, and follows that course to near its junction with the Sieve, a few miles above Florence, from which point its general direction is westward to the sea. From the bend at Arezzo, a depression called the Val di Chiana runs southeastward until it strikes into the valley of the Paglia, a tributary of the Tiber, and thus connects the basin of the latter river with that of the Arno. In the Middle Ages, and down to the eighteenth century, the Val di Chiana was often overflowed and devastated by the torrents which poured down from the highlands, transporting great quantities of slime with their currents, stagnating upon its surface, and gradually converting it into a marshy and unhealthy district, which was at last very greatly reduced in population and productiveness. It had, in fact, become so desolate that even the swallow had deserted it.†

* The fact, that the mixing of salt and fresh water in coast marshes and lagoons is deleterious to the sanitary condition of the vicinity, seems almost universally admitted, though the precise reason why a mixture of both should be more injurious than either alone, is not altogether clear. It has been suggested that the admission of salt water to the lagoons and rivers kills many fresh water plants and animals, while the fresh water is equally fatal to many marine organisms, and that the decomposition of the remains originates poisonous miasmata. Other theories however have been proposed. The whole subject is fully and ably discussed by Dr. Salvagnoli Marchetti in the appendix to his valuable *Rapporto sul Bonificamento delle Maremme Toscane.* See also the *Memorie Economico-Statistiche sulle Maremme Toscane*, of the same author.

† This curious fact is thus stated in the preface to Fossombroni (*Memorie sopra la Val di Chiana*, edition of 1835, p. xiii), from which also I borrow most of the data hereafter given with respect to that valley "It is perhaps not universally known, that the swallows, which come from

The bed of the Arno near Arezzo and that of the Paglia at the southern extremity of the Val di Chiana did not differ

the north [south] to spend the summer in our climate, do not frequent marshy districts with a malarious atmosphere. A proof of the restoration of salubrity in the Val di Chiana is furnished by these aerial visitors, which had never before been seen in those low grounds, but which have appeared within a few years at Forano and other points similarly situated."

Is the air of swamps destructive to the swallows, or is their absence in such localities merely due to the want of human habitations, near which this half-domestic bird loves to breed, perhaps because the house fly and other insects which follow man are found only in the vicinity of his dwellings?

In almost all European countries, the swallow is protected, by popular opinion or superstition, from the persecution to which almost all other birds are subject. It is possible that this respect for the swallow is founded upon ancient observation of the fact just stated on the authority of Fossombroni. Ignorance mistakes the effect for the cause, and the absence of this bird may have been supposed to be the occasion, not the consequence, of the unhealthiness of particular localities. This opinion once adopted, the swallow would become a sacred bird, and in process of time fables and legends would be invented to give additional sanction to the prejudices which protected it. The Romans considered the swallow as consecrated to the Penates, or household gods, and according to Peretti (*Le Serate del Villaggio*, p. 168) the Lombard peasantry think it a sin to kill them, because they are *le gallinelle del Signore*, the chickens of the Lord.

The following little Tuscan *rispetto* from Gradi (*Racconti Popolari*, p. 33) well expresses the feeling of the peasantry toward this bird:

O rondinella che passi lo mare
Torna 'ndietro, vo' dirti du' parole;
Dammi 'na penna delle tue bell' ale,
Vo' scrivere 'na lettera al mi' amore;
E quando l' avrò scritta 'n carta bella,
Ti renderò la penna, o rondinella;
E quando l' avrò scritta 'n carta bianca,
Ti renderò la penna che ti manca;
E quando l' avrò scritta in carta d' oro,
Ti renderò la penna al tuo bel volo.

O swallow, that fliest beyond the sea,
Turn back! I would fain have a word with thee.
A feather oh grant, from thy wing so bright!
For I to my sweetheart a letter would write;
And when it is written on paper fine
I'll give thee, O swallow, that feather of thine;
—On paper so white, and I'll give thee back,
O pretty swallow, the pen thou dost lack;
—On paper of gold, and then I'll restore
To thy beautiful pinion the feather once more.

much in level. The general inclination of the valley was therefore small; it does not appear to have ever been divided into opposite slopes by a true watershed, and the position of the summit seems to have shifted according to the varying amount and place of deposit of the sediment brought down by the lateral streams which emptied into it. The length of its principal channel of drainage, and even the direction of its flow at any given point, were therefore fluctuating. Hence, much difference of opinion was entertained at different times with regard to the normal course of this stream, and, consequently, to the question whether it was to be regarded as properly an affluent of the Tiber or of the Arno.

The bed of the latter river at the bend has been eroded to the depth of thirty or forty feet, and that, apparently, at no very remote period. If it were elevated to what was evidently its original height, the current of the Arno would be so much above that of the Paglia as to allow of a regular flow from its channel to the latter stream, through the Val di Chiana, provided the bed of the valley had remained at the level which excavations prove it to have had a few centuries ago, before it was raised by the deposits I have mentioned. These facts, together with the testimony of ancient geographers which scarcely admits of any other explanation, are thought to prove that all the waters of the Upper Arno were originally discharged through the Val di Chiana into the Tiber, and that a part of them still continued to flow, at least occasionally, in that direction down to the days of the Roman empire, and perhaps for some time later. The depression of the bed of the Arno, and the raising of that of the valley by the deposits of the lateral torrents and of the Arno itself, finally cut off the branch of the river which had flowed to the Tiber, and all its

Popular traditions and superstitions are so closely connected with localities, that, though an emigrant people may carry them to a foreign land, they seldom survive a second generation. The swallow, however, is still protected in New England by prejudices of transatlantic origin; and I remember hearing, in my childhood, that if the swallows were killed, the cows would give bloody milk.

waters were turned into its present channel, though the principal drainage of the Val di Chiana appears to have been in a southeastwardly direction until within a comparatively recent period.

In the sixteenth century, the elevation of the bed of the valley had become so considerable, that in 1551, at a point about ten miles south of the Arno, it was found to be not less than one hundred and thirty feet above that river; then followed a level of ten miles, and then a continuous descent to the Paglia. Along the level portion of the valley was a boatable channel, and lakes, sometimes a mile or even two miles in breadth, had formed at various points farther south. At this period, the drainage of the summit level might easily have been determined in either direction, and the opposite descents of the valley made to culminate at the north or at the south end of the level. In the former case, the watershed would have been ten miles south of the Arno; in the latter, twenty miles, and the division would have been not very unequal.

Various schemes were suggested at this time for drawing off the stagnant waters, as well as for the future regular drainage of the valley, and small operations for those purposes were undertaken with partial success; but it was feared that the discharge of the accumulated waters into the Tiber would produce a dangerous inundation, while the diversion of the drainage into the Arno would increase the violence of the floods to which that river was very subject, and no decisive steps were taken. In 1606, an engineer whose name has not been preserved proposed, as the only possible method of improvement, the piercing of a tunnel through the hills bounding the valley on the west to convey its waters to the Ombrone, but the expense and other objections prevented the adoption of this project.* The fears of the Roman Government for the security of the valley of the Tiber had induced it to construct barriers across that part of the channel which lay within its territory,

* Morozzi, *Dello stato antico e moderno del fiume Arno*, ii, p. 42.

and these obstructions, though not specifically intended for that purpose, naturally promoted the deposit of sediment and the elevation of the bed of the valley in their neighborhood. The effect of this measure and of the continued spontaneous action of the torrents was, that the northern slope, which in 1551 had commenced at the distance of ten miles from the Arno, was found in 1605 to begin nearly thirty miles south of that river, and in 1645 it had been removed about six miles farther in the same direction.*

In the seventeenth century, the Tuscan and Papal Governments consulted Galileo, Torricelli, Castelli, Cassini, Viviani, and other distinguished philosophers and engineers, on the possibility of reclaiming the valley by a regular artificial drainage. Most of these eminent physicists were of opinion that the measure was impracticable, though not altogether for the same reasons; but they seem to have agreed in thinking that the opening of such channels, in either direction, as would give the current a flow sufficiently rapid to drain the lands properly, would dangerously augment the inundations of the river—whether the Tiber or the Arno—into which the waters should be turned. The general improvement of the valley was now for a long time abandoned, and the waters were allowed to spread and stagnate until carried off by partial drainage, infiltration, and evaporation. Torricelli had contended that the slope of a large part of the valley was too small to allow it to be drained by ordinary methods, and that no practicable depth and width of canal would suffice for that purpose. It could be laid dry, he thought, only by converting its surface into an inclined plane, and he suggested that this might be accomplished by controlling the flow of the numerous torrents which pour into it, so as to force them to deposit their sediment at the pleasure of the engineer, and, consequently, to elevate the level of the area over which it should be spread.† This plan

* Morozzi, *Dello stato, etc., dell'Arno*, ii, pp. 39, 40.

† Torricelli thus expressed himself on this point: "If we content ourselves with what nature has made practicable to human industry, we shall endeavor to control, as far as possible, the outlets of these streams, which,

did not meet with immediate general acceptance, but it was soon adopted for local purposes at some points in the southern part of the valley, and it gradually grew in public favor and was extended in application until its final triumph a hundred years later.

In spite of these encouraging successes, however, the fear of danger to the valley of the Arno and the Tiber, and the difficulty of an agreement between Tuscany and Rome—the boundary between which states crossed the Val di Chiana not far from the halfway point between the two rivers—and of reconciling other conflicting interests, prevented the resumption of the projects for the general drainage of the valley until after the middle of the eighteenth century. In the mean time the science of hydraulics had become better understood, and the establishment of the natural law according to which the velocity of a current of water, and of course the proportional quantity discharged by it in a given time, are increased by increasing its mass, had diminished if not dissipated the fear of exposing the banks of the Arno to greater danger from inundations by draining the Val di Chiana into it.

The suggestion of Torricelli was finally adopted as the basis of a comprehensive system of improvement, and it was decided to continue and extend the inversion of the original flow of the waters, and to turn them into the Arno from a point as far to the south as should be found practicable. The conduct of the works was committed to a succession of able engineers who, for a long series of years, were under the general direction of the celebrated philosopher and statesman Fossombroni, and the success has fully justified the expectations of the most sanguine advocates of the scheme. The plan of improvement embraced two branches: the one, the removal of certain obstructions in the bed of the Arno, and, consequently, the further depression of the channel of that river, in certain places, with the view

by raising the bed of the valley with their deposits, will realize the fable of the Tagus and the Pactolus, and truly roll golden sands for him that is wise enough to avail himself of them."—FOSSOMBRONI, *Memorie sopra la Val di Chiana*, p. 219.

of increasing the rapidity of its current; the other, the gradual filling up of the ponds and swamps, and raising of the lower grounds of the Val di Chiana, by directing to convenient points the flow of the streams which pour down into it, and there confining their waters by temporary dams until the sediment was deposited where it was needed. The economical result of these operations has been, that in 1835 an area of more than four hundred and fifty square miles of pond, marsh, and damp, sickly low grounds had been converted into fertile, healthy and well-drained soil, and, consequently, that so much territory has been added to the agricultural domain of Tuscany.

But in our present view of the subject, the geographical revolution which has been accomplished is still more interesting. The climatic influence of the elevation and draining of the soil must have been considerable, though I do not know that an increase or a diminution of the mean temperature or precipitation in the valley has been established by meteorological observation. There is, however, in the improvement of the sanitary condition of the Val di Chiana, which was formerly extremely unhealthy, satisfactory proof of a beneficial climatic change. The fevers, which not only decimated the population of the low grounds but infested the adjacent hills, have ceased their ravages, and are now not more frequent than in other parts of Tuscany. The strictly topographical effect of the operations in question, besides the conversion of marsh into dry surface, has been the inversion of the inclination of the valley for a distance of thirty-five miles, so that this great plain which, within a comparatively short period, sloped and drained its waters to the south, now inclines and sends its drainage to the north. The reversal of the currents of the valley has added to the Arno a new tributary equal to the largest of its former affluents, and a most important circumstance connected with this latter fact is, that the increase of the volume of its waters has accelerated their velocity in a still greater proportion, and, instead of augmenting the danger from its inundations, has almost wholly obviated that source of apprehension. Between the beginning of the fifteenth century and the year 1761, thirty-one destructive floods of the Arno are recorded; between 1761, when the principal streams of the Val di Chiana were diverted into that river, and 1835, not one.*

Improvements in the Tuscan Maremme.

In the improvements of the Tuscan Maremma, more formidable difficulties have been encountered. The territory to be reclaimed was more extensive; the salubrious places of retreat for laborers and inspectors were more remote; the courses of the rivers to be controlled were longer and their natural inclination less rapid; some of them, rising in wooded regions, transported comparatively little earthy matter,† and above all,

* Arrian observes that at the junction of the Hydaspes and the Acesines, both of which are described as wide streams, "one very narrow river is formed of two confluents, and its current is very swift."—ARRIAN, *Alex. Anab.*, vi, 4.

A like example is observed in the Anapus near Syracuse, which, below the junction of its two branches, is narrower, though swifter than either of them, and such cases are by no means unfrequent. The immediate effect of the confluence of two rivers upon the current below depends upon local circumstances, and especially upon the angle of incidence. If the two nearly coincide in direction, so as to include a small angle, the joint current will have a greater velocity than the slower confluent, perhaps even than either of them. If the two rivers run in transverse, still more if they flow in more or less opposite directions, the velocity of the principal branch will be retarded both above and below the junction, and at high water it may even set back the current of the affluent.

On the other hand, the diversion of a considerable branch from a river retards its velocity below the point of separation, and here a deposit of earth in its channel immediately begins, which has a tendency to turn the whole stream into the new bed. "Theory and the authority of all hydrographical writers combine to show that the channels of rivers undergo an elevation of bed below a canal of diversion."—Letter of FOSSOMBRONI, in SALVAGNOLI, *Raccolta di Documenti*, p. 32. See the early authorities and discussions on the principle stated in the text, in FRISI, *Del modo di regolare i Fiumi e i Torrenti*, libro iii, capit. i.

† This difficulty has been remedied as to one important river of the Maremma, the Pecora, by clearings recently executed along its upper

the coast, which is a recent deposit of the waters, is little elevated above the sea, and admits into its lagoons and the mouths of its rivers floods of salt water with every western wind, every rising tide.*

The western coast of Tuscany is not supposed to have been an unhealthy region before the conquest of Etruria by the Romans, but it certainly became so within a few centuries after that event. This was a natural consequence of the neglect or wanton destruction of the public improvements, and especially the hydraulic works in which the Etruscans were so skilful, and of the felling of the upland forests, to satisfy the demand for wood at Rome for domestic, industrial, and military purposes. After the downfall of the Roman empire, the incursions of the barbarians, and then feudalism, foreign domination, intestine wars, and temporal and spiritual tyrannies,

course. "The condition of this marsh and of its affluents are now, November, 1859, much changed, and it is advisable to prosecute its improvement by deposits. In consequence of the extensive felling of the woods upon the plains, hills, and mountains of the territory of Massa and Scarlino, within the last ten years, the Pecora and other affluents of the marsh receive, during the rains, water abundantly charged with slime, so that the deposits within the first division of the marsh are already considerable, and we may now hope to see the whole marsh and pond filled up in a much shorter time than we had a right to expect before 1850. This circumstance totally changes the terms of the question, because the filling of the marsh and pond, which then seemed almost impossible on account of the small amount of sediment deposited by the Pecora, has now become practicable."—SALVAGNOLI, *Rapporto sul Bonificamento delle Maremme Toscane*, pp. li, lii.

The annual amount of sediment brought down by the rivers of the Maremma is computed at more than 12,000,000 cubic yards, or enough to raise an area of four square miles one yard. Between 1830 and 1859 more than three times that quantity was deposited in the marsh and shoal water lake of Castiglione alone.—SALVAGNOLI, *Raccolta di Documenti*, pp. 74, 75.

* The tide rises ten inches on the coast of Tuscany. See Memoir by FANTONI, in the appendix to SALVAGNOLI, *Rapporto*, p. 189.

On the tides of the Mediterranean, see BÖTTGER, *Das Mittelmeer*, p. 190. Not having Admiral Smyth's Mediterranean—on which Böttger's work is founded—at hand, I do not know how far credit is due to the former author for the matter contained in the chapter referred to.

aggravated still more cruelly the moral and physical evils which Tuscany and the other Italian States were doomed to suffer, and from which they have enjoyed but brief respites during the whole period of modern history. The Maremma was already proverbially unhealthy in the time of Dante, who refers to the fact in several familiar passages, and the petty tyrants upon its borders often sent criminals to places of confinement in its territory, as a slow but certain mode of execution. Ignorance of the causes of the insalubrity, and often the interference of private rights,* prevented the adoption of measures to remove it, and the growing political and commercial importance of the large towns in more healthful localities absorbed the attention of Government, and deprived the Maremma of its just share in the systems of physical improvement which were successfully adopted in interior and Northern Italy.

Before any serious attempts were made to drain or fill up the marshes of the Maremme, various other sanitary experiments were tried. It was generally believed that the insalubrity of the province was the consequence, not the cause, of its depopulation, and that, if it were once densely inhabited, the ordinary operations of agriculture, and especially the

* In Catholic countries, the discipline of the church requires a *meagre* diet at certain seasons, and as fish is not flesh, there is a great demand for that article of food at those periods. For the convenience of monasteries and their patrons, and as a source of pecuniary emolument to ecclesiastical establishments and sometimes to lay proprietors, great numbers of artificial fish ponds were created during the Middle Ages. They were generally shallow pools formed by damming up the outlet of marshes, and they were among the most fruitful sources of endemic disease, and of the peculiar malignity of the epidemics which so often ravaged Europe in those centuries. These ponds, in religious hands, were too sacred to be infringed upon for sanitary purposes, and when belonging to powerful lay lords they were almost as inviolable. The rights of fishery were a standing obstacle to every proposal of hydraulic improvement, and to this day large and fertile districts in Southern Europe remain sickly and almost unimproved and uninhabited, because the draining of the ponds upon them would reduce the income of proprietors who derive large profits by supplying the faithful, in Lent, with fish, and with various species of waterfowl which, though very fat, are, ecclesiastically speaking, meagre.

maintenance of numerous domestic fires, would restore it to its ancient healthfulness.* In accordance with these views, settlers were invited from various parts of Italy, from Greece, and, after the accession of the Lorraine princes, from that country also, and colonized in the Maremme. To strangers coming from soils and skies so unlike those of the Tuscan marshes, the climate was more fatal than to the inhabitants of the neighboring districts, whose constitutions had become in some degree inured to the local influences, or who at least knew better how to guard against them. The consequence very naturally was that the experiment totally failed to produce the desired effects, and was attended with a great sacrifice of life and a heavy loss to the treasury of the state.

The territory known as the Tuscan Maremma, *ora maritima*, or Maremme—for the plural form is most generally used—lies upon and near the western coast of Tuscany, and comprises about 1,900 square miles English, of which 500 square miles, or 320,000 acres, are plain and marsh including 45,500 acres of water surface, and about 290,000 acres are forest. One of the mountain peaks, that of Mount Amiata, rises to the height of 6,280 feet. The mountains of the Maremma are healthy, the lower hills much less so, as the malaria is felt at some points at the height of 1,000 feet, and the plains, with the exception of a few localities favorably situated on the seacoast, are in a high degree pestilential. The fixed population is about 80,000, of whom one sixth live on the plains in the winter and about one tenth in the summer. Nine or ten thousand laborers come down from the mountains of the Maremma and the neighboring provinces into the plain, during the latter season, to cultivate and gather the crops.

Out of this small number of inhabitants and strangers, 35,619 were ill enough to require medical treatment between the 1st of June, 1840, and the 1st of June, 1841, and more than one half the cases were of intermittent, malignant, gas-

* Macchiavelli advised the Government of Tuscany "to provide that men should restore the wholesomeness of the soil by cultivation, and purify the air by fires."—SALVAGNOLI, *Memorie*, p. 111.

tric, or catarrhal fever. Very few agricultural laborers escaped fever, though the disease did not always manifest itself until they had returned to the mountains. In the province of Grosseto, which embraces nearly the whole of the Maremma, the annual mortality was 3.92 per cent. the average duration of life but 23.18 years, and 75 per cent. of the deaths were among persons engaged in agriculture.

The filling up of the low grounds and the partial separation of the waters of the sea and the land, which had been in progress since the year 1827, now began to show very decided effects upon the sanitary condition of the population. In the year ending June 1st, 1842, the number of the sick was reduced by more than 2,000, and the cases of fever by more than 4,000. The next year, the cases of fever fell to 10,500, and in that ending June 1st, 1844, to 9,200. The political events of 1848 and the preceding and following years, occasioned the suspension of the works of improvement in the Maremma, but they were resumed after the revolution of 1859, and are now in successful progress.

I have spoken, with some detail, of the improvements in the Val di Chiana and the Tuscan Maremma, because of their great relative importance, and because their history is well known; but like operations have been executed in the territory of Pisa and upon the coast of the duchy of Lucca. In the latter case, they were confined principally to prevention of the intermixing of fresh water with that of the sea. In 1741, sluices or lock gates were constructed for this purpose, and the following year, the fevers, which had been destructive to the coast population for a long time previous, disappeared altogether. In 1768 and 1769, the works having fallen to decay, the fevers returned in a very malignant form, but the rebuilding of the gates again restored the healthfulness of the shore. Similar facts recurred in 1784 and 1785, and again from 1804 to 1821. This long and repeated experience has at last impressed upon the people the necessity of vigilant attention to the sluices, which are now kept in constant repair. The health of the coast is uninterrupted, and Viareggio, the capital town of the district, is now much frequented for its sea baths and its general salubrity, at a season when formerly it was justly shunned as the abode of disease and death.*

It is now a hundred years since the commencement of the improvements in the Val di Chiana, and those of the Maremma have been in more or less continued operation for above a generation. They have, as we have seen, produced important geographical changes in the surface of the earth and in the flow of considerable rivers, and their effects have been not less conspicuous in preventing other changes, of a deleterious character, which would infallibly have taken place if they had not been arrested by the improvements in question. It has been already stated that, in order to prevent the overflow of the valley of the Tiber by freely draining the Val di Chiana into it, the Papal authorities, long before the commencement of the Tuscan works, constructed strong barriers near the southern end of the valley, which detained the waters of the wet season until they could be gradually drawn off into the Paglia. They consequently deposited most of their sediment in the Val di Chiana and carried down comparatively little earth to the Tiber. The lateral streams contributing the largest quantities of sedimentary matter to the Val di Chiana originally flowed into that valley near its northern end; and the change of their channels and outlets in a southern direction, so as to raise that part of the valley by their deposits and thereby reverse its drainage, was one of the principal steps in the process of improvement.

We have seen that the north end of the Val di Chiana near the Arno had been raised by spontaneous deposit of sediment to such a height as to interpose a sufficient obstacle to all flow in that direction. If, then, the Roman dam had not been erected, or the works of the Tuscan Government undertaken, the whole of the earth, which has been arrested by those works and employed to raise the bed

* GIORGINI, *Sur les causes de l'Insalubrité de l'air dans le voisinage des marais, etc., lue à l'Académie des Sciences à Paris,* le 12 Juillet, 1825. Reprinted in SALVAGNOLI, *Rapporto, etc.*, appendice, p. 5, *et seqq.*

and reverse the declivity of the valley, would have been carried down to the Tiber and thence into the sea. The deposit thus created, would, of course, have contributed to increase the advance of the shore at the mouth of that river, which has long been going on at the rate of three mètres and nine tenths (twelve feet and nine inches) per annum.* It is evident that a quantity of earth, sufficient to effect the immense changes I have described in a wide valley more than thirty miles long, if deposited at the outlet of the Tiber, would have very considerably modified the outline of the coast, and have exerted no unimportant influence on the flow of that river, by raising its point of discharge and lengthening its channel.

The sediment washed into the marshes of the Maremme is not less than 12,000,000 cubic yards per annum. The escape of this quantity into the sea, which is now almost wholly prevented, would be sufficient to advance the coast line fourteen yards per year, for a distance of forty miles, computing the mean depth of the sea near the shore at twelve yards. It is true that in this case, as well as in that of other rivers, the sedimentary matter would not be distributed equally along the shore, and much of it would be carried out into deep water, or perhaps transported by the currents to distant coasts. The immediate effects of the deposit, therefore, would not be so palpable as they appear in this numerical form, but they would be equally certain, and would infallibly manifest themselves, first, perhaps, at some remote point, and afterward at or near the outlets of the rivers which produced them.

Obstruction of River Mouths.

The mouths of a large proportion of the streams known to ancient internal navigation are already blocked up by sandbars or fluviatile deposits, and the maritime approaches to river harbors frequented by the ships of Phenicia and Carthage and Greece and Rome are shoaled to a considerable

* See the careful estimates of ROSET, *Moyens de forcer les Torrents, etc.*, pp. 42, 44.

distance out to sea. The inclination of almost every known river bed has been considerably reduced within the historical period, and nothing but great volume of water, or exceptional rapidity of flow, now enables a few large streams like the Amazon, the La Plata, the Ganges, and, in a less degree, the Mississippi, to carry their own deposits far enough out into deep water to prevent the formation of serious obstructions to navigation. But the degradation of their banks, and the transportation of earthy matter to the sea by their currents, are gradually filling up the estuaries even of these mighty floods, and unless the threatened evil shall be averted by the action of geological forces, or by artificial contrivances more efficient than dredging machines, the destruction of every harbor in the world which receives a considerable river must inevitably take place at no very distant date.

This result would, perhaps, have followed in some incalculably distant future, if man had not come to inhabit the earth as soon as the natural forces which had formed its surface had arrived at such an approximate equilibrium that his existence on the globe was possible; but the general effect of his industrial operations has been to accelerate it immensely. Rivers, in countries planted by nature with forests and never inhabited by man, employ the little earth and gravel they transport chiefly to raise their own beds and to form plains in their basins.* In their upper course, where the current is swiftest, they are most heavily charged with coarse rolled or suspended matter, and this, in floods, they deposit on their shores in the mountain valleys where they rise; in their middle course, a lighter earth is spread over the bottom of their

* Rivers which transport sand, gravel, pebbles, heavy mineral matter in short, tend to raise their own beds; those charged only with fine, light earth, to cut them deeper. The prairie rivers of the West have deep channels, because the mineral matter they carry down is not heavy enough to resist the impulse of even a moderate current, and those tributaries of the Po which deposit their sediment in the lakes—the Ticino, the Adda, the Oglio, and the Mincio—flow, in deep cuts, for the same reason.—BAUMGARTEN, l. c., p. 132.

widening basins, and forms plains of moderate extent; the fine silt which floats farther is deposited over a still broader area, or, if carried out to sea, is, in great part quickly swept far off by marine currents and dropped at last in deep water. Man's "improvement" of the soil increases the erosion from its surface; his arrangements for confining the lateral spread of the water in floods compel the rivers to transport to their mouths the earth derived from that erosion even in their upper course; and, consequently, the sediment they deposit at their outlets is not only much larger in quantity, but composed of heavier materials, which sink more readily to the bottom of the sea and are less easily removed by marine currents.

The tidal movement of the ocean, deep sea currents, and the agitation of inland waters by the wind, lift up the sands strewn over the bottom by diluvial streams or sent down by mountain torrents, and throw them up on dry land, or deposit them in sheltered bays and nooks of the coast—for the flowing is stronger than the ebbing tide, the affluent than the refluent wave. This cause of injury to harbors it is not in man's power to resist by any means at present available; but, as we have seen, something can be done to prevent the degradation of high grounds, and to diminish the quantity of earth which is annually abstracted from the mountains, from table lands, and from river banks, to raise the bottom of the sea.

This latter cause of harbor obstruction, though an active agent, is, nevertheless, in many cases, the less powerful of the two. The earth suspended in the lower course of fluviatile currents is lighter than sea sand, river water lighter than sea water, and hence, if a land stream enters the sea with a considerable volume, its water flows over that of the sea, and bears its slime with it until it lets it fall far from shore, or, as is more frequently the case, mingles with some marine current and transports its sediment to a remote point of deposit. The earth borne out of the mouths of the Nile is in part carried over the waves which throw up sea sand on the beach, and deposited in deep water, in part drifted by the current, which sweeps east and north along the coasts of Egypt and Syria, until it finds a resting place in the northeastern angle of the Mediterranean.* Thus the earth loosened by the rude Abyssinian ploughshare, and washed down by the rain from the hills of Ethiopia which man has stripped of their protecting forests, contributes to raise the plains of Egypt, to shoal the maritime channels which lead to the city built by Alexander near the mouth of the Nile, and to fill up the harbors made famous by Phenician commerce.

* "The stream carries this mud, &c., at first farther to the east, and only lets it fall where the force of the current becomes weakened. This explains the continual advance of the land seaward along the Syrian coast, in consequence of which Tyre and Sidon no longer lie on the shore, but some distance inland. That the Nile contributes to this deposit may easily be seen, even by the unscientific observer, from the stained and turbid character of the water for many miles from its mouths. A somewhat alarming phenomenon was observed in this neighborhood in 1801, on board the English frigate Romulus, Captain Culverhouse, on a voyage from Acre to Abukir. Dr. E. D. Clarke, who was a passenger on board this ship, thus describes it:

"'26th July.—To-day, Sunday, we accompanied the captain to the wardroom to dine, as usual, with his officers. While we were at table, we heard the sailors who were throwing the lead suddenly cry out: "Three and a half!" The captain sprang up, was on deck in an instant, and, almost at the same moment, the ship slackened her way, and veered about. Every sailor on board supposed she would ground at once. Meanwhile, however, as the ship came round, the whole surface of the water was seen to be covered with thick, black mud, which extended so far that it appeared like an island. At the same time, actual land was nowhere to be seen—not even from the masthead—nor was any notice of such a shoal to be found on any chart on board. The fact is, as we learned afterward, that a stratum of mud, stretching from the mouths of the Nile for many miles out into the open sea, forms a movable deposit along the Egyptian coast. If this deposit is driven forward by powerful currents, it sometimes rises to the surface, and disturbs the mariner by the sudden appearance of shoals where the charts lead him to expect a considerable depth of water. But these strata of mud are, in reality, not in the least dangerous. As soon as a ship strikes them they break up at once, and a frigate may hold her course in perfect safety where an inexperienced pilot, misled by his soundings, would every moment expect to be stranded.'"—Böttger, *Das Mittelmeer*, pp. 188, 189.

Subterranean Waters.

I have frequently alluded to a branch of geography, the importance of which is but recently adequately recognized—the subterranean waters of the earth considered as stationary reservoirs, as flowing currents, and as filtrating fluids. The earth drinks in moisture by direct absorption from the atmosphere, by the deposition of dew, by rain and snow, by percolation from rivers and other superficial bodies of water, and sometimes by currents flowing into caves or smaller visible apertures.* Some of this humidity is exhaled again by the

* The caves of Carniola receive considerable rivers from the surface of the earth, which cannot, in all cases, be identified with streams flowing out of them at other points, and like phenomena are not uncommon in other limestone countries.

The cases are certainly not numerous where marine currents are known to pour continuously into cavities beneath the surface of the earth, but there is at least one well-authenticated instance of this sort—that of the mill streams at Argostoli in the island of Cephalonia. It had been long observed that the sea water flowed into several rifts and cavities in the limestone rocks of the coast, but the phenomenon has excited little attention until very recently. In 1833, three of the entrances were closed, and a regular channel, sixteen feet long and three feet wide, with a fall of three feet, was cut into the mouth of a larger cavity. The sea water flowed into this canal, and could be followed eighteen or twenty feet beyond its inner terminus, when it disappeared in holes and clefts in the rock.

In 1858, the canal had been enlarged to the width of five feet and a half, and a depth of a foot. The water pours rapidly through the canal into an irregular depression and forms a pool, the surface of which is three or four feet below the adjacent soil, and about two and a half or three feet below the level of the sea. From this pool it escapes through several holes and clefts in the rock, and has not yet been found to emerge elsewhere.

There is a tide at Argostoli of about six inches in still weather, but it is considerably higher with a south wind. I do not find it stated whether water flows through the canal into the cavity at low tide, but it distinctly appears that there is no refluent current, as of course there could not be from a basin so much below the sea. Mousson found the delivery through the canal to be at the rate of 24.88 cubic feet to the second; at what stage of the tide does not appear. Other mills of the same sort have been

soil, some is taken up by organic growths and by inorganic compounds, some poured out upon the surface by springs and either immediately evaporated or carried down to larger streams and to the sea, some flows by subterranean courses into the bed of fresh-water rivers * or of the ocean, and some remains, though even here not in forever motionless repose, to fill deep cavities and underground channels.† In every case

erected, and there appear to be several points on the coast where the sea flows into the land.

Various hypotheses have been suggested to explain this phenomenon, some of which assume that the water descends to a great depth beneath the crust of the earth, but the supposition of a difference of level in the surface of the sea on the opposite sides of the island, which seems confirmed by other circumstances, is the most obvious method of explaining these singular facts. If we suppose the level of the water on one side of the island to be raised by the action of currents three or four feet higher than on the other, the existence of cavities and channels in the rock would easily account for a subterranean current beneath the island, and the apertures of escape might be so deep or so small as to elude observation. See *Aus der Natur*, vol. 19, pp. 129, *et seqq.* See *Appendix*, No. 51.

* "The affluents received by the Seine below Rouen are so inconsiderable, that the augmentation of the volume of that river must be ascribed principally to springs rising in its bed. This is a point of which engineers now take notice, and M. Belgrand, the able officer charged with the improvement of the navigation of the Seine between Paris and Rouen, has devoted much attention to it."—Babinet, *Études et Lectures*, iii, p. 185.

On page 232 of the volume just quoted, the same author observes: "In the lower part of its course, from the falls of the Oise, the Seine receives so few important affluents, that evaporation alone would suffice to exhaust all the water which passes under the bridges of Paris."

This supposes a much greater amount of evaporation than has been usually computed, but I believe it is well settled that the Seine conveys to the sea much more water than is discharged into it by all its superficial branches.

† Girard and Duchatelet maintain that the subterranean waters of Paris are absolutely stagnant. See their report on drainage by artesian wells, *Annales des Ponts et Chaussées*, 1833, 2me sémestre, pp. 313, *et seqq.*

This opinion, if locally true, cannot be generally so, for it is inconsistent with the well-known fact that the very first eruption of water from a boring often brings up leaves and other objects which must have been carried into the underground reservoirs by currents.

the aqueous vapors of the air are the ultimate source of supply, and all these hidden stores are again returned to the atmosphere by evaporation.

The proportion of the water of precipitation taken up by direct evaporation from the surface of the ground seems to have been generally exaggerated, sufficient allowance not being made for moisture carried downward, or in a lateral direction, by infiltration or by crevices in the superior rocky or earthy strata. According to Wittwer, Mariotte found that but one sixth of the precipitation in the basin of the Seine was delivered into the sea by that river, "so that five sixths remained for evaporation and consumption by the organic world." *

Lieutenant Maury—whose scientific reputation, though fallen, has not quite sunk to the level of his patriotism—estimates the annual amount of precipitation in the valley of the Mississippi at 620 cubic miles, the discharge of that river into the sea at 107 cubic miles, and concludes that "this would leave 513 cubic miles of water to be evaporated from this river basin annually." † In these and other like computations, the water carried down into the earth by capillary and larger conduits is wholly lost sight of, and no thought is bestowed upon the supply for springs, for common and artesian wells, and for underground rivers, like those in the great caves of Kentucky, which may gush up in fresh-water currents at the bottom of the Caribbean Sea, or rise to the light of day in the far-off peninsula of Florida.

The progress of the emphatically modern science of geology has corrected these erroneous views, because the observations on which it depends have demonstrated not only the existence, but the movement, of water in nearly all geological formations, have collected evidence of the presence of large reservoirs at greater or less depths beneath surfaces of almost every

* *Physikalische Geographie*, p. 286. It does not appear whether this inference is Mariotte's or Wittwer's. I suppose it is a conclusion of the latter.

† *Physical Geography of the Sea*. Tenth edition. London, 1861, § 274.

character, and have investigated the rationale of the attendant phenomena. The distribution of these waters has been minutely studied with reference to a great number of localities, and though the actual mode of their vertical and horizontal transmission is still involved in much doubt, the laws which determine their aggregation are so well understood, that, when the geology of a given district is known, it is not difficult to determine at what depth water will be reached by the borer, and to what height it will rise.

The same principles have been successfully applied to the discovery of small subterranean collections or currents of water, and some persons have acquired, by a moderate knowledge of the superficial structure of the earth combined with long practice, a skill in the selection of favorable places for digging wells which seems to common observers little less than miraculous. The Abbé Paramelle—a French ecclesiastic who devoted himself for some years to this subject and was extensively employed as a well-finder—states, in his work on Fountains, that in the course of thirty-four years he had pointed out more than ten thousand subterranean springs, and though his geological speculations were often erroneous, the highest scientific authorities in Europe have testified to the great practical value of his methods, and the almost infallible certainty of his predictions.*

Babinet quotes a French proverb, "Summer rain wets nothing," and explains it as meaning that the water of such rains is "almost totally taken up by evaporation." "The rains of summer," he adds, "however abundant they may be, do not penetrate the soil to a greater depth than 15 or 20 centimètres. In summer the evaporating power of the heat is five or six times as great as in winter, and this power is exerted by an atmosphere capable of containing five times as much vapor as in winter." "A stratum of snow which prevents evaporation [from the soil] causes almost all the water that composes it to filter down into the earth, and form a

* PARAMELLE, *Quellenkunde, mit einem Vorwort von* B. COTTA, 1856.

reserve for springs, wells, and rivers which could not be supplied by any amount of summer rain." "This latter—useful, indeed like dew, to vegetation—does not penetrate the soil and accumulate a store to feed springs and to be brought up by them to the open air." * This conclusion, however applicable it may be to the climate and soil of France, is too broadly stated to be accepted as a general truth, and in countries where the precipitation is small in the winter months, familiar observation shows that the quantity of water yielded by deep wells and natural springs depends not less on the rains of summer than on those of the rest of the year, and, consequently, that much of the precipitation of that season must find its way to strata too deep to lose water by evaporation.

The supply of subterranean reservoirs and currents, as well as of springs, is undoubtedly derived chiefly from infiltration, and hence it must be affected by all changes of the natural surface that accelerate or retard the drainage of the soil, or that either promote or obstruct evaporation from it. It has sufficiently appeared from what has gone before, that the spontaneous drainage of cleared ground is more rapid than that of the forest, and consequently, that the felling of the woods, as well as the draining of swamps, deprives the subterranean waters of accessions which would otherwise be conveyed to them by infiltration. The same effect is produced by artificial contrivances for drying the soil either by open ditches or by underground pipes or channels, and in proportion as the sphere of these operations is extended, the effect of them cannot fail to make itself more and more sensibly felt in the diminished supply of water furnished by wells and running springs.†

* *Études et Lectures*, vi, p. 118.

† "The area of soil dried by draining is constantly increasing, and the water received by the surface from atmospheric precipitation is thereby partly conducted into new channels, and, in general, carried off more rapidly than before. Will not this fact exert an influence on the condition of many springs, whose basin of supply thus undergoes a partial or complete transformation? I am convinced that it will, and it is important to collect data for solving the question." BERNHARD COTTA, Preface to PARAMELLE, *Quellenkunde* (German translation), pp. vii, viii. See *Appendix*, No. 52.

It is undoubtedly true that loose soils, stripped of vegetation and broken up by the plough or other processes of cultivation, may, until again carpeted by grasses or other plants, absorb more rain and snow water than when they were covered by a natural growth; but it is also true that the evaporation from such soils is augmented in a still greater proportion. Rain scarcely penetrates beneath the sod of grass ground, but runs off over the surface; and after the heaviest showers a ploughed field will often be dried by evaporation before the water can be carried off by infiltration, while the soil of a neighboring grove will remain half saturated for weeks together. Sandy soils frequently rest on a tenacious subsoil, at a moderate depth, as is usually seen in the pine plains of the United States, where pools of rain water collect in slight depressions on the surface of earth, the upper stratum of which is as porous as a sponge. In the open grounds such pools are very soon dried up by the sun and wind; in the woods they remain unevaporated long enough for the water to diffuse itself laterally until it finds, in the subsoil, crevices through which it may escape, or slopes which it may follow to their outcrop or descend along them to lower strata.

The readiness with which water not obstructed by impermeable strata diffuses itself through the earth in all directions —and, consequently, the importance of keeping up the supply of subterranean reservoirs—find a familiar illustration in the effect of paving the ground about the stems of vines and trees. The surface earth around the trunk of a tree may be made perfectly impervious to water, by flag stones and cement, for a distance greater than the spread of the roots; and yet the tree will not suffer for want of moisture, except in droughts severe enough sensibly to affect the supply in deep wells and springs. Both forest and fruit trees grow well in cities where the streets and courts are closely paved, and where even the lateral access of water to the roots is more or less obstructed by deep cellars and foundation walls. The deep-lying veins and sheets of water, supplied by infiltration from above, send up moisture by capillary attraction, and the pavement prevents the soil beneath it from losing its humidity by evaporation. Hence, city-grown trees find moisture enough for their roots, and though plagued with smoke and dust, often retain their freshness while those planted in the open fields, where sun and wind dry up the soil faster than the subterranean fountains can water it, are withering from drought. Without the help of artificial conduit or of water carrier, the Thames and the Seine refresh the ornamental trees that shade the thoroughfares of London and of Paris, and beneath the hot and reeking mould of Egypt, the Nile sends currents to the extremest border of its valley.*

* See the interesting observations of KRIEGK on this subject, *Schriften zur allgemeinen Erdkunde*, cap. iii, § 6, and especially the passages in RITTER'S *Erdkunde*, vol. i, there referred to.

Laurent, (*Mémoires sur le Sahara Oriental*, pp. 8, 9), in speaking of a river at El-Faid, "which, like all those of the desert, is, most of the time, without water," observes, that many wells are dug in the bed of the river in the dry season, and that the subterranean current thus reached appears to extend itself laterally, at about the same level, at least a kilomètre from the river, as water is found by digging to the depth of twelve or fifteen mètres at a village situated at that distance from the bank.

The most remarkable case of infiltration known to me by personal observation is the occurrence of fresh water in the beach sand on the eastern side of the Gulf of Akaba, the eastern arm of the Red Sea. If you dig a cavity in the beach near the sea level, it soon fills with water so fresh as not to be undrinkable, though the sea water two or three yards from it contains even more than the average quantity of salt. It cannot be maintained that this is sea water freshened by filtration through a few feet or inches of sand, for salt water cannot be deprived of its salt by that process. It can only come from the highlands of Arabia, and it would seem that there must exist some large reservoir in the interior to furnish a supply which, in spite of evaporation, holds out for months after the last rains of winter, and perhaps even through the year. I observed the fact in the month of June.

The precipitation in the mountains that border the Red Sea is not known by pluviometric measurement, but the mass of debris brought down the ravines by the torrents proves that their volume must be large. The proportion of surface covered by sand and absorbent earth, in Arabia Petræa and the neighboring countries, is small, and the mountains drain themselves rapidly into the wadies or ravines where the torrents are formed; but the beds of earth and disintegrated rock at the bottom of the

Artesian Wells.

The existence of artesian wells depends upon that of subterranean reservoirs and rivers, and the supply yielded by borings is regulated by the abundance of such sources. The waters of the earth are, in many cases, derived from superficial currents which are seen to pour into chasms opened, as it were, expressly for their reception; and in others where no apertures in the crust of the earth have been detected, their existence is proved by the fact that artesian wells sometimes bring up from great depths seeds, leaves, and even living fish, which must have been carried down through channels large enough to admit a considerable stream. But in general, the sheets and currents of water reached by deep boring appear to be primarily due to infiltration from highlands where the water is first collected in superficial or subterranean reservoirs. By means of channels conforming to the dip of the strata, these reservoirs communicate with the lower basins, and exert upon them a fluid pressure sufficient to raise a column to the surface, whenever an orifice is opened.* The water delivered by

valleys are of so loose and porous texture, that a great quantity of water is absorbed in saturating them before a visible current is formed on their surface. In a heavy thunder storm, accompanied by a deluging rain, which I witnessed at Mount Sinai in the month of May, a large stream of water poured, in an almost continuous cascade, down the steep ravine north of the convent, by which travellers sometimes descend from the plateau between the two peaks, but after reaching the foot of the mountain, it flowed but a few yards before it was swallowed up in the sands.

* It is conceivable that in large and shallow subterranean basins the superincumbent earth may rest upon the water and be partly supported by it. In such case the weight of the earth would be an additional, if not the sole, cause of the ascent of the water through the tubes of artesian wells. The elasticity of gases in the cavities may also aid in forcing up water.

A French engineer, M. Mullot, invented a simple method of bringing to the surface water from any one of several successive accumulations at different depths, or of raising it, unmixed, from two or more of them at once. It consists in employing concentric tubes, one within the other, leaving a space for the rise of water between them, and reaching each to the sheet from which it is intended to draw.

an artesian well is, therefore, often derived from distant sources, and may be wholly unaffected by geographical or meteorological changes in its immediate neighborhood, while the same changes may quite dry up common wells and springs which are fed only by the local infiltration of their own narrow basins.

In most cases, artesian wells have been bored for purely economical or industrial purposes, such as to obtain good water for domestic use or for driving light machinery, to reach saline or other mineral springs, and recently, in America, to open fountains of petroleum or rock oil. The geographical and geological effects of such abstraction of fluids from the bowels of the earth are too remote and uncertain to be here noticed; * but artesian wells have lately been employed in Algeria for a purpose which has even now a substantial, and may hereafter acquire a very great geographical importance. It was observed by many earlier as well as recent travellers in the East, among whom Shaw deserves special mention, that the Libyan desert, bordering upon the cultivated shores of the Mediterranean, appeared in many places to rest upon a subterranean lake at an accessible distance below the surface. The Moors are vaguely said to have *bored* artesian wells down to this reservoir, to obtain water for domestic use and irrigation, but

* Many more or less probable conjectures have been made on this subject, but thus far I am not aware that any of the apprehended results have been actually shown to have happened. In an article in the *Annales des Ponts et Chaussées* for July and August, 1839, p. 131, it was suggested that the sinking of the piers of a bridge at Tours in France was occasioned by the abstraction of water from the earth by artesian wells, and the consequent withdrawal of the mechanical support it had previously given to the strata containing it. A reply to this article will be found in VIOLLET, *Théorie des Puits Artésiens*, p. 217.

In some instances the water has rushed up with a force which seemed to threaten the inundation of the neighborhood, and even the washing away of much soil; but in these cases the partial exhaustion of the supply, or the relief of hydrostatic or elastic pressure, has generally produced a diminution of the flow in a short time, and I do not know that any serious evil has ever been occasioned in this way.

I do not find such wells described by any trustworthy traveller, and the universal astonishment and incredulity with which the native tribes viewed the operations of the French engineers sent into the desert for that purpose, are a sufficient proof that this mode of reaching the subterranean waters was new to them. They were, however, aware of the existence of water below the sands, and were dexterous in digging wells—square shafts lined with a framework of palm-tree stems—to the level of the sheet. The wells so constructed, though not technically artesian wells, answer the same purpose; for the water rises to the surface and flows over it as from a spring.*

* See a very interesting account of these wells, and of the workmen who clean them out when obstructed by sand brought up with the water, in Laurent's memoir on the artesian wells recently bored by the French Government in the Algerian desert, *Mémoire sur le Sahara Oriental, etc.*, pp. 19, *et seqq.* Some of the men remained under water from two minutes to two minutes and forty seconds. Several officers are quoted as having observed immersions of three minutes' duration, and M. Berbrugger alleges that he witnessed one of five minutes and fifty-five seconds. The shortest of these periods is longer than the best pearl diver can remain below the surface of salt water. The wells of the Sahara are from twenty to eighty mètres deep.

It has often been asserted that the ancient Egyptians were acquainted with the art of boring artesian wells. Parthey, describing the Little Oasis, mentions ruins of a Roman aqueduct, and observes: "It appears from the recent researches of Aim, a French engineer, that these aqueducts are connected with old artesian wells, the restoration of which would render it practicable to extend cultivation much beyond its present limits. This agrees with ancient testimony. It is asserted that the inhabitants of the oases sunk wells to the depth of 200, 300, and even 500 ells, from which affluent streams of water poured out. See OLYMPIODORUS in *Photii Bibl.*, cod. 80, p. 61, l. 17, ed. Bekk."—PARTHEY, *Wanderungen*, ii, p. 528.

In a paper entitled, *Note relative à l'execution d'un Puits Artésien en Egypte sous la XVIII dynastie*, presented to the Académie des Inscriptions et Belles Lettres, on the 12th of November, 1852, M. Lenormant endeavors to show that a hieroglyphic inscription found at Contrapscelcis proves the execution of a work of this sort in the Nubian desert, at the period indicated in the title to his paper. The interpretation of the inscription is a question for Egyptologists; but if wells were actually bored through the rock by the Egyptians after the Chinese or the European fashion, it is singular that among the numerous and minute representa-

These wells, however, are too few and too scanty in supply to serve any other purposes than the domestic wells of other countries, and it is but recently that the transformation of desert into cultivable land by this means has been seriously attempted. The French Government has bored a large number of artesian wells in the Algerian desert within a few years, and the native sheikhs are beginning to avail themselves of the process. Every well becomes the nucleus of a settlement proportioned to the supply of water, and before the end of the year 1860, several nomade tribes had abandoned their wandering life, established themselves around the wells, and planted more than 30,000 palm trees, besides other perennial vegetables.* The water is found at a small depth, generally from 100 to 200 feet, and though containing too large a pro-

tions of their industrial operations, painted or carved on the walls of their tombs, no trace of the processes employed for so remarkable and important a purpose should have been discovered. See *Appendix*, No. 54.

It is certain that artesian wells have been common in China from a very remote antiquity, and the simple method used by the Chinese—where the borer is raised and let fall by a rope, instead of a rigid rod—has been lately been employed in Europe with advantage. Some of the Chinese wells are said to be 3,000 feet deep; that of Neusalzwerk in Silesia—the deepest in Europe—is 2,300. A well was bored at St. Louis, in Missouri, a few years ago, to supply a sugar refinery, to the depth of 2,199 feet. This was executed by a private firm in three years, at the expense of only $10,000. Another has since been bored at the State capitol at Columbus, Ohio, 2,500 feet deep, but without obtaining the desired supply of water.

* "In the anticipation of our success at Oum-Thiour, every thing had been prepared to take advantage of this new source of wealth without a moment's delay. A division of the tribe of the Selmia, and their sheikh, Aïssa ben Shâ, laid the foundation of a village as soon as the water flowed, and planted twelve hundred date palms, renouncing their wandering life to attach themselves to the soil. In this arid spot, life had taken the place of solitude, and presented itself, with its smiling images, to the astonished traveller. Young girls were drawing water at the fountain; the flocks, the great dromedaries with their slow pace, the horses led by the halter, were moving to the watering trough; the hounds and the falcons enlivened the group of party-colored tents, and living voices and animated movement had succeeded to silence and desolation."—LAURENT, *Mémoires sur le Sahara*, p. 85.

portion of mineral matter to be acceptable to a European palate, it answers well for irrigation, and does not prove unwholesome to the natives.

The most obvious use of artesian wells in the desert at present is that of creating stations for the establishment of military posts and halting places for the desert traveller; but if the supply of water shall prove adequate for the indefinite extension of the system, it is probably destined to produce a greater geographical transformation than has ever been effected by any scheme of human improvement. The most striking contrast of landscape scenery that nature brings near together in time or place, is that between the greenery of the tropics, or of a northern summer, and the snowy pall of leafless winter. Next to this in startling novelty of effect, we must rank the sudden transition from the shady and verdant oasis of the desert to the bare and burning party-colored ocean of sand and rock which surrounds it.* The most sanguine believer in

* The variety of hues and tones in the local color of the desert is, I think, one of the phenomena which most surprise and interest a stranger to those regions. In England and the United States, rock is so generally covered with moss or earth, and earth with vegetation, that untravelled Englishmen and Americans are not very familiar with naked rock as a conspicuous element of landscape. Hence, in their conception of a bare cliff or precipice, they hardly ascribe definite color to it, but depict it to their imagination as wearing a neutral tint not assimilable to any of the hues with which nature tinges her atmospheric or paints her organic creations. There are certainly extensive desert ranges, chiefly limestone formations, where the surface is either white, or has weathered down to a dull uniformity of tone which can hardly be called color at all; and there are sand plains and drifting hills of wearisome monotony of tint. But the chemistry of the air, though it may tame the glitter of the limestone to a dusky gray, brings out the green and brown and purple of the igneous rocks, and the white and red and blue and violet and yellow of the sandstone. Many a cliff in Arabia Petræa is as manifold in color as the rainbow, and the veins are so variable in thickness and inclination, so contorted and involved in arrangement, as to bewilder the eye of the spectator like a disk of party-colored glass in rapid revolution.

In the narrower wadies, the mirage is not common; but on broad expanses, as at many points between Cairo and Suez, and in Wadi el Araba, it mocks you with lakes and land-locked bays, studded with islands and

indefinite human progress hardly expects that man's cunning will accomplish the universal fufilment of the prophecy, "the desert shall blossom as the rose," in its literal sense; but sober geographers have thought the future conversion of the sand plains of Northern Africa into fruitful gardens, by means of artesian wells, not an improbable expectation. They have gone farther, and argued that, if the soil were covered witl fields and forests, vegetation would call down moisture from the Libyan sky, and that the showers which are now wasted on the sea, or so often deluge Southern Europe with destructive inundation, would in part be condensed over the arid wastes of Africa, and thus, without further aid from man, bestow abundance on regions which nature seems to have condemned to perpetual desolation.

An equally bold speculation, founded on the well-known fact, that the temperature of the earth and of its internal waters increases as we descend beneath the surface, has suggested that artesian wells might supply heat for industrial and domestic purposes, for hot-house cultivation, and even for the local amelioration of climate. The success with which Count Lardarello has employed natural hot springs for the evaporation of water charged with boracic acid, and other fortunate applications of the heat of thermal sources, lend some countenance to the latter project; but both must, for the present, be ranked among the vague possibilities of science, not regarded as probable future triumphs of man over nature.

Artificial Springs.

A more plausible and inviting scheme is that of the creation of perennial springs by husbanding rain and snow water,

fringed with trees, all painted with an illusory truth of representation absolutely indistinguishable from the reality. The checkered earth, too, is canopied with a heaven as variegated as itself. You see, high up in the sky, rosy clouds at noonday, colored probably by reflection from the ruddy mountains, while near the horizon float cumuli of a transparent ethereal blue, seemingly balled up out of the clear cerulean substance of the firmament, and detached from the heavenly vault, not by color or consistence, but solely by the light and shade of their prominences.

storing it up in artificial reservoirs of earth, and filtering it through purifying strata, in analogy with the operations of nature. The sagacious Palissy—starting from the theory that all springs are primarily derived from precipitation, and reasoning justly on the accumulation and movement of water in the earth—proposed to reduce theory to practice, and to imitate the natural processes by which rain is absorbed by the earth and given out again in running fountains. "When I had long and diligently considered the cause of the springing of natural fountains and the places where they be wont to issue," says he, "I did plainly perceive, at last, that they do proceed and are engendered of nought but the rains. And it is this, look you, which hath moved me to enterprise the gathering together of rain water after the manner of nature, and the most closely according to her fashion that I am able; and I am well assured that by following the formulary of the Supreme Contriver of fountains, I can make springs, the water whereof shall be as good and pure and clear as of such which be natural." * Palissy discusses the subject of the origin of springs at length and with much ability, dwelling specially on infiltration, and, among other things, thus explains the frequency of springs in mountainous regions: "Having well considered the which, thou mayest plainly see the reason why there be more springs and rivulets proceeding from the mountains than from the rest of the earth; which is for no other cause but that the rocks and mountains do retain the water of the rains like vessels of brass. And the said waters falling upon the said mountains descend continually through the earth, and through crevices, and stop not till they find some place that is bottomed with stone or close and thick rocks; and they rest upon such bottom until they find some channel or other manner of issue, and then they flow out in springs or brooks or rivers, according to the greatness of the reservoirs and of the outlets thereof." †

After a full exposition of his theory, Palissy proceeds to

* *Œuvres de Palissy*, *Des Eaux et Fontaines*, p. 157.

† Id., p. 166. See *Appendix*, No. 55.

describe his method of creating springs, which is substantially the same as that lately proposed by Babinet in the following terms: "Choose a piece of ground containing four or five acres, with a sandy soil, and with a gentle slope to determine the flow of the water. Along its upper line, dig a trench five or six feet deep and six feet wide. Level the bottom of the trench, and make it impermeable by paving, by macadamizing, by bitumen, or, more simply and cheaply, by a layer of clay. By the side of this trench dig another, and throw the earth from it into the first, and so on until you have rendered the subsoil of the whole parcel impermeable to rain water. Build a wall along the lower line with an aperture in the middle for the water, and plant fruit or other low trees upon the whole, to shade the ground and check the currents of air which promote evaporation. This will infallibly give you a good spring which will flow without intermission and supply the wants of a whole hamlet or a large chateau." * Babinet states that the whole amount of precipitation on a reservoir of the proposed area, in the climate of Paris, would be about 13,000 cubic yards, not above one half of which, he thinks, would be lost, and, of course, the other half would remain available to supply the spring. I much doubt whether this expectation would be realized in practice, in its whole extent; for if Babinet is right in supposing that the summer rain is wholly evaporated, the winter rains, being much less in quantity, would hardly suffice to keep the earth saturated and give off so large a surplus.

The method of Palissy, though, as I have said, similar in principle to that of Babinet, would be cheaper of execution,

* Babinet, *Études et Lectures sur les Sciences d'Observation*, ii, p. 225. Our author precedes his account of his method with a complaint which most men who indulge in thinking have occasion to repeat many times in the course of their lives. "I will explain to my readers the construction of artificial fountains according to the plan of the famous Bernard de Palissy, who, a hundred and fifty [three hundred] years ago, came and took away from me, a humble academician of the nineteenth century, this discovery which I had taken a great deal of pains to make. It is enough to discourage all invention when one finds plagiarists in the past as well as in the future!" (P. 224.)

and, at the same time, more efficient. He proposes the construction of relatively small filtering receptacles, into which he would conduct the rain falling upon a large area of rocky hillside, or other sloping ground not readily absorbing water. This process would, in all probability, be a very successful, as well as an inexpensive, mode of economizing atmospheric precipitation, and compelling the rain and snow to form perennial fountains at will.

Economizing Precipitation.

The methods suggested by Palissy and by Babinet are of limited application, and designed only to supply a sufficient quantity of water for the domestic use of small villages or large private establishments. Dumas has proposed a much more extensive system for collecting and retaining the whole precipitation in considerable valleys, and storing it in reservoirs, whence it is to be drawn for household and mechanical purposes, for irrigation, and, in short, for all the uses to which the water of natural springs and brooks is applicable. His plan consists in draining both surface and subsoil, by means of conduits differing in construction according to local circumstances, but in the main not unlike those employed in improved agriculture, collecting the water in a central channel, securing its proper filterage, checking its too rapid flow by barriers at convenient points, and finally receiving the whole in spacious covered reservoirs, from which it may be discharged in a constant flow or at intervals as convenience may dictate.*

There is no reasonable doubt that a very wide employment of these various contrivances for economizing and supplying water is practicable, and the expediency of resorting to them is almost purely an economical question. There appears to be no serious reason to apprehend collateral evils from them, and in fact all of them, except artesian wells, are simply indirect methods of returning to the original arrangements of nature, or, in other words, of restoring the fluid circulation of the

* M. G. Dumas, *La Science des Fontaines*, 1857.

globe; for when the earth was covered with the forest, perennial springs gushed from the foot of every hill, brooks flowed down the bed of every valley. The partial recovery of the fountains and rivulets which once abundantly watered the face of the agricultural world seems practicable by such means, even without any general replanting of the forests; and the cost of one year's warfare, if judiciously expended in a combination of both methods of improvcment, would secure, to almost every country that man has exhausted, an amelioration of climate, a renovated fertility of soil, and a general physical improvement, which might almost be characterized as a new creation.

CHAPTER V

THE SANDS.

ORIGIN OF SAND—SAND NOW CARRIED DOWN TO THE SEA—THE SANDS OF EGYPT AND THE ADJACENT DESERT—THE SUEZ CANAL—THE SANDS OF EGYPT—COAST DUNES AND SAND PLAINS—SAND BANKS—DUNES ON COAST OF AMERICA—DUNES OF WESTERN EUROPE—FORMATION OF DUNES—CHARACTER OF DUNE SAND—INTERIOR STRUCTURE OF DUNES—FORM OF DUNES—GEOLOGICAL IMPORTANCE OF DUNES—INLAND DUNES—AGE, CHARACTER, AND PERMANENCE OF DUNES—USE OF DUNES AS BARRIER AGAINST THE SEA—ENCROACHMENTS OF THE SEA—THE LIIMFJORD—ENCROACHMENTS OF THE SEA—DRIFTING OF DUNE SANDS—DUNES OF GASCONY—DUNES OF DENMARK—DUNES OF PRUSSIA—ARTIFICIAL FORMATION OF DUNES—TREES SUITABLE FOR DUNE PLANTATIONS—EXTENT OF DUNES IN EUROPE—DUNE VINEYARDS OF CAPE BRETON—REMOVAL OF DUNES—INLAND SAND PLAINS—THE LANDES OF GASCONY—THE BELGIAN CAMPINE—SANDS AND STEPPES OF EASTERN EUROPE—ADVANTAGES OF RECLAIMING DUNES—GOVERNMENT WORKS OF IMPROVEMENT.

Origin of Sand.

SAND, which is found in beds or strata at the bottom of the sea or in the channels of rivers, as well as in extensive deposits upon or beneath the surface of the dry land, appears to consist essentially of the detritus of rocks. It is not always by any means clear through what agency the solid rock has been reduced to a granular condition; for there are beds of quartzose sand, where the sharp, angular shape of the particles renders it highly improbable that they have been formed by gradual abrasion and attrition, and where the supposition of a crushing mechanical force seems equally inadmissible. In common sand, the quartz grains are the most numerous; but this is not a proof that the rocks from which these particles were derived were wholly, or even chiefly, quartzose in character; for, in many composite rocks, as, for example, in the granitic group, the mica, felspar, and hornblende are more easily decomposed by chemical action, or disintegrated, comminuted, and reduced to an impalpable state by mechanical force, than the quartz. In the destruction of such rocks, therefore, the quartz would survive the other ingredients, and remain unmixed, when they had been decomposed and had entered into new chemical combinations, or been ground to slime and washed away by water currents.

The greater or less specific gravity of the different constituents of rock doubtless aids in separating them into distinct masses when once disintegrated, though there are veined and stratified beds of sand where the difference between the upper and lower layers, in this respect, is too slight to be supposed capable of effecting a complete separation.* In cases where rock has been reduced to sandy fragments by heat, or by obscure chemical and other molecular forces, the sandbeds may remain undisturbed, and represent, in the series of geological strata, the solid formations from which they were derived. The large masses of sand not found in place have been transported and accumulated by water or by wind, the former being generally considered the most important of these agencies; for the extensive deposits of the Sahara, of the deserts of Persia, and of that of Gobi, are commonly supposed to have been swept together or distributed by marine currents, and to have been elevated above the ocean by the same means as other upheaved strata.

* In the curiously variegated sandstone of Arabia Petræa—which is certainly a reaggregation of loose sand derived from particles of older rocks—the contiguous veins frequently differ very widely in color, but not sensibly in specific gravity or in texture; and the singular way in which they are now alternated, now confusedly intermixed, must be explained otherwise than by the weight of the respective grains which compose them. They seem, in fact, to have been let fall by water in violent ebullition or tumultuous mechanical agitation, or by a succession of sudden aquatic or aerial currents flowing in different directions and charged with differently colored matter.

Meteoric and mechanical influences are still active in the reduction of rocks to a fragmentary state; but the quantity of sand now transported to the sea seems to be comparatively inconsiderable, because—not to speak of the absence of diluvial action—the number of torrents emptying directly into the sea is much less than it was at earlier periods. The formation of alluvial plains in maritime bays, by the sedimentary matter brought down from the mountains, has lengthened the flow of such streams and converted them very generally into rivers, or rather affluents of rivers much younger than themselves. The filling up of the estuaries has so reduced the slope of all large and many small rivers, and, consequently, so checked the current of what the Germans call their *Unterlauf*, or lower course, that they are much less able to transport heavy material than at earlier epochs. The slime deposited by rivers at their junction with the sea, is usually found to be composed of material too finely ground and too light to be denominated sand, and it can be abundantly shown that the sandbanks at the outlet of large streams are of tidal, not of fluviatile origin, or, in lakes and tideless seas, a result of the concurrent action of waves and of wind.

Large deposits of sand, therefore, must in general be considered as of ancient, not of recent formation, and many eminent geologists ascribe them to diluvial action. Staring has discussed this question very fully, with special reference to the sands of the North Sea, the Zuiderzee, and the bays and channels of the Dutch coast.* His general conclusion is, that the

* *De Bodem van Nederland*, i, pp. 243, 246–377, *et seqq*. See also the arguments of Brémontier as to the origin of the dune sands of Gascony, *Annales des Ponts et Chaussées*, 1833, 1er sémestre, pp. 158, 161. Brémontier estimates the sand annually thrown up on that coast at five cubic toises and two feet to the running toise (ubi supra, p. 162), or rather more than two hundred and twenty cubic feet to the running foot. Laval, upon observations continued through seven years, found the quantity to be twenty-five mètres per running mètre, which is equal to two hundred and sixty-eight cubic feet to the running foot.—*Annales des Ponts et Chaussées*, 1842, 2me sémestre, p. 229. These computations make the proportion of sand deposited on the coast of Gascony three or four times as great as that

rivers of the Netherlands "move sand only by a very slow displacement of sandbanks, and do not carry it with them as a suspended or floating material." The sands of the German Ocean he holds to be a product of the "great North German drift," deposited where they now lie before the commencement of the present geological period, and he maintains similar opinions with regard to the sands thrown up by the Mediterranean at the mouths of the Nile and on the Barbary coast.*

Sand now carried to the Sea.

There are, however, cases where mountain streams still bear to the sea perhaps relatively small, but certainly absolutely large, amounts of disintegrated rock.† The quantity of

observed by Andresen on the shores of Jutland. Laval estimates the total quantity of sand annually thrown up on the coast of Gascony at 6,000,000 cubic mètres, or more than 7,800,000 cubic yards.

* *De Bodem van Nederland*, i, p. 339.

† The conditions favorable to the production of sand from disintegrated rock, by causes now in action, are perhaps nowhere more perfectly realized than in the Sinaitic Peninsula. The mountains are steep and lofty, unprotected by vegetation or even by a coating of earth, and the rocks which compose them are in a shattered and fragmentary condition. They are furrowed by deep and precipitous ravines, with beds sufficiently inclined for the rapid flow of water, and generally without basins in which the larger blocks of stone rolled by the torrents can be dropped and left in repose; there are severe frosts and much snow on the higher summits and ridges, and the winter rains are abundant and heavy. The mountains are principally of igneous formation, but many of the less elevated peaks are capped with sandstone, and on the eastern slope of the peninsula you may sometimes see, at a single glance, several lofty pyramids of granite, separated by considerable intervals, and all surmounted by horizontally stratified deposits of sandstone often only a few yards square, which correspond to each other in height, are evidently contemporaneous in origin, and were once connected in continuous beds. The degradation of the rock on which this formation rests is constantly bringing down masses of it, and mingling them with the basaltic, porphyritic, granitic, and calcareous fragments which the torrents carry down to the valleys, and, through them, in a state of greater or less disintegration, to the sea. The quantity of sand annually washed into the Red Sea by the larger torrents of the Lesser Peninsula, is probably at least equal to that contributed to the ocean by

sand and gravel carried into the Mediterranean by the torrents of the Maritime Alps, the Ligurian Apennines, the islands of Corsica, Sardinia, and Sicily, and the mountans of Calabria, is apparently great. In mere mass, it is possible, if not probable, that as much rocky material, more or less comminuted, is contributed to the basin of the Mediterranean by Europe, even excluding the shores of the Adriatic and the Euxine, as is washed up from it upon the coasts of Africa and Syria. A great part of this material is thrown out again by the waves on the European shores of that sea. The harbors of Luni, Albenga, San Remo, and Savona west of Genoa, and of Porto Fino on the other side, are filling up, and the coast near Carrara and Massa is said to have advanced upon the sea to a dis-

any streams draining basins of no greater extent. Absolutely considered, then, the mass may be said to be large, but it is apparently very small as compared with the sand thrown up by the German Ocean and the Atlantic on the coasts of Denmark and of France. There are, indeed, in Arabia Petræa, many torrents with very short courses, for the sea waves in many parts of the peninsular coast wash the base of the mountains. In these cases, the debris of the rocks do not reach the sea in a sufficiently comminuted condition to be entitled to the appellation of sand, or even in the form of well-rounded pebbles. The fragments retain their angular shape, and, at some points on the coast, they become cemented together by lime or other binding substances held in solution or mechanical suspension in the sea water, and are so rapidly converted into a singularly heterogeneous conglomerate, that one deposit seems to be consolidated into a breccia before the next winter's torrents cover it with another.

In the northern part of the peninsula there are extensive deposits of sand intermingled with agate pebbles and petrified wood, but these are evidently neither derived from the Sinaitic group, nor products of local causes known to be now in action.

I may here notice the often repeated but mistaken assertion, that the petrified wood of the Western Arabian desert consists wholly of the stems of palms, or at least of endogenous vegetables. This is an error. I have myself picked up in that desert, within the space of a very few square yards, fragments both of fossil palms, and of at least two petrified trees distinctly marked as of exogenous growth both by annular structure and by knots. In ligneous character, one of these almost precisely resembles the grain of the extant beech, and this specimen was wormeaten before it was converted into silex.

tance of 475 feet in thirty-three years.* Besides this, we have no evidence of the existence of deep-water currents in the Mediterranean, extensive enough and strong enough to transport quartzose sand across the sea. It may be added that much of the rock from which the torrent sands of Southern Europe are derived contains little quartz, and hence the general character of these sands is such that they must be decomposed or ground down to an impalpable slime, long before they could be swept over to the African shore.

The torrents of Europe, then, do not at present furnish the material which composes the beach sands of Northern Africa, and it is equally certain that those sands are not brought down by the rivers of the latter continent. They belong to a remote geological period, and have been accumulated by causes which we cannot at present assign. The wind does not stir water to great depths with sufficient force to disturb the bottom,† and

* Böttger, *Das Mittelmeer*, p. 128.

† The testimony of divers and of other observers on this point is conflicting, as might be expected from the infinite variety of conditions by which the movement of water is affected. It is generally believed that the action of the wind upon the water is not perceptible at greater depths than from fifteen feet in ordinary, to eighty or ninety in extreme cases; but these estimates are probably very considerably below the truth. Andresen quotes Brémontier as stating that the movement of the waves sometimes extends to the depth of five hundred feet, and he adds that others think it may reach to six or even seven hundred feet below the surface.—Andresen, *Om Klitformationen*, p. 20.

Many physicists now suppose that the undulations of great bodies of water reach even deeper. But a movement of undulation is not necessarily a movement of translation, and besides, there is very frequently an undertow, which tends to carry suspended bodies out to sea as powerfully as the superficial waves to throw them on shore. Sandbanks sometimes recede from the coast, instead of rolling toward it. Reclus informs us that the Mauvaise, a sandbank near the Point de Grave, on the Atlantic coast of France, has moved five miles to the west in less than a century.—*Revue des Deux Mondes*, for December, 1862, p. 905.

The action of currents may, in some cases, have been confounded with that of the waves. Sea currents, strong enough, possibly, to transport sand for some distance, flow far below the surface in parts of the open ocean, and in narrow straits they have great force and velocity. The

the sand thrown upon the coast in question must be derived from a narrow belt of sea. It must hence, in time, become exhausted, and the formation of new sandbanks and dunes upon the southern shores of the Mediterranean will cease at last for want of material.*

But even in the cases where the accumulations of sand in extensive deserts appear to be of marine formation, or rather aggregation, and to have been brought to their present position by upheaval, they are not wholly composed of material collected or distributed by the currents of the sea; for, in all such regions, they continue to receive some small contributions from the disintegration of the rocks which underlie, or crop out through, the superficial deposits. In some instances, too, as in Northern Africa, additions are constantly made to the mass by the prevalence of sea winds, which transport, or, to

divers employed at Constantinople in 1853 found in the Bosphorus, at the depth of twenty-five fathoms and at a point much exposed to the wash from Galata and Pera, a number of bronze guns supposed to have belonged to a ship of war blown up about a hundred and fifty years before. These guns were not covered by sand or slime, though a crust of earthy matter, an inch in thickness, adhered to their upper surfaces, and the bottom of the strait appeared to be wholly free from sediment. The current was so powerful at this depth that the divers were hardly able to stand, and a keg of nails, purposely dropped into the water, in order that its movements might serve as a guide in the search for a bag of coin accidentally lost overboard from a ship in the harbor, was rolled by the stream several hundred yards before it stopped.

* Few seas have thrown up so much sand as the shallow German Ocean; but there is some reason to think that the amount of this material now cast upon its northern shores is less than at some former periods, though no extensive series of observations on this subject has been recorded. On the Spit of Agger, at the present outlet of the Liimfjord, Andresen found the quantity during ten years, on a beach about five hundred and seventy feet broad, equal to an annual deposit of an inch and a half over the whole surface.—*Om Klitformationen*, p. 56.

This gives seventy-one and a quarter cubic feet to the running foot—a quantity certainly much smaller than that cast up by the same sea on the shores of the Dano-German duchies and of Holland, and, as we have seen, scarcely one fourth of that deposited by the Atlantic on the coast of Gascony. See *ante*, p. 453, note.

speak more precisely, roll the finer beach sand to considerable distances into the interior. But this is a very slow process, and the exaggerations of travellers have diffused a vast deal of popular error on the subject.

Sands of Egypt.

In the narrow valley of the Nile—which, above its bifurcation near Cairo, is, throughout Egypt and Nubia, generally bounded by precipitous cliffs—wherever a ravine or other considerable depression occurs in the wall of rock, one sees what seems a stream of desert sand pouring down, and common observers have hence concluded that the whole valley is in danger of being buried under a stratum of infertile soil. The ancient Egyptians apprehended this, and erected walls, often of unburnt brick, across the outlet of gorges and lateral valleys, to check the flow of the sand streams. In later ages, these walls have mostly fallen into decay, and no preventive measures against such encroachments are now resorted to. But the extent of the mischief to the soil of Egypt, and the future danger from this source, have been much overrated. The sand on the borders of the Nile is neither elevated so high by the wind, nor transported by that agency in so great masses, as is popularly supposed; and of that which is actually lifted or rolled and finally deposited by air currents, a considerable proportion is either calcareous, and, therefore, readily decomposable, or in the state of a very fine dust, and so, in neither case, injurious to the soil. There are, indeed, both in Africa and in Arabia, considerable tracts of fine silicious sand, which may be carried far by high winds, but these are exceptional cases, and in general the progress of the desert sand is by a rolling motion along the surface.* So little is it lifted, and so incon-

* Sand heaps, three and even six hundred feet high, are indeed formed by the wind, but this is effected by driving the particles up an inclined plane, not by lifting them. Brémontier, speaking of the sand hills on the western coast of France, says: "The particles of sand composing them are not large enough to resist wind of a certain force, nor small enough to

siderable is the quantity yet remaining on the borders of Egypt, that a wall four or five feet high suffices for centuries to check its encroachments. This is obvious to the eye of every observer who prefers the true to the marvellous; but the old-world fable of the overwhelming of caravans by the fearful simoom—which even the Arabs no longer repeat, if indeed they are the authors of it—is so thoroughly rooted in the imagination of Christendom that most desert travellers, of the tourist class, think they shall disappoint the readers of their journals if they do not recount the particulars of their escape from being buried alive by a sand storm, and the popular demand for a "sensation" must be gratified accordingly.*

be taken up by it, like dust; they only roll along the surface from which they are detached, and, though moving with great velocity, they rarely rise to a greater height than three or four inches."—*Mémoire sur les Dunes*, *Annales des Ponts et Chaussées*, 1833, 1er sémestre, p. 148.

Andresen says that a wind, having a velocity of forty feet per second, is strong enough to raise particles of sand as high as the face and eyes of a man, but that, in general, it rolls along the ground, and is scarcely ever thrown more than to the height of a couple of yards from the surface. Even in these cases, it is carried forward by a hopping, not a continuous, motion; for a very narrow sheet or channel of water stops the drift entirely, all the sand dropping into it until it is filled up.

The character of the motion of sand drifts is well illustrated by an interesting fact not much noticed hitherto by travellers in the East. In situations where the sand is driven through depressions in rock beds, or over deposits of silicious pebbles, the surface of the stone is worn and smoothed much more effectually than it could be by running water, and you may pick up, in such localities, rounded, irregularly broken fragments of agate, which have received from the attrition of the sand as fine a polish as could be given them by the wheel of the lapidary.

Very interesting observations on the polishing of hard stones by drifting sand will be found in the Geological Report of William P. Blake: *Pacific Railroad Report*, vol. v, pp. 92, 230, 231. The same geologist observes, p. 242, that the sand of the Colorado desert does not rise high in the air, but bounds along on the surface or only a few inches above it.

* Wilkinson says that, in much experience in the most sandy parts of the Libyan desert, and much inquiry of the best native sources, he never saw or heard of any instance of danger to man or beast from the mere accumulation of sand transported by the wind. Chesney's observations in

Another circumstance is necessary to be considered in estimating the danger to which the arable lands of Egypt are exposed. The prevailing wind in the valley of the Nile and its borders is from the north, and it may be said without exaggeration that the north wind blows for three quarters of the year.* The effect of winds blowing up the valley is to drive the sands of the desert plateau which border it, in a direction parallel with the axis of the valley, not transversely to it; and if it ran in a straight line, the north wind would carry no desert sand into it. There are, however, both curves and angles in its course, and hence, wherever its direction deviates from that of the wind, it might receive sand drifts from the desert plain through which it runs. But, in the course of ages, the winds have, in a great measure, bared the projecting points of their ancient deposits, and no great accumulations remain in situations from which either a north or a south wind would carry them into the valley.†

The Suez Canal.

These considerations apply, with equal force, to the supposed danger of the obstruction of the Suez Canal by the drift-

Arabia, and the testimony of the Bedouins he consulted, are to the same purpose. The dangers of the simoom are of a different character, though they are certainly aggravated by the blinding effects of the light particles of dust and sand borne along by it, and by that of the inhalation of them upon the respiration.

* In the narrow valley of the Nile, bounded as it is, above the Delta, by high cliffs, all air currents from the northern quarter become north winds, though of course varying in partial direction, in conformity with the sinuosities of the valley. Upon the desert plateau they incline westward, and have already borne into the valley the sands of the eastern banks, and driven those of the western quite out of the Egyptian portion of the Nile basin.

† "The North African desert falls into two divisions: the Sahel, or western, and the Sahar, or eastern. The sands of the Sahar were, at a remote period, drifted to the west. In the Sahel, the prevailing east winds drive the sand-ocean with a progressive westward motion. The eastern half of the desert is swept clean."—NAUMANN, *Geognosie*, ii, p. 1173.

ing of the desert sands. The winds across the isthmus are almost uniformly from the north, and they swept it clean of flying sands long ages since. The traces of the ancient canal between the Red Sea and the Nile are easily followed for a considerable distance from Suez. Had the drifts upon the isthmus been as formidable as some have feared and others have hoped, those traces would have been obliterated, and Lake Timsah and the Bitter Lakes filled up, many centuries ago. The few particles driven by the rare east and west winds toward the line of the canal, would easily be arrested by plantations or other simple methods, or removed by dredging. The real dangers and difficulties of this magnificent enterprise—and they are great—consist in the nature of the soil to be removed in order to form the line, and especially in the constantly increasing accumulation of sea sand at the southern terminus by the tides of the Red Sea, and at the northern, by the action of the winds. Both seas are shallow for miles from the shore, and the excavation and maintenance of deep channels, and of capacious harbors with easy and secure entrances, in such localities, is doubtless one of the hardest problems offered to modern engineers for practical solution.

Sands of Egypt.

The sand let fall in Egypt by the north wind is derived, not from the desert, but from a very different source—the sea. Considerable quantities of sand are thrown up by the Mediterranean, at and between the mouths of the Nile, and indeed along almost the whole southern coast of that sea, and drifted into the interior to distances varying according to the force of the wind and the abundance and quality of the material. The sand so transported contributes to the gradual elevation of the Delta, and of the banks and bed of the river itself. But just in proportion as the bed of the stream is elevated, the height of the water in the annual inundations is increased also, and as the inclination of the channel is diminished, the rapidity of the current is checked, and the deposition of the slime it holds in suspension consequently promoted. Thus the winds and the water, moving in contrary directions, join in producing a common effect.

The sand, blown over the Delta and the cultivated land higher up the stream during the inundation, is covered or mixed with the fertile earth brought down by the river, and no serious injury is sustained from it. That spread over the same ground after the water has subsided, and during the short period when the soil is not stirred by cultivation or covered by the flood, forms a thin pellicle over the surface as far as it extends, and serves to divide and distinguish the successive layers of slime deposited by the annual inundations. The particles taken up by the wind on the sea beach are borne onward, by a hopping motion, or rolled along the surface, until they are arrested by the temporary cessation of the wind, by vegetation, or by some other obstruction, and they may, in process of time, accumulate in large masses, under the lee of rocky projections, buildings, or other barriers which break the force of the wind.

In these facts we find the true explanation of the sand drifts, which have half buried the Sphinx and so many other ancient monuments in that part of Egypt. These drifts, as I have said, are not primarily from the desert, but from the sea; and, as might be supposed from the distance they have travelled, they have been long in gathering. While Egypt was a great and flourishing kingdom, measures were taken to protect its territory against the encroachment of sand, whether from the desert or from the sea; but the foreign conquerors, who destroyed so many of its religious monuments, did not spare its public works, and the process of physical degradation undoubtedly began as early as the Persian invasion. The urgent necessity, which has compelled all the successive tyrannies of Egypt to keep up some of the canals and other arrangements for irrigation, was not felt with respect to the advancement of the sands; for their progress was so slow as hardly to be perceptible in the course of a single reign, and long experience has shown that, from the natural effect of the inundations, the cultivable soil of the valley is, on the whole, trenching upon the domain of the desert, not retreating before it.

The oases of the Libyan, as well as of many Asiatic deserts, have no such safeguards. The sands are fast encroaching upon them, and threaten soon to engulf them, unless man shall resort to artesian wells and plantations, or to some other efficient means of checking the advance of this formidable enemy, in time to save these islands of the waste from final destruction.

Accumulations of sand are, in certain cases, beneficial as a protection against the ravages of the sea; but, in general, the vicinity, and especially the shifting of bodies of this material, are destructive to human industry, and hence, in civilized countries, measures are taken to prevent its spread. This, however, can be done only where the population is large and enlightened, and the value of the soil, or of the artificial erections and improvements upon it, is considerable. Hence in the deserts of Africa and of Asia, and the inhabited lands which border on them, no pains are usually taken to check the drifts, and when once the fields, the houses, the springs, or the canals of irrigation are covered or choked, the district is abandoned without a struggle, and surrendered to perpetual desolation.*

Sand Dunes and Sand Plains.

Two forms of sand deposit are specially important in European and American geography. The one is that of dune or shifting hillock upon the coast, the other that of barren plain in the interior. The coast dunes are composed of sand washed

* In parts of the Algerian desert, some efforts are made to retard the advance of sand dunes which threaten to overwhelm villages. "At Debila," says Laurent, "the lower parts of the lofty dunes are planted with palms, * * * but they are constantly menaced with burial by the sands. The only remedy employed by the natives consists in little dry walls of crystallized gypsum, built on the crests of the dunes, together with hedges of dead palm leaves. These defensive measures are aided by incessant labor; for every day the people take up in baskets the sand blown over to them the night before and carry it back to the other side of the dune."—*Mémoires sur le Sahara*, p. 14.

up from the depths of the sea by the waves, and heaped in knolls and ridges by the winds. The sand with which many plains are covered, appears sometimes to have been deposited upon them while they were yet submerged, sometimes to have been drifted from the sea coast, and scattered over them by wind currents, sometimes to have been washed upon them by running water. In these latter cases, the deposit, though in itself considerable, is comparatively narrow in extent and irregular in distribution, while, in the former, it is often evenly spread over a very wide surface. In all great bodies of either sort, the silicious grains are the principal constituent, though, when not resulting from the disintegration of silicious rock and still remaining in place, they are generally accompanied with a greater or less admixture of other mineral particles, and of animal and vegetable remains,* and they are, also, usually somewhat changed in consistence by the ever-varying conditions of temperature and moisture to which they have been exposed since their deposit. Unless the proportion of these latter ingredients is so large as to create a certain adhesiveness in the mass—in which case it can no longer properly be called sand—it is infertile, and, if not charged with water, partially agglutinated by iron, lime, or other cement, or confined by alluvion resting upon it, it is much inclined to drift, whenever,

* Organic constituents, such as comminuted shells, and silicious and calcareous exuviæ of infusorial animals and plants, are sometimes found mingled in considerable quantities with mineral sands. These are usually the remains of aquatic vegetables or animals, but not uniformly so, for the microscopic organisms, whose flinty cases enter so largely into the sand-beds of the Mark of Brandenburg, are still living and prolific in the dry earth. See WITTWER, *Physikalische Geographie*, p. 142.

The desert on both sides of the Nile is inhabited by a land snail, and thousands of its shells are swept along and finally buried in the drifts by every wind. Every handful of the sand contains fragments of them. FORCHHAMMER, in LEONHARD und BRONN'S *Jahrbuch*, 1841, p. 8, says of the sand hills of the Danish coast: "It is not rare to find, high in the knolls, marine shells, and especially those of the oyster. They are due to the oyster eater [*Hæmalopus ostralegus*], which carries his prey to the top of the dunes to devour it." See also STARING, *De Bodem van*, N. I. p. 321.

by any chance, the vegetable network which, in most cases, thinly clothes and at the same time confines it, is broken.

Human industry has not only fixed the flying dunes, but, by mixing clay and other tenacious earths with the superficial stratum of extensive sand plains, and by the application of fertilizing substances, it has made them abundantly productive of vegetable life. These latter processes belong to agriculture and not to geography, and, therefore, are not embraced within the scope of the present subject. But the preliminary steps, whereby wastes of loose, drifting barren sands are transformed into wooded knolls and plains, and finally, through the accumulation of vegetable mould, into arable ground, constitute a conquest over nature which precedes agriculture—a geographical revolution—and, therefore, an account of the means by which the change has been effected belongs properly to the history of man's influence on the great features of physical geography. I proceed, then, to examine the structure of dunes, and to describe the warfare man wages with the sand hills, striving on the one hand to maintain and even extend them, as a natural barrier against encroachments of the sea, and, on the other, to check their moving and wandering propensities, and prevent them from trespassing upon the fields he has planted and the habitations in which he dwells.

Coast Dunes.

Coast dunes are oblong ridges or round hillocks, formed by the action of the wind upon sands thrown up by the waves on the beach of seas, and sometimes of fresh-water lakes. On most coasts, the supply of sand for the formation of dunes is derived from tidal waves. The flow of the tide is more rapid, and consequently its transporting power greater, than that of the ebb; the momentum, acquired by the heavy particles in rolling in with the water, tends to carry them even beyond the flow of the waves; and at the turn of the tide, the water is in a state of repose long enough to allow it to let fall much of the solid matter it holds in suspension. Hence, on all low, tide-washed coasts of seas with sandy bottoms, there exist several conditions favorable to the formation of sand deposits along high-water mark.* If the land winds are of greater fre-

* There are various reasons why the formation of dunes is confined to low shores, and this law is so universal, that when bluffs are surmounted by them, there is always cause to suspect upheaval, or the removal of a sloping beach in front of the bluff, after the dunes were formed. Bold shores are usually without a sufficient beach for the accumulation of large deposits; they are commonly washed by a sea too deep to bring up sand from its bottom; their abrupt elevation, even if moderate in amount, would still be too great to allow ordinary winds to lift the sand above them; and their influence in deadening the wind which blows toward them would even more effectually prevent the raising of sand from the beach at their foot.

Forchhammer, describing the coast of Jutland, says that, in high winds, "one can hardly stand upon the dunes, except when they are near the water line and have been cut down perpendicularly by the waves. Then the wind is little or not at all felt—a fact of experience very common on our coasts, observed on all the steep shore bluffs of two hundred feet in height, and, in the Faroe Islands, on precipices two thousand feet high. In heavy gales in those islands, the cattle fly to the very edge of the cliffs for shelter, and frequently fall over. The wind, impinging against the vertical wall, creates an ascending current which shoots somewhat past the crest of the rock, and thus the observer or the animal is protected against the tempest by a barrier of air."—Leonhard und Bronn, *Jahrbuch*, 1841, p. 3.

The calming, or rather diversion, of the wind by cliffs extends to a considerable distance in front of them, and no wind would have sufficient force to raise the sand vertically, parallel to the face of a bluff, even to the height of twenty feet.

It is very commonly believed that it is impossible to grow forest trees on sea-shore bluffs, or points much exposed to strong winds. The observations just cited tend to show that it would not be difficult to protect trees from the mechanical effect of the wind, by screens much lower than the height to which they are expected to grow. Recent experiments confirm this, and it is found that, though the outer row or rows may suffer from the wind, every tree shelters a taller one behind it. Extensive groves have thus been formed in situations where an isolated tree would not grow at all.

Piper, in his *Trees of America*, p. 19, gives an interesting account of Mr. Tudor's success in planting trees on the bleak and barren shore of Nahant. "Mr. Tudor," observes he, "has planted more than ten thousand trees at Nahant, and, by the results of his experiments, has fully demonstrated that trees, properly cared for in the beginning, may be made to grow up to the

quency, duration, or strength than the sea winds, the sands left by the retreating wave will be constantly blown back into the water; but if the prevailing air currents are in the opposite direction, the sands will soon be carried out of the reach of the highest waves, and transported continually farther and farther into the interior of the land, unless obstructed by high grounds, vegetation, or other obstacles.

The tide, though a usual, is by no means a necessary condition for the accumulations of sand out of which dunes are formed. The Baltic and the Mediterranean are almost tideless seas, but there are dunes on the Russian and Prussian coasts of the Baltic, and at the mouths of the Nile and many other points on the shores of the Mediterranean. The vast shoals in the latter sea, known to the ancients as the Greater and Lesser Syrtis, are of marine origin. They are still filling up with sand, washed up from greater depths, or sometimes drifted from the coast in small quantities, and will probably be converted, at some future period, into dry land covered with sand hills. There are also extensive ranges of dunes upon the eastern shores of the Caspian, and at the southern, or rather southeastern extremity of Lake Michigan.* There is no doubt that this latter lake formerly extended much farther in that direction, but its southern portion has gradually shoaled and at last been converted into solid land, in consequence of the prevalence of the northwest winds. These blow over the lake a large part of the year, and create a southwardly set of the currents, which wash up sand from the bed of the lake and throw it on shore. Sand is taken up from the beach at Michigan City by every wind from that quarter, and, after a heavy blow of

very bounds of the ocean, exposed to the biting of the wind and the spray of the sea. The only shelter they require is, at first, some interruption to break the current of the wind, such as fences, houses, or other trees."

* The careful observations of Colonel J. D. Graham, of the United States Army, show a tide of about three inches in Lake Michigan. See "A Lunar Tidal Wave in the North American Lakes," demonstrated by Lieut.-Colonel J. D. Graham, in the fourteenth volume of the *Proceedings of the American Association for the Advancement of Science.*

some hours' duration, sand ridges may be observed on the north side of the fences, like the snow wreaths deposited by a drifting wind in winter. Some of the particles are carried back by contrary winds, but most of them lodge on or behind the dunes, or in the moist soil near the lake, or are entangled by vegetables, and tend permanently to elevate the level. Like effects are produced by constant sea winds, and dunes will generally be formed on all low coasts where such prevail, whether in tideless or in tidal waters.

Jobard thus describes the *modus operandi*, under ordinary circumstances, at the mouths of the Nile, where a tide can scarcely be detected: "When a wave breaks, it deposits an almost imperceptible line of fine sand. The next wave brings also its contribution, and shoves the preceding line a little higher. As soon as the particles are fairly out of the reach of the water they are dried by the heat of the burning sun, and immediately seized by the wind and rolled or borne farther inland. The gravel is not thrown out by the waves, but rolls backward and forward until it is worn down to the state of fine sand, when it, in its turn, is cast upon the land and taken up by the wind." * This description applies only to the common every-day action of wind and water; but just in proportion to the increasing force of the wind and the waves, there is an increase in the quantity of sand, and in the magnitude of the particles carried off from the beach by it, and, of course, every storm in a landward direction adds sensibly to the accumulation upon the shore.

Sand Banks.

Although dunes, properly so called, are found only on dry land and above ordinary high-water mark, and owe their elevation and structure to the action of the wind, yet, upon many shelving coasts, accumulations of sand much resembling dunes are formed under water at some distance from the shore by the oscillations of the waves, and are well known by the

* STARING, *De Bodem van Nederland*, i, p. 327, note.

name of sand banks. They are usually rather ridges than banks, of moderate inclination, and with the steepest slope seaward; and their form differs from that of dunes only in being lower and more continuous. Upon the western coast of the island of Amrum, for example, there are three rows of such banks, the summits of which are at a distance of perhaps a couple of miles from each other; so that, including the width of the banks themselves, the spaces between them, and the breadth of the zone of dunes upon the land, the belt of moving sands on that coast is probably not less than eight miles wide.

Under ordinary circumstances, sand banks are always rolling landward, and they compose the magazine from which the material for the dunes is derived. The dunes, in fact, are but aquatic sand banks transferred to dry land. The laws of their formation are closely analogous, because the action of the two fluids, by which they are respectively accumulated and built up, is very similar when brought to bear upon loose particles of solid matter. It would, indeed, seem that the slow and comparatively regular movements of the heavy, unelastic water ought to affect such particles very differently from the sudden and fitful impulses of the light and elastic air. But the velocity of the wind currents gives them a mechanical force approximating to that of the slower waves, and, however difficult it may be to explain all the phenomena that characterize the structure of the dunes, observation has proved that it is nearly identical with that of submerged sand banks. The differences of form are generally ascribable to the greater number and variety of surface accidents of the ground on which the sand hills of the land are built up, and to the more frequent changes, and wider variety of direction, in the courses of the wind.

Dunes on the Coast of America.

Upon the Atlantic coast of the United States, the prevalence of western or off-shore winds is unfavorable to the formation of dunes, and, though marine currents lodge vast quantities of sand, in the form of banks, on that coast, its shores are proportionally more free from sand hills than some others of lesser extent. There are, however, very important exceptions. The action of the tide throws much sand upon some points of the New England coast, as well as upon the beaches of Long Island and other more southern shores, and here dunes resembling those of Europe are formed. There are also extensive ranges of dunes on the Pacific coast of the United States, and at San Francisco they border some of the streets of the city.

The dunes of America are far older than her civilization, and the soil they threaten or protect possesses, in general, too little value to justify any great expenditure in measures for arresting their progress or preventing their destruction. Hence, great as is their extent and their geographical importance, they have, at present, no such intimate relations to human life as to render them objects of special interest in the point of view I am taking, and I do not know that the laws of their formation and motion have been made a subject of original investigation by any American observer.

Dunes of Western Europe.

Upon the western coast of Europe, on the contrary, the ravages occasioned by the movement of sand dunes, and the serious consequences often resulting from the destruction of them, have long engaged the earnest attention of governments and of scientific men, and for nearly a century persevering and systematic effort has been made to bring them under human control. The subject has been carefully studied in Denmark and the adjacent duchies, in Western Prussia, in the Netherlands, and in France; and the experiments in the way of arresting the drifting of the dunes, and of securing them, and the lands they shelter, from the encroachments of the sea, have resulted in the adoption of a system of coast improvement substantially the same in all these countries. The sands, like the forests, have now their special literature, and the volumes and memoirs, which describe them and the processes employed to subdue them, are full of scientific interest and of practical instruction.*

Formation of Dunes.

The laws which govern the formation of dunes are substantially these. We have seen that, under certain conditions, sand is accumulated above high-water mark on low sea and lake shores. So long as the sand is kept wet by the spray or by capillary attraction, it is not disturbed by air currents, but as soon as the waves retire sufficiently to allow it to dry, it

* The principal special works and essays on this subject known to me are:

BRÉMONTIER, *Mémoire sur les Dunes, etc.*, 1790, reprinted in *Annales des Ponts et Chaussées*, 1833, 1er sémestre, pp. 145–186.

Rapport sur les différents Mémoires de M. Brémontier, par LAUMONT et autres, 1806, same volume, pp. 192, 224.

LEFORT, *Notice sur les Travaux de Fixation des Dunes*, *Annales des Ponts et Chaussées*, 1831, 2me sémestre, pp. 320–332.

FORCHHAMMER, *Geognostische Studien am Meeres Ufer*, in LEONHARD und BRONN, *Jahrbuch, etc.*, 1841, pp. 1, 38.

J. G. KOHL, *Die Inseln und Marschen der Herzogthümer Schleswig und Holstein*, 1846, vol. ii, pp. 112–162, 193–204.

LAVAL, *Mémoire sur les Dunes du Golfe de Gascogne*, *Annales des Ponts et Chaussées*, 1847, 2me sémestre, pp. 218–268.

G. C. A. KRAUSE, *Der Dünenbau auf den Ostsee-Küsten West-Preussens*, 1850, 1 vol. 8vo.

W. C. H. STARING, *De Bodem van Nederland*, 1856, vol. i, pp. 310–341, and 424–431.

Same author, *Voormaals en Thans*, 1858, pages cited.

C. C. ANDRESEN, *Om Klitformationen og Klittens Behandling og Bestyrelse*, 1861, 1 vol. 8vo, x, 392 pp., much the most complete treatise on the subject.

ANDRESEN cites, upon the origin of the dunes: HULL, *Over den Oorsprong en de Geschiedenis der Hollandsche Duinen*, 1838, and GROSS's *Veiledning ved Behandlingen af Sandflugtstrækningerne*, 1847; and upon the improvement of sand plains by planting, PANNEWITZ, *Anleitung zum Anbau der Sandflächen*, 1832. I am not acquainted with either of the latter two works but I have consulted with advantage, on this subject, DELAMARRE, *Historique de la Création d'une Richesse millionaire par la culture des Pins*, 1827; BOITEL, *Mise en valeur des terres pauvres par le Pin maritime*, 1857; and BRINCKEN, *Ansichten über die Bewaldung der Steppen des Europäischen Russlands*, 1854.

becomes the sport of the wind, and is driven up the gently sloping beach until it is arrested by stones, vegetables, or other obstructions, and thus an accumulation is formed which constitutes the foundation of a dune. However slight the elevation thus created, it serves to stop or retard the progress of the sand grains which are driven against its shoreward face, and to protect from the further influence of the wind the particles which are borne beyond it, or rolled over its crest, and fall down behind it. If the shore above the beach line were perfectly level and straight, the grass or bushes upon it of equal height, the sand thrown up by the waves uniform in size and weight of particles as well as in distribution, and if the action of the wind were steady and regular, a continuous bank would be formed, everywhere alike in height and cross section. But no such constant conditions anywhere exist. The banks are curved, broken, unequal in elevation; they are sometimes bare, sometimes clothed with vegetables of different structure and dimensions; the sand thrown up is variable in quantity and character; and the winds are shifting, gusty, vortical, and often blowing in very narrow currents. From all these causes, instead of uniform hills, there rise irregular rows of sand heaps, and these, as would naturally be expected, are of a pyramidal, or rather conical shape, and connected at bottom by more or less continuous ridges of the same material.

On a receding coast, dunes will not attain so great a height as on more secure shores, because they are undermined and carried off before they have time to reach their greatest dimensions. Hence, while at sheltered points in Southwestern France, there are dunes three hundred feet or more in height, those on the Frisic Islands and the exposed parts of the coast of Schleswig-Holstein range only from twenty to one hundred feet. On the western shores of Africa, it is said that they sometimes attain an elevation of six hundred feet. This is one of the very few points known to geographers where desert sands are advancing seaward, and here they rise to the greatest altitude to which sand grains can be carried by the wind.

The hillocks, once deposited, are held together and kept in

shape, partly by mere gravity, and partly by the slight cohesion of the lime, clay, and organic matter mixed with the sand; and it is observed that, from capillary attraction, evaporation from lower strata, and retention of rain water, they are always moist a little below the surface.* By successive accumulations, they gradually rise to the height of thirty, fifty, sixty, or a hundred feet, and sometimes even much higher. Strong winds, instead of adding to their elevation, sweep off loose particles from their surface, and these, with

* "Dunes are always full of water, from the action of capillary attraction. Upon the summits, one seldom needs to dig more than a foot to find the sand moist, and in the depressions, fresh water is met with near the surface."—FORCHHAMMER, in LEONHARD und BRONN, for 1841, p. 5, note.

On the other hand, Andresen, who has very carefully investigated this as well as all other dune phenomena, maintains that the humidity of the sand ridges cannot be derived from capillary attraction. He found by experiment that drift sand was not moistened to a greater height than eight and a half inches, after standing a whole night in water. He states the minimum of water contained by the sand of the dunes, one foot below the surface, after a long drought, at two per cent., the maximum, after a rainy month, at four per cent. At greater depths the quantity is larger. The hygroscopicity of the sand of the coast of Jutland he found to be thirty-three per cent. by measure, or 21.5 by weight. The annual precipitation on that coast is twenty-seven inches, and, as the evaporation is about the same, he argues that rain water does not penetrate far beneath the surface of the dunes, and concludes that their humidity can be explained only by evaporation from below.—*Om Klitformationen*, pp. 106–110.

In the dunes of Algeria, water is so abundant that wells are constantly dug in them at high points on their surface. They are sunk to the depth of three or four mètres only, and the water rises to the height of a mètre in them.—LAURENT, *Mémoire sur le Sahara*, pp. 11, 12, 13.

The same writer observes (p. 14) that the hollows in the dunes are planted with palms which find moisture enough a little below the surface. It would hence seem that the proposal to fix the dunes which are supposed to threaten the Suez Canal, by planting the maritime pine and other trees upon them, is not altogether so absurd as it is thought to be by some of those disinterested philanthropists of other nations who are distressed with fears that French capitalists will lose the money they have invested in that great undertaking.

Ponds of water are often found in the depressions between the sand hills of the dune chains in the North American desert.

others blown over or between them, build up a second row of dunes, and so on according to the character of the wind, the supply and consistence of the sand, and the face of the country In this way is formed a belt of sand dunes, irregularly dispersed and varying much in height and dimensions, and sometimes many miles in breadth. On the Island of Sylt, in the German Sea, where there are several rows, the width of the belt is from half a mile to a mile. There are similar ranges on the coast of Holland, exceeding two miles in breadth, while at the mouths of the Nile they form a zone not less than ten miles wide. The base of some of the dunes in the Delta of the Nile is reached by the river during the annual inundation, and the infiltration of the water, which contains lime, has converted the lower strata into a silicious limestone, or rather a calcareous sandstone, and thus afforded an opportunity of studying the structure of that rock in a locality where its origin and mode of aggregation and solidification are known.

Character of Dune Sand.

"Dune sand," says Staring, "consists of well-rounded grains of quartz, more or less colored by iron, and often mingled with fragments of shells, small indeed, but still visible to the naked eye.* These fragments are not constant constituents of dune sand. They are sometimes found at the very summits of the hillocks, as at Overveen; in the King's Dune,

* According to the French authorities, the dunes of France are not always composed of quartzose sand. "The dune sands" of different characters, says Brémontier, "partake of the nature of the different materials which compose them. At certain points on the coast of Normandy they are found to be purely calcareous; they are of mixed composition on the shores of Brittany and Saintonge, and generally quartzose between the mouth of the Gironde and that of the Adour."—*Mémoire sur les Dunes, Annales des Ponts et Chaussées*, t. vii, 1833, 1er sémestre, p. 146.

In the dunes of Long Island and of Jutland, there are considerable veins composed almost wholly of garnet. For a very full examination of the mechanical and chemical composition of the dune sands of Jutland, see Andresen, *Om Klitformationen*, p. 110.

near Egmond, they form a coarse calcareous gravel very largely distributed through the sand, while the interior dunes between Haarlem and Warmond exhibit no trace of them. It is yet undecided whether the presence or absence of these fragments is determined by the period of the formation of the dunes, or whether it depends on a difference in the process by which different dunes have been accumulated. Land shells, such as snails, for example, are found on the surface of the dunes in abundance, and many of the shelly fragments in the interior of the hillocks may be derived from the same source." *

J. G. Kohl has some poetical thoughts upon the origin and character of the dune sands, which are worth quoting:

"The sand was composed of pure transparent quartz. I could not observe this sand without the greatest admiration. If it is the product of the waves, breaking and crushing flints and fragments of quartz against each other, it is a result which could be brought about only in the course of countless ages. We need not lift ourselves to the stars, to their incalculable magnitudes and distances and numbers, in order to feel the giddiness of astonishment. Here, upon earth, in the simple sand, we find miracle enough. Think of the number of sand grains contained in a single dune, then of all the dunes upon this widely extended coast—not to speak of the innumerable grains in the Arabian, African, and Prussian deserts—this, of itself, is sufficient to overwhelm a thoughtful fancy. How long, how many times must the waves have risen and sunk in order to reduce these vast heaps to powder!

"During the whole time I spent on this coast, I had always some sand in my fingers, was rubbing and rolling it about, examining it on all sides, holding a little shining grain on the tip of my finger, and thinking to myself how, in its corners, its angles, its whole configuration, it might very probably have a history longer than that of the old German nation—possibly longer than that of the human race. Where was the original quartz crystal, of which this is a fragment, first

* *De Bodem van Nederland*, i, p. 323.

formed? To what was it once fixed? What power broke it loose? How was it beaten smaller and ever smaller by the waves? They tossed it, for æons, to and fro upon the beach, rolled it up and down, forced it to make thousands and thousands of daily voyages for millions and millions of days. Then the wind bore it away, and used it in building up a dune; there it lay for centuries, packed in with its fellows, protecting the marshes and cherished by the inhabitants, till, seized again by the pursuing sea, it fell once more into the water, there to begin the endless dance anew—and again to be swept away by the wind—and again to find rest in the dunes, a protection and a blessing to the coast. There is something mysterious about such a grain of sand, and at last I went so far as to fancy a little immortal spark linked with each one, presiding over its destiny, and sharing its vicissitudes. Could we arm our eyes with a microscope, and then dive, like a sparling, into one of these dunes, the pile, which is in fact only a heap of countless little crystal blocks, would strike us as the most marvellous building upon earth. The sunbeams would pass, with illuminating power, through all these little crystalline bodies. We should see how every sand grain is formed, by what multifarious little facets it is bounded, we should even discover that it is itself composed of many distinct particles." *

Sand concretions form within the dunes and especially in the depressions between them. These are sometimes so extensive and impervious as to retain a sufficient supply of water to feed perennial springs, and to form small permanent ponds, and they are a great impediment to the penetration of roots, and consequently to the growth of trees planted, or germinating from self-sown seeds, upon the dunes.†

* J. G. Kohl, *Die Inseln und Marschen der Herzogthümer Schleswig und Holstein*, ii, p. 200.

† Staring, *De Bodem van Nederland*, i, p. 317. See also, Bergsöe, *Reventlov's Virksomhed*, ii, p. 11.

"In the sand-hill ponds mentioned in the text, there is a vigorous growth of bog plants accompanied with the formation of peat, which goes on regularly as long as the dune sand does not drift. But if the surface of the dunes is broken, the sand blows into the ponds, covers the peat, and

Interior Structure of Dunes.

The interior structure of the dunes, the arrangement of their particles, is not, as might be expected, that of an unorganized, confused heap, but they show a strong tendency to stratification. This is a point of much geological interest, because it indicates that sandstone may owe its stratified character to the action of wind as well as of water. The origin and peculiar character of these layers are due to a variety of causes. A southwest wind and current may deposit upon a dune a stratum of a given color and mineral composition, and this may be succeeded by a northwest wind and current, bringing with them particles of a different hue, constitution, and origin.

Again, if we suppose a violent tempest to strew the beach with sand grains very different in magnitude and specific gravity, and, after the sand is dry, to be succeeded by a gentle breeze, it is evident that only the lighter particles will be taken up and carried to the dunes. If, after some time, the wind freshens, heavier grains will be transported and deposited on the former, and a still stronger succeeding gale will roll up yet larger kernels. Each of these deposits will form a stratum. If we suppose the tempest to be followed, after the sand is dry, not by a gentle breeze, but by a wind powerful enough to lift at the same time particles of very various magnitudes and weights, the heaviest will often lodge on the dune while the lighter will be carried farther. This would produce a stratum of coarse sand, and the same effect might result from the blowing away of light particles out of a mixed layer, while the heavier remained undisturbed.* Still another cause of

puts an end to its formation. When, in the course of time, marine currents cut away the coast, the dunes move landward and fill up the ponds, and thus are formed the remarkable strata of fossile peat called Martörv, which appears to be unknown to the geologists of other parts of Europe."—FORCHHAMMER, in LEONHARD und BRONN, 1841, p. 13.

* The lower strata must be older than the superficial layers, and the particles which compose them may in time become more disintegrated, and therefore finer than those deposited later and above them.

stratification may be found in the occasional interposition of a thin layer of leaves or other vegetable remains between successive deposits, and this I imagine to be more frequent than has been generally supposed.

The eddies of strong winds between the hillocks must also occasion disturbances and re-arrangements of the sand layers, and it seems possible that the irregular thickness and the strange contortions of the strata of the sandstone at Petra may be due to some such cause. A curious observation of Professor Forchhammer suggests an explanation of another peculiarity in the structure of the sandstone of Mount Seir. He describes dunes in Jutland, composed of yellow quartzose sand intermixed with black titanian iron. When the wind blows over the surface of the dunes, it furrows the sand with alternate ridges and depressions, ripples, in short, like those of water. The swells, the dividing ridges of the system of sand ripples, are composed of the light grains of quartz, while the heavier iron rolls into the depressions between, and thus the whole surface of the dune appears as if covered with a fine black network.

Form of Dunes.

The sea side of dunes, being more exposed to the caprices of the wind, is more irregular in form than the lee or land side, where the arrangement of the particles is affected by fewer disturbing and conflicting influences. Hence, the stratification of the windward slope is somewhat confused, while the sand on the lee side is found to be disposed in more regular beds, inclining landward, and with the largest particles lowest, where their greater weight would naturally carry them. The lee side of the dunes, being thus formed of sand deposited according to the laws of gravity, is very uniform in its slope, which, according to Forchhammer, varies little from an angle of 30° with the horizon, while the more exposed and irregular weather side lies at an inclination of from 5° to 10°. When, however, the outer tier of dunes is formed so near the water-line as to be exposed to the immediate action of the waves, it is undermined, and the face of the hill is very steep and sometimes nearly perpendicular.

Geological Importance of Dunes.

These observations, and other facts which a more attentive study on the spot would detect, might furnish the means of determining interesting and important questions concerning geological formations in localities very unlike those where dunes are now thrown up. For example, Studer supposes that the drifting sand hills of the African desert were originally coast dunes, and that they have been transported to their present position far in the interior, by the rolling and shifting leeward movement to which all dunes not covered with vegetation are subject. The present general drift of the sands of that desert appears to be to the southwest and west, the prevailing winds blowing from the northeast and east; but it has been doubted whether the shoals of the western coast of Northern Africa, and the sands upon that shore, are derived from the bottom of the Atlantic, in the usual manner, or, by an inverse process, from those of the Sahara. The latter, as has been before remarked, is probably the truth, though observations are wanting to decide the question.* There is nothing violently improbable in the supposition that they may have been first thrown up by the Mediterranean on its Libyan coast, and thence blown south and west over the vast space they now cover. But whatever has been their source and movement, they can hardly fail to have left on their route some sandstone monuments to mark their progress, such, for example, as we have seen are formed from the dune sand at the mouth of the Nile; and it is conceivable that the character of the drifting sands themselves, and of the conglomerates and sandstones to whose formation they have contributed, might furnish satisfac-

* "On the west coast of Africa the dunes are drifting seawards, and always receiving new accessions from the Sahara. They are constantly advancing out into the sea." See *ante*, p. 16, note.—NAUMANN, *Geognosie*, ii, p. 1172. See *Appendix*, No.

tory evidence as to their origin, their starting point, and the course by which they have wandered so far from the sea.*

If the sand of coast dunes is, as Staring describes it, composed chiefly of well-rounded quartzose grains, fragments of shells, and other constant ingredients, it would often be recognizable as coast sand, in its agglutinate state of sandstone. The texture of this rock varies from an almost imperceptible fineness of grain to great coarseness, and affords good facilities for microscopic observation of its structure. There are sandstones, such, for example, as are used for grindstones, where the grit, as it is called, is of exceeding sharpness; others where the angles of the grains are so obtuse that they scarcely act at all on hard metals. The former may be composed of grains of rock, disintegrated indeed, and recemented together, but not, in the meanwhile, much rolled; the latter, of sands long washed by the sea, and drifted by land winds. There is,

* Forchhammer, after pointing out the coincidence between the inclined stratification of dunes and the structure of ancient tilted rocks, says: "But I am not able to point out a sandstone formation corresponding to the dunes. Probably most ancient dunes have been destroyed by submersion before the loose sand became cemented to solid stone, but we may suppose that circumstances have existed somewhere which have preserved the characteristics of this formation."—Leonhard und Bronn, 1841, p. 8, 9.

Such formations, however, certainly exist. I find from Laurent (*Mémoire sur le Sahara, etc.*, p. 12), that in the Algerian desert there exist "sandstone formations" not only "corresponding to the dunes," but actually consolidated within them. "A place called El-Mouia-Tadjer presents a repetition of what we saw at El-Baya; one of the funnels formed in the middle of the dunes contains wells from two mètres to two and a half in depth, dug in a sand which pressure, and probably the presence of certain salts, have cemented so as to form true sandstone, soft indeed, but which does not yield except to the pickaxe. These sandstones exhibit an inclination which seems to be the effect of wind; for they conform to the direction of the sands which roll down a scarp occasioned by the primitive obstacle." See *Appendix*, No. 57.

The dunes near the mouth of the Nile, the lower sands of which have been cemented together by the infiltration of Nile water, would probably show a similar stratification in the sandstone which now forms their base.

indeed, so much resemblance between the effects of driving winds and of rolling water upon light bodies, that there would be difficulty in distinguishing them;* but after all, it is not probable that sandstone, composed of grains thrown up from the salt sea, and long tossed by the winds, would be identical in its structure with that formed from fragments of rock crushed by mechanical force, or disintegrated by heat, and again agglutinated without much exposure to the action of moving water.†

* Forchhammer ascribes the resemblance between the furrowing of the dune sands and the beach ripples, not to the similarity of the effect of wind and water upon sand, but wholly to the action of the former fluid; in the first instance, directly, in the latter, through the water. "The wind ripples on the surface of the dunes precisely resemble the water ripples of sand flats occasionally overflowed by the sea; and with the closest scrutiny, I have never been able to detect the slightest difference between them. This is easily explained by the fact, that the water ripples are produced by the action of light wind on the water which only transmits the air waves to the sand."—LEONHARD und BRONN, 1841, pp. 7, 8.

† American observers do not agree in their descriptions of the form and character of the sand grains which compose the interior dunes of the North American desert. C. C. Parry, geologist to the Mexican Boundary Commission, in describing the dunes near the station at a spring thirty-two miles west from the Rio Grande at El Paso, says: "The separate grains of the sand composing the sand hills are seen under a lens to be angular, and not rounded, as would be the case in regular beach deposits."—*U. S. Mexican Boundary Survey*, *Report of*, vol. i, *Geological Report of C. C. Parry*, p. 10.

In the general description of the country traversed, same volume, p. 47, Colonel Emory says that on an "examination of the sand with a microscope of sufficient power," the grains are seen to be angular, not rounded by rolling in water.

On the other hand, Blake, in *Geological Report*, *Pacific Railroad Rep.*, vol. v, p. 119, observes that the grains of the dune sand, consisting of quartz, chalcedony, carnelian, agate, rose quartz, and probably chrysolite, were much rounded; and on page 241, he says that many of the sand grains of the Colorado desert are perfect spheres.

On page 20 of a report in vol. ii. of the *Pacific Railroad Report*, by the same observer, it is said that an examination of dune sands brought from the Llano Estacado by Captain Pope, showed the grains to be "much rounded by attrition."

The sands described by Mr. Parry and Colonel Emory are not from the

Inland Dunes.

I have met with some observations indicating a structural difference between interior and coast dunes, which might perhaps be recognized in the sandstones formed from these two species of sand hills respectively. In the great American desert between the Andes and the Pacific, Meyen found sand heaps of a perfect falciform shape.* They were from seven to fifteen feet high, the chord of their arc measuring from twenty to seventy paces. The slope of the convex face is described as very small, that of the concave as high as 70° or 80°, and their surfaces were rippled. No smaller dunes were observed, nor any in the process of formation. The concave side uniformly faced the northwest, except toward the centre of the desert, where, for a distance of one or two hundred paces, they gradually opened to the west, and then again gradually resumed the former position.

Pöppig ascribes a falciform shape to the movable, a conical to the fixed dunes, or *medanos*, of the same desert. "The medanos," he observes, "are hillock-like elevations of sand, some having a firm, others a loose base. The former [latter], which are always crescent shaped, are from ten to twenty feet high, and have an acute crest. The inner side is perpendicular, and the outer or bow side forms an angle with a steep inclination downward. When driven by violent winds, the medanos pass rapidly over the plains. The smaller and lighter ones move quickly forward, before the larger; but the latter soon overtake and crush them, whilst they are themselves shivered by the collision. These medanos assume all sorts of extraordinary figures, and sometimes move along the plain in rows forming

same localities as those examined by Mr. Blake, and the difference in their character may denote a difference of origin or of age.

* LAURENT (*Mémoire sur le Sahara*, pp. 11, 12, and elsewhere) speaks of a funnel-shaped depression at a high point in the dunes, as a characteristic feature of the sand hills of the Algerian desert. This seems to be an approximation to the crescent form noticed by Meyen and Pöppig in the inland dunes of Peru.

most intricate labyrinths. * * A plain often appears to be covered with a row of medanos, and some days afterward it is again restored to its level and uniform aspect. * * *

"The medanos with immovable bases are formed on the blocks of rocks which are scattered about the plain. The sand is driven against them by the wind, and as soon as it reaches the top point, it descends on the other side until that is likewise covered; thus gradually arises a conical-formed hill. Entire hillock chains with acute crests are formed in a similar manner. * * * On their southern declivities are found vast masses of sand, drifted thither by the mid-day gales. The northern declivity, though not steeper than the southern, is only sparingly covered with sand. If a hillock chain somewhat distant from the sea extends in a line parallel with the Andes, namely, from S. S. E. to N. N. W., the western declivity is almost entirely free of sand, as it is driven to the plain below by the southeast wind, which constantly alternates with the wind from the south." *

It is difficult to reconcile this description with that of Meyen, but if confidence is to be reposed in the accuracy of either observer, the formation of the sand hills in question must be governed by very different laws from those which determine the structure of coast dunes. Captain Gilliss, of the American navy, found the sand hills of the Peruvian desert to be in general crescent shaped, as described by Meyen, and a similar structure is said to characterize the inland dunes of the Llano Estacado and other plateaus of the North American desert, though these latter are of greater height and other dimensions than those described by Meyen. There is no very obvious explanation of this difference in form between maritime and inland sand hills, and the subject merits investigation.†

* *Travels in Peru*, New York, 1848, chap. ix.

† Notwithstanding the general tendency of isolated coast dunes and of the peaks of the sand ridges to assume a conical form, Andresen states that the hills of the inner or landward rows are sometimes *bow-shaped*, and sometimes undulating in outline.—*Om Klitformationen*, p. 84. He says further that: "Before an obstruction, two or three feet high and con-

Age, Character, and Permanence of Dunes.

The origin of most great lines of dunes goes back past all history. There are on many coasts, several distinct ranges of sand hills which seem to be of very different ages, and to have been formed under different relative conditions of land and water.* In some cases, there has been an upheaval of the coast

siderably longer, lying perpendicularly to the direction of the wind, the sand is deposited with a windward angle of from 6° to 12°, and the bank presents a concave face to the wind, while, behind the obstruction, the outline is convex;" and he lays it down as a general rule, that a slope, *from* which sand is blown, is left with a concavity of about one inch of depth to four feet of distance; a slope, *upon* which sand is dropped by the wind, is convex. It appears from Andresen's figures, however, that the concavity and convexity referred to, apply, not to the *horizontal longitudinal* section of the sand bank, as his language unexplained by the drawings might be supposed to mean, but to the *vertical cross-section*, and hence the dunes he describes, with the exception above noted, do not correspond to those of the American deserts.—*Om Klitformationen*, p. 86.

The dunes of Gascony, which sometimes exceed three hundred feet in height, present the same concavity and convexity of *vertical* cross-section. The slopes of these dunes are much steeper than those of the Netherlands and the Danish coast; for while all observers agree in assigning to the seaward and landward faces of these latter, respectively, angles of from 5° to 12°, and 30° with the horizon, the corresponding faces of the dunes of Gascony present angles of from 10° to 25°, and 50° to 60°.—Laval, *Mémoire sur les Dunes de Gascogne*, *Annales des Ponts et Chaussées*, 1847, 2me sémestre.

* Krause, speaking of the dunes on the coast of Prussia, says: "Their origin belongs to three different periods, in which important changes in the relative level of sea and land have unquestionably taken place. * * * Except in the deep depressions between them, the dunes are everywhere sprinkled, to a considerable height, with brown oxydulated iron, which has penetrated into the sand to the depth of from three to eighteen inches, and colored it red. * * * Above the iron is a stratum of sand differing in composition from ordinary sea sand, and on this, growing woods are always found. * * * The gradually accumulated forest soil occurs in beds of from one to three feet thick, and changes, proceeding upward, from gray sand to black humus." Even on the third or seaward range, the sand grasses appear and thrive luxuriantly, at least on the west coast, though Krause doubts whether the dunes of the east coast were ever thus protected.—*Der Dünenbau*, pp. 8, 11.

line since the formation of the oldest hillocks, and these have become inland dunes, while younger rows have been thrown up on the new beach laid bare by elevation of the sea bed. Our knowledge of the mode of their first accumulation is derived from observation of the action of wind and water in the few instances where, with or without the aid of man, new coast dunes have been accumulated, and of the influence of wind alone in elevating new sand heaps inland of the coast tier, when the outer rows are destroyed by the sea, as also when the sodded surface of ancient sands has been broken, and the subjacent strata laid open to the air.

It is a question of much interest, in what degree the naked condition of most dunes is to be ascribed to the improvidence and indiscretion of man. There are, in Western France, extensive ranges of dunes covered with ancient and dense forests, while the recently formed sand hills between them and the sea are bare of vegetation, and are rapidly advancing upon the wooded dunes, which they threaten to bury beneath their drifts. Between the old dunes and the new, there is no discoverable difference in material or in structure; but the modern sand hills are naked and shifting, the ancient, clothed with vegetation and fixed. It has been conjectured that artificial methods of confinement and plantation were employed by the primitive inhabitants of Gaul; and Laval, basing his calculations on the rate of annual movement of the shifting dunes, assigns the fifth century of the Christian era as the period when these processes were abandoned.*

There is no historical evidence that the Gauls were acquainted with artificial methods of fixing the sands of the coast, and we have little reason to suppose that they were advanced enough in civilization to be likely to resort to such processes, especially at a period when land could have had but a moderate value.

* LAVAL, *Mémoire sur les Dunes de Gascogne*, *Annales des Ponts et Chaussées*, 1847, 2me sémestre, p. 231. The same opinion had been expressed by BRÉMONTIER, *Annales des Ponts et Chaussées*, 1833, 1er sémestre, p. 185.

In other countries, dunes have spontaneously clothed themselves with forests, and the rapidity with which their surface is covered by various species of sand plants, and finally by trees, where man and cattle and burrowing animals are excluded from them, renders it highly probable that they would, as a general rule, protect themselves, if left to the undisturbed action of natural causes. The sand hills of the Frische Nehrung, on the coast of Prussia, were formerly wooded down to the water's edge, and it was only in the last century that, in consequence of the destruction of their forests, they became moving sands.* There is every reason to believe that the dunes of the Netherlands were clothed with trees until after the Roman invasion. The old geographers, in describing these countries, speak of vast forests extending to the very brink of the sea; but drifting coast dunes are first mentioned by the chroniclers of the Middle Ages, and so far as we know they have assumed a destructive character in consequence of the improvidence of man.† The history of the dunes of Michigan,

* "In the Middle Ages," says Willibald Alexis, as quoted by Müller, *Das Buch der Pflanzenwelt* i, p. 16, "the Nehrung was extending itself further, and the narrow opening near Lochstadt had filled itself up with sand. A great pine forest bound with its roots the dune sand and the heath uninterruptedly from Danzig to Pillau. King Frederick William I was once in want of money. A certain Herr von Korff promised to procure it for him, without loan or taxes, if he could be allowed to remove something quite useless. He thinned out the forests of Prussia, which then, indeed, possessed little pecuniary value; but he felled the entire woods of the Frische Nehrung, so far as they lay within the Prussian territory. The financial operation was a success. The king had money, but in the elementary operation which resulted from it, the state received irreparable injury. The sea winds rush over the bared hills; the Frische Haff is half-choked with sand; the channel between Elbing, the sea, and Königsberg is endangered, and the fisheries in the Haff injured. The operation of Herr von Korff brought the king 200,000 thalers. The state would now willingly expend millions to restore the forests again."

† Staring, *Voormaals en Thans*, p. 231. Had the dunes of the Netherlandish and French coasts, at the period of the Roman invasion, resembled the moving sand hills of the present day, it is inconceivable that they could have escaped the notice of so acute a physical geographer as Strabo;

so far as I have been able to learn from my own observation, or that of others, is the same. Thirty years ago, when that region was scarcely inhabited, they were generally covered with a thick growth of trees, chiefly pines, and underwood, and there was little appearance of undermining and wash on the lake side, or of shifting of the sands, except where the trees had been cut or turned up by the roots.*

Nature, as she builds up dunes for the protection of the sea shore, provides, with similar conservatism, for the preservation of the dunes themselves; so that, without the interference of man, these hillocks would be, not perhaps absolutely perpetual, but very lasting in duration, and very slowly altered in form or position. When once covered with the trees, shrubs, and herbaceous growths adapted to such localities, dunes undergo no apparent change, except the slow occasional undermining of

and the absolute silence of Cæsar, Ptolemy, and the encyclopædic Pliny, respecting them, would be not less inexplicable.

The Old Northern language, the ancient tongue of Denmark, though rich in terms descriptive of natural scenery, had no name for dune, nor do I think the sand hills of the coast are anywhere noticed in Icelandic literature. The modern Icelanders, in treating of the dunes of Jutland, call them *klettr*, hill, cliff, and the Danish *klit* is from that source. The word Düne is also of recent introduction into German. Had the dunes been distinguished from other hillocks, in ancient times, by so remarkable a feature as the propensity to drift, they would certainly have acquired a specific name in both Old Northern and German. So long as they were wooded knolls, they needed no peculiar name; when they became formidable, from the destruction of the woods which confined them, they acquired a designation.

* The sands of Cape Cod were partially, if not completely, covered with vegetation by nature. Dr. Dwight, describing the dunes as they were in 1800, says: "Some of them are covered with beach grass; some fringed with whortleberry bushes; and some tufted with a small and singular growth of oaks. * * * The parts of this barrier, which are covered with whortleberry bushes and with oaks, have been either not at all, or very little blown. The oaks, particularly, appear to be the continuation of the forests originally formed on this spot. * * * They wore all the marks of extreme age; were, in some instances, already decayed, and in others decaying; were hoary with moss, and were deformed by branches, broken and wasted, not by violence, but by time."—*Travels*, iii, p. 91.

the outer tier, and accidental destruction by the exposure of the interior, from the burrowing of animals, or the upturning of trees with their roots, and all these causes of displacement are very much less destructive when a vegetable covering exists in the immediate neighborhood of the breach.

Before the occupation of the coasts by civilized and therefore destructive man, dunes, at all points where they have been observed, seem to have been protected in their rear by forests, which served to break the force of the winds in both directions,* and to have spontaneously clothed themselves with a dense growth of the various plants, grasses, shrubs, and trees, which nature has assigned to such soils. It is observed in Europe that dunes, though now without the shelter of a forest country behind them, begin to protect themselves as soon as human trespassers are excluded, and grazing animals denied access to them. Herbaceous and arborescent plants spring up almost at once, first in the depressions, and then upon the surface of the sand hills. Every seed that sprouts, binds together a certain amount of sand by its roots, shades a little ground with its leaves, and furnishes food and shelter for still younger or smaller growths. A succession of a very few favorable seasons suffices to bind the whole surface together with a vegetable network, and the power of resistance possessed by the dunes themselves, and the protection they afford to the fields behind them, are just in proportion to the abundance and density of the plants they support.

The growth of the vegetable covering can, of course, be much accelerated by judicious planting and watchful care, and this species of improvement is now carried on upon a vast scale, wherever the value of land is considerable and the population dense. In the main, the dunes on the coast of the German Sea, notwithstanding the great quantity of often fertile

* Bergsöe (*Reventlovs Virksomhed*, ii, 3) states that the dunes on the west coast of Jutland were stationary before the destruction of the forests to the east of them. The felling of the tall trees removed the resistance to the lower currents of the westerly winds, and the sands have since buried a great extent of fertile soil. See also same work, ii, p. 124.

land they cover, and the evils which result from their movement, are, upon the whole, a protective and beneficial agent, and their maintenance is an object of solicitude with the governments and people of the shores they protect.*

Use of Dunes as a Barrier against the Sea.

Although the sea throws up large quantities of sand on flat lee-shores, there are, as we have seen, many cases where it continually encroaches on those same shores and washes them away. At all points of the shallow North Sea where the agitation of the waves extends to the bottom, banks are forming and rolling eastward. Hence the sea sand tends to accumulate upon the coast of Schleswig-Holstein and Jutland, and were there no conflicting influences, the shore would rapidly extend itself westward. But the same waves which wash the sand to the coast undermine the beach they cover, and still more rapidly degrade the shore at points where it is too high to receive partial protection by the formation of dunes upon it. The earth of the coast is generally composed of particles finer, lighter, and more transportable by water than the sea sand. While, therefore, the billows raised by a heavy west wind may roll up and deposit along the beach thousands of tons of sand, the same waves may swallow up even a larger quantity of fine shore earth. This earth, with a portion of the sand, is swept off by northwardly and southwardly currents, and let fall at other points of the coast, or carried off, altogether,

* "We must, therefore, not be surprised to see the people here deal as gingerly with their dunes, as if treading among eggs. He who is lucky enough to own a molehill of dune pets it affectionately, and spends his substance in cherishing and fattening it. That fair, fertile, rich province, the peninsula of Eiderstädt in the south of Friesland, has, on the point toward the sea, only a tiny row of dunes, some six miles long or so; but the people talk of their fringe of sand hills as if it were a border set with pearls. They look upon it as their best defence against Neptune. They have connected it with their system of dikes, and for years have kept sentries posted to protect it against wanton injury."—J. G. KOHL, *Die Inseln u. Marschen Schleswig-Holsteins*, ii, p. 115.

out of the reach of causes which might bring it back to its former position.

Although, then, the eastern shore of the German Ocean here and there advances into the sea, it in general retreats before it, and but for the protection afforded it by natural arrangements seconded by the art and industry of man, whole provinces would soon be engulfed by the waters. This protection consists in an almost unbroken chain of sand banks and dunes, extending from the northernmost point of Jutland to the Elbe, a distance of not much less than three hundred miles, and from the Elbe again, though with more frequent and wider interruptions, to the Atlantic borders of France and Spain.* So long as the dunes are maintained by nature or by human art, they serve, like any other embankment or dike, as a partial or a complete protection against the encroachments of the sea; and on the other hand, when their drifts are not checked by natural processes, or by the industry of man, they become a cause of as certain, if not of as sudden, destruction as the ocean itself whose advance they retard.

Encroachments of the Sea.

The eastward progress of the sea on the Danish and Netherlandish coast, and on certain shores of the Atlantic, depends so much on local geological structure, on the force and direction of tidal and other marine currents, on the volume and rapidity of coast rivers, on the contingencies of the weather and on other varying circumstances, that no general rate can be assigned to it.

* Sand banks sometimes connect themselves with the coast at both ends, and thus cut off a portion of the sea. In this case, as well as when salt water is enclosed by sea dikes, the water thus separated from the ocean gradually becomes fresh, or at least brackish. The Haffs, or large expanses of fresh water in Eastern Prussia—which are divided from the Baltic by narrow sand banks called Nehrungen, or, at sheltered points of the coast, by fluviatile deposits called Werders—all have one or more open passages, through which the water of the rivers that supply them at last finds its way to the sea.

At Agger, near the western end of the Liimfjord, in Jutland, the coast was washed away, between the years 1815 and 1839, at the rate of more than eighteen feet a year. The advance of the sea appears to have been something less rapid for a century before; but from 1840 to 1857, it gained upon the land no less than thirty feet a year. At other points of the shore of Jutland, the loss is smaller, but the sea is encroaching generally upon the whole line of the coast.*

The Liimfjord.

The irruption of the sea into the fresh-water lagoon of Liimfjord in Jutland, in 1825—one of the most remarkable encroachments of the ocean in modern times—is expressly ascribed to "mismanagement of the dunes" on the narrow neck of land which separated the fjord from the North Sea. At earlier periods, the sea had swept across the isthmus, and even burst through it, but the channel had been filled up again, sometimes by artificial means, sometimes by the operation of natural causes, and on all these occasions effects were produced very similar to those resulting from the formation of the new channel in 1825, which still remains open.† Within comparatively recent historical ages, the Liimfjord has thus been several times alternately filled with fresh and with salt water, and man has produced, by neglecting the dunes, or at least might have prevented by maintaining them, changes identical with those which are usually ascribed to the action of great geological causes, and sometimes supposed to have required vast periods of time for their accomplishment.

"This breach," says Forchhammer, "which converted the Liimfjord into a sound, and the northern part of Jutland into an island, occasioned remarkable changes. The first and most

* Andresen, *Om Klitformationen*, pp. 68–72.

† Id., pp. 231, 232. Andresen's work, though printed in 1861, was finished in 1859. Lyell (*Antiquity of Man*, 1863, p. 14) says: "Even in the course of the present century, the salt waters have made one eruption into the Baltic by the Liimfjord, although they have been now again excluded."

striking phenomenon was the sudden destruction of almost all the fresh-water fish previously inhabiting this lagoon, which was famous for its abundant fisheries. Millions of fresh-water fish were thrown on shore, partly dead and partly dying, and were carted off by the people. A few only survived, and still frequent the shores at the mouth of the brooks. The eel, however, has gradually accommodated itself to the change of circumstances, and is found in all parts of the fjord, while to all other fresh-water fish, the salt water of the ocean seems to have been fatal. It is more than probable that the sand washed in by the irruption covers, in many places, a layer of dead fish, and has thus prepared the way for a petrified stratum similar to those observed in so many older formations.

"As it seems to be a law of nature that animals whose life is suddenly extinguished while yet in full vigor, are the most likely to be preserved by petrification, we find here one of the conditions favorable to the formation of such a petrified stratum. The bottom of the Liimfjord was covered with a vigorous growth of aquatic plants, belonging both to fresh and to salt water, especially *Zostera marina*. This vegetation totally disappeared after the irruption, and, in some instances, was buried by the sand; and here again we have a familiar phenomenon often observed in ancient strata—the indication of a given formation by a particular vegetable species—and when the strata deposited at the time of the breach shall be accessible by upheaval, the period of eruption will be marked by a stratum of *Zostera*, and probably by impressions of fresh-water fishes.

"It is very remarkable that the *Zostera marina*, a sea plant, was destroyed even where no sand was deposited. This was probably in consequence of the sudden change from brackish to salt water. * * It is well established that the Liimfjord communicated with the German Ocean at some former period. To that era belong the deep beds of oyster shells and *Cardium edule*, which are still found at the bottom of the fjord. And now, after an interval of centuries, during which the lagoon contained no salt-water shell fish, it again produces great numbers of *Mytilus edulis.* Could we obtain a deep section of the bottom, we should find beds of *Ostrea edulis* and *Cardium edule*, then a layer of *Zostera marina* with fresh-water fish, and then a bed of *Mytilus edulis.* If, in course of time, the new channel should be closed, the brooks would fill the lagoon again with fresh water; fresh-water fish and shell fish would reappear, and thus we should have a repeated alternation of organic inhabitants of the sea and of the waters of the land.

"These events have been accompanied with but a comparatively insignificant change of land surface, while the formations in the bed of this inland sea have been totally revolutionized in character." *

Coasts of Schleswig-Holstein, Holland, and France.

On the islands on the coast of Schleswig-Holstein, the advance of the sea has been more unequivocal and more rapid. Near the beginning of the last century, the dunes which had protected the western coast of the island of Sylt began to roll to the east, and the sea followed closely as they retired. In 1757, the church of Rantum, a village upon that island, was obliged to be taken down in consequence of the advance of the sand hills; in 1791, these hills had passed beyond its site, the waves had swallowed up its foundations, and the sea gained so rapidly, that, fifty years later, the spot where they lay was seven hundred feet from the shore.†

The most prominent geological landmark on the coast of Holland is the Huis te Britten, *Arx Britannica*, a fortress built by the Romans, in the time of Caligula, on the main land near the mouth of the Rhine. At the close of the seventeenth century, the sea had advanced sixteen hundred paces beyond it. The older Dutch annalists record, with much parade of numerical accuracy, frequent encroachments of the sea

* FORCHHAMMER, *Geognostische Studien am Meeres-Ufer*, LEONHARD und BRONN, *Jahrbuch*, 1841, pp. 11, 13.

† ANDRESEN, *Om Klitformationen*, pp. 68, 72.

upon many parts of the Netherlandish coast. But though the general fact of an advance of the ocean upon the land is established beyond dispute, the precision of the measurements which have been given is open to question. Staring, however, who thinks the erosion of the coast much exaggerated by popular geographers, admits a loss of more than a million and a half acres, chiefly worthless morass;* and it is certain that but for the resistance of man, but for his erection of dikes and protection of dunes, there would now be left of Holland little but the name. It is, as has been already seen, still a debated question among geologists whether the coast of Holland now is, and for centuries has been, subsiding. I believe most investigators maintain the affirmative; and if the fact is so, the advance of the sea upon the land is, in part, due to this cause. But the rate of subsidence is at all events very small, and therefore the encroachments of the ocean upon the coast are mainly to be ascribed to the erosion and transportation of the soil by marine waves and currents.

The sea is fast advancing at several points of the western coast of France, and unknown causes have given a new impulse to its ravages since the commencement of the present century. Between 1830 and 1842, the Point de Grave, on the north side of the Gironde, retreated one hundred and eighty mètres, or about fifty feet per year; from the latter year to 1846, the rate was increased to more than three times that quantity, and the loss in those four years was above six hundred feet. All the buildings at the extremity of the peninsula have been taken down and rebuilt farther landward, and the lighthouse of the Grave now occupies its third position. The sea attacked the base of the peninsula also, and the Point de Grave and the adjacent coasts have been for twenty years the scene of one of the most obstinately contested struggles between man and the ocean recorded in the annals of modern engineering.

It cannot, indeed, be affirmed that human power is able to arrest altogether the incursions of the waves on sandy coasts,

* *Voormaals en Thans*, pp. 126, 170.

by planting the beach, and clothing the dunes with wood. On the contrary, both in Holland and on the French coast, it has been found necessary to protect the dunes themselves by piling and by piers and sea walls of heavy masonry. But experience has amply shown that the processes referred to are entirely successful in preventing the movement of the dunes, and the drifting of their sands over cultivated lands behind them ; and that, at the same time, the plantations very much retard the landward progress of the waters.*

Drifting of Dune Sands.

Besides their importance as a barrier against the inroads of the ocean, dunes are useful by sheltering the cultivated ground behind them from the violence of the sea wind, from salt spray, and from the drifts of beach sand which would otherwise overwhelm them. But the dunes themselves, unless their surface sands are kept moist, and confined by the growth of plants, or at least by a crust of vegetable earth, are constantly rolling inward; and thus, while, on one side, they lay bare the traces of ancient human habitations or other evidences of the social life of primitive man, they are, on the other, burying fields, houses, churches, and converting populous districts into barren and deserted wastes.

Especially destructive are they when, by any accident, a cavity is opened into them to a considerable depth, thereby giving the wind access to the interior, where the sand is thus first dried, and then scooped out and scattered far over the neighboring soil. The dune is now a magazine of sand, no longer a rampart against it, and mischief from this source seems more difficult to resist than from almost any other drift, because the supply of material at the command of the wind, is more abundant and more concentrated than in its original thin and widespread deposits on the beach. The burrowing of

* See a very interesting article entitled "Le Littoral de la France," by Élisée Reclus, in the *Revue des Deux Mondes*, for December, 1862, pp. 901, 936.

conies in the dunes is, in this way, not unfrequently a cause of their destruction and of great injury to the fields behind them. Drifts, and even inland sand hills, sometimes result from breaking the surface of more level sand deposits, far within the range of the coast dunes. Thus we learn from Staring, that one of the highest inland dunes in Friesland owes its origin to the opening of the drift sand by the uprooting of a large oak.*

Great as are the ravages produced by the encroachment of the sea upon the western shores of continental Europe, they have been in some degree compensated by spontaneous marine deposits at other points of the coast, and we have seen in a former chapter that the industry of man has reclaimed a large territory from the bosom of the ocean. These latter triumphs are not of recent origin, and the incipient victories which paved the way for them date back perhaps as far as ten centuries. In the mean time, the dunes had been left to the operation of the laws of nature, or rather freed, by human imprudence, from the fetters with which nature had bound them, and it is scarcely three generations since man first attempted to check their destructive movements. As they advanced, he unresistingly yielded and retreated before them, and they have buried under their sandy billows many hundreds of square miles of luxuriant cornfields and vineyards and forests.

Dunes of Gascony.

On the west coast of France, a belt of dunes, varying in width from a quarter of a mile to five miles, extends from the Adour to the estuary of the Gironde, and covers an area of three hundred and seventy-five square miles. When not fixed by vegetable growths, they advance eastward at a mean rate of about one rod, or sixteen and a half feet, a year. We do not know historically when they began to drift, but if we suppose their motion to have been always the same as at present, they would have passed over the space between the sea coast

* *De Bodem van Nederland*, i, p. 425. See *Appendix*, No.

and their eastern boundary, and covered the large area above mentioned, in fourteen hundred years. We know, from written records, that they have buried extensive fields and forests and thriving villages, and changed the courses of rivers, and that the lighter particles carried from them by the winds, even where not transported in sufficient quantities to form sand hills, have rendered sterile much land formerly fertile.* They have also injuriously obstructed the natural drainage of the maritime districts by choking up the beds of the streams, and forming lakes and pestilential swamps of no inconsiderable extent. In fact, so completely do they embank the coast, that between the Gironde and the village of Mimizan, a distance of one hundred miles, there are but two outlets for the discharge of all the waters which flow from the land to the sea; and the eastern front of the dunes is bordered by a succession of stagnant pools, some of which are more than six miles in length and breadth.†

The Dunes of Denmark and Prussia.

In the small kingdom of Denmark, inclusive of the duchies of Schleswig and Holstein, the dunes cover an area of more than two hundred and sixty square miles. The breadth of the

* The movement of the dunes has been hardly less destructive on the north side of the Gironde. See the valuable article of ÉLISÉE RECLUS already referred to, in the *Revue des Deux Mondes*, for December, 1862, entitled "Le Littoral de la France."

† LAVAL, *Mémoire sur les Dunes du Golfe de Gascogne*, *Annales des Ponts et Chaussées*, 1847, p. 223. The author adds, as a curious and unexplained fact, that some of these pools, though evidently not original formations but mere accumulations of water dammed up by the dunes, have, along their western shore, near the base of the sand hills, a depth of more than one hundred and thirty feet, and hence their bottoms are not less than eighty feet below the level of the lowest tides. Their western banks descend steeply, conforming nearly to the slope of the dunes, while on the northeast and south the inclination of their beds is very gradual. The greatest depth of these pools corresponds to that of the sea ten miles from the shore. Is it possible that the weight of the sands has pressed together the soil on which they rest, and thus occasioned a subsidence of the surface extending beyond their base? See *Appendix*, No. 59.

chain is very various, and in some places it consists only of a single row of sand hills, while in others, it is more than six miles wide. The general rate of eastward movement of the drifting dunes is from three to twenty-four feet per annum. If we adopt the mean of thirteen feet and a half for the annual motion, the dunes have traversed the widest part of the belt in about twenty-five hundred years. Historical data are wanting as to the period of the formation of these dunes and of the commencement of their drifting; but there is recorded evidence that they have buried a vast extent of valuable land within three or four centuries, and further proof is found in the fact that the movement of the sands is constantly uncovering ruins of ancient buildings, and other evidences of human occupation, at points far within the present limits of the uninhabitable desert. Andresen estimates the average depth of the sand deposited over this area at thirty feet, which would give a cubic mile and a half for the total quantity.*

The drifting of the dunes on the coast of Prussia commenced not much more than a hundred years ago. The Frische Nehrung is separated from the mainland by the Frische Haff, and there is but a narrow strip of arable land along its eastern borders. Hence its rolling sands have covered a comparatively small extent of dry land, but fields and villages have been buried and valuable forests laid waste by them. The loose coast row has drifted over the inland ranges, which, as was noticed in the description of these dunes on a former page, were protected by a surface of different composition, and the sand has thus been raised to a height which it could not have reached upon level ground. This elevation has enabled it to advance upon and overwhelm woods, which, upon a plain, would have checked its progress, and, in one instance, a forest of many hundred acres of tall pines was destroyed by the drifts between 1804 and 1827.

Control of Dunes by Man.

There are three principal modes in which the industry of

* Andresen, *Om Klitformationem*, pp. 56, 79, 82.

man is brought to bear upon the dunes. First, the creation of them, at points where, from changes in the currents or other causes, new encroachments of the sea are threatened; second, the maintenance and protection of them where they have been naturally formed; and third, the removal of the inner rows where the belt is so broad that no danger is to be apprehended from the loss of them.

Artificial Formation of Dunes.

In describing the natural formation of dunes, it was said that they began with an accumulation of sand around some vegetable or other accidental obstruction to the drifting of the particles. A high, perpendicular cliff, which deadens the wind altogether, prevents all accumulation of sand; but, up to a certain point, the higher and broader the obstruction, the more sand will heap up in front of it, and the more will that which falls behind it be protected from drifting farther. This familiar observation has taught the inhabitants of the coast that an artificial wall or dike will, in many situations, give rise to a broad belt of dunes. Thus a sand dike or wall, of three or four miles in length, thrown in 1610 across the Koegras, a tide-washed flat between the Zuiderzee and the North Sea, has occasioned the formation of rows of dunes a mile in breadth, and thus excluded the sea altogether from the Koegras. A similar dike, called the Zijperzeedijk, has produced another scarcely less extensive belt in the course of two centuries.

A few years since, the sea was threatening to cut through the island of Ameland, and, by encroachment on the southern side and the blowing off of the sand from a low flat which connected the two higher parts of the island, it had made such progress, that in heavy storms the waves sometimes rolled quite across the isthmus. The construction of a breakwater and a sand dike have already checked the advance of the sea, and a large number of sand hills has been formed, the rapid growth of which promises complete future security against both wind and wave. Similar effects have been produced by the erection of plank fences, and even of simple screens of wattling and reeds.*

Protection of Dunes.

The dunes of Holland are sometimes protected from the dashing of the waves by a *revêtement* of stone, or by piles; and the lateral high-water currents, which wash away their base, are occasionally checked by transverse walls running from the foot of the dunes to low-water mark; but the great expense of such constructions has prevented their adoption on a large scale.† The principal means relied on for the protection of the sand hills are the planting of their surfaces and the exclusion of burrowing and grazing animals. There are grasses, creeping plants, and shrubs of spontaneous growth, which flourish in loose sand, and, if protected, spread over considerable tracts, and finally convert their face into a soil capable of cultivation, or, at least, of producing forest trees. Krause enumerates one hundred and seventy-one plants as native to the coast sands of Prussia, and the observations of Andresen in Jutland carry the number of these vegetables up to two hundred and thirty-four.

Some of these plants, especially the *Arundo arenaria* or *arenosa*, or *Psamma* or *Psammophila arenaria*—Klittetag, or Hjelme in Danish, helm in Dutch, Dünenhalm, Sandschilf, or Hügelrohr in German, gourbet in French, and marram in English—are exclusively confined to sandy soils, and thrive

* Staring, *De Bodem van Nederland*, i, pp. 329–331. Id., *Voormaals en Thans*, p. 163. Andresen, *Om Klitformationen*, pp. 280, 295.

The creation of new dunes, by the processes mentioned in the text, seems to be much older in Europe than the adoption of measures for securing them by planting. Dr. Dwight mentions a case in Massachusetts, where a beach was restored, and new dunes formed, by planting beach grass. "Within the memory of my informant, the sea broke over the beach which connects Truro with Province Town, and swept the body of it away for some distance. The beach grass was immediately planted on the spot; in consequence of which the beach was again raised to a sufficient height, and in various places into hills."—*Dwight's Travels*, iii, p. 93.

† Staring, i, pp. 310, 332.

well only in a saline atmosphere.* The arundo grows to the height of about twenty-four inches, but sends its strong roots with their many rootlets to a distance of forty or fifty feet. It has the peculiar property of flourishing best in the loosest soil, and a sand shower seems to refresh it as the rain revives the thirsty plants of the common earth. Its roots bind together the dunes, and its leaves protect their surface. When the sand ceases to drift, the arundo dies, its decaying roots fertilizing the sand, and the decomposition of its leaves forming a layer of vegetable earth over it. Then follows a succession of other plants which gradually fit the sand hills, by growth and decay, for forest planting, for pasturage, and sometimes for ordinary agricultural use.

But the protection and gradual transformation of the dunes is not the only service rendered by this valuable plant. Its leaves are nutritious food for sheep and cattle, its seeds for poultry; † cordage and netting twine are manufactured from its fibres, it makes a good material for thatching, and its dried roots furnish excellent fuel. These useful qualities, unfortunately, are too often prejudicial to its growth. The peasants feed it down with their cattle, cut it for rope making, or dig it up for fuel, and it has been found necessary to resort to severe legislation to prevent them from bringing ruin upon themselves by thus improvidently sacrificing their most effectual safeguard against the drifting of the sands.‡

In 1539, a decree of Christian III, king of Denmark, imposed a fine upon persons convicted of destroying certain species of sand plants upon the west coast of Jutland. This ordinance was renewed and made more comprehensive in 1558,

* There is some confusion in the popular use of these names, and in the scientific designations of sand plants, and they are possibly applied to different plants in different places. Some writers style the gourbet *Calamagrostis arenaria*, and distinguish it from the Danish Klittetag or Hjelme.

† Bread, not indeed very palatable, has been made of the seeds of the arundo, but the quantity which can be gathered is not sufficient to form an important economical resource.—ANDRESEN, *Om Klitformationen*, p. 160.

‡ BERGSÖE, *Reventlovs Virksomhed*, ii, p. 4.

and in 1569 the inhabitants of several districts were required, by royal rescript, to do their best to check the sand drifts, though the specific measures to be adopted for that purpose are not indicated. Various laws against stripping the dunes of their vegetation were enacted in the following century, but no active measures were taken for the subjugation of the sand drifts until 1779, when a preliminary system of operation for that purpose was adopted. This consisted in little more than the planting of the *Arundo arenaria* and other sand plants, and the exclusion of animals destructive to these vegetables.*

* Measures were taken for the protection of the dunes of Cape Cod, in Massachusetts, during the colonial period, though I believe they are now substantially abandoned. A hundred years ago, before the valley of the Mississippi, or even the rich plains of Central and Western New York, were opened to the white settler, the value of land was relatively much greater in New England than it is at present, and consequently some rural improvements were then worth making, which would not now yield sufficient returns to tempt the investment of capital. The money and the time required to subdue and render productive twenty acres of sea sand on Cape Cod, would buy a "section" and rear a family in Illinois. The son of the Pilgrims, therefore, abandons the sand hills, and seeks a better fortune on the fertile prairies of the West.

Dr. Dwight, who visited Cape Cod in the year 1800, after describing the "beach grass, a vegetable bearing a general resemblance to sedge, but of a light bluish-green, and of a coarse appearance," which "flourishes with a strong and rapid vegetation on the sands," observes that he received "from a Mr. Collins, formerly of Truro, the following information:" "When he lived at Truro, the inhabitants were, under the authority of law, regularly warned in the month of April, yearly, to plant beach grass, as, in other towns of New England, they are warned to repair highways. It was required by the laws of the State, and under the proper penalties for disobedience; being as regular a public tax as any other. The people, therefore, generally attended and performed the labor. The grass was dug in bunches, as it naturally grows; and each bunch divided into a number of smaller ones. These were set out in the sand at distances of three feet. After one row was set, others were placed behind it in such a manner as to shut up the interstices; or, as a carpenter would say, so as to break the joints. * * * When it is once set, it grows and spreads with rapidity. * * * The seeds are so heavy that they bend down the heads of the grass; and when ripe, drop directly down by its side, where they immediately vegetate. Thus in a short time the ground is covered.

Ten years later, plantations of forest trees, which have since proved so valuable a means of fixing the dunes and rendering them productive, were commenced, and have been continued ever since.* During this latter period, Brémontier, without any knowledge of what was doing in Denmark, experimented upon the cultivation of forest trees on the dunes of Gascony, and perfected a system, which, with some improvements in

"Where this covering is found, none of the sand is blown. On the contrary, it is accumulated and raised continually as snow gathers and rises among bushes, or branches of trees cut and spread upon the earth. Nor does the grass merely defend the surface on which it is planted; but rises, as that rises by new accumulations; and always overtops the sand, however high that may be raised by the wind."—*Dwight's Travels in New England and New York*, ii, p. 92, 93.

This information was received in 1800, and it relates to a former state of things, probably more than twenty years previous, and earlier than 1779, when the Government of Denmark first seriously attempted the conquest of the dunes.

The depasturing of the beach grass—a plant allied in habits, if not in botanical character, to the arundo—has been attended with very injurious effects in Massachusetts. Dr. Dwight, after referring to the laws for its propagation, already cited, says: "The benefit of this useful plant, and of these prudent regulations, is, however, in some measure lost. There are in Province Town, as I was informed, one hundred and forty cows. These animals, being stinted in their means of subsistence, are permitted to wander, at times, in search of food. In every such case, they make depredations on the beach grass, and prevent its seeds from being formed. In this manner the plant is ultimately destroyed."—*Travels*, iii, p. 94.

On page 101 of the same volume, the author mentions an instance of great injury from this cause. "Here, about one thousand acres were entirely blown away to the depth, in many places, of ten feet. * * * Not a green thing was visible except the whortleberries, which tufted a few lonely hillocks rising to the height of the original surface and prevented by this defence from being blown away also. These, although they varied the prospect, added to the gloom by their strongly picturesque appearance, by marking exactly the original level of the plain, and by showing us in this manner the immensity of the mass which had been thus carried away by the wind. The beach grass had been planted here, and the ground had been formerly enclosed; but the gates had been left open, and the cattle had destroyed this invaluable plant."

* ANDRESEN, *Om Klitformationen*, pp. 237, 240.

matters of detail, is still largely pursued on those shores. The example of Denmark was soon followed in the neighboring kingdom of Prussia, and in the Netherlands; and, as we shall see hereafter, these improvements have been everywhere crowned with most flattering success.

Under the administration of Reventlov, a little before the close of the last century, the Danish Government organized a regular system of improvement in the economy of the dunes. They were planted with the arundo and other vegetables of similar habits, protected against trespassers, and at last partly covered with forest trees. By these means much waste soil has been converted into arable ground, a large growth of valuable timber obtained, and the further spread of the drifts, which threatened to lay waste the whole peninsula of Jutland, to a considerable extent arrested.

In France, the operations for fixing and reclaiming the dunes—which began under the direction of Brémontier about the same time as in Denmark, and which are, in principle and in many of their details, similar to those employed in the latter kingdom—have been conducted on a far larger scale, and with greater success, than in any other country. This is partly owing to a climate more favorable to the growth of suitable forest trees than that of Northern Europe, and partly to the liberality of the Government, which, having more important landed interests to protect, has put larger means at the disposal of the engineers than Denmark and Prussia have found it convenient to appropriate to that purpose. The area of the dunes already secured from drifting, and planted by the processes invented by Brémontier and perfected by his successors, is about 100,000 acres.* This amount of productive soil, then, has been added to the resources of France, and a still greater quantity

* "These plantations, perseveringly continued from the time of Brémontier now cover more than 40,000 hectares, and compose forests which are not only the salvation of the department, but constitute its wealth."—Clavé, *Études Forestières*, p. 254.

Other authors have stated the plantations of the French dunes to be much more extensive.

of valuable land has been thereby rescued from the otherwise certain destruction with which it was threatened by the advance of the rolling sand hills.

The improvements of the dunes on the coast of West Prussia began in 1795, under Sören Björn, a native of Denmark, and, with the exception of the ten years between 1807 and 1817, they have been prosecuted ever since. The methods do not differ essentially from those employed in Denmark and France, though they are modified by local circumstances, and, with respect to the trees selected for planting, by climate. In 1850, between the mouth of the Vistula and Kahlberg, 6,300 acres, including about 1,900 acres planted with pines and birches, had been secured from drifting; between Kahlberg and the eastern boundary of West-Prussia, 8,000 acres; and important preliminary operations had been carried on for subduing the dunes on the west coast.*

Trees suited to Dune Plantations.

The tree which has been found to thrive best upon the sand hills of the French coast, and at the same time to confine the sand most firmly and yield the largest pecuniary returns, is the maritime pine, *Pinus maritima*, a species valuable both for its timber and for its resinous products. It is always grown from seed, and the young shoots require to be protected for several seasons, by the branches of other trees, planted in rows, or spread over the surface and staked down, by the growth of the *Arundo arenaria* and other small sand plants, or by wattled hedges. The beach, from which the sand is derived, has been generally planted with the arundo, because the pine does not thrive well so near the sea; but it is thought that a species of tamarisk is likely to succeed in that latitude even better than the arundo. The shade and the protection offered by the branching top of this pine are favorable to the growth of deciduous trees, and, while still young, of shrubs and smaller plants, which contribute more rapidly to the formation of vegetable

* KRUSE, *Dünenbau*, pp. 34, 38, 40.

mould, and thus, when the pine has once taken root, the redemption of the waste is considered as effectually secured.

In France, the maritime pine is planted on the sands of the interior as well as on the dunes of the sea coast, and with equal advantage. This tree resembles the pitch pine of the Southern American States in its habits, and is applied to the same uses. The extraction of turpentine from it begins at the age of about twenty years, or when it has attained a diameter of from nine to twelve inches. Incisions are made up and down the trunk, to the depth of about half an inch in the wood, and it is insisted that if not more than two such slits are cut, the tree is not sensibly injured by the process. The growth, indeed, is somewhat checked, but the wood becomes superior to that of trees from which the turpentine is not extracted. Thus treated, the pine continues to flourish to the age of one hundred or one hundred and twenty years, and up to this age the trees on a hectare yield annually 350 kilogrammes of essence of turpentine, and 280 kilogrammes of resin, worth together 110 francs. The expense of extraction and distillation is calculated at 44 francs, and a clear profit of 66 francs per hectare, or more than five dollars per acre, is left.* This is exclusive of the value of the timber, when finally cut, which, of course, amounts to a very considerable sum.

In Denmark, where the climate is much colder, hardier conifers, as well as the birch and other northern trees, are found to answer a better purpose than the maritime pine, and

* These processes are substantially similar to those employed in the pineries of the Carolinas, but they are better systematized and more economically conducted in France. In the latter country, all the products of the pine, even to the cones, find a remunerating market, while, in America, the price of resin is so low, that in the fierce steamboat races on the great rivers, large quantities of it are thrown into the furnaces to increase the intensity of the fires. In a carefully prepared article on the Southern pineries published in an American magazine—I think Harper's—a few years ago, it was stated that the resin from the turpentine distilleries was sometimes allowed to run to waste; and the writer, in one instance, observed a mass, thus rejected as rubbish, which was estimated to amount to two thousand barrels. See *Appendix*, No. 60.

it is doubtful whether this tree would be able to resist the winter on the dunes of Massachusetts. Probably the pitch pine of the Northern States, in conjunction with some of the American oaks, birches, and poplars, and especially the robinia or locust, would prove very suitable to be employed on the sand hills of Cape Cod and Long Island. The ailanthus, now coming into notice as a sand-loving tree, may, perhaps, serve a better purpose than any of them.

Extent of Dunes in Europe.

The dunes of Denmark, as we have seen, cover an area of two hundred and sixty square miles, or one hundred and sixty-six thousand acres; those of the Prussian coast are vaguely estimated at from eighty-five to one hundred and ten thousand acres; those of Holland at one hundred and forty thousand acres; * those of Gascony at about three hundred thousand acres.† I do not find any estimate of their extent in other provinces of France, in the duchies of Schleswig and Holstein, or in the Baltic provinces of Russia, but it is probable that the entire quantity of dune land upon the eastern shores of the Atlantic and the Baltic does not fall much short of a million of acres.‡ This vast deposit of sea sand extends along

* ANDRESEN, *Om Klitformationen*, pp. 78, 262, 275.

† LAVAL, *Mémoire sur les Dunes du Golfe de Gascogne*, *Annales des Ponts et Chaussées*, 1847, 2me sémestre, p. 261. See *Appendix*, No, .

‡ There are extensive ranges of dunes on various parts of the coasts of the British Islands, but I find no estimate of their area. Pannewitz (*Anleitung zum Anbau der Sandflächen*), as cited by Andresen (*Om Klitformationen*, p. 45), states that the drifting sands of Europe, including, of course, sand plains as well as dunes, cover an extent of 21,000 square miles. This is, perhaps, an exaggeration, though there is, undoubtedly, much more desert land of this description on the European continent than has been generally supposed. There is no question that most of this waste is capable of reclamation by simple planting, and no mode of physical improvement is better worth the attention of civilized Governments than this.

There are often serious objections to extensive forest planting on soils capable of being otherwise made productive, but they do not apply to sand

the coast for a distance of several hundred miles, and from the time of the destruction of the forests which covered it, to the year 1789, the whole line was rolling inward and burying the soil beneath it, or rendering the fields unproductive by the sand which drifted from it. At the same time, as the sand hills moved eastward, the ocean was closely following their retreat and swallowing up the ground they had covered, as fast as their movement left it bare.

Planting the dunes has completely prevented the surface sands from blowing over the soil to the leeward of the plantations, and though it has not, in all cases, arrested the encroachments of the sea, it has so greatly retarded the rapidity of their advance, that sandy coasts, when once covered with forests, may be considered as substantially secure, so long as proper measures are taken for the protection of the woods.

Dune Vineyards of Cap Breton.

In the vicinity of Cap Breton in France, a peculiar process is successfully employed, both for preventing the drifting of dunes, and for rendering the sands themselves immediately productive; but this method is applicable only in exceptional cases of favorable climate and exposure. It consists in planting vineyards upon the dunes, and protecting them by hedges of broom, *Erica scoparia*, so disposed as to form rectangles about thirty feet by forty. The vines planted in these enclosures thrive admirably, and the grapes produced by them are among the best grown in France. The dunes are so far from being an unfavorable soil for the vine, that fresh sea-sand is regularly employed as a fertilizer for it, alternating every other season with ordinary manure. The quantity of sand thus applied every second year, raises the surface of the vineyard about four or five inches. The vines are cut down every year to three or four shoots, and the raising of the soil rapidly cov-

wastes, which, until covered by woods, are not only a useless incumbrance, but a source of serious danger to all human improvements in the neighborhood of them.

ers the old stocks. As fast as buried, they send out new roots near the surface, and thus the vineyard is constantly renewed, and has always a youthful appearance, though it may have been already planted a couple of generations. This practice is ascertained to have been followed for two centuries, and is among the oldest well-authenticated attempts of man to resist and vanquish the dunes.*

Removal of Dunes.

The artificial removal of dunes, no longer necessary as a protection, does not appear to have been practised upon a large scale except in the Netherlands, where the numerous canals furnish an easy and economical means of transporting the sand, and where the construction and maintenance of sea and river dikes, and of causeways and other embankments and fillings, create a great demand for that material. Sand is also employed in Holland, in large quantities, for improving the consistence of the tough clay bordering upon or underlying diluvial deposits, and for forming an artificial soil for the growth of certain garden and ornamental vegetables. When the dunes are removed, the ground they covered is restored to the domain of industry; and the quantity of land, recovered in the Netherlands by the removal of the barren sands which encumbered it, amounts to hundreds and perhaps thousands of acres.†

Inland Sand Plains.

The inland sand plains of Europe are either derived from the drifting of dunes or other beach sands, or consist of diluvial deposits. As we have seen, when once the interior of a dune is laid open to the wind, its contents are soon scattered far and wide over the adjacent country, and the beach sands, no longer checked by the rampart which nature had constrained them to build against their own encroachments, are also carried to con-

* BOITEL, *Mise en valeur des Terres pauvres par le Pin maritime*, pp. 212, 218.

† See *Appendix*, No.

siderable distances from the coast. Few regions have suffered so much from this cause in proportion to their extent, as the peninsula of Jutland. So long as the woods, with which nature had planted the Danish dunes, were spared, they seem to have been stationary, and we have no historical evidence, of an earlier date than the sixteenth century, that they had become in any way injurious. From that period, there are frequent notices of the invasions of cultivated grounds by the sands; and excavations are constantly bringing to light proof of human habitation and of agricultural industry, in former ages, on soils now buried beneath deep drifts from the dunes and beaches of the sea coast.*

Extensive tracts of valuable plain land in the Netherlands and in France have been covered in the same way with a layer of sand deep enough to render them infertile, and they can be restored to cultivation only by processes analogous to those employed for fixing and improving the dunes.† Diluvial sand plains, also, have been reclaimed by these methods in the Duchy of Austria, between Vienna and the Semmering ridge, in Jutland, and in the great champaign country of Northern Germany, especially the Mark Brandenburg, where artificial forests can be propagated with great ease, and where, consequently, this branch of industry has been pursued on a great scale, and with highly beneficial results, both as respects the supply of forest products and the preparation of the soil for agricultural use.

As a general rule, inland sands are looser, dryer, and more inclined to drift, than those of the sea coast, where the moist and saline atmosphere of the ocean keeps them always more or less humid and cohesive. No shore dunes are so movable as the medanos of Peru described in a passage quoted from Pöp-

* For details, consult Andresen, *Om Klitformationen*, pp. 223, 236.

† When the deposit is not very deep, and the adjacent land lying to the leeward of the prevailing winds is covered with water, or otherwise worthless, the surface is sometimes freed from the drifts by repeated harrowings, which loosen the sand, so that the wind takes it up and transports it to grounds where accumulations of it are less injurious.

pig on a former page, or as the sand hills of Poland, both of which seem better entitled to the appellation of sand waves than those of the Sahara or of the Arabian desert. The sands of the valley of the Lower Euphrates—themselves probably of submarine origin, and not derived from dunes—are advancing to the northwest with a rapidity which seems fabulous when compared with the slow movement of the sand hills of Gascony and the Low German coasts. Loftus, speaking of Niliyya, an old Arab town a few miles east of the ruins of Babylon, says that, "in 1848, the sand began to accumulate around it, and in six years, the desert, within a radius of six miles, wás covered with little, undulating domes, while the ruins of the city were so buried that it is now impossible to trace their original form or extent." * Loftus considers this sand flood as the "vanguard of those vast drifts which, advancing from the southeast, threaten eventually to overwhelm Babylon and Baghdad."

An observation of Layard, cited by Loftus, appears to me to furnish a possible explanation of this irruption. He "passed two or three places where the sand, issuing from the earth like water, is called 'Aioun-er-rummal,' sand springs." These "springs" are very probably merely the drifting of sand from the ancient subsoil, where the protecting crust of aquatic deposit and vegetable earth has been broken through, as in the case of the drift which arose from the upturning of an oak mentioned on a former page. When the valley of the Euphrates was regularly irrigated and cultivated, the underlying sands were bound by moisture, alluvial slime, and vegetation; but now, that all improvement is neglected, and the surface, no longer watered, has become parched, powdery, and naked, a mere accidental fissure in the superficial stratum may soon be enlarged to a wide opening, that will let loose sand enough to overwhelm a province.

The Landes of Gascony.

The most remarkable sand plain of France lies at the southwestern extremity of the empire, and is generally known as

* *Travels and Researches in Chaldæa*, chap. ix.

the Landes, or heaths, of Gascony. Clavé thus describes it: "Composed of pure sand, resting on an impermeable stratum called *alios*, the soil of the Landes was, for centuries, considered incapable of cultivation. Parched in summer, drowned in winter, it produced only ferns, rushes, and heath, and scarcely furnished pasturage for a few half-starved flocks. To crown its miseries, this plain was continually threatened by the encroachments of the dunes. Vast ridges of sand, thrown up by the waves, for a distance of more than fifty leagues along the coast, and continually renewed, were driven inland by the west wind, and, as they rolled over the plain, they buried the soil and the hamlets, overcame all resistance, and advanced with fearful regularity. The whole province seemed devoted to certain destruction, when Brémontier invented his method of fixing the dunes by plantations of the maritime pine." *

Although the Landes had been almost abandoned for ages, they show numerous traces of ancient cultivation and prosperity, and it is principally by means of the encroachments of the sands that they have become reduced to their present desolate condition. The destruction of the coast towns and harbors, which furnished markets for the products of the plains, the damming up of the rivers, and the obstruction of the smaller channels of natural drainage by the advance of the dunes, were no doubt very influential causes; and if we add the drifting of the sea sand over the soil, we have at least a partial explanation of the decayed agriculture and diminished population of this great waste. When the dunes were once arrested, and the soil to the east of them was felt to be secure against invasion by them, experiments, in the way of agricultural improvement, by drainage and plantation, were commenced, and they have been attended with such signal success, that the complete recovery of one of the dreariest and most extensive wastes in Europe may be considered as both a probable and a near event.†

* *Études Forestières*, p. 253.

† LAVERGNE, *Économie Rurale de la France*, p. 300, estimates the area of the Landes of Gascony at 700,000 hectares, or about 1,700,000 acres.

The Belgian Campine.

In the northern part of Belgium, and extending across the confines of Holland, is another very similar heath plain, called the Campine. This is a vast sand flat, interspersed with marshes and inland dunes, and, until recently, considered wholly incapable of cultivation. Enormous sums have been expended in reclaiming it by draining and other familiar agricultural processes, but without results at all proportional to the capital invested. In 1849, the unimproved portion of the Campine was estimated at little less than three hundred and fifty thousand acres. The example of France has prompted experiments in the planting of trees, especially the maritime pine, upon this barren waste, and the results have been such as to show that its sands may both be fixed and made productive, not only without loss, but with positive pecuniary advantage.*

Sands and Steppes of Eastern Europe.

There are still unsubdued sand wastes in many parts of interior Europe not familiarly known to tourists or even geographers. "Olkucz and Schiewier in Poland," says Naumann,

The same author states (p. 304), that when the Moors were driven from Spain by the blind cupidity and brutal intolerance of the age, they demanded permission to establish themselves in this desert; but political and religious prejudices prevented the granting of this liberty. At this period the Moors were a far more cultivated people than their Christian persecutors, and they had carried many arts, that of agriculture especially, to a higher pitch than any other European nation. But France was not wise enough to accept what Spain had cast out, and the Landes remained a waste for three centuries longer. See *Appendix*, No. .

The forest of Fontainebleau, which contains above 40,000 acres, is not a plain, but its soil is composed almost wholly of sand, interspersed with ledges of rock. The sand forms not less than ninety-eight per cent. of the earth, and, as it is almost without water, it would be a drifting desert but for the artificial propagation of forest trees upon it.

* *Économie Rurale de la Belgique, par* EMILE DE LAVELEYE, *Revue des Deux Mondes*, Juin, 1861, pp. 617–644.

"lie in true sand deserts, and a boundless plain of sand stretches around Ozenstockau, on which there grows neither tree nor shrub. In heavy winds, this plain resembles a rolling sea, and the sand hills rise and disappear like the waves of the ocean. The heaps of waste from the Olkucz mines are covered with sand to the depth of four fathoms." * No attempts have yet been made to subdue the sands of Poland, but when peace and prosperity shall be restored to that unhappy country, there is no reasonable doubt that the measures, which have proved so successful on similar formations in Germany, may be employed with advantage in the Polish deserts.

There are sand drifts in parts of the steppes of Russia, but in general the soil of those vast plains is of a different, though very varied, composition, and is covered with vegetation. The steppes, however, have many points of analogy with the sand plains of Northern Germany, and if they are ever fitted for civilized occupation, it must be by the same means, that is, by planting forests. It is disputed whether the steppes were ever wooded. They were certainly bare of forest growth at a very remote period; for Herodotus describes the country of the Scythians between the Ister and the Tanais as woodless, with the exception of the small province of Xylæa between the Dnieper and the Gulf of Perekop. They are known to have been occupied by a large nomade and pastoral population down to the sixteenth century, though these tribes are now much reduced in numbers. The habits of such races are scarcely less destructive to the forest than those of civilized life. Pastoral tribes do not employ much wood for fuel or for construction, but they carelessly or recklessly burn down the forests, and their cattle effectually check the growth of young trees wherever their range extends.

At present, the furious winds which sweep over the plains, the droughts of summer, and the rights and abuses of pasturage, constitute very formidable obstacles to the employment of measures which have been attended with so valuable results on the sand wastes of France and Germany. The Russian Gov-

* *Geognosie*, ii, p. 1173.

ernment has, however, attempted the wooding of the steppes, and there are thriving plantations in the neighborhood of Odessa, where the soil is of a particularly loose and sandy character.* The trees best suited to this locality, and, as there is good reason to suppose, to sand plains in general, is the *Ailanthus glandulosa*, or Japan varnish tree.† The remarkable success which has crowned the experiments with the ailanthus at Odessa, will, no doubt, stimulate to similar trials elsewhere, and it seems not improbable that the arundo and the maritime pine, which have fixed so many thousand acres of drifting sands in Western Europe, will be, partially at least, superseded by the tamarisk and the varnish tree.

Advantages of Reclaiming the Sands.

If we consider the quantity of waste land which has been made productive by the planting of the sand hills and plains, and the extent of fertile soil, the number of villages and other human improvements, and the value of the harbors, which the same process has saved from being buried under the rolling dunes, and at last swallowed up forever by the invasions of the sea, we shall be inclined to rank Brémontier and Reventlov among the greatest benefactors of their race. With the excep-

* According to HOHENSTEIN, *Der Wald*, pp. 228, 229, an extensive plantation of pines—a tree new to Southern Russia—was commenced in 1842, on the barren and sandy banks of the Ingula, near Elisabethgrod, and has met with very flattering success. Other experiments in sylviculture at different points on the steppes promise valuable results.

† "Sixteen years ago," says an Odessa landholder, "I attempted to fix the sand of the steppes, which covers the rocky ground to the depth of a foot, and forms moving hillocks with every change of wind. I tried acacias and pines in vain; nothing would grow in such a soil. At length I planted the varnish tree, or *ailanthus*, which succeeded completely in binding the sand." This result encouraged the proprietor to extend his plantations over both dunes and sand steppes, and in the course of sixteen years this rapidly growing tree had formed real forests. Other landowners have imitated his example with great advantage.—RENTSCH, *Der Wald*, p. 44, 45.

tion of the dikes of the Netherlands, their labors are the first deliberate and direct attempts of man to make himself, on a great scale, a geographical power, to restore natural balances which earlier generations had disturbed, and to atone, by acts guided by foreseeing and settled purpose, for the waste which thoughtless improvidence had created.

Government Works.

There is an important political difference between these latter works and the diking system of the Netherlandish and German coasts. The dikes originally were, and in modern times very generally have been, private enterprises, undertaken with no other aim than to add a certain quantity of cultivable soil to the former possessions of their proprietor, or sometimes of the state. In short, with few exceptions, they have been merely a pecuniary investment, a mode of acquiring land not economically different from purchase. The planting of the dunes, on the contrary, has always been a public work, executed, not with the expectation of reaping a regular direct percentage of income from the expenditure, but dictated by higher views of state economy—by the same governmental principles, in fact, which animate all commonwealths in repelling invasion by hostile armies, or in repairing the damages that invading forces may have inflicted on the general interests of the people. The restoration of the forests in the southern part of France, as now conducted by the Government of that empire, is a measure of the same elevated character as the fixing of the dunes. In former ages, forests were formed or protected simply for the sake of the shelter they afforded to game, or for the timber they yielded; but the recent legislation of France, and of some other Continental countries, on this subject, looks to more distant as well as nobler ends, and these are among the public acts which most strongly encourage the hope that the rulers of Christendom are coming better to understand the true duties and interests of civilized government.

CHAPTER VI.

PROJECTED OR POSSIBLE GEOGRAPHICAL CHANGES BY MAN.

CUTTING OF MARINE ISTHMUSES—THE SUEZ CANAL—CANAL ACROSS ISTHMUS OF DARIEN—CANALS TO THE DEAD SEA—MARITIME CANALS IN GREECE—CANAL OF SAROS—CAPE COD CANAL—DIVERSION OF THE NILE—CHANGES IN THE CASPIAN—IMPROVEMENTS IN NORTH AMERICAN HYDROGRAPHY—DIVERSION OF RHINE—DRAINING OF THE ZUIDERZEE—WATERS OF THE KARST—SUBTERRANEAN WATERS OF GREECE—SOIL BELOW ROCK—COVERING ROCKS WITH EARTH—WADIES OF ARABIA PETRÆA—INCIDENTAL EFFECTS OF HUMAN ACTION—RESISTANCE TO GREAT NATURAL FORCES—EFFECTS OF MINING—ESPY'S THEORIES—RIVER SEDIMENT—NOTHING SMALL IN NATURE.

Cutting of Marine Isthmuses.

BESIDES the great enterprises of physical transformation of which I have already spoken, other works of internal improvement or change have been projected in ancient and modern times, the execution of which would produce considerable, and, in some cases, extremely important, revolutions in the face of the earth. Some of the schemes to which I refer are evidently chimerical; others are difficult, indeed, but cannot be said to be impracticable, though discouraged by the apprehension of disastrous consequences from the disturbance of existing natural or artificial arrangements; and there are still others, the accomplishment of which is ultimately certain, though for the present forbidden by economical considerations.

When we consider the number of narrow necks or isthmuses which separate gulfs and bays of the sea from each other, or from the main ocean, and take into account the time and cost, and risks of navigation which would be saved by executing channels to connect such waters, and thus avoiding the necessity of doubling long capes and promontories, or even continents, it seems strange that more of the enterprise and money which have been so lavishly expended in forming artificial rivers for internal navigation should not have been bestowed upon the construction of maritime canals. Many such have been projected in early and in recent ages, and some trifling cuts between marine waters have been actually made, but no work of this sort, possessing real geographical or even commercial importance, has yet been effected.

These enterprises are attended with difficulties and open to objections, which are not, at first sight, obvious. Nature guards well the chains by which she connects promontories with mainlands, and binds continents together. Isthmuses are usually composed of adamantine rock or of shifting sands—the latter being much the more refractory material to deal with. In all such works there is a necessity for deep excavation below low-water mark—always a matter of great difficulty; the dimensions of channels for sea-going ships must be much greater than those of canals of inland navigation; the height of the masts or smoke pipes of that class of vessels would often render bridging impossible, and thus a ship canal might obstruct a communication more important than that which it was intended to promote; the securing of the entrances of marine canals and the construction of ports at their termini would in general be difficult and expensive, and the harbors and the channel which connected them would be extremely liable to fill up by deposits washed in from sea and shore. Besides all this, there is, in many cases, an alarming uncertainty as to the effects of joining together waters which nature has put asunder. A new channel may deflect strong currents from safe courses, and thus occasion destructive erosion of shores otherwise secure, or promote the transportation of sand or slime to block up important harbors, or it may furnish a powerful enemy with dangerous facilities for hostile operations along the coast.

Nature sometimes mocks the cunning and the power of man by spontaneously performing, for his benefit, works which he shrinks from undertaking, and the execution of which by him she would resist with unconquerable obstinacy. A dangerous sand bank, that all the enginery of the world could not dredge out in a generation, may be carried off in a night by a strong river flood, or a current impelled by a violent wind from an unusual quarter, and a passage scarcely navigable by fishing boats may be thus converted into a commodious channel for the largest ship that floats upon the ocean. In the remarkable gulf of Liimfjord in Jutland, nature has given a singular example of a canal which she alternately opens as a marine strait, and, by shutting again, converts into a fresh-water lagoon. The Liimfjord was doubtless originally an open channel from the Atlantic to the Baltic between two islands, but the sand washed up by the sea blocked up the western entrance, and built a wall of dunes to close it more firmly. This natural dike, as we have seen, has been more than once broken through, and it is perhaps in the power of man, either permanently to maintain the barrier, or to remove it and keep a navigable channel constantly open. If the Liimfjord becomes an open strait, the washing of sea sand through it would perhaps block up some of the belts and small channels now important for the navigation of the Baltic, and the direct introduction of a tidal current might produce very perceptible effects on the hydrography of the Cattegat.

The Suez Canal.

If the Suez Canal—the greatest and most truly cosmopolite physical improvement ever undertaken by man—shall prove successful, it will considerably affect the basins of the Mediterranean and of the Red Sea, though in a different manner, and probably in a less degree than the diversion of the current of the Nile from the one to the other—to which I shall presently refer—would do. It is, indeed, conceivable, that if a free channel be once cut from sea to sea, the coincidence of a high tide and a heavy south wind might produce a hydraulic force that would convert the narrow canal into an open strait. In such a case, it is impossible to estimate, or even to foresee, the consequences which might result from the unobstructed mingling of the flowing and ebbing currents of the Red Sea with the almost tideless waters of the Mediterranean. There can be no doubt, however, that they would be of a most important character as respects the simply geographical features and the organic life of both. But the shallowness of the two seas at the termini of the canal, the action of the tides of the one and the currents of the other, and the nature of the intervening isthmus, render the occurrence of such a cataclysm in the highest degree improbable. The obstruction of the canal by sea sand at both ends is a danger far more difficult to guard against and avert, than an irruption of the waters of either sea.

There is, then, no reason to expect any change of coast lines or of natural navigable channels as a direct consequence of the opening of the Suez Canal, but it will, no doubt, produce very interesting revolutions in the animal and vegetable population of both basins. The Mediterranean, with some local exceptions—such as the bays of Calabria, and the coast of Sicily so picturesquely described by Quatrefages *—is comparatively poor in marine vegetation, and in shell as well as in fin fish. The scarcity of fish in some of its gulfs is proverbial, and you may scrutinize long stretches of beach on its northern shores, after every south wind for a whole winter, without finding a dozen shells to reward your search. But no one who has not looked down into tropical or subtropical seas can conceive the amazing wealth of the Red Sea in organic life. Its bottom is carpeted or paved with marine plants, with zoophytes and with shells, while its waters are teeming with infinitely varied forms of moving life. Most of its vegetables and its animals, no doubt, are confined by the laws of their organization to warmer temperatures than that of the Mediterranean, but among them there must be many, whose habitat is of a wider range, many whose powers of accommodation would enable them to acclimate themselves in a colder sea.

* *Souvenirs d'un Naturaliste*, i, pp. 204 *et seqq.*

We may suppose the less numerous aquatic fauna and flora of the Mediterranean to be equally capable of climatic adaptation, and hence, when the canal shall be opened, there will be an interchange of the organic population not already common to both seas. Destructive species, thus newly introduced, may diminish the numbers of their proper prey in either basin, and, on the other hand, the increased supply of appropriate food may greatly multiply the abundance of others, and at the same time add important contributions to the aliment of man in the countries bordering on the Mediterranean.

A collateral feature of this great project deserves notice as possessing no inconsiderable geographical importance. I refer to the conduit or conduits constructed from the Nile to the isthmus, *primarily to supply* fresh water to the laborers on the great canal, and ultimately to serve as aqueducts for the city of Suez, and for the irrigation and reclamation of a large extent of desert soil. In the flourishing days of the Egyptian empire, the waters of the Nile were carried over important districts east of the river. In later ages, most of this territory relapsed into a desert, from the decay of the canals which once fertilized it. There is no difficulty in restoring the ancient channels, or in constructing new, and thus watering not only all the soil that the wisdom of the Pharaohs had improved, but much additional land. Hundreds of square miles of arid sand waste would thus be converted into fields of perennial verdure, and the geography of Lower Egypt would be thereby sensibly changed. If the canal succeeds, considerable towns will grow up at once at both ends of the channel, and at intermediate points, all depending on the maintenance of aqueducts from the Nile, both for water and for the irrigation of the neighboring fields which are to supply them with bread. Important interests will thus be created, which will secure the permanence of the hydraulic works and of the geographical changes produced by them, and Suez, or Port Said, or the city at Lake Timsah, may become the capital of the government which has been so long established at Cairo.

Canal across the Isthmus of Darien.

The most colossal project of canalization ever suggested, whether we consider the physical difficulties of its execution, the magnitude and importance of the waters proposed to be united, or the distance which would be saved in navigation, is that of a channel between the Gulf of Mexico and the Pacific, across the Isthmus of Darien. I do not now speak of a lock canal, by way of the Lake of Nicaragua or any other route—for such a work would not differ essentially from other canals, and would scarcely possess a geographical character—but of an open cut between the two seas. It has been by no means shown that the construction of such a channel is possible, and, if it were opened, it is highly probable that sand bars would accumulate at both entrances, so as to obstruct any powerful current through it. But if we suppose the work to be actually accomplished, there would be, in the first place, such a mixture of the animal and vegetable life of the two great oceans as I have stated to be likely to result from the opening of the Suez Canal between two much smaller basins. In the next place, if the channel were not obstructed by sand bars, it might sooner or later be greatly widened and deepened by the mechanical action of the current through it, and consequences, not inferior in magnitude to any physical revolution which has taken place since man appeared upon the earth, might result from it.

What those consequences would be is in a great degree matter of pure conjecture, and there is much room for the exercise of the imagination on the subject; but, as more than one geographer has suggested, there is one possible result which throws all other conceivable effects of such a work quite into the shade. I refer to changes in the course of the two great oceanic rivers, the Gulf Stream and the corresponding current on the Pacific side of the isthmus. The warm waters which the Gulf Stream transports to high latitudes and then spreads out, like an expanded hand, along the eastern shores of the Atlantic, give out, as they cool, heat enough to raise the mean temperature of Western Europe several degrees. In fact, the Gulf

Stream is the principal cause of the superiority of the climate of Western Europe over those of Eastern America and Eastern Asia in the corresponding latitudes. All the meteorological conditions of the former region are in a great measure regulated by it, and hence it is the grandest and most beneficent of all purely geographical phenomena. We do not yet know enough of the laws which govern the movements of this mighty flood of warmth and life to be able to say whether its current would be perceptibly affected by the severance of the Isthmus of Darien; but as it enters and sweeps round the Gulf of Mexico, it is possible that the removal of the resistance of the land which forms the western shore of that sea, might allow the stream to maintain its original westward direction, and join itself to the tropical current of the Pacific.

The effect of such a change would be an immediate depression of the mean temperature of Western Europe to the level of that of Eastern America, and perhaps the climate of the former continent might become as excessive as that of the latter, or even a new "ice period" be occasioned by the withdrawal of so important a source of warmth from the northern zones. Hence would result the extinction of vast multitudes of land and sea plants and animals, and a total revolution in the domestic and rural economy of human life in all those countries from which the New World has received its civilized population. Other scarcely less startling consequences may be imagined as possible; but the whole speculation is too dreary, distant, and improbable to deserve to be long indulged in.*

* "If we suppose the narrow isthmus of Central America to be sunk in the ocean, the warm equatorial current would no longer follow its circuitous route around the Gulf of Mexico, but pour itself through the new opening directly into the Pacific. We should then lose the warmth of the Gulf Stream, and cold polar currents flowing farther southward would take its place and be driven upon our coasts by the western winds. The North Sea would resemble Hudson's Bay, and its harbors be free from ice at best only in summer. The power and prosperity of its coasts would shrivel under the breath of winter, as a medusa thrown on shore shrinks to an insignificant film under the influence of the destructive atmosphere. Commerce, industry, fertility of soil, population, would disappear, and the vast

Canals to the Dead Sea.

The project of Captain Allen for opening a new route to India by cuts between the Mediterranean and the Dead Sea, and between the Dead Sea and the Red Sea, presents many interesting considerations.* The hypsometrical observations of Bertou, Roth, and others, render it highly probable, if not certain, that the watershed in the Wadi-el-Araba between the Dead Sea and the Red Sea is not less than three hundred feet above the mean level of the latter, and if this is so, the execution of a canal from the one sea to the other is quite out of the question. But the summit level between the Mediterranean and the Jordan, near Jezreel, is believed to be little, if at all, more than one hundred feet above the sea, and the distance is so short that the cutting of a channel through the dividing ridge would probably be found by no means an impracticable undertaking. Although, therefore, we have no reason to believe it possible to open a navigable channel to the east by way of the Dead Sea, there is not much doubt that the basin of the latter might be made accessible from the Mediterranean.

The level of the Dead Sea lies 1,316.7 feet below that of the ocean. It is bounded east and west by mountain ridges, rising to the height of from 2,000 to 4,000 feet above the ocean. From its southern end, a depression called the Wadi-el-Araba extends to the Gulf of Akaba, the eastern arm of the Red Sea. The Jordan empties into its northern extremity, after having passed through the Lake of Tiberias at an elevation of 663.4 feet above the Dead Sea, or 653.3 below the Mediterranean, and drains a considerable valley north of the lake, as well as the plain of Jericho, which lies between the lake and the sea. If the waters of the Mediterranean were admitted

waste—a new Labrador—would become a worthless appendage of some clime more favored by nature."—HARTWIG, *Das Leben des Meeres*, p. 70.

* I know nothing of Captain Allen's work but its title and its subject. Very probably he may have anticipated many of the following speculations, and thrown light on points upon which I am ignorant.

freely into the basin of the Dead Sea, they would raise its surface to the general level of the ocean, and consequently flood all the dry land below that level within the basin.

I do not know that accurate levels have been taken in the valley of the Jordan above the Lake of Tiberias, and our information is very vague as to the hypsometry of the northern part of the Wadi-el-Araba. As little do we know where a contour line, carried around the basin at the level of the Mediterranean, would strike its eastern and western borders. We cannot, therefore, accurately compute the extent of now dry land which would be covered by the admission of the waters of the Mediterranean, or the area of the inland sea which would be thus created. Its length, however, would certainly exceed one hundred and fifty miles, and its mean breadth, including its gulfs and bays, could scarcely be less than fifteen, perhaps even twenty. It would cover very little ground now occupied by civilized or even uncivilized man, though some of the soil which would be submerged—for instance, that watered by the Fountain of Elisha and other neighboring sources—is of great fertility, and, under a wiser government and better civil institutions, might rise to importance, because, from its depression, it possesses a very warm climate, and might supply Southeastern Europe with tropical products more readily than they can be obtained from any other source. Such a canal and sea would be of no present commercial importance, because they would give access to no new markets or sources of supply; but when the fertile valleys and the deserted plains east of the Jordan shall be reclaimed to agriculture and civilization, these waters would furnish a channel of communication which might become the medium of a very extensive trade.

Whatever might be the economical results of the opening and filling of the Dead Sea basin, the creation of a new evaporable area, adding not less than 2,000 or perhaps 3,000 square miles to the present fluid surface of Syria, could not fail to produce important meteorological effects. The climate of Syria would be tempered, its precipitation and its fertility increased, the courses of its winds and the electrical condition of its atmosphere modified. The present organic life of the valley would be extinguished, and many tribes of plants and animals would emigrate from the Mediterranean to the new home which human art had prepared for them. It is possible, too, that the addition of 1,300 feet, or forty atmospheres, of hydrostatic pressure upon the bottom of the basin might disturb the equilibrium between the internal and the external forces of the crust of the earth at this point of abnormal configuration, and thus produce geological convulsions the intensity of which cannot be even conjectured.

Maritime Canals in Greece.

A maritime canal executed and another projected in ancient times, the latter of which is again beginning to excite attention, deserve some notice, though their importance is of a commercial rather than a geographical character. The first of these is the cut made by Xerxes through the rock which connects the promontory of Mount Athos with the mainland; the other, a navigable canal through the Isthmus of Corinth. In spite of the testimony of Herodotus and Thucydides, the Romans classed the canal of Xerxes among the fables of "mendacious Greece," and yet traces of it are perfectly distinct at the present day through its whole extent, except at a single point where, after it had become so choked as to be no longer navigable, it was probably filled up to facilitate communication by land between the promontory and the country in the rear of it.

If the fancy kingdom of Greece shall ever become a sober reality, escape from its tutelage and acquire such a moral as well as political status that its own capitalists—who now prefer to establish themselves and employ their funds anywhere else rather than in their native land—have any confidence in the permanency of its institutions, a navigable channel will no doubt be opened between the gulfs of Lepanto and Ægina. The annexation of the Ionian Islands to Greece will make such a work almost a political necessity, and it would not only furnish valuable facilities for domestic intercourse, but become an important channel of communication between the Levant and the countries bordering on the Adriatic, or conducting their trade through that sea.

As I have said, the importance of this latter canal and of a navigable channel between Mount Athos and the continent would be chiefly commercial, but both of them would be conspicuous instances of the control of man over nature in a field where he has thus far done little to interfere with her spontaneous arrangements. If they were constructed upon such a scale as to admit of the free passage of the water through them, in either direction, as the prevailing winds should impel it, they would exercise a certain influence on the coast currents, which are important as hydrographical elements, and also as producing abrasion of the coast and a drift at the bottom of seas, and hence would be entitled to a higher rank than simply as artificial means of transit.

Canal of Saros.

It has been thought practicable to cut a canal across the peninsula of Gallipoli from the outlet of the Sea of Marmora into the Gulf of Saros. It may be doubted whether the mechanical difficulties of such a work would not be found insuperable; but when Constantinople shall recover the important political and commercial rank which naturally belongs to her, the execution of such a canal will be recommended by strong reasons of military expediency, as well as by the interests of trade. An open channel across the peninsula would divert a portion of the water which now flows through the Dardanelles, diminish the rapidity of that powerful current, and thus in part remove the difficulties which obstruct the navigation of the strait. It would considerably abridge the distance by water between Constantinople and the northern coast of the Ægean, and it would have the important advantage of obliging an enemy to maintain two blockading fleets instead of one.

Cape Cod Canal.

The opening of a navigable cut through the narrow neck which separates the southern part of Cape Cod Bay in Massachusetts from the Atlantic, was long ago suggested, and there are few coast improvements on the Atlantic shores of the United States which are recommended by higher considerations of utility. It would save the most important coasting trade of the United States the long and dangerous navigation around Cape Cod, afford a new and safer entrance to Boston harbor for vessels from Southern ports, secure a choice of passages, thus permitting arrivals upon the coast and departures from it at periods when wind and weather might otherwise prevent them, and furnish a most valuable internal communication in case of coast blockade by a foreign power. The difficulties of the undertaking are no doubt formidable, but the expense of maintenance and the uncertainty of the effects of currents setting through the new strait are still more serious objections.

Diversion of the Nile.

Perhaps the most remarkable project of great physical change, proposed or threatened in earlier ages, is that of the diversion of the Nile from its natural channel, and the turning of its current into either the Libyan desert or the Red Sea. The Ethiopian or Abyssinian princes more than once menaced the Memlouk sultans with the execution of this alarming project, and the fear of so serious an evil is said to have induced the Moslems to conciliate the Abyssinian kings by large presents, and by some concessions to the oppressed Christians of Egypt.* Indeed, Arabic historians affirm that in the tenth

* "Some haue writtē, that by certain kings inhabiting aboue, the *Nilus* should there be stopped; & at a time prefixt, let loose vpon a certaine tribute payd them by the *Aegyptians.* The error springing perhaps frō a truth (as all wandring reports for the most part doe) in that the *Sultan* doth pay a certaine annuall summe to the *Abissin* Emperour for not diuerting the course of the Riuer, which (they say) he may, or impouerish it at the least."—GEORGE SANDYS, *A Relation of a Journey, etc.*, p. 98.

century the Ethiopians dammed the river, and, for a whole year, cut off its waters from Egypt. The probable explanation of this story is to be found in a season of extreme drought, such as have sometimes occurred in the valley of the Nile. About the beginning of the sixteenth century, Albuquerque the "Terrible" revived the scheme of turning the Nile into the Red Sea, with the hope of destroying the transit trade through Egypt by way of Kosseir. In 1525 the King of Portugal was requested by the Emperor of Abyssinia to send him engineers for that purpose; a successor of that prince threatened to attempt the project about the year 1700, and even as late as the French occupation of Egypt, the possibility of driving out the intruder by this means was suggested in England.

It cannot be positively affirmed that the diversion of the waters of the Nile to the Red Sea is impossible. In the chain of mountains which separates the two valleys, Brown found a deep depression or wadi, extending from the one to the other, at no great elevation above the bed of the river. The Libyan desert is so much higher than the Nile below the junction of the two principal branches at Khartum, that there is no reason to believe a new channel for their united waters could be found in that direction; but the Bahr-el-Abiad flows through, if it does not rise in, a great table land, and some of its tributaries are supposed to communicate in the rainy season with branches of great rivers flowing in quite another direction. Hence it is probable that a portion at least of the waters of this great arm of the Nile—and perhaps a quantity the abstraction of which would be sensibly felt in Egypt—might be sent to the Atlantic by the Niger, lost in the inland lakes of Central Africa, or employed to fertilize the Libyan sand wastes.

Admitting the possibility of turning the whole river into the Red Sea, let us consider the probable effect of the change. First and most obvious is the total destruction of the fertility of Middle and Lower Egypt, the conversion of that part of the valley into a desert, and the extinction of its imperfect civilization, if not the absolute extirpation of its inhabitants. This is the calamity threatened by the Abyssinian princes and the ferocious Portuguese warrior, and feared by the sultans of Egypt. Beyond these immediate and palpable consequences neither party then looked; but a far wider geographical area, and far more extensive and various human interests, would be affected by the measure. The spread of the Nile during the annual in undation covers, for many weeks, several thousand square miles with water, and at other seasons of the year pervades the same and even a larger area with moisture by infiltration. The abstraction of so large an evaporable surface from the southern shores of the Mediterranean could not but produce important effects on many meteorological phenomena, and the humidity, the temperature, the electrical condition and the atmospheric currents of Northeastern Africa might be modified to a degree that would sensibly affect the climate of Europe.

The Mediterranean, deprived of the contributions of the Nile, would require a larger supply, and of course a stronger current, of water from the Atlantic through the Straits of Gibraltar; the proportion of salt it contains would be increased, and the animal life of at least its southern borders would be consequently modified; the current which winds along its southern, eastern, and northeastern shores would be diminished in force and volume, if not destroyed altogether, and its basin and its harbors would be shoaled by no new deposits from the highlands of inner Africa.

In the much smaller Red Sea, more immediately perceptible, if not greater, effects, would be produced. The deposits of slime would reduce its depth, and perhaps, in the course of ages, divide it into an inland and an open sea; its waters would be more or less freshened, and its immensely rich marine fauna and flora changed in character and proportion, and, near the mouth of the river, perhaps even destroyed altogether; its navigable channels would be altered in position and often quite obstructed; the flow of its tides would be modified by the new geographical conditions; the sediment of the river would form new coast lines and lowlands, which would be covered with vegetation, and probably thereby produce sensible climatic changes.

Changes in the Caspian.

The Russian Government has contemplated the establishment of a nearly direct water communication between the Caspian Sea and the Sea of Azoff, partly by natural and partly by artificial channels, and there are now navigable canals between the Don and the Volga; but these works, though not wanting in commercial and political interest, do not possess any geographical importance. It is, however, very possible to produce appreciable geographical changes in the basin of the Caspian by the diversion of the great rivers which flow from Central Russia. The surface of the Caspian is eighty-three feet below the level of the Sea of Azoff, and its depression has been explained upon the hypothesis that the evaporation exceeds the supply derived, directly and indirectly, from precipitation, though able physicists now maintain that the sinking of this sea is due to a subsidence of its bottom from geological causes. At Tsaritsin, the Don, which empties into the Sea of Azoff, and the Volga, which pours into the Caspian, approach each other within ten miles. Near this point, by means of open or subterranean canals, the Don might be turned into the Volga, or the Volga into the Don. If we suppose the whole or a large proportion of the waters of the Don to be thus diverted from their natural outlet and sent down to the Caspian, the equilibrium between the evaporation from that sea and its supply of water might be restored, or its level even raised above its ancient limits. If the Volga were turned into the Sea of Azoff, the Caspian would be reduced in dimensions until the balance between loss and gain should be reëstablished, and it would occupy a much smaller area than at present. Such changes in the proportion of solid and fluid surface would have some climatic effects in the territory which drains into the Caspian, and on the other hand, the introduction of a greater quantity of fresh water into the Sea of Azoff would render that gulf less saline, affect the character and numbers of its fish, and perhaps be not wholly without sensible influence on the water of the Black Sea.

Improvements in North American Hydrography.

We are not yet well enough acquainted with the geography of Central Africa, or of the interior of South America, to conjecture what hydrographical revolutions might there be wrought; but from the fact that many important rivers in both continents drain extensive table lands, of very moderate inclination, there is reason to suppose that important changes in the course of rivers might be accomplished. Our knowledge of the drainage of North America is much more complete, and it is certain that there are numerous points where the courses of great rivers, or the discharge of considerable lakes, might be completely diverted, or at least partially directed into different channels.

The surface of Lake Erie is 565 feet above that of the Hudson at Albany, and it is so near the level of the great plain lying east of it, that it was found practicable to supply the western section of the canal, which unites it with the Hudson, with water from the lake, or rather from the Niagara which flows out of it. Hence a channel might be constructed, which would draw off into the valley of the Genesee any desirable proportion of the water naturally discharged by the Niagara. The greatest depth of water yet sounded in Lake Erie is but two hundred and seventy feet, the mean depth one hundred and twenty. Open canals parallel with the Niagara, or directly toward the Genesee, might be executed upon a scale which would exercise an important influence on the drainage of the lake, if there were any adequate motive for such an undertaking. Still easier would it be to create additional outlets for the waters of Lake Superior at the Saut St. Mary—where the river which drains the lake descends twenty-two feet in a single mile—and thus produce incalculable effects, both upon that lake and upon the great chain of inland waters which communicate with it.

The summit level between Lake Michigan and the Des

Plaines, a tributary of the Mississippi, is only twenty-seven feet above the lake, and the intervening distance is but a very few miles. It has often been proposed to cut an open channel across this ridge, and there is no doubt of the practicability of the project. Were this accomplished, although such a cut would not, of itself, form a navigable canal, a part of the waters of Lake Michigan would be contributed to the Gulf of Mexico, instead of to that of St. Lawrence, and the flow might be so regulated as to keep the Illinois and the Mississippi at flood at all seasons of the year. The increase in the volume of these rivers would augment their velocity and their transporting power, and consequently, the erosion of their banks and the deposit of slime in the Gulf of Mexico, while the introduction of a larger body of cold water into the beds of these rivers would very probably produce a considerable effect on the animal life that peoples them. The diversion of water from the common basin of the great lakes through a new channel, in a direction opposite to their natural discharge, would not be absolutely without influence on the St. Lawrence, though probably the effect would be too small to be in any way perceptible.

Diversion of the Rhine.

The interference of physical improvements with vested rights and ancient arrangements, is a more formidable obstacle in old countries than in new, to enterprises involving anything approaching to a geographical revolution. Hence such projects meet with stronger opposition in Europe than in America, and the number of probable changes in the face of nature in the former continent is proportionally less. I have noticed some important hydraulic improvements as already executed or in progress in Europe, and I may refer to some others as contemplated or suggested. One of these is the diversion of the Rhine from its present channel below Ragatz, by a cut through the narrow ridge near Sargans, and the consequent turning of its current into the Lake of Wallenstadt. This would be an extremely easy undertaking, for the ridge is but twenty feet above the level of the Rhine, and hardly two hundred yards wide. There is no present adequate motive for this diversion, but it is easy to suppose that it may become advisable within no long period. The navigation of the Lake of Constance is rapidly increasing in importance, and the shoaling of the eastern end of that lake by the deposits of the Rhine may require a remedy which can be found by no other so ready means as the discharge of that river into the Lake of Wallenstadt. The navigation of this latter lake is not important, nor is it ever likely to become so, because the rocky and precipitous character of its shores renders their cultivation impossible. It is of great depth, and its basin is capacious enough to receive and retain all the sediment which the Rhine would carry into it for thousands of years.

Draining of the Zuiderzee.

I have referred to the draining of the Lake of Haarlem as an operation of great geographical as well as economical and mechanical interest. A much more gigantic project, of a similar character, is now engaging the attention of the Netherlandish engineers. It is proposed to drain the great salt-water basin called the Zuiderzee. This inland sea covers an area of not less than two thousand square miles, or about one million three hundred thousand acres. The seaward half, or that portion lying northwest of a line drawn from Enkhuizen to Staveren, is believed to have been converted from a marsh to an open bay since the fifth century after Christ, and this change is ascribed, partly if not wholly, to the interference of man with the order of nature. The Zuiderzee communicates with the sea by at least six considerable channels, separated from each other by low islands, and the tide rises within the basin to the height of three feet. To drain the Zuiderzee, these channels must first be closed and the passage of the tidal flood through them cut off. If this be done, the coast currents will be restored approximately to the lines they followed fourteen or fifteen centuries ago, and there can be little doubt that an appreciable effect will thus be produced upon all the tidal phenomena of that coast, and, of course, upon the maritime geography of Holland.

A ring dike and canal must then be constructed around the landward side of the basin, to exclude and carry off the fresh-water streams which now empty into it. One of these, the Ijssel, a considerable river, has a course of eighty miles, and is, in fact, one of the outlets of the Rhine, though augmented by the waters of several independent tributaries. These preparations being made, and perhaps transverse dikes erected at convenient points for dividing the gulf into smaller portions, the water must be pumped out by machinery, in substantially the same way as in the case of the Lake of Haarlem. No safe calculations can be made as to the expenditure of time and money required for the execution of this stupendous enterprise, but I believe its practicability is not denied by competent judges, though doubts are entertained as to its financial expediency. The geographical results of this improvement would be analogous to those of the draining of the Lake of Haarlem, but many times multiplied in extent, and its meteorological effects, though perhaps not perceptible on the coast, could hardly fail to be appreciable in the interior of Holland.

Waters of the Karst.

The singular structure of the Karst, the great limestone plateau lying to the north of Trieste, has suggested some engineering operations which might be attended with sensible effects upon the geography of the province. I have described this table land as, though now bare of forests, and almost of vegetation, having once been covered with woods, and as being completely honeycombed by caves through which the drainage of that region is conducted. Schmidl has spent years in studying the subterranean geography and hydrography of this singular district, and his discoveries, and those of earlier cave-hunters, have led to various proposals of physical improvement of a novel character. Many of the underground water courses of the Karst are without visible outlet, and, in some instances at least, they, no doubt, send their waters, by deep channels, to the Adriatic.* The city of Trieste is very insufficiently provided with fresh water. It has been thought practicable to supply this want by tunnelling through the wall of the plateau, which rises abruptly in the rear of the town, until some subterranean stream is encountered, the current of which can be conducted to the city. More visionary projectors have gone further, and imagined that advantage might be taken of the natural tunnels under the Karst for the passage of roads, railways, and even navigable canals. But however chimerical these latter schemes may seem, there is every reason to believe that art might avail itself of these galleries for improving the imperfect drainage of the champaign country bounded by the Karst, and that stopping or opening the natural channels might very much modify the hydrography of an extensive region.

Subterranean Waters of Greece.

There are parts of continental Greece which resemble the Karst and the adjacent plains in being provided with a natural subterranean drainage. The superfluous waters run off into limestone caves called *catavothra* (καταβόθρα). In ancient times, the entrances to the catavothra were enlarged or partially closed as the convenience of drainage or irrigation required, and there is no doubt that similar measures might be adopted at the present day with great advantage both to the salubrity and the productiveness of the regions so drained.

* The Recca, a river with a considerable current, has been satisfactorily identified with a stream flowing through the cave of Trebich, and with the Timavo—the Timavus of Virgil and the ancient geographers—which empties through several mouths into the Adriatic between Trieste and Aquileia. The distance from Trieste to a suitable point in the grotto of Trebich is thought to be less than three miles, and the difficulties in the way of constructing a tunnel do not seem formidable. The works of Schmidl, *Die Höhlen des Karstes*, and *Der unterirdische Lauf der Recca*, are not common out of Germany, but the reader will find many interesting facts derived from them in two articles entitled *Der unterirdische Lauf der Recca*, in *Aus der Natur*, xx, pp. 250–254, 263–266.

Soil below Rock.

One of the most singular changes of natural surface effected by man is that observed by Beechey and by Barth at Lîn Tefla, and near Gebel Genûnes, in the district of Ben Gâsi, in Northern Africa. In this region the superficial stratum originally consisted of a thin sheet of rock covering a layer of fertile earth. This rock has been broken up, and, when not practicable to find use for it in fences, fortresses, or dwellings, heaped together in high piles, and the soil, thus bared of its stony shell, has been employed for agricultural purposes.* If we remember that gunpowder was unknown at the period when these remarkable improvements were executed, and of course that the rock could have been broken only with the chisel and wedge, we must infer that land had at that time a very great pecuniary value, and, of course, that the province, though now exhausted, and almost entirely deserted by man, had once a dense population.

Covering Rock with Earth.

If man has, in some cases, broken up rock to reach productive ground beneath, he has, in many other instances, covered bare ledges, and sometimes extensive surfaces of solid stone, with fruitful earth, brought from no inconsiderable distance. Not to speak of the Campo Santo at Pisa, filled, or at least coated, with earth from the Holy Land, for quite a different purpose, it is affirmed that the garden of the monastery of St. Catherine at Mount Sinai is composed of Nile mud, transported on the backs of camels from the banks of that river. Parthey and older authors state that all the productive soil of the Island of Malta was brought over from Sicily.† The accuracy

* BARTH, *Wanderungen durch die Küsten des Mittelmeeres*, i, p. 353. In a note on page 380, of the same volume, Barth cites Strabo as asserting that a similar practice prevailed in Iapygia; but it may be questioned whether the epithet τραχεῖα, applied by Strabo to the original surface, necessarily implies that it was covered with a continuous stratum of rock.

† PARTHEY, *Wanderungen durch Sicilien und die Levante*, i, p. 404.

of the information may be questioned in both cases, but similar practices, on a smaller scale, are matter of daily observation in many parts of Southern Europe. Much of the wine of the Moselle is derived from grapes grown on earth carried high up the cliffs on the shoulders of men. In China, too, rock has been artificially covered with earth to an extent which gives such operations a real geographical importance, and the accounts of the importation of earth at Malta, and the fertilization of the rocks on Mount Sinai with slime from the Nile, may be not wholly without foundation.

Wadies of Arabia Petræa.

In the latter case, indeed, river sediment might be very useful as a manure, but it could hardly be needed as a soil; for the growth of vegetation in the wadies of the Sinaitic Peninsula shows that the disintegrated rock of its mountains requires only water to stimulate it to considerable productiveness. The wadies present, not unfrequently, narrow gorges, which might easily be closed, and thus accumulations of earth, and reservoirs of water to irrigate it, might be formed which would convert many a square mile of desert into flourishing date gardens and cornfields. Not far from Wadi Feiran, on the most direct route to Wadi Esh-Sheikh, is a very narrow pass called by the Arabs El Bueb (El Bab) or, The Gate, which might be securely closed to a very considerable height, with little labor or expense. Above this pass is a wide and nearly level expanse, containing a hundred acres, perhaps much more. This is filled up to a certain regular level with deposits brought down by torrents before the Gate, or Bueb, was broken through, and they have now worn down a channel in the deposits to the bed of the wadi. If a dam were constructed at the pass, and reservoirs built to retain the winter rains, a great extent of valley might be rendered cultivable.

Incidental Effects of Human Action.

I have more than once alluded to the collateral and unsought consequences of human action as being often more momentous than the direct and desired results. There are cases where such incidental, or, in popular speech, accidental, consequences, though of minor importance in themselves, serve to illustrate natural processes; others, where, by the magnitude and character of the material traces they leave behind them, they prove that man, in primary or in more advanced stages of social life, must have occupied particular districts for a longer period than has been supposed by popular chronology. "On the coast of Jutland," says Forchhammer, "wherever a bolt from a wreck or any other fragment of iron is deposited in the beach sand, the particles are cemented together, and form a very solid mass around the iron. A remarkable formation of this sort was observed a few years ago in constructing the sea wall of the harbor of Elsineur. This stratum, which seldom exceeded a foot in thickness, rested upon common beach sand, and was found at various depths, less near the shore, greater at some distance from it. It was composed of pebbles and sand, and contained a great quantity of pins, and some coins of the reign of Christian IV, between the beginning and the middle of the seventeenth century. Here and there, a coating of metallic copper had been deposited by galvanic action, and the presence of completely oxydized metallic iron was often detected. An investigation undertaken by Councillor Reinhard and myself, at the instance of the Society of Science, made it in the highest degree probable that this formation owed its origin to the street sweepings of the town, which had been thrown upon the beach, and carried off and distributed by the waves over the bottom of the harbor." * These and other familiar observations of the like sort show that a sandstone reef, of no inconsiderable magnitude, might

* *Geognostische Studien am Meeres Ufer*, LEONHARD und BRONN, *Jahrbuch*, 1841, pp. 25, 26.

originate from the stranding of a ship with a cargo of iron,* or from throwing the waste of an establishment for working metals into running water which might carry it to the sea.

Parthey records a singular instance of unforeseen mischief from an interference with the arrangements of nature. A landowner at Malta possessed a rocky plateau sloping gradually toward the sea, and terminating in a precipice forty or fifty feet high, through natural openings in which the sea water flowed into a large cave under the rock. The proprietor attempted to establish salt works on the surface, and cut shallow pools in the rock for the evaporation of the water. In order to fill the salt pans more readily, he sank a well down to the cave beneath, through which he drew up water by a windlass and buckets. The speculation proved a failure, because the water filtered through the porous bottom of the pans, leaving little salt behind. But this was a small evil, compared with other destructive consequences that followed. When the sea was driven into the cave by violent west or northwest winds, it shot a *jet d'eau* through the well to the height of sixty feet, the spray of which was scattered far and wide over the neighboring gardens and blasted the crops. The well was now closed with stones, but the next winter's storms hurled them out again, and spread the salt spray over the grounds in the vicinity as before. Repeated attempts were made to stop the orifice, but at the time of Parthey's visit the sea had thrice burst through, and it was feared that the evil was without remedy.†

I have mentioned the great exent of the heaps of oyster and other shells left by the American Indians on the Atlantic coast of the United States. Some of the Danish kitchen-middens, which closely resemble them, are a thousand feet long, from one hundred and fifty to two hundred wide, and from six to ten high. These piles have an importance as geological witnesses, independent of their bearing upon human

* Kohl, *Schleswig-Holstein*, ii, p. 45.

† *Wanderungen durch Sicilien und die Levante*, i, p. 406.

history. Wherever the coast line appears, from other evidence, to have remained unchanged in outline and elevation since they were accumulated, they are found near the sea, and not more than about ten feet above its level. In some cases they are at a considerable distance from the beach, and in these instances, so far as yet examined, there are proofs that the coast has advanced in consequence of upheaval or of fluviatile or marine deposit. Where they are altogether wanting, the coast seems to have sunk or been washed away by the sea. The constancy of these observations justifies geologists in arguing, where other evidence is wanting, the advance of land or sea respectively, or the elevation or depression of the former, from the position or the absence of these heaps alone.

Every traveller in Italy is familiar with Monte Testaccio, the mountain of potsherds, at Rome; but this deposit, large as it is, shrinks into insignificance when compared with masses of similar origin in the neighborhood of older cities. The cast-away pottery of ancient towns in Magna Græcia composes strata of such extent and thickness that they have been dignified with the appellation of the ceramic formation. The Nile, as it slowly changes its bed, exposes in its banks masses of the same material, so vast that the population of the world during the whole historical period would seem to have chosen this valley as a general deposit for its broken vessels.

The fertility imparted to the banks of the Nile by the water and the slime of the inundations, is such that manures are little employed. Hence much domestic waste, which would elsewhere be employed to enrich the soil, is thrown out into vacant places near the town. Hills of rubbish are thus piled up which astonish the traveller almost as much as the solid pyramids themselves. The heaps of ashes and other household refuse collected on the borders and within the limits of Cairo were so large, that the removal of them by Ibrahim Pacha has been looked upon as one of the great works of the age.

The soil near cities, the street sweepings of which are spread upon the ground as manure, is perceptibly raised by them and by other effects of human industry, and in spite of all efforts to remove the waste, the level of the ground on which large towns stand is constantly elevated. The present streets of Rome are twenty feet above those of the ancient city. The Appian way between Rome and Albano, when cleared out a few years ago, was found buried four or five feet deep, and the fields along the road were elevated nearly or quite as much. The floors of many churches in Italy, not more than six or seven centuries old, are now three or four feet below the adjacent streets, though it is proved by excavations that they were built as many feet above them.

Resistance to Great Natural Forces.

I have often spoken of the greater and more subtile natural forces, and especially of geological agencies, as powers beyond human guidance or resistance. This is no doubt at present true in the main, but man has shown that he is not altogether impotent to struggle with even these mighty servants of nature, and his unconscious as well as his deliberate action may in some cases have increased or diminished the intensity of their energies. It is a very ancient belief that earthquakes are more destructive in districts where the crust of the earth is solid and homogeneous, than where it is of a looser and more interrupted structure. Aristotle, Pliny the elder, and Seneca believed that not only natural ravines and caves, but quarries, wells, and other human excavations, which break the continuity of the terrestrial strata and facilitate the escape of elastic vapors, have a sensible influence in diminishing the violence and preventing the propagation of the earth waves. In all countries subject to earthquakes this opinion is still maintained, and it is asserted that, both in ancient and in modern times, buildings protected by deep wells under or near them have suffered less from earthquakes than those the architects of which have neglected this precaution.*

* LANDGREBE, *Naturgeschichte der Vulkane*, ii, pp. 19, 20.

If the commonly received theory of the cause of earthquakes is true—that, namely, which ascribes them to the elastic force of gases accumulated or generated in subterranean reservoirs—it is evident that open channels of communication between such reservoirs and the atmosphere might serve as a harmless discharge of gases that would otherwise acquire destructive energy. The doubt is whether artificial excavations can be carried deep enough to reach the laboratory where the elastic fluids are distilled. There are, in many places, small natural crevices through which such fluids escape, and the source of them sometimes lies at so moderate a depth that they pervade the superficial soil and, as it were, transpire from it, over a considerable area. When the borer of an ordinary artesian well strikes into a cavity in the earth, imprisoned air often rushes out with great violence, and this has been still more frequently observed in sinking mineral-oil wells. In this latter case, the discharge of a vehement current of inflammable fluid sometimes continues for hours and even longer periods. These facts seem to render it not wholly improbable that the popular belief of the efficacy of deep wells in mitigating the violence of earthquakes is well founded.

In general, light, wooden buildings are less injured by earthquakes than more solid structures of stone or brick, and it is commonly supposed that the power put forth by the earth wave is too great to be resisted by any amount of weight or solidity of mass that man can pile up upon the surface. But the fact that in countries subject to earthquakes many very large and strongly constructed palaces, temples, and other monuments have stood for centuries, comparatively uninjured, suggests a doubt whether this opinion is sound. The earthquake of the first of November, 1755, which was felt over a twelfth part of the earth's surface, was probably the most violent of which we have any clear and distinct account, and it seems to have exerted its most destructive force at Lisbon. It has often been noticed as a remarkable fact, that the mint, a building of great solidity, was almost wholly unaffected by the shock which shattered every house and church in the city, and its escape from the common ruin can hardly be accounted for except upon the supposition that its weight, compactness, and strength of material enabled it to resist an agitation of the earth which overthrew all weaker structures. On the other hand, a stone pier in the harbor of Lisbon, on which thousands of people had taken refuge, sank with its foundations to a great depth during the same earthquake; and it is plain that where subterranean cavities exist, at moderate depths, the erection of heavy masses upon them would tend to promote the breaking down of the strata which roof them over.

No physicist, I believe, has supposed that man can avert the eruption of a volcano or diminish the quantity of melted rock which it pours out of the bowels of the earth; but it is not always impossible to divert the course of even a large current of lava. "The smaller streams of lava near Catania," says Ferrara, in describing the great eruption of 1669, "were turned from their course by building dry walls of stone as a barrier against them. * * * It was proposed to divert the main current from Catania, and fifty men, protected by hides, were sent with hooks and iron bars to break the flank of the stream near Belpasso.* When the opening was made,

* Soon after the current issues from the volcano, it is covered above and at its sides, and finally in front, with scoriæ, formed by the cooling of the exposed surface, which bury and conceal the fluid mass. The stream rolls on under the coating, and between the walls of scoriæ, and it was the lateral crust which was broken through by the workmen mentioned in the text.

The distance to which lava flows, before its surface begins to solidify, depends on its volume, its composition, its temperature and that of the air, the force with which it is ejected, and the inclination of the declivity over which it runs. In most cases it is difficult to approach the current at points where it is still entirely fluid, and hence opportunities of observing it in that condition are not very frequent. In the eruption of February, 1850, on the east side of Vesuvius, I went quite up to one of the outlets. The lava shot out of the orifice upward with great velocity, like the water from a spring, in a stream eight or ten feet in diameter, throwing up occasionally volcanic bombs, but it immediately spread out on the declivity down which it flowed, to the width of several yards. It continued red hot in broad daylight, and without a particle of scoriæ on its surface, for a fluid lava poured forth and flowed rapidly toward Paterno; but the inhabitants of that place, not caring to sacrifice their own town to save Catania, rushed out in arms and put a stop to the operation." * In the eruption of Vesuvius in 1794, the viceroy saved from impending destruction the town of Portici, and the valuable collection of antiquities then deposited there but since removed to Naples, by employing several thousand men to dig a ditch above the town, by which the lava current was carried off in another direction.†

Effects of Mining.

The excavations made by man, for mining and other purposes, may sometimes occasion disturbance of the surface by the subsidence of the strata above them, as in the case of the mine of Fahlun, but such accidents must always be too inconsiderable in extent to deserve notice in a geographical point of view. Such excavations, however, may interfere materially with the course of subterranean waters, and it has even been conjectured that the removal of large bodies of metallic ore

course of at least one hundred yards. At this distance, the suffocating, sulphurous vapors became so dense that I could follow the current no farther. The undulations of the surface were like those of a brook swollen by rain. I estimated the height of the waves at five or six inches by a breadth of eighteen or twenty. To the eye, the fluidity of the lava seemed as perfect as that of water, but masses of cold lava weighing ten or fifteen pounds floated upon it like cork.

The heat emitted by lava currents seems extremely small when we consider the temperature required to fuse such materials and the great length of time they take in cooling. I saw at Nicolosi ancient oil jars, holding a hundred gallons or more, which had been dug out from under a stream of old lava above that town. They had been very slightly covered with volcanic ashes before the lava flowed over them, but the lead with which holes in them had been plugged was not melted. The current that buried Mompiliere in 1669 was thirty-five feet thick, but marble statues, in a church over which the lava formed an arch, were found uncalcined and uninjured in 1704. See SCROPE, *Volcanoes*, chap. VI. § 6.

* FERRARA, *Descrizione dell' Etna*, p. 108.

† LANDGREBE, *Naturgeschichte der Vulkane*, ii, p. 82.

from their original deposits might, at least locally, affect the magnetic and electrical condition of the earth's crust to a sensible degree.

Accidental fires in mines of coal or lignite sometimes lead to consequences not only destructive to large quantities of valuable material, but may, directly or indirectly, produce results important in geography. The coal occasionally takes fire from the miners' lights or other fires used by them, and, if long exposed to air in deserted galleries, may be spontaneously kindled. Under favorable circumstances, a stratum of coal will burn till it is exhausted, and a cavity may be burnt out in a few months which human labor could not excavate in many years. Wittwer informs us that a coal mine at St. Etienne in Dauphiny has been burning ever since the fourteenth century, and that a mine near Duttweiler, another near Epterode, and a third at Zwickau, have been on fire for two hundred years. Such conflagrations not only produce cavities in the earth, but communicate a perceptible degree of heat to the surface, and the author just quoted cites cases where this heat has been advantageously employed in forcing vegetation.*

* *Physikalische Geographie*, p. 168. Beds of peat, accidentally set on fire, sometimes continue to burn for months. I take the following account of a case of this sort from a recent American journal:

"A Curious Phenomenon.—When the track of the railroad between Brunswick and Bath was being graded, in crossing a meadow near the populous portion of the latter city, the 'dump' suddenly took on a sinking symptom, and down went the twenty feet fill of gravel, clay, and broken rocks, out of sight, and it was a long, *long* time before dirt trains could fill the capacious stomach that seemed ready to receive all the solid material that could be turned into it. The difficulty was at length overcome, but all along the side of the sinkage the earth was thrown up, broken into yawning chasms, and the surface was thus elevated above its old watery level. Since that time this ground, thus slightly elevated, has been cultivated, and has yielded enormously of whatever the owner seemed disposed to plant upon it. Some three months ago, by some means unknown to us, the underlying peat took fire, and for weeks, as we had occasion to pass it, we noticed the smoke arising from the smouldering combustion beneath the surface. Rains fell, but the fire burned, and the smoke continued to arise. Monday we had occasion to pass the spot, and though nearly a

Espy's Theories.

Espy's well known suggestion of the possibility of causing rain artificially, by kindling great fires, is not likely to be turned to practical account, but the speculations of this able meteorologist are not, for that reason, to be rejected as worthless. His labors exhibit great industry in the collection of facts, much ingenuity in dealing with them, remarkable insight into the laws of nature, and a ready perception of analogies and relations not obvious to minds less philosophically constituted. They have unquestionably contributed very essentially to the advancement of meteorological science. The possibility that the distribution and action of electricity may be considerably modified by long lines of iron railways and telegraph wires, is a kindred thought, and in fact rests much on the same foundation as the belief in the utility of lightning rods, but such influence is too obscure and too small to have been yet detected.

River Sediment.

The manifestation of the internal heat of the earth at any given point is conditioned by the thickness of the crust at such point. The deposits of rivers tend to augment that thickness at their estuaries. The sediment of slowly flowing rivers emptying into shallow seas is spread over so great a surface that we can hardly imagine the foot or two of slime they let fall over a wide area in a century to form an element among even the infinitesimal quantities which compose the terms of the equations of nature. But some swift rivers, rolling mountains of fine earth, discharge themselves into deeply scooped gulfs or bays, and in such cases the deposit amounts, in the course of a few years, to a mass the transfer of which from the surface of a large basin, and its accumulation at a single point, may be

week's rain had been drenching the ground, and though the surface was whitened with snow, and though pools of water were standing upon the surface in the immediate neighborhood, still the everlasting subterranean fire was burning, and the smoke arising through the snow."

supposed to produce other effects than those measurable by the sounding line. Now, almost all the operations of rural life, as I have abundantly shown, increase the liability of the soil to erosion by water. Hence, the clearing of the valley of the Ganges by man must have much augmented the quantity of earth transported by that river to the sea, and of course have strengthened the effects, whatever they may be, of thickening the crust of the earth in the Bay of Bengal. In such cases, then, human action must rank among geological influences.

Nothing Small in Nature.

It is a legal maxim that "the law concerneth not itself with trifles," *de minimus non curat lex;* but in the vocabulary of nature, little and great are terms of comparison only; she knows no trifles, and her laws are as inflexible in dealing with an atom as with a continent or a planet.* The human opera-

* One of the sublimest, and at the same time most fearful suggestions that have been prompted by the researches of modern science, was made by Babbage in the ninth chapter of his *Ninth Bridgewater Treatise.* I have not the volume at hand, but the following explanation will recall to the reader, if it does not otherwise make intelligible, the suggestion I refer to.

No atom can be disturbed in place, or undergo any change of temperature, of electrical state, or other material condition, without affecting, by attraction or repulsion or other communication, the surrounding atoms. These, again, by the same law, transmit the influence to other atoms, and the impulse thus given extends through the whole material universe. Every human movement, every organic act, every volition, passion, or emotion, every intellectual process, is accompanied with atomic disturbance, and hence every such movement, every such act or process affects all the atoms of universal matter. Though action and reaction are equal, yet reaction does not restore disturbed atoms to their former place and condition, and consequently the effects of the least material change are never cancelled, but in some way perpetuated, so that no action can take place in physical, moral, or intellectual nature, without leaving all matter in a different state from what it would have been if such action had not occurred. Hence, to use language which I have employed on another occasion: there exists, not alone in the human conscience or in the omniscience of the Creator, but in external material nature, an ineffaceable, imperishable

tions mentioned in the last few paragraphs, therefore, do act in the ways ascribed to them, though our limited faculties are at present, perhaps forever, *incapable* of weighing their immediate, still more their ultimate consequences. But our inability to assign definite values to these causes of the disturbance of natural arrangements is not a reason for ignoring the existence of such causes in any general view of the relations between man and nature, and we are never justified in assuming a force to be insignificant because its measure is unknown, or even because no physical effect can now be traced to it as its origin. The collection of phenomena must precede the analysis of them, and every new fact, illustrative of the action and reaction between humanity and the material world around it, is another step toward the determination of the great question, whether man is of nature or above her.

record, possibly legible even to created intelligence, of every act done, every word uttered, nay, of every wish and purpose and thought conceived by mortal man, from the birth of our first parent to the final extinction of our race; so that the physical traces of our most secret sins shall last until time shall be merged in that eternity of which not science, but religion alone, assumes to take cognizance.

APPENDIX.

No. 1 (page 19, *note*). It may be said that the cases referred to in the note on p. 19—and indeed all cases of a supposed acclimation consisting in physiological changes—are instances of the origination of new varieties by natural selection, the hardier maize, tomato, and other vegetables of the North, being the progeny of seeds of individuals endowed, exceptionally, with greater power of resisting cold than belongs in general to the species which produced them. But, so far as the evidence of change of climate, from a difference in vegetable growth, is concerned, it is immaterial whether we adopt this view or maintain the older and more familiar doctrine of a local modification of character in the plants in question.

No. 2 (page 24, *note*). The adjectives of direction in *-erly* are not unfrequently used to indicate, in a loose way, the course of winds blowing from unspecified points between N. E. and S. E.; S. E. and S. W.; S. W. and N. W. or N. W. and N. E. If the employment of these words were understood to be limited to thus expressing a direction nearer to the cardinal point from whose name the adjective is taken than to any other cardinal point, they would be valuable elements of English meteorological nomenclature.

No. 3 (page 31). I find a confirmation of my observations on the habits of the beaver as a geographical agency, in a report of the proceedings of the British Association, in the London Athenæum of October 8, 1864, p. 469. It is there stated that Viscount Milton and Dr. Cheadle, in an expedition across the Rocky Mountains by the Yellow Head, or Leather Pass, observed that "a great portion of the country to the east of the mountains" had been "completely changed in character by the agency of the beaver, which formerly existed here in enormous numbers. The shallow valleys were formerly traversed by rivers and chains of lakes which, dammed up along their course at numerous points, by the work of those animals, have become a series of marshes in various stages of consolidation. So complete has this change been, that hardly a stream is found for a distance of two hundred miles, with the exception of the large rivers. The animals have thus destroyed, by their own labors, the waters necessary to their own existence."

When the process of "consolidation" shall have been completed, and the forest reëstablished upon the marshes, the water now diffused through them will be collected in the lower or more yielding portions, cut new channels for their flow, become running brooks, and thus restore the ancient aspect of the surface.

No. 4 (page 33, *note*). The lignivorous insects that attack living trees almost uniformly confine their ravages to trees already unsound or diseased in growth from the depredations of leaf-eaters, such as caterpillars and the like, or from other causes. The decay of the tree, therefore, is the cause not the consequence of the invasions of the borer. This subject has been discussed by Perris in the *Annales de la Société Entomologique de la France* for 1851 (?), and his conclusions are confirmed by the observations of Samanos, who quotes, at some length, the views of Perris. "Having, for fifteen years," says the latter author, "incessantly studied the habits of lignivorous insects in one of the best wooded regions of France, I have observed facts enough to feel myself warranted in expressing my conclusions, which are: that insects in general—I am not speaking of those which confine their voracity to the leaf—do not attack trees in sound health, and they assail those only whose normal conditions and functions have been by some cause impaired."

See, more fully, Samanos, *Traité de la Culture du Pin Maritime*, Paris, 1864, pp. 140–145.

No. 5 (page 34, *note*). Very interesting observations, on the agency of the squirrel and other small animals in planting and in destroying nuts and other seeds of trees, may be found in a paper on the Succession of Forests in Thoreau's *Excursions*, pp. 135 *et seqq.*

I once saw several quarts of beech-nuts taken from the winter quarters of a family of flying squirrels in a hollow tree. The kernels were neatly stripped of their shells and carefully stored in a dry cavity.

No. 6 (page 40, *note*). Schroeder van der Kolk, in *Het Verschil tusschen den Psychischen Aanleg van het Dier en van den Mensch*, cites from Burdach and other authorities many interesting facts respecting instincts lost, or newly developed and become hereditary, in the lower animals, and he quotes Aristotle and Pliny as evidence that the common quadrupeds and fowls of our fields and our poultry yards were much less perfectly domesticated in their times than long, long ages of servitude have now made them.

Perhaps the half-wild character ascribed by P. Læstadius and other Swedish writers to the reindeer of Lapland, may be in some degree due to the comparative shortness of the period during which he has been partially tamed. The domestic swine bred in the woods of Hungary and the buffaloes of Southern Italy are so wild and savage as to be very dangerous to all but their keepers. The former have relapsed into their original condition, the latter have not yet been reclaimed from it.

Among other instances of obliterated instincts, Schroeder van der Kolk states that in Holland, where, for centuries, the young of the cow has been usually taken from the dam at birth and fed by hand, calves, even if left with the mother, make no attempt to suck; while in England, where calves are not weaned until several weeks old, they resort to the udder as naturally as the young of wild quadrupeds.—*Ziel en Ligchaam*, p. 128, *n.*

No. 7 (page 60, *first note*). At Piè di Mulera, at the outlet of the Val Anzasca, near the principal hotel, is a vine measuring thirty-one inches in circumference. The door of the chapter-hall in the cloister of the church of San Giovanni, at Saluzzo, is of vine wood, and the boards of which the panels were made could not have been less than ten inches wide. Statues and other objects of considerable dimensions, of vine wood, are mentioned by ancient writers.

No. 8 (page 6, *second note*). Cartier, A. D. 1535–'6, mentions "vines, great melons, cucumbers, gourds [courges], pease, beans of various colors, but not like ours," as common among the Indians of the banks of the St. Lawrence.—*Bref Recit*, etc., reprint. Paris, 1863, pp. 13, a; 14, b; 20, b; 31, a.

No. 8 (page 65, *second paragraph*). It may be considered very highly probable, if not certain, that the undiscriminating herbalists of the sixteenth century must have overlooked many plants native to this island. An English botanist, in an hour's visit to Aden, discovered several species of plants on rocks always reported, even by scientific travellers, as absolutely barren. But after all, it appears to be well established that the original flora of St. Helena was extremely limited, though now counting hundreds of species.

No. 9 (page 66, *first note*). Although the vine *genus* is very catholic and cosmopolite in its habits, yet particular *varieties* are extremely fastidious and exclusive in their requirements as to soil and climate. The stocks of many celebrated vineyards lose their peculiar qualities by transplantation, and the most famous wines are capable of production only in certain well-defined, and for the most part narrow districts. The Ionian vine which bears the little stoneless grape known in commerce as the Zante currant, has resisted almost all efforts to naturalize it elsewhere, and is scarcely grown except in two or three of the Ionian islands and in a narrow territory on the northern shores of the Morea.

No. 10 (page 68, *first note*). In most of the countries of Southern Europe, sheep and beeves are wintered upon the plains, but driven in the summer to mountain pastures at many days' distance from the homesteads of their owners. They transport seeds in their coats in both directions, and hence Alpine plants often shoot up at the foot of the mountains, the grasses of the plain on the borders of the glaciers; but in both cases, they usually fail to propagate themselves by ripening their seed. This explains the scattered tufts of common clover, with pale and flaccid blossoms, which are sometimes seen at heights exceeding 7,000 feet above the sea.

No. 11 (page 78, *last paragraph*). The poisonous wild parsnip, which is very common in New England, is popularly believed to be identical with the garden parsnip, and differenced only by conditions of growth, a richer soil depriving it, it is said, of its noxious properties. Many wild medicinal plants, such as pennyroyal for example, are so much less aromatic and powerful, when cultivated in gardens, than when self-sown on meagre soils, as to be hardly fit for use.

No. 12 (page 74, *second note*). See in Thoreau's *Excursions*, an interesting description of the wild apple-trees of Massachusetts.

No. 13 (page 86, *first paragraph*). It is said at Courmayeur that a very few ibexes of a larger variety than those of the Cogne mountains, still linger about the Grande Jorasse.

No. 14 (page 92, *first note*). In Northern and Central Italy, one often sees hillocks crowned with grove-like plantations of small trees, much resembling large arbors. These serve to collect birds, which are entrapped in nets in great numbers. These plantations are called *ragnaje*, and the reader will find, in Bindi's edition of Davanzati, a very pleasant description of a ragnaja, though its authorship is not now ascribed to that eminent writer.

No. 15 (page 93, *second note*). The appearance of the dove-like grouse, *Tetrao paradoxus*, or *Syrrhaptes Pallasii*, in various parts of Europe, in 1859 and the following years, is a noticeable exception to the law of regularity which seems to govern the movements and determine the habitat of birds. The proper home of this bird is the steppes of Tartary, and it is not recorded to have been observed in Europe, or at least west of Russia, until the year abovementioned, when many flocks of twenty or thirty, and even a hundred individuals, were seen in Bohemia, Germany, Holland, Denmark, England, Ireland, and France. A considerable flock frequented the Frisian island of Borkum for more than five months. It was hoped they would breed and remain permanently in the island, but this expectation has been disappointed, and the steppe-grouse seems to have disappeared again altogether.

No. 16 (page 94, *note*). From an article by A. Esquiros, in the *Revue des De x Mondes* for Sept. 1, 1864, entitled, *La vie Anglaise*, p. 119, it appears that such occurrences as that stated in the note are not unfrequent on the British coast.

No. 17 (page 100, *first paragraph*). I cannot learn that caprification is now practised in Italy, but it is still in use in Greece.

No. 18 (page 112, *first note*). The recent great multiplication of vipers in some parts of France, is a singular and startling fact.

Toussenel, quoting from official documents, states, that upon the offer of a reward of fifty centimes, or ten cents, a head, *twelve thousand* vipers were brought to the prefect of a single department, and that in 1859 fifteen hundred snakes and twenty quarts of snakes' eggs were found under a farm-house hearthstone. The granary, the stables, the roof, the very beds swarmed with serpents, and the family were obliged to abandon its habitation. Dr. Viaugrandmarais, of Nantes, reported to the prefect of his department more than two hundred recent cases of viper bites, twenty-four of which proved fatal.—*Tristia*, p. 176 *et seqq*.

No. 19 (page 121, *first note*). The Beduins are little given to the chase, and seldom make war on the game birds and quadrupeds of the desert. Hence the wild animals of Arabia are less timid than those of Europe. On one occasion, when I was encamped during a sand storm of some violence in Arabia Petræa, a wild pigeon took refuge in one of our tents which had not been blown down, and remained quietly perched on a boy in the midst of four or five persons, until the storm was over, and then took his departure, *insalutato hospite*.

No. 20 (page 122). It is possible that time may modify the habits of the fresh water fish of the North American States, and accommodate them to the new physical conditions of their native waters. Hence it may be hoped that nature, even unaided by art, will do something toward restoring the ancient plenty of our lakes and rivers. The decrease of our fresh water fish cannot be ascribed alone to exhaustion by fishing, for in the waters of the valleys and flanks of the Alps, which have been inhabited and fished ten times as long by a denser population, fish are still very abundant, and they thrive and multiply under circumstances where no American species could live at all. On the southern slope of those mountains, trout are caught in great numbers, in the swift streams which rush from the glaciers, and where the water is of icy coldness, and so turbid with particles of fine-ground rock, that you cannot see an inch below the surface. The glacier streams of Switzerland, however, are less abundant in fish.

No. 21 (page 131, *note*). Vaupell, though agreeing with other writers as to the injury done to the forest by most domestic animals—which he illustrates in an interesting way in his posthumous work, *The Danish Woods*—thinks, nevertheless, that at the season when the mast is falling swine are rather useful than otherwise to forests of beech and oak, by treading into the ground and thus sowing beechnuts and acorns, and by destroying moles and mice.—*De Danske Skove*, p. 12.

No. 22 (page 135, *note*). The able authors of Humphreys and Abbot's most valuable Report on the Physics and Hydraulics of the Mississippi, conclude that the delta of that river began its encroachments on the Gulf of Mexico not more than 4,400 years ago, before which period they suppose the Mississippi to have been "a comparatively clear stream," conveying very little sediment to the sea. The present rate of advance of the delta is 262 feet a year, and there are reasons for thinking that the amount of deposit has long been approximately constant.—*Report*, pp. 435, 436.

The change in the character of the river must, if this opinion is well founded, be due to some geological revolution, or at least convulsion, and the hypothesis of the former existence of one or more great lakes in its upper valley, whose bottoms are occupied by the present prairie region, has been suggested. The shores of these supposed lakes have not, I believe, been traced, or even detected, and we cannot admit the truth of this hypothesis without supposing changes much more extensive than the mere bursting of the barrier which confined the waters.

No. 23 (page 143, *note*). See on this subject a paper by J. Jamin, in the *Revue des Deux Mondes* for Sept. 15, 1864; and, on the effects of human industry on the atmosphere, an article in *Aus der Natur*, vol. 29, 1864, pp. 443, 449, 465 *et seqq.*

No. 24 (page 159, *second paragraph*). All evergreens, even the broad-leaved trees, resist frosts of extraordinary severity better than the deciduous trees of the same climates. Is not this because the vital processes of trees of persistent foliage are less interrupted during winter than those of trees which annually shed their leaves, and therefore more organic heat is developed?

No. 25 (page 191, *first paragraph*). In discussing the influence of mountains on precipitation, meteorologists have generally treated the popular belief, that mountains "attract" to them clouds floating within a certain distance from them, as an ignorant prejudice, and they ascribe the appearance of clouds about high peaks solely to the condensation of the humidity of the air carried by atmospheric currents up the slopes of the mountain to a colder temperature. But if mountains do not really draw clouds and invisible vapors to them, they are an exception to the universal law of attraction. The attraction of the small Mount Shehallien was found sufficient to deflect from the perpendicular, by a measurable quantity, a plummet weighing but a few ounces. Why, then, should not greater masses attract to them volumes of vapor weighing hundreds of tons, and floating freely in the atmosphere within moderate distances of the mountains?

No. 26 (page 198, *note*). Élisée Redus ascribes the diminution of the ponds which border the dunes of Gascony to the absorption of their water by the trees which have been planted upon the sands.—*Revue des Deux Mondes*, 1 Aug., 1863, p. 694.

No. 27 (page 219, *note*). The waste of wood in European carpentry was formerly enormous, the beams of houses being both larger and more numerous than permanence or stability required. In examining the construction of the houses occupied by the eighty families which inhabit the village of Faucigny, in Savoy, in 1854, the forest inspector found that *fifty thousand* trees had been employed in building them. The builders "seemed," says Hudry-Menos, "to have tried to solve the problem of piling upon the walls the largest quantity of timber possible without crushing them." —*Revue des Deux Mondes*, 1 June, 1864, p. 601.

No. 28 (page 231, *note*). In a remarkable pamphlet, to which I shall have occasion to refer more than once hereafter, entitled *Avant-projet pour la création d'un sol fertile à la surface des Landes de Gascogne*, Duponchel argues with much force, that the fertilizing properties of river-slime are generally due much more to its mineral than to its vegetable constituents.

No. 29 (page 265, *note*). Even the denser silicious stones are penetrable by fluids and the coloring matter they contain, to such an extent that agates and other forms of silex may be artificially stained through their substance. This art was known to and practised by the ancient lapidaries, and it has been revived in recent times.

No. 30 (page 268). There is good reason for thinking that many of the earth and rock slides in the Alps occurred at an earlier period than the origin of the forest vegetation which, in later ages, covered the flanks of those mountains. See *Bericht über die Untersuchung der Schweizerischen Hochgebirgswaldungen*. 1862. P. 61.

Where more recent slides have been again clothed with woods, the trees, shrubs, and smaller plants which spontaneously grow upon them are usually of different species from those observed upon soil displaced at remote periods. This difference is so marked that the site of a slide can often be recognized at a great distance by the general color of the foliage of its vegetation.

No. 31 (page 286, *note*). It should have been observed that the venomous principle of poisonous mushrooms is not decomposed and rendered innocent by the process described in the note. It is merely extracted by the acidulated or saline water employed for soaking the plants, and care should be taken that this water be thrown away out of the reach of mischief.

No. 32 (page 293, *note*). Gaudry estimates the ties employed in the railways of France at thirty millions, to supply which not less than two millions of large trees have been felled. These ties have been, upon the average, at least once renewed, and hence we must double the number of ties and of trees required to furnish them.—*Revue des Deux Mondes*, 15 July, 1863, p. 425.

No. 33 (page 294, *second paragraph of note*). After all, the present consumption of wood and timber for fuel and other domestic and rural purposes, in many parts of Europe, seems incredibly small to an American. In rural Switzerland, the whole supply of firewood, fuel for small smitheries, dairies, breweries, brick and lime kilns, distilleries, fences, furniture, tools, and even house building—exclusive of the small quantity derived from the trimmings of fruit trees, grape vines and hedges, and from decayed fences and buildings—does not exceed an average of *two hundred and thirty cubic feet*, or less than two cords, a year per household. The average consumption of wood in New England for domestic fuel alone, is from five to ten times as much as Swiss families require for all the uses above enumerated. But the existing habitations of Switzerland are sufficient for a population which increases but slowly, and in the peasants' houses but a single room is usually heated. See *Bericht über die Untersuchung der Schweiz. Hochgebirgswaldungen*, pp. 85–89.

No. 34 (page 304). Among more recent manuals may be mentioned: *Les Etudes de Maitre Pierre.* Paris, 1864. 12mo; BAZELAIRE, *Traité de Reboisement.* 2d edition, Paris, 1864; and, in Italian, SIEMONI, *Manuale teorico-pratico d'arte Forestale.* Firenze, 1864. 8vo. A very important work has lately been published in France by Viscount de Courval, which is known to me only by a German translation published at Berlin, in 1864, under the title, *Das Aufästen der Waldbäume.* The principal feature of De Courval's very successful system of sylviculture, is a mode of trimming which compels the tree to develop the stem by reducing the lateral ramification. Beginning with young trees, the buds are rubbed off from the stems, and superfluous lateral shoots are pruned down to the trunk. When large trees are taken in hand, branches which can be spared, and whose removal is necessary to obtain a proper length of stem, are very smoothly cut off quite close to the trunk, and the exposed surface is *immediately* brushed over with mineral-coal tar. When thus treated, it is said that the healing of the wound is perfect, and without any decay of the tree.

No. 35 (page 313). The most gorgeous autumnal coloring I have observed in the vegetation of Europe, has been in the valleys of the Durance and its tributaries in Dauphiny. I must admit that neither in variety nor in purity and brilliancy of tint, does this coloring fall much, if at all, short of that of the New England woods. But there is this difference: in Dauphiny, it is only in small shrubs that this rich painting is seen, while in North America the foliage of large trees is dyed in full splendor. Hence the American woodland has fewer broken lights and more of what painters call breadth of coloring. Besides this, the arrangement of the leafage in large globular or conical masses, affords a wider scale of light and shade, thus aiding now the gradation, now the contrast of tints, and gives the American October landscape a softer and more harmonious tone than marks the humble shrubbery of the forest hill-sides of Dauphiny.

Thoreau—who was not, like some very celebrated landscape critics of the present day, an outside spectator of the action and products of natural forces, but, in the old religious sense, an *observer* of organic nature, living, more than almost any other descriptive writer, among and with her children—has a very eloquent paper on the "Autumnal Tints" of the New England landscape.—See his *Excursions*, pp. 215 *et seqq.*

Few men have personally noticed so many facts in natural history accessible to unscientific observation as Thoreau, and yet he had never seen that very common and striking spectacle, the phosphorescence of decaying wood, until, in the latter years of his life, it caught his attention in a bivouac in the forests of Maine. He seems to have been more excited by this phenomenon than by any other described in his works. It must be a capacious eye that takes in all the visible facts in the history of the most familiar natural object.—*The Maine Woods*, p. 184.

"The luminous appearance of bodies projected against the sky adjacent to the rising" or setting sun, so well described in Professor Necker's Letter to Sir David Brewster, is, as Tyndall observes, "hardly ever seen by either guides or travellers, though it would seem, *prima facie*, that it must be of frequent occurrence." See TYNDALL, *Glaciers of the Alps.* Part I. Second ascent of Mont Blanc.

Judging from my own observation, however, I should much doubt whether this brilliant phenomenon can be so often seen in perfection as would be expected; for I have frequently sought it in vain at the foot of the Alps, under conditions apparently otherwise identical with those where, in the elevated Alpine valleys, it shows itself in the greatest splendor.

No. 36 (page 324). European poets, whose knowledge of the date palm is not founded on personal observation, often describe its trunk as not only slender, but particularly *straight.* Nothing can be farther from the truth. When the Orientals compare the form of a beautiful girl to the stem of the palm, they do not represent it as rigidly straight, but on the contrary as made up of graceful curves, which seem less like permanent outlines than like flowing motion. In a palm grove, the trunks, so far from standing planted upright like the candles of a chandelier, bend in a vast variety of curves, now leaning towards, now diverging from, now crossing, each other, and among a hundred you will hardly see two whose axes are parallel.

No. 37 (page 316, *first note*). Charles Martin ascribes the power of reproduction by shoots from the stump to the cedar of Mount Atlas, which appears to be identical with the cedar of Lebanon.—*Revue des Deux Mondes*, 15 July, 1864, p. 315.

No. 38 (page 332). In an interesting article on recent internal improvements in England, in the London Quarterly Review for January, 1858, it is related that in a single rock cutting on the Liverpool and Manchester railway, 480,000 cubic yards of stone were removed; that the earth excavated and removed in the construction of English railways up to that date, amounted to a hundred and fifty million cubic yards, and that at the Round Down Cliff, near Dover, a single blast of nineteen thousand pounds of powder blew down a thousand million tons of chalk, and covered fifteen acres of land with the fragments.

No. 39 (page 339). According to Reventlov, whose work is one of the best sources of information on the subject of diking-in tide-washed flats, *Salicornia herbacea* appears as soon as the flat is raised high enough to be dry for three hours at ordinary ebb tide, or, in other words, where the ordinary flood covers it to a depth of not more than two feet. At a flood depth of one foot, the *Salicornia* dies and is succeeded by various sand plants. These are followed by *Poa distans* and *Poa maritima* as the ground is raised by further deposits, and these plants finally by common grasses. The *Salicornia* is preceded by *confervæ*, growing in deeper water, which spread over the bottom, and when covered by a fresh deposit of slime reappear above it, and thus vegetable and alluvial strata alternate until the flat is raised sufficiently high for the growth of *Salicornia*.—*Om Marskdannelsen paa Vestkysten af Hertugdömmet Slesvig*, pp. 7, 8.

No. 40 (page 348, *note*). The drijftil employed for the ring dike of the Lake of Haarlem, was in part cut in sections fifty feet long by six or seven wide, and these were navigated like rafts to the spot where they were sunk to form the dike.—Emile de Laveleye, *Revue des Deux Mondes*, 15 Sept., 1863, p. 285.

No. 41 (page 352, *last paragraph*). See on the influence of the improvements in question on tidal and other marine currents, Staring, *De Bodem van Nederland*, I. p. 279.

Although the dikes of the Netherlands and the adjacent states have protected a considerable extent of coast from the encroachments of the sea, and have won a large tract of cultivable land from the dominion of the waters, it has been questioned whether a different method of accomplishing these objects might not have been adopted with advantage. It has been suggested that a system of inland dikes and canals, upon the principle of those which, as will be seen in a subsequent part of the chapter on the waters, have been so successfully employed in the Val di Chiana and in Egypt, might have elevated the low grounds above the ocean tides, by spreading over them the sediment brought down by the Rhine, the Maes, and the Scheld. If this process had been introduced in the Middle Ages and constantly pursued to our times, the superficial and coast geography, as well as the hydrography of the countries in question, would undoubtedly have presented an aspect very different from their present condition; and by combining the process with a system of maritime dikes, which would have been necessary, both to resist the advance of the sea and to retain the slime deposited by river overflows, it is possible that the territory of those states would have been as extensive as it now is, and, at the same time, more elevated by several feet. But it must be borne in mind that we do not know the proportions in which the marine deposits that form the polders have been derived from materials brought down by these rivers or from other more remote sources. Much of the river slime has no doubt been transported by marine currents quite beyond the reach of returning streams, and it is uncertain how far this loss has been balanced by earth washed by the sea from distant shores and let fall on the coasts of the Netherlands and other neighboring countries.

We know little or nothing of the quantity of solid matter brought down by the rivers of Western Europe in early ages, but, as the banks of those rivers are now generally better secured against wash and abrasion than in former centuries, the sediment transported by them must be less than at periods nearer the removal of the primitive forests of their valleys. Klöden states the quantity of sedimentary matter now annually brought down by the Rhine at Bonn to be sufficient only to cover a square English mile to the depth of a little more than a foot.—*Erdkunde*, I. p. 384.

No. 42 (page 358, *first paragraph*). Meteorological observations have been regularly recorded at Zwanenburg, near the north end of the Lake of Haarlem, for more than a century, and since 1845 a similar register has been kept at the Helder, forty or fifty miles farther north. In comparing these two series of observations, it is found that about the end of the year 1852, when the drawing off of the waters of the Lake of Haarlem was completed, and the preceding summer had dried the grounds laid bare so as greatly to reduce the evaporable surface, a change took place in the relative temperature of the two stations. Taking the mean of every successive period of five days from 1845 to 1852, the temperature at Zwanenburg was thirty-three hundredths of a centigrade degree *lower* than at the Helder. Since the end of 1852, the thermometer at Zwanenburg has stood, from the 11th of April to the 20th of September inclusive, twenty-two hundredths of a degree *higher* than at the Helder, but from the 14th of October to the 17th of March, it has averaged one-tenth of a degree *lower* than its mean between the same dates before 1853.

There is no reasonable doubt that these differences are due to the draining of the lake. There has been less refrigeration from evaporation in summer, and the ground has absorbed more solar heat at the same period, while in the winter it has radiated more warmth then when it was covered with water. Doubtless the quantity of humidity contained in the atmosphere has also been affected by the same cause, but observations do not appear to have been made on that point. See Krecke, *Het Klimaat van Nederland*, II. 64.

No. 43 (page 358, *note*). In the course of the present year (1864), there have been several land slips on the borders of the Lake of Como, and in one instance the grounds of a villa lying upon the margin of the water suffered a considerable displacement. If the lake should be lowered to any considerable extent, in pursuance of the plan mentioned in the note on page 358, there is ground to fear that the steep shores of the lake might, at some points, be deprived of a lateral pressure requisite to their stability, and slide into the water as on the Lake of Lungern. See p. 356.

No. 44 (page 369, *last paragraph but one of note*). In like manner, while the box, the cedar, the fir, the oak, the pine, "beams," and "timber," are very frequently mentioned in the Old Testament, not one of these words is found in the New, *except* the case of the "beam in the eye," in the parable in Matthew and Luke.

No. 45 (page 375, *note*). In all probability, the real change effected by human art in the superficial geography of Egypt, is the conversion of pools and marshes into dry land, by a system of transverse dikes, which compelled the flood water to deposit its sediment on the banks of the river instead of carrying it to the sea. The *colmate* of modern Italy were thus anticipated in ancient Egypt.

No. 46 (page 378). We have seen in *Appendix*, No. 40, *ante*, that the mean temperature of a station on the borders of the Lake of Haarlem—a sheet of water formerly covering sixty-two and a half square English miles—for the period between the 11th of April and the 20th of September, had been raised not less than a degree of Fahrenheit by the draining of that lake; or, to state the case more precisely, that the formation of the lake, which was a consequence of man's improvidence, had reduced the temperature one degree F. below the natural standard. The artificially irrigated lands of France, Piedmont, and Lombardy, taken together, are fifty times as extensive as the Lake of Haarlem, and they are situated in climates where evaporation is vastly more rapid than in the Netherlands. They must therefore, no doubt, affect the local climate to a far greater extent than has been observed in connection with the draining of the lake in question. I do not know that special observations have been made with a view to measure the climatic effects of irrigation, but in the summer I have often found the *morning* temperature, when the difference would naturally be least perceptible, on the watered plains of Piedmont, nine miles south of Turin, several degrees lower than that recorded at an observatory in the city.

No. 47 (page 391, *note*). The Roman aqueduct known as the Pont du Gard, near Nismes, was built, in all probability, nineteen centuries ago. The bed of the river Gardon, a rather swift stream, which flows beneath it, can have suffered but a slight depression since the piers of the aqueduct were founded.

No. 48 (page 393, *first note*). Duponchel makes the following remarkable statement: "The river Herault rises in a granitic region, but soon reaches calcareous formations, which it traverses for more than sixty kilometres, rolling through deep and precipitous ravines, into which the torrents are constantly discharging enormous masses of pebbles belonging to the hardest rocks of the Jurassian period. These debris, continually renewed, compose, even below the exit of the gorge where the river enters into a regular channel cut in a tertiary deposit, broad beaches, prodigious accumulations of rolled pebbles, extending several kilometres down the stream, but they diminish in size and weight so rapidly that above the mouth of the river, which is at a distance of thirty or thirty-five kilometres from the gorge, every trace of calcareous matter has disappeared from the sands of the bottom, which are exclusively silicious."—*Avant-projet pour la création d'un sol fertile*, etc., p. 20.

No. 49 (page 404, *first paragraph of second note*). The length of the lower course of the Po having been considerably increased by the filling up of the Adriatic with its deposits, the velocity of the current ought, *prima facie*, to have been diminished and its bed raised in proportion. There are grounds for believing that this has happened in the case of the Nile, and one reason why the same effect has not been more sensibly perceptible in the Po is, that the confinement of the current by continuous embankments gives it a high-water velocity sufficient to sweep out deposits let fall at lower stages and slower movements of the water. Torrential streams tend first to excavate, then to raise, their beds. No general law on this point can be stated in relation to the middle and lower course of rivers. The conditions which determine the question of the depression or elevation of a river bed are too multifarious, variable, and complex to be subjected to formulæ, and they can scarcely even be enumerated. See, however, note on p. 431.

No. 50 (page 406, *first paragraph*). The system proposed in the text is substantially the Egyptian method, the Nile dikes having been constructed rather to retain than to exclude the water. The waters of rivers which flow down planes of gentle inclination, deposit in their inundations the largest proportion of their sediment as soon as, by overflowing their banks, they escape from the swift current of the channel, and consequently the immediate banks of such rivers become higher than the grounds lying farther from the stream. In the "intervals," or "bottoms," of the great North American rivers, the alluvial banks are elevated and dry, the flats more remote from the river lower and swampy. This is generally observable in Egypt, though less so than in the valley of the Mississippi, where, below Cape Girardeau, the alluvial banks constitute natural glacis descending as you recede from the river, at an average of seven feet in the first mile.—HUMPHREYS AND ABBOT'S *Report*, pp. 96, 97.

The Egyptian crossdikes, by retaining the water of the inundations, compel it to let fall its remaining slime, and hence the elevation of the remoter land goes on at a rate not very much slower than that of the immediate banks. Probably transverse embankments would produce the same effect in the Mississippi valley. In the great floods of this river, it is observed that, at a certain distance from the channel, the bottoms, though lower than the banks, are flooded to a less depth. See cross sections in Plate IV. of Humphreys and Abbot's Report. This apparently anomalous fact is due, I suppose, to the greater swiftness of the current of the overflowing water in the low grounds, which are often drained through the channels of rivers whose beds lie at a lower level than that of the Mississippi, or by the bayous which are so characteristic a feature of the geography of that valley. A judicious use of dikes would probably convert the swamps of the lower Mississippi valley into a region like Egypt.

No. 51 (*second note*). The mean discharge of the Mississippi is 675,000 cubic feet per second, and, accordingly, that river contributes to the sea about eleven times as much water as the Po, and more than six and a half times as much as the Nile. The discharge of the Mississippi is estimated at one-fourth of the precipitation in its basin, certainly a very large proportion, when we consider the rapidity of evaporation in many parts of the basin, and the probable loss by infiltration.—HUMPHREYS AND ABBOT'S *Report*, p. 93.

No. 52 (page 423, *first paragraph*). Artificially directed currents of water have been advantageously used in civil engineering for displacing and transporting large quantities of earth, and there is no doubt that this agency might be profitably employed to a far greater extent than has yet been attempted. Some of the hydraulic works in California for washing down masses of auriferous earth are on a scale stupenduous enough to produce really important topographical changes.

No. 53 (page 435, *first note*). I have lately been informed by a resident of the Ionian Islands, who is familiar with this phenomenon, that the sea flows uninterruptedly into the sub-insular cavities, at all stages of the tide.

No. 54 (page 438, *note*). It is observed in Cornwall that deep mines are freer from water in artificially well-drained, than in undrained agricultural districts.—ESQUIROS, *Revue des Deux Mondes*, Nov. 15, 1863, p. 430.

No. 55 (page 441). See, on the Artesian wells of the Sahara, and especially on the throwing up of living fish by them, an article entitled, *Le Sahara*, etc., by Charles Martins, in the *Revue des Deux Mondes* for August 1, 1864, pp. 618, 619.

No. 56 (page 441, *first note*). From the article in the *Rev. des Deux Mondes*, referred to in the preceding note, it appears that the wells discovered by Ayme were truly artesian. They were bored in rock, and provided at the outlet with a pear-shaped valve of stone, by which the orifice could be closed or opened at pleasure.

No. 57 (page 477, *second note*). Hull ingeniously suggests that, besides other changes, fine sand intermixed with or deposited above a coarser stratum, as well as the minute particles resulting from the disintegration of the latter, may be carried by rain in the case of dunes, or by the ordinary action of sea water in that of subaqueous sandbanks, down through the interstices in the coarser layer, and thus the relative position of fine sand and gravel may be more or less changed.—*Oorsprong der Hollandsche Duinen*, p. 103.

No. 58 (page 479). It appears from Laurent, that marine shells, of extant species, are found in the sands of the Sahara, far from the sea, and even at considerable depths below the surface.—*Mémoires sur le Sahara Oriental*, p. 62.

This observation has been confirmed by late travellers, and is an important link in the chain of evidence which tends to prove that the upheaval of the Libyan desert is of comparatively recent date.

No. 59 (p. 480). "At New Quay [in England] the dune sands are converted to stone by an oxyde of iron held in solution by the water which pervades them. This stone, which is formed, so to speak, under our eye, has been found solid enough to be employed for building."—Esquiros, *L'Angleterre et la vie Anglaise*, *Revue des Deux Mondes*, 1 March, 1864, pp. 44, 45.

No. 60 (page 496, *first paragraph*). In Ditmarsh, the breaking of the surface by the manœuvering of a corps of cavalry let loose a sand-drift which did serious injury before it was subdued.—Kohl, *Inseln u. Marschen.* etc., III. p. 282.

Similar cases have occurred in Eastern Massachusetts, from equally slight causes.—See Thoreau, *A Week on the Concord and Merrimack Rivers*, pp. 151–208.

No. 61 (page 497, *last note*). A more probable explanation of the fact stated in the note is suggested by Èlisée Reclus, in an article entitled, *Le Littoral de la France*, in the *Revue des Deux Mondes* for Sept. 1, 1864, pp. 193, 194. This able writer believes such pools to be the remains of ancient maritime bays, which have been cut off from the ocean by gradually accumulated sand banks raised by the waves and winds to the character of dunes.

No. 62 (page 506, *note*). The statement in the note is confirmed by Olmsted: "There is not a sufficient demand for rosin, except of the first qualities, to make it worth transporting from the inland distilleries; it is ordinarily, therefore, conducted off to a little distance, in a wooden trough, and allowed to flow from it to waste upon the ground. At the first distillery I visited, which had been in operation but one year, there lay a congealed pool of rosin, estimated to contain over three thousand barrels."—*A Journey in the Seaboard Slave States*, 1863, p. 345.

No. 63 (page 507). In an article on the dunes of Europe, in Vol. 29 (1864) of *Aus der Natur*, p. 590, the dunes are estimated to cover, on the islands and coasts of Schleswig Holstein, in Northwest Germany, Denmark, Holland, and France, one hundred and eighty-one German, or nearly four thousand English square miles; in Scotland, about ten German, or two hundred and ten English miles; in Ireland, twenty German, or four hundred and twenty English miles; and in England, one hundred and twenty German, or more than twenty-five hundred English miles.

No. 64 (page 512, *last paragraph*). For a brilliant account of the improvement of the Landes, see Edmond About, *Le Progrès*, Chap. VII.

In the memoir referred to in *Appendix*, No. 46, *ante*, Duponchel proposes the construction of artificial torrents to grind calcareous rock to slime by rolling and attrition in its bed, and, at the same time, the washing down of an argillaceous deposit which is to be mixed with the calcareous slime and distributed over the Landes by watercourses constructed for the purpose. By this means, he supposes that a highly fertile soil could be formed on the surface, which would also be so raised by the process as to admit of freer drainage. That nothing may be wanting to recommend this project, Duponchel suggests that, as some of the rivers of Western France are auriferous, it is probable that gold enough may be collected from the washings to reduce the cost of the operations materially.

No. 65 (page 528, *first paragraph*). The opening of a channel across Cape Cod would have, though perhaps to a smaller extent, the same effects in interchanging the animal life of the southern and northern shores of the isthmus, as in the case of the Suez canal; for although the breadth of Cape Cod does not anywhere exceed twenty miles, and is in some places reduced to one, it appears from the official reports on the Natural History of Massachusetts, that the population of the opposite waters differs widely in species.

Not having the original documents at hand, I quote an extract from the *Report on the Invertebrate Animals of Mass.*, given by Thoreau, *Excursions*, p. 69: "The distribution of the marine shells is well worthy of notice as a geological fact. Cape Cod, the right arm of the Commonwealth, reaches out into the ocean some fifty or sixty miles. It is nowhere many miles wide; but this narrow point of land has hitherto proved a barrier to the migration of many species of mollusca. Several genera and numerous species, which are separated by the intervention of only a few miles of land, are effectually prevented from mingling by the Cape, and do not pass from one side to the other * * * * Of the one hundred and ninety-seven marine species, eighty-three do not pass to the south shore, and fifty are not found on the north shore of the Cape."

Probably the distribution of the species of mollusks is affected by unknown local conditions, and therefore an open canal across the Cape might not make every species that inhabits the waters on one side common to those of the other; but there can be no doubt that there would be a considerable migration in both directions.

The fact stated in the report may suggest an important caution in drawing conclusions upon the relative age of formations from the character of their fossils. Had a geological movement or movements upheaved to different levels the bottoms of waters thus separated by a narrow isthmus, and dislocated the connection between those bottoms, naturalists, in after ages, reasoning from the character of the fossil faunas, might have assigned them to different, and perhaps very widely distant, periods.

No. 65 (page 548, *first paragraph*). To the geological effects of the thickening of the earth's crust in the Bay of Bengal, are to be added those of thinning it on the highlands where the Ganges rises. The same action may, as a learned friend suggests to me, even have a cosmical influence. The great rivers of the earth, taken as a whole, transport sediment from the polar regions in an equatorial direction, and hence tend to increase the equatorial diameter, and at the same time, by their inequality of action, to a continual displacement of the centre of gravity, of the earth. The motion of the globe and of all bodies affected by its attraction, is modified by every change of its form, and in this case we are not authorized to say that such effects are in any way compensated.

INDEX.

THE END.

www.ingramcontent.com/pod-product-compliance
Lightning Source LLC
LaVergne TN
LVHW101438200926
848483LV00004B/398
* 9 7 8 1 4 1 8 1 4 8 1 7 1 *